REF
QC173.59
.S65
C6
cop. 1

FORM 125 M

SCIENCE DIVISION

The Chicago Public Library

Received_____ JUL 8 1977

SYNTHESE LIBRARY

MONOGRAPHS ON EPISTEMOLOGY,

LOGIC, METHODOLOGY, PHILOSOPHY OF SCIENCE,

SOCIOLOGY OF SCIENCE AND OF KNOWLEDGE,

AND ON THE MATHEMATICAL METHODS OF

SOCIAL AND BEHAVIORAL SCIENCES

Managing Editor:

JAAKKO HINTIKKA, *Academy of Finland and Stanford University*

Editors:

ROBERT S. COHEN, *Boston University*

DONALD DAVIDSON, *Rockefeller University and Princeton University*

GABRIËL NUCHELMANS, *University of Leyden*

WESLEY C. SALMON, *University of Arizona*

VOLUME 74

BOSTON STUDIES IN THE PHILOSOPHY OF SCIENCE

EDITED BY ROBERT S. COHEN AND MARX W. WARTOFSKY

VOLUME XXII

THE CONCEPTS OF SPACE AND TIME

Their Structure and Their Development

Edited by

MILIČ ČAPEK

D. REIDEL PUBLISHING COMPANY

DORDRECHT-HOLLAND / BOSTON-U.S.A.

Library of Congress Cataloging in Publication Data

Main entry under title:

The Concepts of space and time.

    (Boston studies in the philosophy of science ; 22)
(Synthese library ; 74)
    Bibliography: p.
    1. Space and time. I. Čapek, Milič. II. Series.
Q174.B67 vol. 22 [QC173.59.S65] 510s [115' .4]
ISBN 90-277-0355-8   73-75761
ISBN 90-277-0375-2 pbk.

Published by D. Reidel Publishing Company,
P.O. Box 17, Dordrecht, Holland

Sold and distributed in the U.S.A., Canada, and Mexico
by D. Reidel Publishing Company, Inc.
Lincoln Building, 160 Old Derby Street, Hingham,
Mass. 02043, U.S.A.

REF
QC
173.59
.S65
C6
cop. 1

All Rights Reserved
This selection copyright © 1976 by D. Reidel Publishing Company, Dordrecht, Holland
No part of the material protected by this copyright notice may be reproduced or
utilized in any form or by any means, electronic or mechanical,
including photocopying, recording or by any informational storage and
retrieval system, without written permission from the copyright owner

Printed in The Netherlands by D. Reidel, Dordrecht

PREFACE

Professor Milič Čapek is distinguished as philosopher and as historian of scientific ideas, whose life-time studies center on what is known as the philosophy of nature and its history. What is distinctive to his approach within this field is that he is greatly appreciative of Bergson, James, Peirce and Whitehead. His two previous books, *The Philosophical Impact of Contemporary Physics* of 1961, [Van Nostrand, N.Y., reprinted with two appendices in 1969], and *Bergson and Modern Physics*, [volume VII of these *Boston Studies*, 1971], reveal both his critical attitude towards, and the influence of, these thinkers – an influence tempered by his understanding of the philosophical import that contemporary physics brings into our picture of the world.

What Čapek has set out to present here, in the form of selections (which are secondary and expository in the case of the distant nebulous past, but primary otherwise), is that parts of our views of nature greatly and mutually influence other parts, and that our conception of the world keeps evolving. Thus, ideas of time intertwine with ideas of space, and both with ideas of matter and force. But it is the breadth and scope of this selection that should catch the reader's attention: writers from antiquity to the present day and age, metaphysical and scientific, better known and not so well known, including some who surprisingly remain well-mentioned but not well-read, such as Pierre Gassendi and Pierre Duhem, some fascinating pages of whom are here offered in English translation for the first time. We are grateful to David A. and Mary-Alice Sipfle, Walter Emge, and Professor Čapek for their translations.

Of time and space, there is no end – at least of discussion, theory, argument, and belief. Nor is there any lack of useful anthologies and surveys: J. J. C. Smart's collection *Problems of Space and Time* (Collier, N.Y., 1964), J. T. Fraser's symposium of contemporary essays, *The Voices of Time* (Braziller, N.Y., 1966), Richard Gale's anthology *The Philosophy of Time* (Doubleday, N.Y., 1967), the international conference proceedings *The Study of Time* (Springer, N.Y. and Berlin, 1972), and Max Jammer's noted *Concepts of Space* (2nd ed., Harvard University Press, 1969) come to mind at once. We believe this book with Čapek's lucid introduction will serve students and scholars too, and for many years to come, as a coherent, perceptive, and manageably brief historical entry to the issues and the texts of Western conceptions of space and time.

ROBERT S. COHEN
MARX W. WARTOFSKY

Center for Philosophy and History of Science
Boston University
September 1975

ACKNOWLEDGMENTS

The editor wishes to thank the following editors, publishers and authors for permission to reprint articles of which they hold the copyright.

George Allen and Unwin, Ltd. (London), for F. M. Cornford 'The Invention of Space' from *Essays in Honour of Gilbert Murray* (1936); for B. Russell's 'Early Defense of Newton's Absolute Space', 'On Zeno's Paradoxes', and 'On Change, Time and Motion' from *The Principles of Mathematics* (World rights exc. U.S.A.); for E. Meyerson 'The Elimination of Time in Classical Science' from *Identity and Reality* (1931) (World rights exc. U.S.A.); and from H. Bergson 'On Zeno's Paradoxes' from *Matter and Memory* (1911).

The Aristotelian Society (London), for A. N. Whitehead 'Comment on the Paradox of the Twins' from 'The Problem of Simultaneity' in *Aristotelian Society Supplement* **3** (1923), © 1923 The Aristotelian Society.

W. A. Benjamin, Inc. (Menlo Park, California), for D. Bohm 'Comment on the Paradox of the Twins' from *The Special Theory of Relativity* (© 1965).

The Bobbs-Merrill Co., Inc. (Indianapolis), for H. Bergson 'Discussion with Becquerel of the Paradox of the Twins' from *Duration and Simultaneity* (© 1965, tr. L. Jacobson) (Rights for U.S.A. its dependencies and territories.)

George Braziller, Inc. (New York), for M. Čapek 'The Inclusion of Becoming in the Physical World' from 'Time in Relativity Theory: Arguments for a Philosophy of Becoming', from J. T. Fraser (ed.), *Voices of Time* (© 1966) (slightly enlarged).

Cambridge University Press (New York), for A. N. Whitehead 'The Inapplicability of the Concept of Instant on the Quantum Level' from *Science and the Modern World* (1926) (World rights exc. U.S.A.); for A. S. Eddington 'The Arrow of Time, Entropy and the Expansion of the Universe' from *New Pathways of Science* (1935); for R. Descartes 'View of Space as Plenum' from *Philosophical Works of Descartes* (1931); for

ACKNOWLEDGMENTS

A. A. Robb 'The Conical Order of Time-Space' from *The Absolute Relations of Time and Space* (1921).

The Clarendon Press (Oxford), for C. Bailey's 'Matter and the Void According to Leucippus', 'The Continuity and Infinity of Space According to Epicurus and Lucretius', and 'The Relational Theory of Time in Ancient Atomism' from *The Greek Atomists and Epicurus* (1928).

J. M. Dent & Sons Ltd. (London), for B. Pascal 'The Relativity of Magnitude' from *Pensées*, Everyman's Library (1931) (World rights exc. U.S.A.).

Dover Publications, Inc. (New York), for H. Minkowski 'The Union of Space and Time' from 'Space and Time' from A. Sommerfeld (ed.), *The Principle of Relativity* (n.d.); for H. Reichenbach's 'The Principle of Equivalence' and 'Comment on the Clock Paradox' from *Philosophy of Space and Time*, tr. Maria Reichenbach and John Freund (1956).

E. P. Dutton & Co., Inc. (New York), for B. Pascal 'The Relativity of Magnitude' from *Pensées*. Retranslated from the Lafuma text by John Warrington. Trans. © 1960 by J. M. Dent & Sons, Ltd. Everyman's Library Edition. Published by E. R. Dutton & Co., Inc. and used with their permission.

Les Éditions Payot (Paris), for E. Meyerson's 'On Various Interpretations of the Relativistic Time' and 'The Relativistic Explanation of Gravitation' from *La déduction relativiste* (1925).

A. Grünbaum for 'The Exclusion of Becoming from the Physical World' from his paper 'The Meaning of Time' in *Basic Issues in the Philosophy of Time*, ed. E. Freeman and W. Sellars, Open Court, La Salle, Ill., 1971, pp. 196–227.

Hafner Publishing Co. (New York) for St. Augustine's 'On the Beginning of Time' from *The City of God* (© 1948; trans. and edited by Marcus Dods, D. D.).

Harvard University Press (Cambridge, Mass.), for M. Jammer 'Gradual Emancipation from Aristotle: from Crescas to Gilbert' from *Concepts of Space*, 2nd edition (© 1954 and 1969 by the President and Fellows of Harvard College); for J. B. Stallo 'Criticism of Newton, Euler, Kant and Neumann' from P. W. Bridgman (ed.), *The Concepts and Theories of Modern Physics* (The Belknap Press of Harvard University Press, © 1960 by the President and Fellows of Harvard College).

Harvard University Press (Cambridge, Mass.), Heinemann Ltd. (Lon-

ACKNOWLEDGMENTS

don), and The Loeb Classical Library for Plotinus's 'Criticism of the Relational Theories of Time' from *Ennead III* (© 1967 by the President and Fellows of Harvard College; translated by H. Armstrong); for Aristotle 'On Time, Motion and Change' from *Physics*, Book IV, (© 1957 by the President and Fellows of Harvard College; trans. by P. H. Wicksteed and F. M. Cornford); for St. Augustine's 'Views on Time' from *The Confessions*.

Hermann, Éditeurs (Paris), for P. Duhem's 'Plato's Theory of Space and the Geometrical Composition of the Elements', 'Space and the Void According to Aristotle', 'Place and the Void According to John Philopon', 'Absolute Frame of Reference According to St. Thomas', 'The Empyrean as the Place of the Universe' and 'The Problem of the Absolute Clock' from *Le système du monde*, vol. I (n.d.), vol. VII (1956); for P. Frank 'Is the Future Already Here?' from *Interpretations and Misinterpretations of Modern Physics* (1938).

Holt, Rinehart and Winston (New York), for H. Bergson's 'On Zeno's Paradoxes' from Creative Evolution (© 1911).

Humanities Press, Inc. (Atlantic Highlands, N.J.), for F. M. Cornford 'The Elimination of Time by Parmenides' from *Plato and Parmenides* (1950) (Rights for U.S.A. and territories.)

The Johns Hopkins Press (Baltimore, Md.), for A. Koyré 'The Finite World of Copernicus' from *From the Closed World to the Infinite Universe* (1957).

Alfred A. Knopf, Inc. (New York), for W. K. Clifford's 'On the Bending of Space' and 'On the Space-Theory of Matter' from J. R. Newman (ed.), *The Common Sense of the Exact Sciences* (© 1946 and renewed 1974).

Librairie Philosophique J. Vrin (Paris), for H. More 'On the Difference between Extension and Matter' from Ch. Adam and Paul Tannery (eds.), Descartes, *Oeuvres*, vol. 5 (1956); for A. Koyré 'The Infinite Space in the Fourteenth Century' from *Études d'Histoire de la Pensée Philosophique* (Cahiers des Annales #19), (1961).

The Library of Living Philosophers, Inc. (La Salle, Ill.), for H. P. Robertson, 'Geometry as a Branch of Physics', being Chapter 11 (pp. 315–332) in Paul A. Schilpp (ed.), *Albert Einstein: Philosopher-Scientist* (vol. 7 in *The Library of Living Philosophers*), La Salle, Ill., Open Court, 3rd edition, 1970; for Kurt Gödel, 'A Remark About the Relationship

ACKNOWLEDGMENTS

Between Relativity Theory and Idealistic Philosophy', being Chapter 21 (pp. 557–562) in Paul A. Schilpp (ed.), *Albert Einstein: Philosopher-Scientist* (vol. 7 in *The Library of Living Philosophers*), La Salle, Ill., Open Court, 3rd edition, 1970; for Albert Einstein, 'Reply to Criticisms', pp. 687–688 in Paul A. Schilpp (ed.), *Albert Einstein: Philosopher-Scientist* (vol. 7 in *The Library of Living Philosophers*), La Salle, Ill., Open Court, 3rd edition, 1970. The translation from Einstein's original German by Paul A. Schilpp.

Manchester University Press (Manchester, England), for G. W. Leibniz and Samuel Clarke's 'Discussion on the Nature of Space and Time' from H. G. Alexander (ed.), *Leibniz-Clarke Correspondence* (1956).

Methuen (London), for A. Einstein 'The Inadequacy of Classical Models of Ether' from *Sidelights on Relativity* (1922).

The M.I.T. Press (Cambridge, Mass.) for N. Wiener 'Spatio-Temporal Continuity, Quantum Theory and Music' from *I am a Mathematician* (© 1964 by Norbert Wiener).

Macmillan Publishing Co. (New York), for H. Höffding 'Establishment and Extension of the New World Scheme: Giordano Bruno' from *History of Modern Philosophy* (© 1900); for S. Sambursky's 'The Stoic Idea of Space' and 'The Stoic Doctrine of Eternal Recurrence' from *The Physical World of the Greeks* (© 1956 by Merton Dagut) (Rights for U.S.A. and its dependencies.); for A. N. Whitehead 'The Inapplicability of the Concept of Instant on the Quantum Level' from *Science and the Modern World* (© 1925, renewed 1953 by Evelyn Whitehead) (Rights for U.S.A. and its dependencies).

Thomas Nelson and Sons, Ltd. (Sunbury-on-the-Thames, Middlesex) for G. Berkeley 'Criticism of Newton' from A. A. Luce (ed.), *The Works of George Berkeley*, vol. 4, (tr. from *De Motu*, 1721).

W. W. Norton & Co., Inc. (New York), for B. Russell's 'Early Defense of Newton's Absolute Space', 'On Zeno's Paradoxes', and 'On Change, Time and Motion' from *Principles of Mathematics* (Rights for U.S.A. and dependencies) (© 1938, 1966s. All rights reserved).

Open Court Publishing Co. (La Salle, Ill.), for E. Mach 'Criticism of Newton's Concept of Absolute Space' from *Science of Mechanics* (1942).

Presses Universitaires de France (Paris), for A. Calinon 'Geometrical Spaces' from 'Les espaces géométriques' from *Revue Philosophique de la France et de l'étranger*, vol. 27 (1889), for A. Einstein 'Comment on

Meyerson's "la déduction relativiste"' from *Revue Philosophique* **105** (1928).

Routledge & Kegan Paul Ltd. (London), for F. M. Cornford 'The Elimination of Time by Parmenides' from *Plato and Parmenides* (1939) (World rights exc. U.S.A.), for A. Schopenhauer 'On the Necessary Attributes of Time and Space' from *The World as Will and Idea*, for S. Sambursky's 'The Stoic Idea of Space' and 'The Stoic Doctrine of Eternal Recurrence' from *The Physical World of the Greeks* (1956) (World rights exc. U.S.A.), for D. Bohm 'Inadequacy of Laplacean Determinism' from *Causality and Chance in Modern Physics* (1957).

S. Sambursky for 'The Stoic Views of Time' from his *Physics of the Stoics* (Routledge and Kegan Paul, London 1959).

University of California Press (Berkeley, Calif.), for I. Newton's 'On Absolute Space and Absolute Motion' and 'On Time' from *Sir Isaac Newton's Mathematical Principles of Natural Philosophy and His System of the World (Principia)* (rev. by Florian Cajori, trans. by Andrew Motte). Originally published 1934, 1962, by the University of California Press; reprinted by permission of the Regents of the University of California.

G. J. Whitrow for '"Becoming" and the Nature of Time' from his *Natural Philosophy of Time* (Thomas Nelson, London and Edinburgh 1961).

John Wiley and Sons, Inc. (New York), for V. Lenzen 'Geometrical Physics' from *The Nature of Physical Theory* (© 1931 by V. Lenzen), for R. B. Lindsay and H. Margenau 'Time: Continuous or Discrete' from *Foundations of Physics* (© 1936 by R. B. Lindsay and H. Margenau).

Yale University Press (New Haven, Conn.), for H. Weyl 'The Open World' from *The Open World* (© 1932).

TABLE OF CONTENTS

PREFACE V

ACKNOWLEDGMENTS VII

INTRODUCTION XVII

PART 1 / ANCIENT AND CLASSICAL IDEAS OF SPACE

F. M. CORNFORD / The Invention of Space 3
C. BAILEY / Matter and the Void According to Leucippus 17
P. DUHEM / Plato's Theory of Space and the Geometrical Composition of the Elements 21
P. DUHEM / Space and the Void According to Aristotle 27
S. SAMBURSKY / The Stoic Idea of Space 31
C. BAILEY / The Continuity and Infinity of Space According to Epicurus and Lucretius 33
P. DUHEM / Place and the Void According to John Philopon 39
P. DUHEM / Absolute Frame of Reference According to St. Thomas 41
P. DUHEM / The Empyrean as the Place of the Universe 43
A. KOYRÉ / The Infinite Space in the Fourteenth Century 47
A. KOYRÉ / The Finite World of Copernicus 51
H. HÖFFDING / Establishment and Extension of the New World Scheme: Giordano Bruno 57
M. JAMMER / Gradual Emancipation from Aristotle: from Crescas to Gilbert 65
R. DESCARTES / View of Space as Plenum 73
H. MORE / On the Difference between Extension and Matter (From his First Letter to René Descartes) 85

B. PASCAL / The Relativity of Magnitude 89
P. GASSENDI / The Reality of Infinite Void 91
I. NEWTON / On Absolute Space and Absolute Motion 97
J. LOCKE / On Infinite Space and its Difference from Matter 107
L. EULER / Argument for the Reality of Absolute Space 113
J. C. MAXWELL / On Absolute Space 121
C. NEUMANN / On the Necessity of the Absolute Frame of Reference 125
B. RUSSELL / Early Defense of Newton's Absolute Space 129

PART 2 / THE CLASSICAL AND ANCIENT CONCEPTS OF TIME

F. M. CORNFORD / The Elimination of Time by Parmenides 137
C. BAILEY / The Relational Theory of Time in Ancient Atomism 143
ARISTOTLE / On Time, Motion and Change 147
S. SAMBURSKY / The Stoic Views of Time 159
S. SAMBURSKY / The Stoic Doctrine of Eternal Recurrence 167
PLOTINUS / Criticism of the Relational Theories of Time 173
ST. AUGUSTINE / Views on Time 179
P. DUHEM / The Problem of the Absolute Clock 185
B. TELESIO / Independence of Time from Motion 187
G. BRUNO / Hesitations between Absolute and Relational Theory of Time 189
P. GASSENDI / Reality of Absolute Time 195
I. BARROW / Absolute Time 203
I. NEWTON / On Time 209
J. LOCKE / On Succession and Duration 211
R. J. BOSCOVICH / On the Relativity of Temporal Intervals 225
A. SCHOPENHAUER / On the Necessary Attributes of Time and Space 227
J. C. MAXWELL / Absolute Time and the Order of Nature 231
C. NEUMANN / On the Definition of the Equality of Successive Intervals of Time 233

B. RUSSELL / On Zeno's Paradoxes	235
H. BERGSON / On Zeno's Paradoxes	245
B. RUSSELL / On Change, Time and Motion	251
E. MEYERSON / The Elimination of Time in Classical Science	255

PART 3 / MODERN VIEWS OF SPACE AND TIME AND THEIR ANTICIPATIONS

G. BERKELEY / Criticism of Newton	267
G. W. LEIBNIZ and S. CLARKE / Discussion on the Nature of Space and Time	273
R. J. BOSCOVICH / Criticism of Newton's Alleged Proof of Absolute Motion	289
W. K. CLIFFORD / On the Bending of Space	291
W. K. CLIFFORD / On the Space-Theory of Matter	295
A. CALINON / Geometrical Spaces	297
J. B. STALLO / Criticism of Newton, Euler, Kant and Neumann	305
E. MACH / Criticism of Newton's Concept of Absolute Space	309
H. POINCARÉ / The Measure of Time	317
A. EINSTEIN / The Inadequacy of Classical Models of Aether	329
H. MINKOWSKI / The Union of Space and Time	339
E. MEYERSON / On Various Interpretations of the Relativistic Time	353
A. EINSTEIN / Comment on Meyerson's 'La déduction relativiste'	363
A. A. ROBB / The Conical Order of Time-Space	369
P. FRANK / Is the Future Already Here?	387
H. REICHENBACH / The Principle of Equivalence	397
H. P. ROBERTSON / Geometry as a Branch of Physics	409
E. MEYERSON / The Relativistic Explanation of Gravitation	425
V. LENZEN / Geometrical Physics	431
H. BERGSON / Discussion with Becquerel of the Paradox of the Twins	433
A. N. WHITEHEAD / Comment on the Paradox of the Twins	441

H. REICHENBACH / Comment on the Clock Paradox	447
D. BOHM / Comment on the Paradox of the Twins	451
K. GÖDEL / Static Interpretation of Space-Time	455
A. EINSTEIN / Comment on Gödel	458
A. S. EDDINGTON / The Arrow of Time, Entropy and the Expansion of the Universe	463
A. GRÜNBAUM / The Exclusion of Becoming from the Physical World	471
M. ČAPEK / The Inclusion of Becoming in the Physical World	501
G. J. WHITROW / 'Becoming' and the Nature of Time	525
R. B. LINDSAY and H. MARGENAU / Time: Continuous or Discrete	533
A. N. WHITEHEAD / The Inapplicability of the Concept of Instant on the Quantum Level	535
N. WIENER / Spatio-Temporal Continuity, Quantum Theory and Music	539
D. BOHM / Inadequacy of Laplacean Determinism and Irreversibility of Time	547
H. WEYL / The Open World	561
INDEX OF NAMES	567

INTRODUCTION

Plan of the Book

This book of selections consists of three parts. The first deals with the ancient and classical views of space. I purposely separated these from the pre-relativistic views of time which the second part treats for the following reasons. First, when space and time are treated together, there is often the tendency to exaggerate their similarities and to play down their specific, differentiating features. Second, time, when treated together with space, is quite often dealt with much more briefly – even in a cursory and appendix-like fashion. This can be clearly seen in comparing Newton's own sections on space and time. They are usually quoted together and in most selections they are included under a single heading 'Newton's views of space, time and motion.' When separated – as in this book – the differences in Newton's treatment of space and time become apparent. Newton is far more concerned about the empirical significance of absolute space than that of absolute time. Absolute space is for him the absolute frame of reference by which absolute motions can be differentiated from the relative ones; time is for him the absolute *immaterial* clock of which the material clocks – whether man-made or the natural periodic motions – are imperfect imitations. Newton tries hard to establish experimentally the difference between the absolute and relative frames of reference (his rotating bucket experiment, and the experiment with the two connected spheres revolving around their common center of gravity), but he does not attempt anything of this sort for time. He candidly concedes – and seems not to be disturbed by it – that no uniform motion, that is no uniform material clock, exists in nature. As we shall see, he was not the first who suspected or explicitly stated it. But what is interesting in the present context is the different lengths with which he treats space and time.

So much for the separation of Part 1 from Part 2. In choosing the selections for these two parts I did not always use the primary sources. Sometimes it seemed preferable to use sections from classical historical works like those of Pierre Duhem, Max Jammer, Alexander Koyré, Harald Höffding, etc. The sections from these works present the ancient and medieval views more comprehensively and in a better perspective than the texts which have artificially to be carved out of the primary sources.

Part 3 deals with modern views of space and time and also with their anticipations. Its main, though not exclusive, contents are the problems related to relativistic space-time. But also included are texts dealing with some consequences of quantum theory, especially those related to the problem of spatio-temporal continuity. Among the materials anticipating or at least foreshadowing some contemporary views, various criticisms of Newton from Berkeley, Leibniz and Boscovich to Stallo and Mach are included along with the remarkable anticipations of the physical significance of non-Euclidian geometries by Clifford and Calinon. But more about this in the following sections of this Introduction.

I

I do not know of any more appropriate introduction to the problem of space than Professor Cornford's excellent, though relatively little known essay, 'The Invention of Space'. Its author combines historical scholarship with a clear awareness of the changes which the concept of space underwent in this century. He raises the following question: "Did the Euclidian era, from which we are now emerging, stretch back, with no definable limit, through all recorded history into the darkness of the Stone Age?" His answer is negative; the infinity of space which Euclidian geometry requires, is *not* a part of immemorial common sense. In truth, a large portion of early Greek philosophy can be understood as a gradual and laborious transition from what Cornford calls 'pre-Euclidian common sense' to the consistent Euclidian thinking. The concept of infinite space was invented by the atomists; their infinite void was the space of Euclid, 'credited with physical existence'.

Cornford does not share the view of many writers on the history of philosophy that Anaximander's '*apeiron*' should be understood as 'Boundless', that is, in the sense of three-dimensional infinity; nor does he think that it means 'qualitatively indeterminate.' His original interpretation of this term is based on interesting philological evidence that, although understood in the sense of 'boundless', this term was so only in the sense of the *two-dimensional* boundlessness of a circle; in the latter sense the term '*apeiron*' was applied to a ring. This was precisely a part of the pre-Euclidian common sense; its space was psychological or physiological which by its own nature was bounded, geocentric (or, rather, 'body-centric') and heterogeneous.

Cornford shows how the struggle between this pre-Euclidian common sense and the concept of Euclidian – and consequently infinite – space pervaded the whole of Greek philosophy. Archytas of Tarentum stated the second postulate of Euclid – that a straight line can be extended in either direction – in a concrete and picturesque way when he asked the famous question, later repeated by Lucretius, Bruno and Locke: "If I am at the extremity of the heaven of the stars, can I stretch outwards my hand or staff? It is absurd to suppose that I could not; and if I can, what is outside must be either body or space." Yet, Parmenides did not even ask this question when he proposed his theory of spherical Plenum; but the question certainly occurred to Melissus who – rather than the atomists – should be credited with the invention of spatial infinity. Aristotle was well aware of this question because he was acquainted with the philosophies of his predecessors; but he still insisted that it did not have a sense. Indeed, it did not have a sense, as long as his definition of place was accepted. The outermost cosmic sphere is *not* in any other place; it contains everything without being contained in anything – *omnia continens sed a nullo alio contenta* – if we use the language by which the medieval cosmology characterized the Christian Empyraeum. This view was a refined and sophisticated version of the pre-Euclidian finite space. How this view of the finite, geocentric and heterogeneous space, hierarchically differentiated into the concentric regions of 'natural places', influenced Plato's and Aristotle's cosmology is explained in the selections from the first volume of Pierre Duhem's monumental *Le système du monde*.

This opposition 'the finite versus infinite space' was closely related to the dispute between the relational and absolutist theory of space. This may sound at first surprising. Was not the conflict between the relational and absolutist theory of space initiated by Leibniz' controversy with Samuel Clarke? But we should not be deceived by the absence of the Newtonian terminology before Newton: not only were there important predecessors of Newton in the sixteenth century, but a concept of space closely similar to that of Newton was indeed present, at least virtually, in the Greek atomists. Einstein, although not a historian, saw it quite clearly when he pointed out, in his preface to Max Jammer's *Concepts of Space*, that the infinite void of atomists is indistinguishable from Newton's absolute space. As soon as matter was defined as plenum, that is, as *occupied* space, the distinction was established between the immutable homogeneous container from its changing, movable physical content. As Lucretius never tired of repeating, the void was as absolute and primary a reality as matter.

Under the combined influence of the *Timaeus* and of the Aristotelian cosmology this concept of an infinite, homogeneous, three-dimensional container was pushed into the background. But it has never disappeared completely. Thus the Stoics still accepted infinite space in which they lodged their finite universe. They accepted, with Aristotle, the finiteness of the world, but not the finiteness of space; they accepted his theory of plenum, but only within the boundaries of their material world. Outside of their world, there was the void – a predecessor of the so called 'imaginary space' of the late Middle Ages from which, as we shall see, the concept of absolute space developed. The Stoic universe was strangely similar to that of Rankine in the middle of the last century: his aether – like the Stoic *pneuma* – did not fill infinite space completely; its boundaries acted like 'reflecting walls' which prevented the unceasing dissipation of the radiant energy in the unlimited depth of space. In the sixth century A.D., John Philoponus, though adhering loyally to the Aristotelian denial of the void, still upheld the distinction between the spatial container and its material content; indeed, by his insistence on the immutability and incorporeality of this container, he foreshadowed the two important features of Newton's space. His thought shows an interesting tension between loyalty to the traditional cosmology of Aristotle and new insights, anticipating modern views. He claims that beyond the spherical surface of the world, that is, beyond the outermost sphere, there is only 'space conceived by reason only.' This concept was clearly a compromise between the Aristotelian denial of any space beyond the sphere of the fixed stars and the uneasy, uncomfortable feeling – uneasy for the Aristotelians – that there *must* be something beyond it! (the Archytas-Lucretius-Bruno question.) In any case, like the Stoic outer void, Philoponus' space conceived 'by reason only' was a predecessor of 'imaginary space' which, in the later Middle Ages, was gradually acquiring the connotation of *actually existing infinite space*, as our text from Koyré clearly shows.

It was only an apparent paradox that the Aristotelian and medieval relational theory of space did not exclude the idea of *absolute place*. This idea did not follow from the inseparability of space from matter which is the essence of any relational theory. But it did follow from the assumption of the boundaries of the universe which for Aristotle and his followers were the limits *of* space, not *in* space. To use modern terminology, the absolute frame of reference was represented by the system 'Earth – outermost sphere'. A certain difficulty which was more of an aesthetic or theological kind was that this privileged system – 'place' – should be immobile – like Newton's absolute place which was a portion of absolute space. But immobility was always regarded by the Greeks and their medieval followers as possessing a more dignified rank than the things in motion. Now in the system 'Earth-heavens' the situation was curiously reversed; the heavens, despite their closeness to the Prime Mover, were moving, that is, revolving around the truly motionless Earth! Thus the most noble attribute of motionlessness belonged to the body which was farthest away from the action of the Unmoved Mover and which was the lowest place in the sub-lunar world of corruption and decay. It is interesting to observe St. Thomas' concern about what he regarded as an anomaly: in order to save the immobility of the place for the sphere of the fixed stars he drew the distinction between its *material* place which is in perpetual rotation, and its *rational* place (*ratio loci*) which is truly immobile. The immobility of the Earth is then of secondary, derivative, nature; it belongs to the Earth only in virtue of the fact that its body is at the center of the truly motionless *rational place* which contains the revolving celestial sphere. Thus both in Philoponus and St. Thomas there is a discernible tendency toward a conceptual separation of the container-like space from its material content; true motionlessness belongs to the former, never to anything *in* space. We shall see that this distinction was fundamental for classical physics and not challenged before the coming of the general theory of relativity.

It is true that alongside of the concept of 'rational place' there was a persistent belief in the existence of the body which was *truly motionless* and truly physical, even though made of different stuff than the perishable elements: this was the tenth sphere, Empyraeum, the motionless outermost limit of the universe. (The ninth sphere was postulated in order to explain slight irregularities in the alleged rotation of the eighth sphere of the fixed stars due to the precession of the equinoxes.) Thus the alliance of theology and cosmology apparently saved the concept of absolute, physically identifiable frame of reference; the motionless 'tenth sphere' played a similar role in the Middle Ages to the motionless aether or 'Body Alpha' in nineteenth century physics: the physical substrate of *absolute rest*.

The distinction between space as a homogeneous container and its physical content was increasingly stressed in the sixteenth century by Campanella, Telesio, Patrizzi, etc., even though the Aristotelian cosmology was still retained. Even Copernicus adhered to it; his universe was literally *heliocentric*, the sun being not only at the center of the solar system, but also at the center of the spherical universe. It was only Bruno who swept away the last sphere of the fixed stars and opened the universe in all directions. There has recently been a distinct tendency to play down the significance of Bruno; after all, it is said he was not a scientist; he did not make systematic observations or experiments; he was only a metaphysician and his view of the universe was a poetic vision, not a scientific discovery. Such an approach completely disregards the importance of the history of *ideas*; it overlooks the fact that the most significant scientific discoveries were made possible by great imaginative efforts by which traditional concepts were eliminated.

The idea of the cosmic sphere was such a concept and our selections from Koyré, Jammer and Höffding show how the transition 'from the closed world to the infinite universe' was achieved, as well as the central role which Bruno played in it. One century before Newton he cleared the ground for the concept of homogeneous space, free of its limits, free of any intrinsic differentiation into 'natural places' in which the five elements of Aristotle were supposed to reside. In this way not only the foundations of Aristotle's cosmology, but also those of his physics were undermined and the principle of the unity of nature was fully proclaimed for the first time since the times of Greek atomists.

The boldness of Bruno's anticipations will stand out if we compare his universe not only with the thoroughly medieval 'Elizabethan universe' of his own time, but also with the views of Nicolas Cusanus, Palingenius and Thomas Digges; their universe, though limitless, remained still differentiated into the heaven and the sublunar world, and the partition between them by the celestial sphere was still retained. It was retained even by Kepler, according to whom it is 'two German miles thick';[1] nothing illustrates better the contrast between a patient, but philosophically conservative and unimaginative scientist and the scientifically minded philosopher, Bruno, who, though too impatient for doing calculations and experimental research, had a better anticipatory grasp of the direction in which science was moving. In truth, it is fair to speak of an anti-experimental, but not anti-empirical bias of Bruno; he certainly had a far better grasp of what became known as the law of inertia than Kepler. In his book *Cena delle ceneri* (1584), he correctly answered the objection of the Aristotelians against Copernicus that a stone dropped from the mast of the moving ship should fall behind it if the Earth is really moving. Bruno in this thought-experiment correctly anticipated that actual experiment performed by Gassendi in the harbour of Marseille more than a half century later (1641). It was also Bruno, not Kepler, who denounced the Osiander preface introducing Copernicus' work as a mere mathematical exercise and not a physical hypothesis.

It is true that this reinstatement of infinite space was a conscious return to the thought of the atomists; without the rediscovery of Lucretius' poem, neither Bruno's nor Gassendi's thought would have been possible. This is true of Gassendi more than Bruno; by returning to the atomistic assertion of a void both outside and inside of the material world, Gassendi, in a conscious opposition to Aristotle, reasserted the immutability of space and its independence from matter. It is true that the immobility of space was held by Campanella, Telesio, Patrizzi and Bruno; but Gassendi did not hesitate to draw the ultimate consequences from the immutability of space – its uncreatibility and indestructibility. Space, according to him, existed prior to the creation of the world, a view which would have appeared heretical to a number of theologians in the Middle Ages – but not to all of them, as Koyré's essay on 'the concept of imaginary space' shows. This term appears occasionally even in Gassendi; God was present in it before he filled it by creating matter in it and is still present in 'imaginary space' outside the material universe. God acts, as Xenophanes already thought, by his own presence everywhere and not by moving around through space.

This association of infinite spatiality with divine omnipresence may be traced considerably farther back into the past; Jammer traced it through medieval Hebrew thought back to Biblical times when it found a magnificent poetic expression in Psalm 139. '*Makom*' is the name by which both space and God were designated. Thus Newton's view of space as '*sensorium Dei*', which was regarded by the 19th century positivistic commentators on Newton as a shocking intrusion of 'mysticism' into Newton's thought, was merely a continuation of the centuries-old tradition. In *Enchiridion metaphysicum*, Henry More compared the attributes traditionally assigned to the Supreme Being with the attributes of space and found them the same: space, like God is "one, simple, immobile, eternal, complete, independent, existing by itself, incorruptible, necessary, measureless, uncreated, unbounded, omnipresent, incorporeal", etc. For this reason More rejected the Cartesian identification of matter with space; space is different from matter and the void is precisely that *incorporeal something* which prevents the walls of the vessel from collapsing. This association of space with spirituality led Koyré to the remark that "by a strange irony of history, the vacuum of the godless atomists became for Henry More God's own extension, the very condition of his action in the world."

But the paradox is only apparent and it lies in the very core of atomism; as both John Burnet and Cyril Bailey observed, it was the founder of materialism who claimed that "a thing may be real without being a body." The incorporeal void alongside with matter was the primary reality for Democritus. The movements of ideas sometimes take an unexpected course and give rise to strange and paradoxical cross-currents; yet, a deeper analysis reveals that they were hardly accidental. It was the immateriality of space which yielded so easily to its divinization by More, Gassendi and Newton.

The Cartesian identification of space with matter is usually regarded as an instance of the relational theory of space. And it is true that Descartes defined motion in respect to the neighbouring bodies, that is, in the relational sense. But was his thought completely consistent in this respect? Certainly not; otherwise he could not have the law of conservation of quantity of motion, mv, as the fundamental principle of nature. For the constancy of the quantity of motion, mv, is meaningful only within a definite frame of reference; in other frames of reference moving with the opposite velocity $(-v)$ this quantity would be zero. By insisting on the absolute constancy of the total momentum mv, Descartes implicitly assumed an absolute frame of reference undistinguishable from the Newtonian space. Thus the denial of the void is not a necessary guarantee of a consistent relational theory.

This is even more obvious in the thought of Bruno. Like Descartes after him, Bruno filled his space with a subtle matter; but unlike Descartes, he differentiated his *aër* (his term for aether) from space which is filled by it; *quod non est aer vacuum ipsum, sed primum, cui vacuum replere convenit*. The distinction between the container and the content is thus fully preserved and only to the container-like space does true immobility belong. This is the meaning of Bruno's words that *aër* would be identical with place, with the void, *if it were not moving (ipseque esset vacuum, si non moveretur)*.

We shall speak of Newton's alleged experimental proofs of absolute space in a comment on Part 3. In the second half of the seventeenth century all the features of the classical concept of space were brought into a clear focus: infinity, infinite divisibility, immutability, causal inefficacy, homogeneity. These features are not logically independent; they are all contained in one feature – *homogeneity*, provided it is understood in the following sense: (a) relativity of place, (b) relativity of magnitude. The first feature follows directly from the homogeneity of space and was labelled by Russell the 'axiom of free mobility': neither the size nor the shape of the bodies are affected by their displacement – all regions of space are equivalent, space is causally inert; in this sense space is physically *nothing* – '*nihil*' in the terminology of Gilbert and Otto von Guericke. From this, the impossibility of external and internal boundaries of space follows; in other words, space must be both infinite and continuous. But while the relativity of place holds in the spaces of constant curvature, that is, besides the space of Euclid also in that of Lobachevsky and Riemann – the relativity of magnitude holds only in Euclidian geometry; only in it can any geometrical figure be constructed on any scale. The famous section from Pascal's *Pensées* shows the concrete applications of this principle: the minute world of microcosmos differs from our world *only* by its dimensions – and so does 'megacosmos' from our own world. This was the basic idea of Swift's fantasies as well as of all physical models which regarded the atom as a miniature of a billiard ball or the double star as a giant molecule. Thus both principles – the relativity of place and of magnitude – affirmed *the unity of nature in space*; nature is basically the same everywhere and on any level of magnitude. This belief was very much alive even at the beginning of this century when Bohr's planetary model of the atom was hailed as a fresh proof of the basic similarity of the microcosmos and of the world of the middle dimensions; it took some time before it was realized that the alleged analogies between the atom and the planetary system are not only very limited, but seriously misleading.

All this was realized only after the Newtonian era was definitely over. But until then, the Newtonian view of space remained practically unchallenged. This is clear from the remaining selections included in Part 1: from Euler's defense of absolute space; from Maxwell's treatment of space which is rigorously Newtonian; from C. Neumann's postulation of the absolute frame of reference ('Body Alpha'); and from Russell's defense of Newton against Mach.

Russell's rejection of Mach's criticism of the rotating bucket experiment shows how conservative was this thinker who always prided himself to be free of traditional influences. This is even more significant because this defense of Newton appeared in 1903 – only two years before Einstein's formulation of the special principle of relativity.

II

Meditation on the nature of time began with radical doubt about its very existence. This is the most striking difference between the development of the concept of time and that of space: doubts about the objective reality of space hardly began before the coming of modern idealism. Is it accidental that the reality of succession was denied by Parmenides at the very dawn of Western thought? I do not think it is; time is far more elusive than space and far more bound up with introspective experience which, in the development of the individual as well as in the development of mankind, comes later than sensory perception. It is true that sensory experience also has its temporal aspects; but these aspects are associated with other, nontemporal features which tend to obscure the true nature of time. Hence the perennial 'Eleatism' of human intellect to which Bergson and, after him, Meyerson and Whitehead called our attention. From Zeno to Russell and some contemporary misinterpretations of relativity, the fallacy of 'spatialization of time' is one of the most persistent features of our intellectual tradition.

Yet, Parmenides' denial of succession and of becoming appeared too radical to the empirically minded atomists. While they retained the Eleatic principle of the immutability of Being, they modified it in such a way as to make it compatible with experience. Change was not denied, but merely reduced to the displacement of the atoms, each of which was the Parmenidean *plenum* on a microscopic scale: uncreated, indestructible, immutable, impenetrable. What was the place of time in such system? Since only two basic principles were posited – matter and the void – time was not included among them; it was a mere 'appearance' (Democritus), 'accident of accidents' (Epicurus); it has 'no being by itself' (Lucretius), – it is a mere function of the changing configurations of the immutable particles. Thus the relational theory of time was born.

In truth, this theory in a confused form can be found already in early Pythagoreans who identified time with the celestial sphere or, more probably, with its revolving motion. This clearly indicated their incapacity to separate time from its concrete content. The reference to the celestial sphere and its revolving motion had a far reaching effect on the subsequent development of the concept of time: (a) it focussed the attention of philosophers on the regular periodicity of the celestial motions by which time can be measured, and thus it deepened the distinction between the qualitative content of time and its metrical aspects; (b) the correlation of time with spatial motion was, as we have just seen, the source of the relational theory of time; (c) finally, the alleged inseparability of time from spatial displacements created the tendency to exaggerate the analogy between space and time and, eventually, to spatialize time altogether, and thus eliminate it entirely. Although this extreme tendency appeared only in Parmenides and Zeno, its influence persisted through the whole history of Western thought; time and change, without being denied completely, were consistently excluded from the realm of 'true reality'. This was the meaning of Plato's view that 'time is a moving image of eternity', created by a *Demiurge* along with the world; the true reality, the Ideas, are uncreated and beyond time. In Aristotle's philosophy the timeless Ideas of Plato were, so to speak, compressed into one single entity – God, the immovable source of every motion. As in Plato, Aristotle's time is coextensive with the world history which however, unlike in the *Timaeus*, is without beginning.

Aristotle denies that time is identical with any movement, even though it is inseparable from it. This insistence on the inseparability of time from movement makes the Aristotelian theory *relational*; but it is significantly different from the relational theory of the atomists. For by 'movement' Aristotle means not only a change in position, but *change in general*. In truth, even local motion is regarded by him as a qualitative change, a transition from potentiality to actuality; thus a stone does not become true stone until it reaches its natural, 'home' place. Thus time is correlated with becoming and qualitative change rather than with quantifiable displacement. The broader connotation of the term 'movement' comes clearly to the fore in Aristotle's concern about the relation of time to the 'movement' of soul: "For when we are aware of movement, we are thereby aware of time, since, even if it were dark and we were conscious of no bodily sensations, but something were 'going on' in our minds, we should, from that very experience, recognize the passage of time." (*Physics*, IV, 11). Without awareness of change – at least of *psychological* change – there would be no awareness of time. This was later criticized by pre-Newtonian absolutists like Bruno, Gassendi and Barrow, as the confusion of the *perception* of time with its existence. But this criticism was not quite fair. After defining time as the "numerable aspect of motion with respect to its successive parts", Aristotle raised the question whether time could exist *without* the counting activity of mind – and his answer was affirmative: time is *numerus numerabilis* (ἀριθμός ἀριθούμενος), i.e. an objective reality *susceptible* of being counted, but *independent* of the act of counting, and consequently independent of the existence of the counting mind.

This is not the only objectivist feature of Aristotle's theory of time which foreshadows the absolutist theory of Newton. In saying that 'time is absolutely the same', when various motions of different speed occur simultaneously, Aristotle clearly formulated the doctrine of the *unity of time* underlying the diversity of motions; time is present even if motion is absent since it is not only the measure of motion, but also the measure of rest; 'for all rest is in time'. The all-embracing unity of time implies absolute simultaneity since 'time is everywhere alike simultaneously' (*Phys.* IV, 12). It is true that the objective reality of time is viewed by him as embodied in the perfectly uniform rotation of the sphere of the fixed stars which represents the absolutely uniform cosmic clock; in this sense, Aristotle's theory remains relational. But his emphasis on the introspective aspects of time struck a new note which has never disappeared from philosophical reflections about the nature of time, and which continued through

Plotinus and St. Augustine to Bergson and Whitehead. This aspect of Aristotle's theory pointed in a direction very different from the relational theory of Epicurus – in the direction of *absolute becoming* rather than of absolute time in the sense of Newton. This was consonant with his emphasis on the genuine ambiguity ('openness') of the future in his famous discussion of 'future contingents'.

Sambursky's texts indicate how modern were certain problems which the Stoics faced. Like Aristotle, they speculated about the paradoxical nature of the present moment; while Xenocrates suggested the theory of atomic temporal intervals, thus curiously anticipating the modern hypothesis of 'chronons', favored by some physicists today, Chrysippus's view was more complex. While he also denied the knife-edge mathematical present, his 'Now', unlike the static 'atom of time', did not have sharp boundaries, resembling in this respect James' 'fringe' or 'drops of time' in his *Pluralistic Universe*. One implication of their theory remained unnoticed by them and is not mentioned by Sambursky either. As noted before, their universe was an island of matter floating in the limitless space; since they insisted on the inseparability of time from events, there should be no time outside the limits of the world. Yet, the extramundane void *temporally coexists* with the world; what else could this mean but that there is one all-embracing time which 'flows' simultaneously both within the world and in the outer vacuum? The Stoics apparently did not face this question; but it reappeared in the late Middle Ages when the concept of ultramundane 'imaginary space' necessarily led to that of 'imaginary time' in which 'imaginary space' endures. As soon as we admit that time 'flows in the void', its independence from the physical content is affirmed and we are departing from the strictly relational theory.

Far better known is the Stoic cosmogonic theory of the eternal return. According to this theory, at the end of each cosmic cycle the universe will be dissolved in the original fire. This will coincide with the beginning of another cycle in which the events of the previous cycle will be reconstituted in all their details and in the same order. But the Stoics – unlike Nietzsche later – did not draw all the consequences from a rigorously circular theory of becoming; thus they believed that while Socrates' life and personality will be *exactly the same* in each successive cycle, the successive Socrateses will not be *numerically* identical since numerical identity would imply an uninterrupted existence. This has a serious consequence for the theory:

for if the successive Socrateses are differentiated only numerically, that is, by their 'different positions in time', is this not a surreptitious return to the idea of irreversible time in which successive identical cycles (identical but for their positions in time) are contained? The doctrine of circular time shows how much the spatial and kinematic analogies obscure the true nature of time. Evidently, if we believe that time is adequately represented by a geometrical line, there does not seem to be any cogent reason why this line should not be curved or even closed. This is basically the same fallacy of spatialization as that committed by early Pythagoreans who identified time with the celestial sphere – the view which Aristotle already called 'childish'.

Although Plotinus' Ineffable One had all the features of the Eleatic Being, he retained change and time on the 'lower level'. It is beyond the scope of this essay to explain how he tried to make the relation between the temporal and timeless levels more intelligible by his idea of emanation by which the lower degrees of reality proceed from the higher ones. Suffice it to say that change and succession appear on the second level of emanation with the World Soul in which individual souls are contained. Unlike Divine Intellect on the first level of emanation, the souls are unable to grasp the timeless truth instantaneously, but only gradually, step by step, by a process of discursive reasoning. In this sense, time, as in Plato, is a 'moving image of eternity'; but 'movement' is here understood in a psychological sense, as 'movement of the soul'. Without this life of the soul, time and movement would disappear: "Suppose that Life, then to revert – an impossibility – to perfect unity: time whose existence is in that Life and Heavens, no longer maintained by that Life, would end at once." (*Enneads*, III, 7.12) From this correlation of mind and life with time it follows that wherever there is time, there is some psychism, at least in a rudimentary form; and *vice versa*. In this respect Plotinus' thought is close to modern temporalistic panpsychism; but by his basic ontological outlook he stands at the opposite pole.

Plotinus' metaphysical doctrine accounts for his distrust of the relational theories of time. For if time is something closely akin to our inner life, something *invisible*, then it is fallacious to identify it with something perceivable by senses. Plotinus considered critically the three theories of time: (a) those which identify time with movement or movements; (b) those which identify it with the thing moved; (c) finally, those which regard it as something belonging to movement. But time cannot be movement since movement itself is in time; furthermore, movement can stop, time cannot. This objection apparently does not apply to the continuous rotating movement of the last sphere, but Plotinus points out that the very language which we use shows that time is something different from this motion. For the revolving motion of the universe returns to the same point in space, *but not to the same moment in time*. Using the same argument as Aristotle, Plotinus points out that only motions can have different speeds (i.e. covering different distances *in time*), but not time itself. If time is not the rotation of the outermost sphere, even less can it be the sphere itself. In truth, time cannot be identified with anything merely spatial; it cannot be identical with the distance covered by a moving body since 'this is not time, but space'. Briefly, the basic error of the relational theories is that either they try to reduce time to something non-temporal, or they move in a vicious circle. Such circularity exists in Aristotle's definition of time as "the number of movement with respect to 'before' and 'after'." Plotinus points out that time cannot result from the addition of the measuring number to motion; for motion itself, as long as it exists, contains 'before' and 'after' and therefore it must be *in* time, unless we take the terms 'before' and 'after' in a spatial sense; but then we would be dealing not with motion, but with its trajectory in space. In reading Plotinus, we begin to understand why Bergson in several places acknowledged his debt to Plotinus, who criticized the fallacy of spatialization long before him.

In Book VI of *The City of God*, St. Augustine anticipated the objection raised so many centuries later by Leibniz against the existence of absolute beginning in time: if the world was created *in* time, what particular reason moved God to choose this particular moment of creation rather than another? In view of the perfect homogeneity of time, there was no sufficient reason to choose one over another. Moreover, the assumption of eternal divine idleness prior to the creation of the world was always uncomfortable to theologians. St. Augustine's answer, that "the world was made not *in* time, but *with* time" ("*Non in tempore, sed cum tempore finxit Deus mundum*"), was later criticized by Isaac Barrow who held it to be incompatible with the doctrine of absolute time. St. Augustine's theory was clearly relational since it regarded time as coextensive with the (finite) history of the world; thus the question of divine idleness before time began is nonsensical since the relation 'before' is meaningful only *in* time.

But while St. Augustine insisted on the correlation of time with change, he was opposed to the mechanistic (i.e. corpuscular-kinetic) relational theory of Lucretius which has so many modern defenders and which correlates time with spatial displacement. St. Augustine rejects this theory by the arguments which were already used by Aristotle and Plotinus. He refers to the same example as Gassendi did much later – to the battle of Joshua which continued even when the sun stopped in its path across the sky; neither does time stop if a potter's wheel or any particular motion stops. We see here the preliminary shades of Newton! But while Newton correlated time with the everlasting duration of the divine mind – and this implied the pre-existence of time before the creation of the universe – St. Augustine correlated it with psychological duration of a finite human mind while excluding it from God altogether. Thus time is not identical with any motion; it is 'distension of the soul'. Augustine's Chapter XI of his *Confessions* belongs to the subtlest introspective analysis of the awareness of time.

While the time of the Aristotelian and medieval cosmology was relational, it was still *uniform* and in this sense universal; it was embodied in the uniform rotation of the last celestial sphere which represented the absolute cosmic clock. Even prior to the removal of this cosmic clock by Giordano Bruno, there were serious doubts about its existence. The fact of the precessional motion, known already to the Greek astronomers, made it necessary to postulate an additional sphere beyond the eighth sphere; only to this sphere – and not to that of 'the fixed stars' – did the truly uniform motion belong. Since some Arabic astronomers doubted the uniform rotation of the ninth sphere and thus were led to the hypothesis of the uniformly revolving tenth sphere, the question naturally emerged whether in this frustrating search for the truly uniform motion we are not involved in an infinite regress and whether any uniformly running cosmic clock exists in nature at all. Pierre Duhem's section 'The Problem of the Absolute Clock' deals with doubts of this kind; Nicolas Bonnet and Gradazei d'Ascoli insisted that the existence of mathematical time is *independent* of the existence of the cosmic clock. Bernardino Telesio (1570), although he retained the Aristotelian cosmology, held against Aristotle that time is logically *prior* to change and motion; while motion cannot exist without time, time can exist *by itself (per se)*, i.e. without motion.

But even after the definitive removal of the celestial clockwork, the concept of absolute time was arrived at only gradually and after some hesitations. This is visible in the thought of Giordano Bruno in the selection included in this book. Certain passages show that Bruno was leaning toward the relational theory of time as, for instance, where he claims that there are as many times as there are stars (*tot tempora, quot astra*); or where he speaks of 'different rates of the flow of time'. On the other hand, he explicitly insisted on the independence of time from motion; time would flow even if all things were at rest; its flow is absolutely uniform which is not true of any celestial motion. Guided by the analogy of the infinite space of which particular spaces are mere parts, he speaks of universal time (*tempus universale*) of which particular durations are finite portions. Against Aristotle (or, rather against what he believed to be Aristotle's view) Bruno held that change is a necessary condition of *the perception of time*, not of its *existence*.

Bruno's main merit was the definite removal of the celestial clock. In this way the unity and uniformity of time had been greatly compromised as long *as time was still regarded as being inseparable from motion*. For what becomes of the unity and uniformity of time if there is no uniformly running clock in which its flow is, so to speak, embodied? There were only two ways out of this difficulty: either to accept fully the consequences of the relational theory and to admit that without any privileged clock there are as many times as there are motions – *tot tempora, quot motus*; or to give up the relational theory altogether and to insist that time is altogether independent of *any* motion and of *any* concrete change. It was this second solution to which Bruno was leaning and which was adopted by Gassendi in his *Syntagma philosophicum*.

But there are some ambiguities even in his thought. In his polemic against Descartes (*Disquisitio metaphysica*, 1644) Gassendi clearly formulated the absolutist theory of time: "Whether things exist or not, whether they move or are at rest, time always flows at an equal rate." *(sive res sint, sive non sint, sive moveantur, sive quiescant, eodem tenore fluit tempus.)* This sentence occurs almost *verbatim* in the passage of Barrow (see our selection) and this led Bernard Rochot to conjecture the direct influence of Gassendi on him.[2] On the other hand, in his *Philosophiae Epicuri Syntagma* written only five years before his death (Pars I,

C. XVI), he apparently accepts the relational theory of time proposed by Epicurus. Rest and motion are, according to Gassendi, 'accidents of things' (*accidentia rerum*) and since time itself depends on these motions, it is an 'accident of accidents' (*accidens accidentium*) which is, so to speak, *twice* removed from objective reality. Gassendi's words are quite explicit in this respect; he says that "time itself results from all the events and accidents and is, so to speak, superadded by our mind to them" *(Tempus ipse omnia Eventa seu accidentia consequitur, ipsisque, Mente intervenente quasi supervenit, superaccidit)*. This strongly subjectivistic element is characteristic of every relational theory of time; it will lead Boltzmann, two centuries later, to say that on the cosmic scale the two directions of time are as indistinguishable as in cosmic space where there is no distinction between 'up' and 'down'.

Only in his *Syntagma philosophicum* is Gassendi's absolutist theory of time freed of the ambiguities which were due to his original orthodox allegiance to Epicurus. The independence of time from its physical content is clearly and explicitly asserted; time was flowing even before the creation of the world and would flow even if God would annihilate everything; furthermore, if God, after obliterating the whole physical universe, would create it again, there would be an interval of empty time which would separate the moment in which the previous universe had been destroyed from the moment in which it was created again out of nothing. It is interesting to note that the reality of such an empty interval of duration between two successive worlds was also defended a few years later by Henry More against Descartes: "If God would destroy the whole universe and much later would create it again, this inter-world (*intermundus*), or the absence of the world, would have its own duration which would be measured by so many days, years, centuries."[3] This dispute between Descartes and More developed from their previous dispute whether the vessel would collapse, if it were really empty: Descartes claimed it would, More disagreed. Here we see how closely the problem of absolute time is related to that of absolute space; as soon as we believe that the void *persists* when emptied of its physical content, we accept the 'temporal vacuum' through which it endures and which represents an interval of empty, i.e. absolute time. Yet, this 'temporal vacuum' is not absolutely empty in a metaphysical sense; it is filled by the everlasting divine duration which cannot be interrupted; in a similar sense, divine duration filled the spatio-temporal vacuum prior to the creation of the world. Hence the association with the divinity, not only of absolute space, but also of absolute time, in the thought of More, Gassendi, Barrow and Newton.

Some sections of Gassendi's chapter have a modern ring; for instance, his criticism of 'the knife-edge present' which is still cherished today by mathematical physicists even after serious doubts which quantum theory and, in particular, the second form of Heisenberg's principle raised about the existence of mathematical instants; or his objections against the tendency to exaggerate the analogies between space and time. In other respects, however, his thought, like that of Newton, was a continuation and even the culmination of the classical thought. He and Newton clearly and explicitly formulated the concept of absolute simultaneity: "any moment of time is the same in all places: (Gassendi); "every indivisible moment of duration is everywhere" (Newton).[4]

Nothing essentially new was added to the classical concept of time in the period roughly bounded by the years 1700 and 1900. John Locke, as Newton's contemporary, called duration 'fleeting extension' or 'perishing distance'; its awareness stems 'from reflection on the train of our ideas', but its existence does not depend on it; nor does it depend on motion. "Duration in itself, is to be considered as going on in one constant, equal, uniform course: but none of the measures of it, which we make use of, can be known to do so." (*An Essay Concerning Human Understanding*, Ch. 14, No. 21.) It was impossible to use a more Newtonian language. There are no limits to 'those boundless oceans of eternity and immensity'; it is meaningful to speak of duration *before* the year 5639 B.C. (the alleged date of the creation of the world) since there is no limit – or beginning – to God's duration. (Ch. 16, No. 12) What was relatively new in Locke was his interest in the introspective basis of our awareness of time. From his time on, the distinction between subjective, psychological and objective, physical time gradually became common. Since our psychological changes go on at certain rate, the perception of very slow and very quick motions is impossible; but although we cannot perceive very short intervals of time, it does not mean that time is absent from these intervals. This means that "all the parts of duration, are duration" (Ch. 16, No. 8), i.e. divisible *ad infinitum* in the same way as "all the parts of extension, are extension." This anticipates almost *verbatim* Kant's analysis of perception: "Space therefore consist only of spaces, time solely of times."

This concept of infinite divisibility of time implies the *relativity of temporal magnitude*: no interval of time is intrinsically smallest or largest – another consequence of the homogeneity of time. It was brought up in a very concrete way by Boscovich in one of his comments on Benedict Stay's didactic poem. It merely draws all the consequences from Locke's claim that time is 'flowing' even in its very minute intervals. In truth, the adjective 'minute' is relative; what appears very short to us, may appear as a very long history to the animals with a much shorter 'specious present'; today we would say with a much faster 'biological clock'. There is a note of fine irony in the concluding lines of Boscovich's passage about the temporal perspective of the worms in a cake of cheese and his comparison with the time-perspective of mankind. This passage is counterpart of Pascal's passage in Part I; to the idea of 'the worlds within the worlds' corresponds the idea of 'histories within histories'.

Kant's view of time was as Newtonian as his view of space. We should not be confused by Kant's de-objectivation or, rather de-reification of space and time which, according to him should not be regarded as 'two eternal and self-subsistent entities' since they are merely the *a priori* forms of sensory perception. But for physics this epistemological distinction does not make any difference; all that a Kantian physicist would do in writing a textbook of physics, would be to add a preface in which he would explain that space and time are not things-in-themselves, that the physical world is phenomenal and should not be confused with noumena. But the content of the textbook itself would be as Newtonian as any other textbook of the last century. In truth, it would be even more so; for, since time and space are not empirical generalizations, but *necessary* conditions of *every* experience, the Newtonian model of the universe with both its components – Euclidian space and mathematically continuous time – is beyond the danger of being challenged by *any* further experience. No wonder that quite a number of neo-Kantians were not happy when this bold prediction of their master clashed with the new trends in physics and geometry. On the other hand, Kant's phenomenalization of time had a lasting effect, especially in German thought, both idealistic and positivistic, native and exported. It strengthened the ancient myth of timeless, noumenal 'realm of Being', underlying our illusory 'stream of experience' unstained by change and unbroken by succession. We shall see how much this metaphysical myth influenced certain misinterpretations of relativistic physics. In view of this, we found it unnecessary to include a section from the *Transcendental Aesthetics* into this anthology, especially since Schopenhauer's 'Predicabilia A Priori' is a remarkably accurate and concise summary of the basic features of the classical as well as the Kantian views of space and time. Similarities and differences of both are clearly stated. The immovability of Newtonian space is mentioned as explicitly (No. 13) as absolute simultaneity (No. 19: "Every part of time is everywhere, i.e. in all space, at once.") The same is true of the infinity, infinite divisibility and unity of both space and time. The Kantian view that it is the *a priori* character of time and space which makes possible arithmetic and geometry respectively is found in Nos. 7 and 27.

A brief section from Maxwell's *Matter and Motion* stresses the container-like character of absolute time which Isaac Barrow called appropriately "space of Motion" (*spatium motus*).[5] A mere passage of time is as much without causal efficacy as a mere displacement in space. Thus the causal order of nature, expressed by the maxim 'The same causes produce the same effects' is based on the causal inertness of both time and space; in other words, an event which differs from the previous one *only* by its position in absolute time is its exact replica, which justifies the usage of the word 'same'.

Carl Neumann was concerned not only about the existence of an absolute frame of reference, but also about the absolute uniform clock; without them, he claimed, the law of inertia is meaningless. In a way similar to his postulating the 'Body Alpha', he postulated the 'inertial clock' by which the equality of two successive intervals can be perceived: "Two material particles in inertial motion endure through the same intervals of time when the equal trajectories of one correspond to the equal trajectories of another." There are a number of difficulties in Neumann's solution. First, there are no purely inertial motions; even if we could have only two particles, there would be gravitational interaction between them which would destroy the strictly inertial character of their motion. Second, even assuming the idealized situation of two 'purely' inertial motions, what we would perceive would be the equality of *trajectories*, not the equality of times; the latter would be *defined* by means of the former, but never directly observed. This leads us to the basic difficulty of which Locke, anticipating both Poincaré and Bergson, was already aware: it is intrinsically impossible to bring two successive intervals together; what is successive, cannot be made simultaneous. In Russell's words, 'the axiom of free mobility' which makes measurements in space possible, cannot be applied to time since the standard measuring unit cannot be 'moved in time'.[6]

The remaining selections might be placed in Part 3 with equal justification; this shows how dim are the boundaries in the continuous development of these ideas. Russell's and Bergson's comments on Zeno's paradoxes show two radically different approaches to the problem of time. Briefly, their basic disagreements can be reduced to the following point: Russell believes that time is adequately symbolized by a geometrical line; Bergson does not. According to Bergson, mathematical continuity and infinite divisibility belong only to the spatial trajectory which motion, so to speak, 'left behind itself' in space; Zeno's paradoxes arise from the confusion of this static trace with motion itself (*mobilité*) which can be grasped only in its process, as *fait accomplissant*, in its present participle, never as *fait accompli*. There are no *faits accomplissants* in the world of Russell; his world contains only static points and instants. According to him, Zeno was right; "we live in an unchanging world" and "the arrow, at every point of its flight, is at rest." Then comes the following mysterious sentence of Russell: "The only point where Zeno probably erred was in inferring (if he did infer) that, because there is no change, therefore the world must be in the same state at one time as at another." It is difficult to see how successive states of the world can be different without change. The only explanation of what appears to be a glaring contradiction is that by 'change' Russell meant dynamic passage, transition, overflow of one moment into the subsequent one in the sense of Bergsonian *durée réelle*; he rejected change understood in this sense since it was incompatible with the mutual externality of instants in his mathematically continuous time. That this explanation is correct is clear from the second selection from Russell where he says: "It is merely the fact that different terms are related to different times that makes the difference between what exists at one time and what exists at another."... "Motion consists *merely* [Russell's italics] in the occupation of different places at different times, subject to continuity. *There is no transition from place to place, no consecutive moment or consecutive position,* no such thing as velocity except in the sense of a real number which is the limit of a certain set of quotients" [Italics added.]. With this, Whitehead later agreed;

velocity, acceleration and, more generally, any physical state does not exist at any instant; but this precisely led him – unlike Russell – to reject the durationless instant as an unreal fiction, and to side with Bergson. From Bergson's own point of view, Russell's 'continuum of instants' is nothing but a juxtaposition of *points*, even though Russell still calls it 'time'; he would have regarded it as another vindication of his claim that a mere arithmetic or geometric treatment converts time into a static manifold of no less static instants. Russell himself admitted it when he wrote: "But every term is *eternal, timeless, immutable*; the relation it may have to parts of time are *equally immutable*" [Italics added]. There was, indeed, no chance of agreement between Bergson (and the later Whitehead) and Russell; however, a more attentive study of other Russell's writings would show that he was sometimes closer to Bergson than he realized.[7]

The final essay of Part 2 is Émile Meyerson's documented study about the elimination of time in classical science. His central thesis is that explanation in its traditional, rationalistic sense tends toward the elimination of diversity both in time and space; when diversity is reluctantly conceded, it is admitted in its most 'innocuous' form as a mere difference of position, either in time or in space. Thus lurking behind this search for identity is the Eleatic ideal, usually present only implicitly and half-consciously, of single and undifferentiated Being. Thus identification in time underlay the classical maxim *causa aequat effectum* which found its concrete embodiment in the conservation laws. There is no need to dwell on Meyerson's documented thesis which was so little understood by scientists without knowledge of the history of science and philosophy; suffice it to say that the full meaning of Meyerson's essay could be grasped only within the context of the other chapters of his *Identity and Reality*. A supplementary reading of his other books, in particular, of his untranslated and lavishly documented *De l'Explication dans les Sciences* adds to the convincingness of his thesis.

III

Among the anticipations of modern views, I included four pre-relativistic critics of the experiments by which Newton tried to establish the reality of absolute space and motion. Two of them (Berkeley and Mach) are well known. Berkeley rejected the Newtonian reification of absolute space on both epistemological and metaphysical grounds: we perceive only *relative* places and *relative* motions; absolute places and absolute motions, i.e. the places and motions *without* reference to concrete physical bodies are unperceivable, and consequently unreal. The rotation of two connected spheres, or of the bucket full of water in Newton's experiment, take place *not* with respect to unperceivable absolute space, but with respect to 'the heaven of the fixed stars'; this is the first reference to 'the great stellar masses' which reappeared later in the criticism of Mach and Stallo. But even if we grant the reality of fictitious absolute space, the motion of the bucket would not be circular, but rather a very complicated motion whose other components are the motion of the earth around its axis, around the sun, etc. Berkeley's argument is strictly kinematic; he acknowledges the fact that centrifugal forces arise in certain situations and not in other, but does not explain the difference. Yet, it was this observed difference which Newton stressed.

In contrast to Berkeley, Boscovich's criticism of Newton is based on dynamical considerations. He points out that the experiment with two connected spheres revolving around the common center of gravity would take place *in exactly the same way* in *any* system moving with a constant vectorial velocity with respect to the frame of reference in which the experiment had been originally performed. Thus Newton's experiment does not give us any information as to which of an infinite number of inertial systems coincides with absolute space. Although this objection is based on a consequence drawn from the dynamical equivalence of *all* inertial systems (i.e. Galileo's principle of relativity), it was not raised either by Stallo or by Mach. The only possible answer which Newton could have given to this objection of Boscovich would be as follows: by means of *optical* experiments it is possible to determine which inertial system coincides with absolute space. And this is what Michelson tried to find out – without any success.

Stallo's argument is similar to that of Berkeley; it is directed against Euler, Kant and Carl Neumann who defended Newton's absolute space. Against both Euler and Neumann, he points out that we cannot meaningfully speak of rotation and, more generally, of *any* motion unless we refer it to some other body or bodies. In other words, to speak of the motion of a body which would be completely solitary in the universe (whether the motion is translatory or rotary) is utterly meaningless. So far Stallo's argument is the same as that of Berkeley; it is based on strictly kinematic considerations. But it goes beyond Berkeley when he claims that not only position and motion, *but the very existence of a body* is relational: "A body cannot survive the system of relations in which alone it has its being; its *presence* or *position* in space is no more possible without reference to other bodies than its *change of position or presence* is possible without such reference." Stallo was unquestionably right; all physical properties of a body, even its inertial mass – which Newton called '*vis insita*' residing *hic et nunc* in a certain region of space at certain time – can exist only in dynamical interaction with other masses and cannot be conceived without it. This is especially clear from Newton's third law as Maxwell pointed out;[8] since inertia manifests itself in the resistance against an accelerating force and since this force is accompanied by an equal force in an opposite direction which has its point of application in *another* mass, we see that we need at least *two* masses in order to have their interaction. Thus the concept of solitary mass which is acted upon by a disembodied force is a result of what Whitehead later called the 'fallacy of simple location' or 'misplaced concreteness'. Thus Stallo, implicitly at least, anticipated Mach's famous saying that in the principle of inertia there is 'an abbreviated reference to the whole universe.' In other words, without the great stellar masses not only would there be no centrifugal forces in the solitary, allegedly rotating body, but the body itself would disappear altogether because of the relational character of *all* its properties; "the neglecting of the rest of the world is impossible" (Mach).

Mach's well-known criticism of Newton appeared about one and a half years after the publication of Stallo's book, but entirely independently of it. (Only later were Mach and Stallo corresponding, and then Mach wrote an extensive introduction to the German translation of Stallo's book in 1901.) They both reached the same conclusion, that in the experiments described by Newton we should say that the rotation takes place not with respect to absolute space, but with respect to the great stellar masses. They used different thought experiments to reach the same conclusion – that inertia (of which centrifugal forces are a manifestation) is of a *relational* nature. While Stallo used the method which Mill called the method of difference in claiming that the centrifugal forces would vanish if the rest of the universe were removed, Mach used 'the method of concomitant variations': he anticipates the rise of centrifugal forces in the liquid rotating *relatively*, with respect to the vessel, if the walls of the vessel were gradually increased to the thickness of several miles. Obviously, not only an orthodox Newtonian, but also a positivistic physicist at that time could answer that neither the abolition of the universe nor the construction of the giant vessel with the walls several miles thick is experimentally feasible and that we cannot base physics on mere 'ifs' in *Gedankenexperimente*. But today we know that Stallo's and Mach's thought experiments paved the way for Einstein's principle of equivalence. We also know that neither Mach nor Stallo even remotely suspected how extensive would be the conceptual reconstruction necessary for this purpose.

The Clarke-Leibniz correspondence has so many comments that hardly anything new can be added. Clarke defended Newton's absolute space and time which, as we pointed out, was also the view of Gassendi.[9] He claimed that things *can* differ *numerically* only, by their position in space or time; Leibniz rejected it on the basis of the principle of sufficient reason, dressed in theological terms. But, as Stallo pointed out (*The Concepts and Theories*, p. 205) Leibniz surreptitiously admitted the existence of absolute space in his fifth letter when he conceded, in the light of Newton's rotating bucket experiment, the difference between absolute and relative motions. Hans Reichenbach in his historical study also conceded the weakness of Leibniz' answer to Clarke. Leibniz – like Berkeley – was unable to account for the dynamical difference between 'absolute' and 'relative' rotations.[10] An even more serious discrepancy occurs in Leibniz' *Initia rerum mathematicarum metaphysica*, written about the same time as his letters to Clarke: although he upheld the relational theory in calling space '*ordo coexistendi*', he speaks of absolute space (*spatium absolutum*) as 'the place of all places' (*locus omnium locurum*)! It was impossible to use a more Newtonian language.

The articles of Clifford and Calinon belong to one and the same category. They deal with the psychological conditions of our perception of space and with possible applications of non-Euclidian geometries to physical space. In this sense, they are resolutely anti-Kantian; the axioms of Euclid, especially the fifth postulate concerning parallel lines, are no *a priori* necessities of thought since both Riemann's and Lobačevski's geometry (each of which is based on a different *denial* of this postulate) are free of contradiction. Our Euclidian geometry, instead of being an *a priori* property of mind, is of empirical origin, resulting from the long pressure of our biological surrounding which is certainly *very* approximately

Euclidian. But, as Clifford and Calinon stressed, this does not mean that space *on a very large scale* must be Euclidian; as Gauss suggested, the parallax of very distant stars may disclose deviations from the Euclidian 'zero-curvature'. In the same way as a small region of the spherical surface is approximately flat, a small volume of either Riemannian or Lobačevsky's space is approximately Euclidian.

While Helmholtz had considered only spaces of constant curvature, Clifford and Calinon went still further in considering not only spaces with a locally variable curvature, but with a curvature changing through time. Clifford's 'space theory of matter' is an amazing philosophical anticipation of the relativistic fusion of matter with space, 'matière resorbée dans l'espace', as Émile Meyerson called it later. While his anticipation was already justly appreciated by Eddington in 1920 (*Space, Time and Gravitation*, p. 192), Calinon's work remained largely ignored; it was only Robertson who recently stressed its significance.

It is true that it is always possible to introduce a certain complication into a physical theory while leaving Euclidian geometry intact. This was the view of Stallo and Poincaré. Although Poincaré's argument was based on the alleged 'greater simplicity' of Euclidian geometry, he was unconsciously influenced by the traditional distinction between homogeneous, immutable, container-like space and changing physical content. The same was true of Stallo despite his nominal protests against the 'reification' of space. As Schlick pointed out in his comment on Helmholtz's writings, Poincaré's error was to consider the simplicity only of geometry, not the simplicity of the whole conceptual framework of which 'physics' and 'geometry' in their traditional senses are mere parts.[10a] Robertson contrasted Poincaré's 'pontifical pronouncement' that "Euclidian geometry has nothing to fear of fresh experiments in physics" with Calinon's remarkable and prophetic insight concerning the physical possibility of those spaces whose constant ('curvature') varies with time. Eddington's 'expanding universe' is precisely such a space. It is interesting to recall that Calinon's anticipation was dismissed in an equally pontifical way by young Bertrand Russell in 1898 as 'grossest absurdities'.[11]

Poincaré was on a far safer ground in his analysis of time measurements. Like Locke before him, he knew that the equality of two successive intervals cannot be directly perceived; from this, his contemporaries, Bergson and Russell, concluded that time is intrinsically immeasurable. He was also aware, like Newton before him, that there is no uniform motion or process in nature; even the rate of the earth's rotation slightly varies. Consequently, the equality of successive time intervals can be only *defined*; and it must be defined in such a way that the equations of mechanics remain 'as simple as possible'. When he wrote in 1905, Poincaré still had in mind the mechanics of Newton; in truth, he mentions Newton's gravitation law explicitly when he speaks of the slowing down of the terrestrial rotation.

In dealing with succession, Poincaré considers the causal theory of time, – later accepted by Carnap, Reichenbach, Robb and Mehlberg – which may be summarized by the words *'propter hoc, ergo post hoc'*. But Poincaré asked how do we recognize the causal antecedent except by the fact that it is temporally *prior*? Are we not moving in a vicious circle?

The difficulty concerning the simultaneity of distant events is no less serious. Such simultaneity cannot be directly perceived, but only computed; but this can be done only if we know the velocity of the luminous signals, which is impossible without measuring it. But we can measure it only if we make some more or less arbitrary assumptions, e.g., that the clocks are not affected by their displacements or that light is being propagated with equal velocity in all directions. (The last assumption was quite improbable at that time since it was assumed that there is a relative motion of the earth with respect to aether.) Poincaré could have pointed out the circularity inherent in the determination of simultaneity much more strongly; for simultaneity of distant events is presupposed in the very concept of spatial distance, since all the points constituting such distance, including its extremities, *co-exist* by definition and are in this sense simultaneous. Consequently, the 'simultaneity of distant events' is nothing but 'simultaneity of the events which are simultaneous'. But this circle became obvious only when the special theory of relativity showed that spatial distance – in classical physics a segment of an Euclidian straight line – lost its originally unique significance. In Whitehead's words, since the advent of relativity "spatial distances stretch through time"; in other words, there are no purely spatial, only spatio-temporal distances.

The significance of Poincaré's essay is not so much in the solutions proposed as in the formulation of the problems relevant for the subsequent development of physics. Within the context of Michelson's vain search for the absolute motion of the earth, Einstein rejected any absolute frame of reference which would be a substrate of absolutely simultaneous events. If we want to retain the concept of aether, Einstein wrote, we must deprive it not only of the mechanical, but even of the most basic *kinematic* properties; it is neither in motion nor at rest! Thus two closely correlated concepts were rejected: immobility of the Newtonian space and simultaneity of distant events. It was only within the framework of these two negations that the constant velocity of electromagnetic signals ceased to be absurd.

Minkowski's well known treatise explains how this necessitated a radical conceptual reconstruction of the relation between space and time: this reconstruction led to the revolutionary rebuilding of the foundations of physics. There is nothing controversial about the mathematical aspect of Minkowski's paper; but what is very much disputed is its philosophical or ontological meaning. Although Minkowski explicitly stated that "three-dimensional geometry becomes a chapter in four-dimensional physics", some physicists and even more philosophers interpreted it in the sense that "three-dimensional physics becomes a part of four-dimensional geometry." Minkowski's space-time was thus frequently conceived as a four-dimensional hyperspace of which time is merely one dimension. Yet, there is no preferential treatment of space by Minkowski; in truth, he said quite explicitly that besides the radical modification of the classical concept of time, a corresponding violation of the classical concept of space is 'indispensable'. But by calling the union of space and time the 'absolute world', Minkowski unintentionally misled those who uphold a static interpretation of space-time; for the term 'world' certainly has a *static* connotation. But a far decisive factor was that such an interpretation was consonant with the centuries-old intellectual tradition reaching back to the very dawn of Western thought; we dealt with it earlier.

This is also pointed out in Meyerson's essay. His criticism of the 'geometrization of time' was endorsed by Einstein in his review of Meyerson's book. But Einstein's own position, as can be seen from his almost sympathetic comment on Gödel's essay, which regards the relativistic physics as a vindication of the idealistic denial of time, was far from being unambiguous.[12] The same is true of Weyl's view, as Meyerson pointed out; he was always careful to speak of 'the three-plus-one-dimensional' instead of 'four-dimensional' space-time, thus indicating that the privileged character of the temporal dimension is preserved even in its relativistic union with space. Yet, on the other hand, his view that "the objective world *is*, it does not become" and that succession exists only for our "blind-folded consciousness creeping along the world-line of its body" influenced those who, like Grünbaum hold that "coming into being is only coming into awareness." Such becomingless view is the very opposite of the 'open world' – the truly Bergsonian term which Weyl used as the title of his later book. As the concluding sentence of Weyl's cited writing indicates, it was the end of the Laplacean determinism, suggested by Heisenberg's principle, which induced him eventually to accept the dynamic world.

Eddington, unlike Weyl, was always consistently opposed to the static interpretation. What Meyerson quotes ("Events do not happen; they are just there, and we come across them...") was not Eddington's view, but the one which he rejected as the context clearly shows.[13] In the same passage Eddington recognized, like Whitrow in his essay, the close connection of rigorous determinism with the static view, another circumstance which induced a number of thinkers to accept the view now still held by Grünbaum. But Grünbaum himself, strangely enough, denies any such connection. It is difficult to see how any kind of indeterminism – that is, any genuine ambiguity of the future – could be reconciled with the becomingless space-time which, as A. A. Robb justly remarked, is another version of what William James called a 'block universe'. In this respect Hilary Putnam was more consistent when he insisted on the strict correlation of rigorous determinism and the static, 'dimensional' view of space-time.[14]

A. A. Robb's writings are not known today as they should be. Yet, not many writers showed so lucidly as he did the fundamental character of the 'before-after' relation in relativistic space-time. Equally rare is his insistence that the simultaneity of distant events is not only relativized, but *eliminated altogether*. (This, by the way was also Einstein's last view.) He expressed it in a rather arresting way by saying that "the present instant, properly speaking, does not extend beyond here"; or that "an instant cannot be properly in two places at once."[15] We are so much used to the Newtonian 'Everywhere-Now' that this strikes us as unbelievable, even though since the times of Olaf Römer we know that 'Now' and 'Seen Now' are two different things. Furthermore, we tend to confuse 'co-instantaneous' and 'contemporary'; the relativity theory denies the former, not the latter since it recognizes 'simultaneity of temporal intervals' as a perfectly meaningful concept. We shall see that the alleged 'paradox of the twins' cannot be even phrased without it.

My view of the nature of relativistic time-space is contained in the paper included here as well as in some other places.[16] It holds that Minkowski's fusion of space with time can be more accurately characterized as dynamization of space rather than 'spatialization of time'. This interpretation and its corollary – the physical emptiness of the future – follows from an attentive reading of Minkowski's formula for the constancy of the world interval and of all its implications. How easily can the implications of this formula be overlooked even by an outstanding physicist-philosopher is shown by the case of Philipp Frank to which I refer (in Note 10, paper herein). The static, becomingless view, represented in these selections by Grünbaum, Gödel and James Jeans (who is criticized in Frank's paper) leads to an intolerable dualism of two realms – the subjective one, to which becoming because of its 'mind-dependence' is confined, and the becomingless world of physics. Such a sharp metaphysical dichotomy creates even greater difficulties than the traditional Cartesian dualism: for, according to Descartes, both the mental and physical realms, despite their profound differences, share at least their *temporal* character; they both belong to the realm of change, i.e. becoming. But in the doctrine of 'mind-dependence of becoming' we have two realms which have *nothing* in common and whose relations and interactions remain completely unintelligible.

Grünbaum is clearly aware of these difficulties and in struggling with them he defends himself along two main lines. First, he claims that the subjectivity of becoming and of 'now' (he correctly recognizes that these two terms are correlated) is no more mysterious than the subjectivity of sensory qualities. Second, he draws a very controversial distinction between time and becoming, and indignantly denies that his becomingless universe is timeless. Finally, in one unguarded moment he concedes that the physical world is not becomingless after all. Let us analyze his utterances in detail.

Sensory qualities such as color, sound, warmth, cold, scent etc. are obviously subjective and nobody – with a curious exception of neo-realists and some phenomenologists – endow them with an objective, i.e. mind-independent status. The objective stimulus of color, for instance, is a certain electromagnetic frequency, altogether different from the sensory quality of color; why should not the objective becomingless world also be altogether different from our stream of experience? A close analysis shows how limited this analogy is. In the first place, it is extremely improbable that our sensory perception, including our passing psychological present, is entirely deceptive. For no matter how dissimilar the sensory qualities are from the physical stimuli, they both occur *in succession* and thus exhibit basically the same temporal order. It is not necessary to go as far as Helmholtz who claimed that the time relations of perceptions furnish a 'true copy' of the time relations of physical events. This is, strictly speaking, true only of the temporal relations of those physical stimuli directly affecting the surface of our bodies; the time relations of *other* physical events (i.e. those taking place at some distance from our bodies) are always inferred, never immediately perceived. It is, however, true that for small distances the perceived and the inferred temporal order practically coincide.

Furthermore, the sequences of our perceptions are 'true copies' of only those physical events which succeed each other at a rate not too different from the temporal rhythm of our consciousness; in other words, to be recognized as distinctly successive, they must not succeed too quickly – otherwise they are fused into the spurious simultaneity of our psychological present. But even if we take all these qualifications into account, it is impossible to claim that there is no objective, physical counterpart to what we experience as becoming and as the psychological present. The biological function of perception is to *select* and *simplify*, but *not to deceive*. Thus when we perceive a single quality of touch instead of thousands of molecular impacts which are its physical counterpart, or when we perceive a single quality of tone instead of thousands of successive air vibrations, it is because of the *economically selective* and *economically simplifying* character of our perception; on this the biological usefulness of sensory information is based. The only difference between the temporal order of our consciousness and that of the physical reality is that the latter is far more complex and more finely grained. But temporal they both are. To my own psychological present corresponds a certain small interval in the history of my own body – and for practical purposes of the whole earth – *and no other interval whether in the past or in the future*. We certainly do not live either in the Cretaceous period or in the year 2000! Only McTaggart for whom "reality is one timeless whole, in which all that appears successive is really co-existent" could brood over the question: Why are we not living in the time of George the Third?

Grünbaum's view becomes *psychologically* more understandable – without becoming more convincing – if we consider what may be called his *second* line of defense: his distinction between 'becoming' and 'time'. When one reads his repeated protests against characterizing his becomingless universe as timeless, one begins to wonder whether the whole dispute between two conflicting interpretations of Minkowski's continuum – spatialization of time and dynamization of space – is not merely semantic. This is what Professor John Graves suggested.[17] For it is extremely difficult to see how these two terms – 'becoming' and 'time' – can be logically separated unless the disputants understand at least one of these terms in a very different sense. Now what does Grünbaum exactly mean by the word 'time'? This is clear from his claim that the becomingless universe does not exclude 'temporal *separations*.' In using the term 'separation' he – whether unconsciously or deliberately – *spatializes* time. More specifically, he represents the *succession* of two events by a geometrical separation of two *juxtaposed* points located on a line which he still calls 'time'. There is no harm in using this spatial symbolism as long as it is understood as a mere symbolism, i.e. as a static translation of genuinely successive terms into spatial imagery; but it becomes a vicious distortion of the true nature of time as soon as it is taken literally. Yet, this is what he does when he eliminates becoming. For only becoming provides the dynamical feature which differentiates the spatial 'before-after' relation from the genuinely temporal succession. There is no genuine succession in a bare spatial pattern; if we say that a point A on the left side of the point B is 'before' B, both points still remain simultaneous, being *co-existing* parts of the same static diagram. Similarly, as long as we interpret Minkowski's word in a static becomingless sense, the events in it are only verbally successive; they really *coexist* together, *totum simul,* in spite of their 'separation' which is nothing but juxtaposition. Only in this way can becoming be excluded from the physical reality and be confined to the subjective realm. Only in this sense can becoming be called 'mind-dependent'.

But does Grünbaum *really* believe it? Consider the following sentence of his: "The mind-dependence thesis does deny that physical events themselves happen in the tensed sense of *coming into being apart from anyone's awareness of them.*" But then he continues: "But this thesis clearly avows that physical events *do happen independently of any mind* in the tenseless sense of merely occurring at certain clock times *in the context of objective relations of earlier and later.*" (Italics mine.) Thus he concedes that physical events do happen independently of any mind after all and that they occur in objective succession: this is exactly what the thesis of mind-independence of becoming holds! No semantic fog, however thick, can ever disguise this fact. In vain does he try to conceal his concession by using the term 'tenseless occurence' which clearly has no intelligible physical meaning; nor does he realize that in rejecting explicitly *totum simul* he gives up his becomingless view in all but name.

But in addition to this Grünbaum makes his position even more difficult by his commitment to physicalism. In the concluding paragraph he writes: "But in characterizing becoming as mind-dependent, I allow fully that the mental events on which it depends themselves require a biochemical physical base or possibly a physical basis involving cybernetic hardware." If this means that the mental events are totally reducible to the brain processes, then he should speak of 'brain-dependence' rather than 'mind-dependence' of becoming. But the brain itself is a part of the becomingless physical reality: where does Grünbaum then obtain any *locus* for becoming – even for the *illusion* of becoming? In the dualistic or Kantian framework it is still conceivable in principle to confine becoming – or rather its illusion – to the mental realm which is regarded as distinct from the external world or 'things-in-themselves.' But in the physicalistic framework there is no 'mental realm' or – which is the same – this allegedly different realm is incorporated into the becomingless physical world. In this context the words of Lotze, quoted by G. J. Whitrow are very relevant: "We must either admit Becoming or else explain the becoming of an unreal appearance of Becoming" – and this, as Whitrow observed, is impossible without an implicit appeal to Becoming.

There is a definite affinity between Grünbaum's view and the philosophy of the British idealist J. M. E. McTaggart whose much commented article 'Unreality of Time' appeared in *Mind* in 1908. Grünbaum's distinction between the past-present-future relation and the earlier-later relations corresponds to McTaggart's distinction between the A-series and the B-series. But unlike Grünbaum, McTaggart correctly recognized that the B-series is impossible without the A-series. From this McTaggart inferred the unreality of time while Grünbaum claims the very opposite; but ultimately their views are the same since Grünbaum's 'temporal separation' of events is only a poorly disguised juxtaposition. In a broader perspective of the history of ideas, Grünbaum belongs to the perennial tradition worshiping timeless Being, the tradition which from Parmenides and Plato up to Laplace, Bradley and McTaggart influenced and largely dominated Western thought. All these views face the question raised in Whitrow's essay – the same which was raised by William James as early as in 1882: "If the future history of the universe pre-exists timelessly (or, as it is fashionable to say, 'tenselessly') in its totality, why is it not already present?" To this our modern neo-Eleatics give no answer.

Among the selections dealing with, or related to, general relativity, the first group deals with the so-called "geometrization of matter." Émile Meyerson, faithful to his method, tried to place the general theory of relativity into a proper historical perspective and concluded that it is a continuation and even a fulfilment of the Cartesian project of 'geometrization of matter.' Einstein in his otherwise laudatory comment on Meyerson denied that such a phrase is meaningful, – unless we cease to regard geometry as an *a priori* science, independent from experience. But in that case geometry becomes, in Robertson's words, 'a branch of physics' and it is far more appropriate to speak with Eddington of 'mechanization of geometry' rather than of 'geometrization of mechanics'. It is certainly useful to use two-dimensional analogies to illustrate certain features of the relativistic physics which our imagination finds hard to accept; for instance, that a straight line is merely a special case of a geodetic line, that finiteness and limitlessness are logically compatible features, or that a space-time diagram of accelerated motion is a curve. But precisely the last illustration shows that we cannot obtain even kinematics – much less dynamics – from mere geometry without reference to time. A merely geometrical space cannot give us any physical diversity nor any physical change as long as it remains homogeneous and immutable. Such was the space of Euclid which Descartes accepted and this was the reason why he failed in his attempt to geometrize matter. In truth, even the Riemannian space without the Cliffordian 'humps' would be physically empty; the presence of local irregularities of curvature would provide us with physical diversity, but *not* with change, since only the activating presence of the temporal dimension can transform the static physical diversity into physical motions or changes of the field. If Victor Lenzen still speaks of the physics thus obtained as 'physical geometry', the emphasis is more on 'physical' than on 'geometry'; he is fully aware how much such 'geometrization' – including Weyl's 'geometrization' of the electromagnetic field – differs from the original Cartesian dream.

The first Reichenbach selection is a lucid introduction to the general theory of relativity; it shows how two phenomena occurring in the gravitational fields – a non-Euclidian behavior of the light rays and an effective slowing down of the clocks (whether natural or artificial) – follow from the principle of equivalence. This will provide us with a better insight into the much discussed twin paradox. Bergson was correct against Becquerel in stressing that since all inertial systems are equivalent, there is a perfect reciprocity of appearances in them; and since the clock in one system cannot be both slower and faster than that in the other one, the so-called 'dilatation of time' has only a referential, perspective-like meaning. But Bergson's error was not to see that in dealing with this 'paradox' we are not on the ground of the special theory any longer; for one of the twins – that who departed from the earth – is subject to an enormous change of velocity at the moment when he starts his return journey. Bergson's objection that acceleration is equally relative and that we have thus an equal right to say that it was the earth which was first moving away in an opposite direction and returned to the rocket traveller, is invalid *since it overlooks the different relations of the two systems to the rest of the universe.* While the traveller's ship is accelerated with respect to the earth *and the great stellar masses*, the earth (and the terrestrial observer) is not. Again, as Mach wrote in a different context, "the neglecting of the rest of the universe is impossible." Bergson's error was excusable because Becquerel phrased his argument exclusively in the language of the special theory; this is so much more curious because in his book he carefully distinguished between apparent dilatation of time in the special theory and effective dilatation in the general theory.[19]

Becquerel's and Bergson's errors are avoided by the other three papers. Whitehead ends his exposition by his interesting, though controversial, attack against conventionalism in chronometry; this would require a lengthy and separate comment. Bohm draws an interesting analogy between the metrical diversity of concomitant temporal series and the variable rhythms of psychological time. Reichenbach explicitly points out that it is the gravitational-accelerational field at the point when the traveller begins his return journey which slows down the traveller's clock; and both Whitehead and Bohm equally stress the crucial importance of the physical change at that point. But if this is so, we face one consequence which is generally ignored by the textbooks on relativity: that no effective slowing down of time occurs as long as the motion of the traveller is inertial. This was the point Bergson made. If we do *not* accept this consequence, we admit that there are at least *some* inertial frames which are *not* dynamically equivalent – the very opposite of what the special theory holds! It would mean that the rest of the universe cannot be neglected even as far as the inertial motions are concerned. But this would not mean a return to Newton's absolute space; the negative result of Michelson's experiment still holds and thus 'the rest of the universe' must not be understood in the Newtonian sense. Yet, it would be enough, as Bridgman pointed out,[20] to make the special theory only *approximately* valid. At present it is simpler not to believe it; the slowing down of the atomic clocks and of the decay of mesons occurs when large accelerations are present, whether in the gravitational field of the Sirius satellite or in the magnetic field of the earth.

It is hardly necessary to add that *no dislocation in time* occurs in the alleged 'paradox' above; it is clear from the diagram that the travelling twin does not visit the future in a Wellsian fashion to return 'back to the present.' The metrically discordant temporal series are *contemporary* and the reunion of the twins – provided an enormous longevity of the terrestrial twin – would constitute another 'passing present', another definite local 'Here-Now' in the world history. Yet, what Kurt Gödel proposes in his paper is another Wellsian trip, this time to the past and then 'back' to the present. Needless to say that a self-intersection of any world-line not only contradicts special relativity; not only leads to the strangest causal anomalies on the level of the weirdest television 'time-tunnel' stories; but it cannot even be phrased in a self-consistent language. The idea of uni-directional time re-emerges in the very formulation which purports to eliminate it. Gödel's time-traveller, by his capacity to interact causally with past events would make envious even the medieval God whose omnipotence was restricted by St. Thomas by his incapacity to undo the past: '*Praeterita autem non fuisse contradictionem implicat*'. (*Summa theologica*,

Q. 25, art. 4.) This particular difficulty is absent in the cyclical theory of time (Stoics, Nietzsche, Abel Rey) according to which the *whole past moment of the universe* will identically and *unchangeably* be *repeated*; but it has other difficulties, mentioned above in connection with the Stoic theory of eternal return. Furthermore, this theory is too closely tied to the whole complex of classical and now obsolete concepts as I have tried to show elsewhere.[21]

Eddington explores the relation of 'time's arrow' to the increase of entropy; to him the former is *defined* by the latter. But he is aware of the *statistical* character of the second law of thermodynamics as well as of the existence of fluctuations in the Brownian motion; does it mean that in such cases time 'flows backwards'? His thought is hazy on this point; his strong belief in irreversibility clashes with his definition of the time direction. He admits the possibility – a very small probability – that the same configuration of atoms will re-occur; he says that the discovery of the expansion of the universe makes this occurrence even more *improbable*, – but *not impossible*! He could have easily pointed out that the concept of 'the same configuration of particles' presupposes the idea of corpuscular entity persisting self-identically through time; as David Bohm shows, in his selection, and as it follows from the discovery of the event-like character of all microphysical particles, there are no such things. In Bohm's words, "because all of the infinity of factors determining what any given thing is are always changing with time, *no such a thing can ever remain identical with itself* as time passes" [Bohm's italics.]. Hence Bohm's deep conviction about the reality and irreversibility of becoming. Eddington suspected the inadequacy of the kinetic-corpuscular interpretation of entropy already before then;[22] in this section he conjectures its close relation to the expansion of the universe.

Eddington's theory of the expanding universe shows how far is the present concept of space from its classical counterpart; it is neither immutable nor infinite, though it is still limitless; but, unlike Einstein's universe, its radius of curvature is still growing. Doubt about the beginningless eternity arises, if we extrapolate the observed expansion pastwards, for there must have been a time when the radius of the spherical space was, if not zero, then at least at its minimum. Does it mean that there was then 'zero-time' of which Lemaître and Gamow speak? There is nothing contradictory in such a concept since the question, 'What was *before* it?' is as illegitimate as that which asks what is *behind* the Riemannian space; as we have seen, St. Augustine points out, 'before' is a relation which is meaningful only *in* time, not outside of it. The alleged necessity to raise this question is due to the inveterate mental habit of symbolizing time by a Euclidian line which can be extended in either direction. Furthermore, this theory can be avoided by the theory of an 'oscillating universe' whose evolution consists in a succession of expansions and contractions.

The selections from Lindsay and Margenau, Whitehead, and Norbert Wiener show that even the last classical feature – that of spatio-temporal continuity – is now questionable as far as its applicability to the microcosmos is concerned. The relativity theory, which is essentially macroscopic in its nature (even though some of its spectacular verifications have been in the microphysical domain), did not raise any doubt about it; only in the light of the quantum phenomena did some physicists begin to ask whether spatio-temporal continuity is not "an exorbitant, enormous extrapolation"[23] of our macroscopic experience. This was already hinted at in the concluding sentence of Victor Lenzen's paper. Hence the speculations about the minimum time, 'chronon', and minimum length 'hodon'. The main defect of these speculations is that they apparently assume what they pretend to deny; does not the allegedly negated 'instant' reappear as a boundary separating two successive chronons? Does not the vanished point reappear as the boundary of two minimum lengths? This difficulty remains insurmountable as long as we continue to use geometrico-visual representations; it disappears or at least is significantly reduced when auditory models are used. This is suggested by two thinkers who are considerably different in other respects, A. N. Whitehead and Norbert Wiener; they both come to the conclusion that 'nature at an instant' is as illegitimate and inapplicable a concept on the microphysical level as the 'world-wide instant' is on the megacosmic level.

Our final selection, from Weyl's essay has already been referred to. In giving an objectivist interpretation of quantum mechanical indeterminacy, he rejects the static Laplacean determinism as resolutely as did David Bohm, even though Bohm is not satisfied with the way this rejection is formulated by the majority of physicists today.

NOTES

[1] Cf. J. L. E. Dreyer, *A History of Astronomy from Thales to Kepler*, 2nd ed., Dover, 1953, p. 411.

[2] Bernard Rochot, 'Sur les notions de temps et d'espace chez quelques auteurs du XVIIe siècle, notamment Gassendi et Barrow', in *Revue d'histoire des sciences et de leurs applications* **6** (1956), pp. 94–106.

[3] H. More's letter to Descartes March 5, 1649 in: R. Descartes, *Œuvres*, V, 302 (ed. by C. Adam and P. Tannery).

[4] *Syntagma philosophicum*, I, p. 224: "quodlibet temporis momentum idem est in omnibus locis"; Newton, *Opera* (ed. by Horsley), vol. III, p. 72. "unumcumque durationis indivisibile momentum ubique."

[5] I. Barrow, 'Lectiones mathematicae', Lect. X *The Mathematical Works of Isaac Barrow* (ed. by W. Whewell), Cambridge University Press, 1860, p. 165.

[6] *An Essay on the Foundations of Geometry*, p. 156: "No day can be brought into temporal coincidence with any other day, to show that the two exactly cover each other."

[7] On this point cf. Appendix II, 'Russell's Hidden Bergsonism', in my book *Bergson and Modern Physics* (*Boston Studies in the Philosophy of Science*, vol. VII, Reidel, Dordrecht, 1971).

[8] *Matter and Motion*, New York, Dover, n.d., pp. 40–42.

[9] Clarke quoted Gassendi's section on time in *Syntagma philosophicum* in support of his view that the divine duration is, contrary to the scholastic view, *not* reducible to the instantaneous Now – *Nunc stans*. (*A Discourse Concerning the Being and Attributes of God*, London, 1719, pp. 43–44.)

[10] H. Reichenbach, 'Die Bewegungslehre bei Newton, Leibniz und Huyghens', *Kantstudien* **19** (1924), pp. 428–9. ['The Theory of Motion According to Newton, Leibniz and Huyghens', tr. Maria Reichenbach, in Hans Reichenbach, *Modern Philosophy of Science*, London, Routledge and Kegan Paul and N.Y., Humanities Press, 1959, and in *Hans Reichenbach – Selected Essays*, Vienna Circle Collection, Dordrecht and Boston, D. Reidel Pub. Co., in press 1976]. This inconsistency can be apparently removed, if we bear in mind that, as Reichenbach correctly points out, Leibniz in this context meant by 'absolute motion' not a displacement in space, but 'inner, dynamic becoming which is in itself beyond space' (*ein dynamisches Geschehen, das an sich ausserraumlich ist*). Now since this change belongs to the inner of the monad and not to the physical world which is merely 'a well founded phenomenon' (*phaenomenon bene fundatum*), there would be seemingly no contradiction between Leibniz' relativism of motion in the physical world and his absoluteness of qualitative change on the metaphysical level of the monads. But this would not do; for Leibniz makes an appeal to the metaphysical, qualitative change for a *physical* purpose – to account for the rise of centrifugal forces in 'absolute rotations'.

[10a] Cf. M. Schlick's notes in H. v. Helmholtz, *Epistemological Writings*, (*Boston Studies in the Philosophy of Science*, vol. 37) ed. and tr. M. Lowe, Y. Elkana and R. S. Cohen from the 1921 edition of P. Hertz and M. Schlick (Reidel, Dordrecht and Boston, 1975).

[11] B. Russell, *An Essay on the Foundations of Geometry*, New York, Dover, 1956, pp. 112–13; B. Russell, 'Les axiomes propres à Euclide sont-ils empiriques?', *Revue de metaphysique et de morale* **6** (1898), p. 773.

[12] In the last years of his life Einstein was still concerned about this problem. According

to Carnap's testimony he felt (around 1952) that the character of 'Now' and the difference between the past and the future are not adequately treated by physicists. Cf. R. Carnap, 'Intellectual Biography' in *The Philosophy of Rudolf Carnap* in *The Library of Living Philosophers*, vol XI (ed. Paul A. Schilpp), Evanston 1963, pp. 37–63. (I am indebted to Professor Abner Shimony for calling my attention to this particular passage.)

[13] *Space, Time and Gravitation*, p. 51. Meyerson clearly overlooked the context.

[14] Cf. Hilary Putnam, 'Time and Physical Geometry', in *The Journal of Philosophy* **64** (1967), 240–247. Like Donald Williams ('The Myth of Passage', *J. Phil.* **48** (1951), 457–72, Putnam entirely overlooks the fact that 'now-lines' never intersect the frontward cone of the causal future.

[15] A. A. Robb, *Geometry of Space and Time*, Cambridge University Press, 1936, p. 15, A. Einstein, 'Autobiographical Notes' in *Albert Einstein, Philosopher-Scientist* (ed. by Paul A. Schilpp), Evanston, 1949, pp. 60–61.

[16] Cf. my articles 'Relativity and the Status of Space', *The Review of Metaphysics* **9** (1955), 169–199; 'The Myth of Frozen Passage', in *Boston Studies in the Philosophy of Science*, vol. II, pp. 441–463; *The Philosophical Impact of Contemporary Physics* (1969). Ch. 9, 16, 17 and Appendices I and II.

[17] John C. Graves, *The Conceptual Foundations of Contemporary Relativity Theory*, MIT Press, Cambridge, Mass., 1971, p. 249.

[18] A. Grünbaum, *Philosophical Problems of Space and Time* (*Boston Studies in the Philosophy of Science*, vol. XII) rev. and enlarged 2nd edition, p. 329 [Reidel, Dordrecht and Boston, 1974].

[19] *Le Principe de relativité et la théorie de la gravitation*, Paris, 1922, p. 240.

[20] P. W. Bridgman, *The Logic of Modern Physics*, MacMillan, 1932, pp. 178–185.

[21] 'The Theory of Eternal Recurrence in Modern Philosophy of Science with Special Reference to C. S. Peirce', *J. Phil.* **57** (1960), 289–296.

[22] Cf. *The Nature of the Physical World*, p. 95.

[23] Schrödinger's words in *Science and Humanism*, Cambridge University Press, 1952, pp. 30–31.

PART 1

ANCIENT AND CLASSICAL IDEAS OF SPACE

F. M. CORNFORD

THE INVENTION OF SPACE*

...When I was taught geometry, geometry and Euclid were synonymous terms; and it never occurred to me to doubt that I lived and moved in Euclidean space, extending, quite obviously, in all its three dimensions, without limit. I suspect that, if we look into our minds, all but a few accomplished mathematicians will find the old framework of space and time still unshaken. Common sense lags a good way behind the reasonings of revolutionary thinkers. We have not yet readjusted our perspective and redrawn our map to accommodate such statements as these, which follow in the President's address:

> Neither space nor time is found to exist in its own right, but only as a way of cutting up something more comprehensive – the space-time continuum. Thus we find that space and time cannot be classified as realities of nature, and the generalized theory of relativity shows that the same is true of their product, the space-time continuum. This can be crumpled and twisted and warped as much as we please without becoming one whit less true to nature – which, of course, can only mean that it is not itself part of nature. Space and time, and also their space-time product, fall into their places as mere mental frameworks of our own construction.

So what we took for the steel structure of the universe turns out to be less like steel than india-rubber; and the india-rubber itself exists only as an arbitrary figment of the human brain. It will be some time yet before common sense assimilates this doctrine and begins to think easily in terms of its concepts.

I am not now concerned with a problem that might perplex the simple mind: How can a ray of light be sure of travelling all round a space that can be twisted and crumpled at the mathematician's pleasure? And if it cannot, what becomes of the astronomer's hope that, if he can only wait long enough, he will see the back of his own head through "a sufficiently powerful telescope"? But there remains a question of interest to those who still care to know something of our inheritance from the past. How did the illusion of the steel framework, as an external fact, come to be imposed upon common sense? If the infinite extent of three-dimensional space is no more than a construction of the human brain and only one of many possible alternatives, all equally agreeable to nature, when and by whom was it constructed? Did the Euclidean era, from which we are now emerging, stretch back, with no definable limit, through all recorded history into the darkness of the Stone Age? Was the geometry set forth by Euclid in the ordered steps of logical deduction simply an explicit formulation of what common sense, from the dawn of human life, had always implicitly conceived? That might naturally be assumed so long as Euclidean space was taken either as a given fact of external nature or as an equally given fact in the constitution of our own minds, needing only to be discovered and displayed in rational argument. But now that it is being replaced by an arbitrary fabric of non-existent india-rubber, the assumption may be questioned.

By whom, then, was the framework created? Professor Eddington's suspicions fall upon Euclid himself.

"The only thing," he writes, "that can be urged against spherical space is that more than twenty centuries ago a certain Greek published a set of axioms which (inferentially) stated that spherical space is impossible. He had, perhaps, more excuse, but no more reason, for his statement than those who repeat it to-day."[1] Euclid was teaching at Alexandria round about 300 B.C. Probably he had been trained in Athens by Plato's pupils at the Academy. But he was, in the main, only codifying a geometry which had been built up piecemeal in scattered theorems by Greek mathematicians of the preceding three centuries. The work was begun in the sixth century by Thales and Pythagoras, the first parents of the two parallel traditions of philosophy. I seek to show that the belief in infinite space as a physical fact can be traced back to the Greek philosophers of the three centuries between Thales and Euclid, but no farther. Granted that we are dealing here with a fabrication of human brains, the brains in question were active between 600 and 300 B.C. Their figment came to be finally imposed on science in the Euclidean era now ending, and to be so deeply ingrained in common sense that we shall find it hard to assimilate the india-rubber substitute.

If this is true, we are concerned with a product of Western civilization in its Hellenic phase. It would help my thesis if I were in a position to show that the Indians or the Chinese, before Western science spread all round the globe, had some different scheme of conception. This seems to me likely, but ignorance confines my argument to the history of Western thought and to the documents that I can read and hope to understand.

Within these limits we start from the known fact that Euclidean geometry was constructed, from beginning to end, by the Greeks of those three centuries. How did they arrive at the notion of that familiar space in which straight lines travel on for ever farther and farther from their starting-point? If, as we are told, such a space does not exist, of its own right, in nature, the construction does not come immediately from observation. Nor can we fall back on the assumption that the mathematicians were simply formulating the implicit conceptions of immemorial common sense; for the ordinary man, no more than the philosophic geometer, could observe what was not there. The inference is that the belief in the infinite extent of space, implied by this geometry, was not implicit in the mind of the Homeric or pre-Homeric Greek. There was a pre-Euclidean common sense, whose conception of the world in space had to be transformed into the Euclidean conception, just as our Euclidean common sense has now to be transformed into the post-Euclidean scheme of relativity.

The evidence for that earlier transformation is to be found in the philosophic literature of our three centuries. As present experience shows, a readjustment of this order cannot be made suddenly; for several generations the old ideas may persist, while common sense lags behind the fresh discoveries of the most advanced minds. In antiquity knowledge spread slowly. When Democritus came to Athens, which lies about 220 miles from his native city, he complained that no one had ever heard his name. A revolution of thought such as may now take one or two generations, might well take a couple of centuries in ancient Greece. The literature of the great creative period preserves abundant traces of the resistance offered by pre-Euclidean common sense to the then revolutionary doctrine of infinite space.

For our purpose the essential property of Euclidean space is that it had no centre and no circumference. In its full abstraction, as conceived by the mathematician, it was an immeasurable blank field, on which the mind could describe all the perfect figures of geometry, but which had no inherent shape of its own. For the physicist it was the frame of the material universe, partly occupied by visible or tangible bodies, whose number and extent were again without definable limit.

Now this physical frame figures in the atomistic systems of antiquity as the Void. The illimitable inane of Lucretius is taken from Epicurus, the contemporary of Euclid. Epicurus took it from the earlier atomists, Leucippus and Democritus, who were at work in the second half of the fifth century. It was these atomists who maintained the existence of an unlimited Void, as a fact in nature. I would suggest that, in so doing, they were endowing the abstract space implied in Greek geometry with physical existence.

As I read the story, what happened was briefly this. As geometry developed, mathematicians were unconsciously led to postulate the infinite space required for the construction of their geometrical figures – that space in which parallel straight lines can be produced 'indefinitely' without meeting or reverting to their starting-point. In the sixth and fifth centuries no distinction was yet drawn between the space demanded by the theorems of geometry and the space which frames the physical world. We know from Aristotle that the earlier Pythagoreans did not even distinguish the solid figures of geometry from the bodies we daily see and handle. Hence the considerations which led mathematicians to recognize infinite space in their science simultaneously led some physicists to recognize an unlimited Void in nature. These were the atomists, whose system was the final outcome of a tradition inspired by Pythagorean mathematics. The atomists broke down the ancient boundaries of the universe and set before mankind, for the first time, the abhorrent and really unimaginable picture of a limitless Void.

If this summary account is correct, the space framework finally accepted by physical science in the Euclidean era is simply the Void of Lucretius. We asked how that framework came to be constructed and imposed upon common sense. It remains to substantiate in more detail the answer suggested: that it was constructed by the reasoning of Greek geometers and imposed by the atomists.

Consider first the progress of geometry towards its final form in the thirteen books of Euclid. We find it presented there as a rigid chain of logical deduction, starting from a number of indemonstrable premisses – definitions, postulates, common notions – and proceeding to more and more complex theorems, in which every step is guaranteed by some previous conclusion. But this form gives no picture of the process by which the various parts of the structure were first discovered. The proposition that the square on the hypotenuse of a right-angled triangle is equal to the squares on the other two sides is said to be due to Pythagoras himself, one of the founders of the whole science. There is no reason to doubt the tradition, and I hope it is also true that Pythagoras sacrificed an ox in the joy of his discovery. Now in Euclid this theorem stands as the last but one in the first book, preceded by forty-six prior propositions and by all the ultimate premisses. But Pythagoras lighted upon it as an isolated truth, and had no idea that he ought to demonstrate forty-six other propositions before he would be warranted in sacrificing his ox. Geometry, in fact, was discovered piecemeal by many independent minds, who attacked particular problems or hit upon particular theorems without co-ordinating their results. So might an unexplored country be mapped by a number of surveyors, each working outwards from a different point and covering as large an area as he could manage. Later some geographer might piece these fragments together. He would find gaps needing to be filled in, and only then would he see the outlines of the country as a whole. In geometry this work of co-ordination was partly done in the fourth century by Plato's colleagues and pupils, and it was triumphantly completed by Euclid. The task involved working backwards, as well as forwards, along the chain of deduction. Among the last things to be established would be those which stand first in the final presentation, the irreducible collection of premisses on which the whole structure depends. In the *Republic* Plato complains that the examination of first premisses had been neglected. It was left for his own school to undertake the task and to carry it, as they supposed, to completion.

I suggest – though I cannot directly prove this – that geometrical space itself may be compared to the outline of a country revealed for the first time to the co-ordinating geographer. It was not realized from the first that the figures employed in the scattered theorems demanded a space of infinite extent. If we suppose this discovery to date from about the middle of the fifth century, then, since the theorems seemed to be established beyond doubt, we can understand why the space they implied was accepted by the atomists as the framework of reality.

I am led to this conjecture by the history of the Void – a curious history traceable through the philosophic writings of our period. I have suggested that the infinite Void is simply Euclidean space credited, as a matter of course, with physical existence. But its existence was maintained by the atomists only in the teeth of very considerable opposition. At the end of our period it was still denied by Aristotle; and his immense authority, fortified by ecclesiastical prejudice, held atomism at bay until physics began to move forward again at the Renaissance. Here, however, we are concerned with the question why the infinite Void met with opposition in the fifth and fourth centuries. The answer lies, at least partly, in the resistance of pre-Euclidean common sense, persisting in the minds of most philosophers. It follows that the space of Euclidean geometry was not the accepted framework of nature before the geometers had mapped it out.

Our first glimpse of the Void in philosophic literature we owe to a passage in Aristotle's *Physics*, recording a feature of the primitive Pythagorean cosmology:

The Pythagoreans too asserted that Void exists and that it enters the Heaven itself, which, as it were, breathes in from the boundless a sort of breath which is at the same time the Void. This keeps things apart, as if it constituted a sort of separation or distinction between things that are next to each other. This holds primarily in the case of numbers; for it is the Void that distinguishes their nature (*Phys.*, IV, 6, 213*b*, 23).

The very obscurity of this statement is witness enough to the archaic character of the system described, which must go back to the sixth century. We are to imagine a spherical universe called 'the Heaven,' a living creature, whose breath is drawn in from the boundless air enveloping it outside. The important point is that 'the Void' is another name for this air or breath. As Aristotle notes in the neighbouring context, we still speak of a vessel as empty when in fact it is filled with air. Within the Heaven, the function of this or vacancy is to keep apart the solid bodies we see and to give them room to move in. Thanks to these vacant intervals, body or matter is not one solid immovable block, but a plurality of discrete things that can move about. The physical picture is not hard to imagine.

What baffles us at first sight is Aristotle's last sentence: "This holds primarily in the case of numbers; for it is the Void that distinguishes their nature." The Pythagoreans represented numbers by patterns of dots or pebbles or counters, arranged in squares, triangles, and other figures, as on our dice and dominoes. Hence we still speak of 'square numbers,' 'cubes,' and so forth. The Void which distinguishes their nature is the blank intervals between these units, or the gaps separating the terms in the series of natural integers.[2] Moreover, since these units were disposed in regular geometrical shapes, the Void is also the blank field (χώρα) marked off by the boundary lines of geometrical figures.³ Under this aspect the Void was the space of geometry.

To our minds it is barely possible to confuse the empty gaps between terms in a series of numbers with the physical air or vacancy that keeps solid bodies apart, or even with geometrical space. But we are dealing here with the most primitive form of atomism, older by perhaps half a century than the atomism of Leucippus. These Pythagoreans simply identified the units of number with geometrical points having position in actual space and indivisible magnitude. They held that physical bodies actually *are* numbers – a number being defined as a plurality of units. Visible and tangible bodies are built up of these monads, which are the units of arithmetic, the points of geometry, and the atoms of body, all at once. The monads are preserved in discrete plurality by intervals of vacancy – gaps between terms in a numerical series, space between the boundaries of geometrical figures, air between the atoms of body and between the surfaces of different bodies.

The essential function of the Void in this system is to keep things apart inside the spherical Heaven and give them room to move in. For that purpose the living Heaven breathes in the vacant air that laps it round. Obviously this internal Void (as we may call it) raises no question of infinite extension. The arguments in defence of the Void, reviewed and criticized by Aristotle in the context, are arguments for the internal Void. It was alleged that bodies could not move without empty spaces to move into; that bodies could not otherwise expand or contract; and that the growth of animals could occur only if the substances they eat could find vacant spaces to occupy.⁴ The whole controversy about the existence of such vacant intervals could be carried on without raising the question of an infinite extent of Void outside the Heaven.

But there is something outside the Heaven; beyond the circumference is the enveloping air which the world breathes. So far there is no occasion for an infinite extent of Void or air outside. The life on our planet is sustained by an envelope of air a few miles in depth; we do not need that air shall spread to the limits of the galaxy and beyond it for ever. Must we suppose that the air round the Pythagorean Heaven was strictly unlimited?

When Aristotle speaks of the Heaven inhaling its breath "from the boundless" (ἐκ τοῦ ἀπείρου), some would take that word as meaning just that unlimited extent which belongs to the Void in later atomism. Were that so, my thesis would fall to the ground; because here, at the threshold of Greek cosmology, we should find infinite space already established in physical existence. Everything hangs on the meaning and implications of 'the boundless' in sixth-century Greek. 'The boundless' figures in the still older Milesian systems of Anaximander and his successor Anaximenes. Anaximander taught that our cosmos was formed of materials drawn from a 'boundless' body which at all times encompasses the world. Anaximenes identified this body with air – that very air which the Pythagorean Heaven breathes. Scholars have debated whether the word ἄπειρον meant 'indefinite in quality' or 'unlimited in extent,' or both. Some who hold that quantity, rather than quality, is in question have assumed that 'boundless' implies the infinite and shapeless extent of Euclidean space or the Lucretian Void. But the word by no means excludes the idea of shape. On the contrary, it is frequently and specially used of circular or spherical shape, because on the circumference of the circle or the sphere there is no beginning or end, no boundary separating one part from another. A scholium on the *Iliad* (Ξ, 200), discussing the phrase 'the limits of the earth,' quotes Porphyry to the following effect. The circumference of the circle and the surface of the sphere are the only figures that can be called in every way uniform. Hence the ancients with good reason described the circle and the sphere as 'boundless'. Thus Aristophanes has the phrase 'Wearing a boundless bronze finger-ring,' meaning a ring having no juncture, no limit as beginning or end. Rings having a bezel with an inset gem are not 'boundless,' as not being uniform. Aeschylus again speaks of women standing round an altar "in a boundless company," meaning the circular arrangement. Euripides calls the seamless tunic "a boundless texture," and speaks of the ether as boundless because it is round (κυκλοτερής) and embraces the earth in its arms.

The last quotation is specially significant. Porphyry takes 'boundless ether' to mean the round encompassing sky, which the speaker in Euripides identifies with Zeus or god. So Anaximander described his 'boundless,' encompassing the world, as 'the divine'; and Empedocles calls his divine universe "a rounded sphere altogether boundless".[5] Thus the word 'boundless' in itself affords no reason to suppose that the enveloping air of Anaximenes or Pythagoras was of infinite extent. There is, on the contrary, some ground for thinking that it actually implied spherical shape.

This interpretation becomes still more probable when we consider the slightly later system of Parmenides. The whole of being, he declares, "since it has a furthest limit, is complete on every side, like the mass of a rounded sphere, equally poised from the centre in every direction." We naturally ask, what is outside this finite sphere of being? Parmenides does not raise that question; apparently it did not occur to him that such a question could be asked. On the other hand, we are left in no doubt as to the answer by what he says about the internal Void of the Pythagoreans. He flatly denied its existence. The Void, to his mind, is simply "nothing," and what is nothing can have no existence. The internal Void, as we saw, was to keep things apart and to provide room for motion. Parmenides accepted the consequence: since the Void, being nothing, cannot exist, a plurality of separate things and motion is impossible. Reality becomes one solid immovable block. The appearance of plurality and motion must be somehow illusory. Now, if nothing cannot exist inside the world, neither can it exist outside. If we do ask the question he ignores, the only possible answer is: Outside the One Being there can be neither something (for all being is inside) nor nothing (for nothing cannot exist or even be conceived).[6] If we find this answer baffling, the fault lies in our own Euclidean preconception that space must extend without limit; therefore, beyond a finite sphere containing all being there must be an endless waste of nothing. The difficulty vanishes when we realize that, at Parmenides' date, no one had seen any reason why there should be an infinity of unoccupied space.

It appears, then, that in these earliest cosmologies the universe of being was finite and spherical, with no endless stretch of emptiness beyond. Space had the form of that which filled space – the form of a sphere with centre and circumference. The point in dispute was, whether the sphere was entirely compact with body or there were vacant intervals inside. Parmenides' denial of these intervals shows that the distinction between air, which is something, and the true Void, which is nothing, was beginning to be drawn. The true Void had its origin in the mathematical Void invoked by the Pythagoreans to separate the units of number. As mathematics became more independent of physics, the confusion of intervals between numbers with the air keeping apart physical bodies could not long persist. So the true Void came to be distinguisted from air. Anaxagoras, in the fifth century, demonstrated by experiments with 'empty' wineskins and waterclocks that the spaces we call empty are really filled with air, which is something, since it resists pressure. By proving the substantial existence of air he thought he was disproving the existence of any true Void.

After Parmenides, the first task of physics was to restore the possibility of a plurality of things and of motion in space. Atomism was revived in a less questionable form by Leucippus and Democritus. They met Parmenides' denial of the Void with a bold reply. "I admit," said Leucippus, "that the Void – sheer emptiness – is 'nothing' or 'not-being'. All the same, this nothing does exist no less than the something, the compact being, we call body." Thus the internal Void was reasserted, no longer as air, but as the true Void. By this time, if my hypothesis is correct, geometers were realizing that their science called for a space of unlimited extent. Geometry, moreover, was detaching itself from arithmetic. It was now denied that space is made up of points that could be identified with the units of number separated by intervals of emptiness, the discontinuous arithmetical Void of the Pythagoreans. Geometrical space was seen to be continuous, not a pattern of empty gaps interrupted by solid things; it penetrates the solids that partly occupy its single continuous medium. At the same time the theorems of geometry were seen to require that parallel straight lines shall travel on for ever through this medium without meeting or returning upon themselves. The atomists now take the revolutionary step of ascribing to a physical Void, outside the visible Heaven, the infinite extent of this geometrical space.[7] Atoms, they held, must be illimitable in number and therefore demand an unlimited extent of space.

The consequences were far more outrageous to pre-Euclidean common sense than we, who have assimilated infinite space, can easily realize. Space was now robbed of its circumference, and therefore of any centre. The immemorial claim of the Earth to be at the centre of the universe was impiously denied. The Earth might still be at the centre of our finite world; but our world has been cut adrift in a limitless waste that has no centre. And now, for the first time,[8] appears the consequent belief in innumerable worlds (*cosmoi*) scattered over endless space. At all times some are coming into being, others passing away. These other worlds were not the same as any stars, or clusters of stars, that we can see: all the stars belong to our world. They were what might now be called 'island universes,' whose existence, entirely beyond the range of observation, was asserted on *a priori* grounds as a reasonable probability. Thus Metrodorus of Chios argued:

That there should be only one world in the infinite would be as strange as that a single ear of corn should grow in a large plain.

Epicurus pointed out that no limited number of worlds could exhaust the unlimited supply of atoms, and Lucretius followed. This doctrine could not arise until the ancient boundaries of spherical space had been broken down and the belief in its strictly infinite extent had deprived space of any centre for our Earth to occupy.

Thus atomism created the picture of illimitable and shapeless vacancy with its sparse population of unnumbered worlds. Once drawn, the picture could never be forgotten so long as men could read Lucretius. It remained to be accepted by the physicists who revived atomism in modern times. Meanwhile its truth was strenuously denied by philosophers who clung to the spherical image of pre-Euclidean common sense. Commentators on the *Timaeus* have doubted whether Plato admits even the internal Void. Aristotle denied any Void, whether internal or external. He demonstrates that there cannot be more than one world, and that the encompassing Heaven is necessarily spherical. Outside the Heaven, he says, there can be "no place or void or time". The Void had been defined as that in which the presence of body, though not actual, is possible. Body cannot exist outside the Heaven. Therefore there is no external Void (*de caelo* i, 9).

It only remained to point out that the space of geometry, if it really required infinite extension, was not, after all, the same thing as physical space. For Plato the objects and truths of mathematics belong to an intelligible realm; the physical world is no more than an imperfect copy or reflection. Geometrical space could be disposed of as an object of thought, not of the senses, or as in some way imaginary. The geometer may claim that his straight lines can be produced 'indefinitely'; but no one can actually draw a line of infinite length. All the mathematician needs, says Aristotle, is a finite line produced as far as he pleases. Aristotle's theory, Dr. Ross remarks,

is here somewhat obscure. He holds strongly that the physical world is a sphere of finite size. The mathematician cannot have a straight line greater than the diameter of this sphere present to him in sensation, and the meaning must be that he is free to *imagine* such a line if he chooses, and if he can.[9]

So tenacious was the resistance of pre-Euclidean common sense. The Greek mind recoiled in horror from the boundless vacancy its own reasonings had conjured into existence. It would be fantastic to suggest that a sound instinct held it back. But listen once more to Sir James Jeans:[10]

Are there any limits at all to the extent of space?

Even a generation ago I think most scientists would have answered this question in the negative. They would have argued that space could be limited only by the presence of something which is not space. We, or rather our imaginations, could only be prevented journeying for ever through space by running up against a wall of something different from space. And, hard though it may be to imagine space extending for ever, it is far harder to imagine a barrier of something different from space which could prevent our imaginations from passing into a further space beyond.

So the Euclidean has argued ever since Archytas, Plato's contemporary, reasoned thus:

If I am at the extremity of the heaven of the fixed stars, can I stretch outwards my hand or staff? It is absurd to suppose that I could not; and if I can, what is outside must be either body or space. We may then in the same way get to the outside of that again, and so on; and if there is always a new place to which the staff may be held out, this clearly involves extension without limit (Eudemus, frag. 30).

So, too, Lucretius:

If for the moment all existing space be held to be bounded, supposing a man runs forward to its outside borders and stands on the utmost verge and then throws a winged javelin, do you choose that when hurled with vigorous force it shall fly to a distance, or do you decide that something can get in its way and stop it? for you must admit and adopt one of the two suppositions; either of which shuts you out from all escape and compels you to grant that the universe stretches without end (I, 968, transl. Munro).

Some two thousand years after Archytas, John Locke repeats his argument:

If body be not supposed infinite, which I think no one will affirm, I would ask, Whether, if God placed a man at the extremity of corporeal beings, he could not stretch out his hand beyond his body? If he could, then he would put his arm where there was before space without body.... If he could not stretch out his hand, it must be because of some external hindrance ... and then I ask, Whether that which hinders his hand be substance or accident, something or nothing? ... I would fain meet with that thinking man, that can in his thoughts set any bounds to space, more than he can to duration; or by thinking hope to arrive at the end of either (*Essay*, ii, 13, 21).

But this argument, Sir James Jeans continues, is not a sound one.

For instance, the earth's surface is of limited extent, but there is no barrier which prevents us from travelling on and on as far as we please. A traveller who did not understand that the earth's surface is spherical would naturally expect that longer and longer journeys would for ever open up new tracts of country awaiting exploration. Yet, as we know, he would necessarily be reduced in time to repeating his own tracks. As a result of its curvature, the earth's surface, although unlimited, is finite in extent. Through his theory of relativity, Einstein claims to have established that space also, although unlimited, is finite in extent. The total volume of space in the universe is of finite amount, just as the surface of the earth is of finite amount, and for the same reason; both bend back on themselves and close up. ... As a consequence of space bending back into itself, a projectile or a ray of light can travel on for ever without going outside space into something which is not space, and yet it cannot go on for ever without repeating its own tracks. For this reason it is probable that light can travel round the whole of space and return to its starting-point, so that if we pointed a sufficiently powerful telescope in the right direction in the midnight sky, we should see the sun and its neighbours in space by light which had made the circuit of the universe.

This post-Euclidean finite but unbounded space takes us back to the pre-Euclidean finite but boundless sphere of Anaximander, Parmenides, and Empedocles. These philosophers did not know as much mathematics as Einstein; but they had the advantage over Newton in knowing much less mathematics than Euclid. They had not been misled by geometry into projecting its infinite space into the external world under the name of the Void. The Euclidean era thus presents itself as a period of aberration, in which common sense was reluctantly lured away from the position that it has now, with no less reluctance, to regain. The whirligig of Time has brought in his revenges upon the impious assailants of spherical Space.

Tantum irreligio potuit suadere malorum

NOTES

* From *Essays in Honor of Gilbert Murray*, Allen & Unwin, London, 1936, pp. 215–235.
[1] *The Expanding Universe* (1933), p. 40.
[2] Cf. Simplicius, *Phys.*, 652, 4 (on this passage): "For what else separates 1 from 2 or 2 from 3 but the void, there being no existence between them?"
[3] Proclus (on *Euclid*, I, Def. xiii, p. 136, Friedlein) remarks that the geometrical term 'boundary' (ὅρος) belongs of right to the primitive 'land-measurement' (γεωμετρία), whereby areas of land were measured and their boundaries kept distinct.
[4] *Physics*, IV, 6; cf. Lucretius, i, 329ff.
[5] Emped., fr. 28, πάμπαν ἀπείρων Σφαῖρος κυκλοτερής. Ar. *Met.*, 1074, bI, "The ancients from the most remote ages have handed down to posterity a mythical tradition that these (the heavenly bodies) are gods and that the divine encompasses the whole of nature."
[6] Cf. Plato, *Theaet*. 180 E. Parmenides declared that the One Being is at rest within its own limits, "having no room in which it moves". There is no vacant space outside, in which it could move about.
[7] Simplicius, *Phys.*, 648, 11, "These asserted the actual existence of an interval between bodies which prevents their being continuous, as Democritus and Leucippus held, who

declared that there is a void, not only inside the cosmos, but also outside – a thing which clearly will not be 'place', but something with an independent existence."
[8] In the minds of most modern writers the whole question of infinite space has been prejudiced by Burnet's attribution of the doctrine of innumerable worlds to Anaximander and other pre-atomists. I have argued (*Classical Quarterly* 38 (1934) 1ff.) that the evidence he relied on is worthless.
[9] *Aristotle* (1923), p. 85. Ar. *Phys.*, 207*b*, 27.
[10] *The Universe Around Us*, p. 70.

C. BAILEY

MATTER AND THE VOID ACCORDING TO LEUCIPPUS*

...But there was yet another objection to which this theory of the existence of matter in the form of infinite discrete particles was liable. If they are discrete, there must be something to separate them (διάστημα): if they are to move – and without motion they cannot combine to form things or shift their position so as to change things – there must be something external to them for them to move in. What is this something? The Pythagoreans, who with their doctrine of the infinitely divisible had been confronted with this problem, had thought of air as lying between the particles of matter, but since the theory of Empedocles had shown that air was an element, as corporeal in substance as earth or fire or water, this answer was no longer possible. Parmenides had seen that the only answer could be 'empty space', but, profoundly convinced as he was that the only existence was that of body, he had denied the existence of empty space altogether: it was 'nothing' (οὐδέν). The world was a corporeal *plenum*, there was no division between parts of matter but all was a continuous whole, neither was there any possibility of motion. Melissus had more recently enforced this position as part of his attack on pluralism:

Nor is there anything empty (κενεόν): for the empty is nothing and that which is nothing cannot be: nor does it (*sc.* the world) move: for it has nowhere to withdraw to (ὑποχωρῆσαι), but it is full. For if there were anything empty, it would withdraw into the empty: but as the empty is not, it has nowhere to withdraw to.

Leucippus was prepared to meet this difficulty perfectly boldly: he affirmed, using Melissus' own terms, that "the elements are the full and the empty", or, as Diogenes puts it, "the universe is the full and the empty", that is, that it is as necessary to conceive of unoccupied space separating the particles of matter as it is to postulate the particles of matter themselves. How then did he answer the metaphysical difficulty of the Milesians? His reply was extremely subtle, and if it seems to us something of a quibble, this is due not so much to Leucippus, as to the persistence of the philosophical tradition which had hitherto refused to acknowledge reality, except in the form of matter. He admitted, as is made clear in Aristotle's exposition, that empty space is 'not real' (μὴ ὄν): the only fully real existence is matter, 'the completely full' (παμπλῆρες). But, he said, though empty space is not in this sense a real thing, it none the less exists: "the real exists not a whit more than the not real, empty space no less than body." In effect Leucippus had introduced a new conception of reality: in the old sense empty space is not real, for it has not the most elementary attribute of matter, it cannot touch or be touched. But it none the less exists: we must form a new idea of existence, something non-corporeal, whose sole function is to be where the fuller reality is not, an existence in which the full reality, matter, can move and have its being. He thus established the necessary requirement for his conception of a discrete infinity of matter; the particles of body were indeed separated by the empty. At the same time, it must be observed, he had protected himself against a possible charge of dualism: if he were taxed with having, like the materialistic pluralists, destroyed the fundamental unity of the universe by the admission of two 'existences', matter and space, he could reply that he still held that there was but one full reality: the other 'existence' was, as it were, but the negation of this reality (τὸ μὴ ὄν), but it did nevertheless exist. For the moment his success was to have conquered in the dialectical debate with the Eleatics, but for the permanent progress of thought he had achieved the far greater triumph of establishing the conception of non-corporeal existence: it is, as Burnet has pointed out, a strange achievement for the founder of the great materialist school of antiquity to have been "the first to say distinctly that a thing might be real without being a body".

In all these different ways Leucippus had shown himself the mediator, and had reconciled the oppositions which had played so large a part in previous debate. He had shown how the One might also be the Many, he had shown that there might be a discrete infinity and yet a limit to division, he had exhibited the existence of empty space without refusing to acknowledge matter as the single reality. But he had of course done much more than this, for he had discovered a theory of the Universe, which was not only in itself remarkable in its penetration, but was to prove capable of development and application with a completeness which was impossible to any of its predecessors. To realize this it is necessary to leave the negative conception of Leucippus as a mediator and to inquire into the positive doctrines of his theory and its application as an explanation of the world.

MATTER AND THE VOID ACCORDING TO LEUCIPPUS

The main principles of the theory of Leucippus have so far established the conception of small discrete particles of homogeneous body separated by and situated in empty space. It is a long step from this to the infinitely complex and varied world of phenomena, and it is necessary at this point to ask how he conceived the nature of the two constituents and by what process they were enabled so to combine as to create the things of experience.

First, both atoms and space are infinite, the atoms in number, space in extent. This might be deduced from the belief that the universe itself was infinite, in which Leucippus followed the Ionians and Melissus, who had revolted on this point from the strict conception of Parmenides. But it would appear that Leucippus[1] argued rather from the infinite variety of phenomena to the infinite number of the atoms, and from that the infinite extent of space follows directly: for in a limited space, infinity of particles could only be obtained by infinite subdivision.

Of the nature of space there was but little to be said: all that Leucippus could predicate of it is that it is 'empty' (κενόν), or as he seems to have put it in one of those quaint terms which he invented for the Atomic theory, 'porous' (μανόν). The use of this term is instructive, for it makes it clear that he was thinking always of the intervals (διαστήματα) between particles of matter. Later on there is some confusion between two possible senses of 'space'. The mathematical sense of extension, though it is sometimes attributed to the Atomists, was always too abstract a conception for them, but they did vary between thinking of space as the whole extent of the universe, some parts of which were occupied by matter, and on the other hand as the 'empty' parts, the intervals between body. But it is clear that by Leucippus at any rate only the latter meaning is intended: 'space' is the sum total of those parts of the whole, which at any given moment are not occupied by matter. Body and space are mutually exclusive, and together they make up the sum total of the universe....

NOTE

* From *The Greek Atomists and Epicurus*, The Clarendon Press, Oxford, 1928, pp. 74–77.

[1] This is certainly the argument which Leucippus adduced for the infinite shapes of the atoms and therefore *a fortiori* for their infinite number, and its method is supported by the argument from sensation noticed above in favour of a limit to division.

P. DUHEM

PLATO'S THEORY OF SPACE AND THE GEOMETRICAL COMPOSITION OF THE ELEMENTS*

The physics of Plato's *Timaeus* appears to be closely related to the physics of Leucippus and Democritus.[1] The role which they attribute to *non-being*, to *nothing-at-all*, to the *void* (τὸ μὴ ὄν, τὸ μηδέν, τὸ κενόν), Plato attributes to what he calls *space* (ἡ χώρα).[2]

As we have seen, at the apex of reality Plato places the ideas of things, ideas which are susceptible to neither generation nor change nor destruction, ideas which are not sensed, which can be known only by rational intuition (νόησις). These ideas properly constitute unchanging being or simply *being* (τὸ ὄν).

At the lowest degree of reality, on the contrary, are changing beings, which are born and perish, and which are known by sense perception (αἴσθησις); Plato often calls this group of beings *generation* (ἡ γένεσις).

To these two categories of beings, Plato adds a third, constituted by *space* (ἡ χώρα). "Here, in short," he says, "is my opinion: being, space and generation exist, and these three exist in three different ways."

Why is it necessary to admit the existence of this space? Because whatever is subject to generation and corruption "is continually in local movement; it comes into being in a certain place and then vanishes from this same place." Such local movement which for a changing being is the beginning of its existence at a certain place followed by its subsequent disappearance from the same place, presupposes a place which persists while this movement is taking place. This place cannot be provided by absolute and ideal being; indeed, unchanging being "never receives into itself any other being from outside, any more than it ever penetrates any other being." This place, then, can be found only in a third kind of being, in space. Like absolute and ideal being, space is immune to destruction; but it is not impenetrable by other beings as ideal being is; it offers a place to all those beings which are born and die.

How can we know this space? Plato tells us that it does not fall under the senses by which we perceive changing and corruptible beings; and, although he does not say so, he doubtless admits that space is not, like pure beings, contemplated by intellectual intuition. "It can be reached," the author of the *Timaeus* continues, "only by a sort of hybrid reasoning." One might think that Plato means geometrical reasoning, which involves both νόησις, and, at the same time, by virtue of the imagination which accompanies it, αἴσθησις. We are hardly convinced of the reality of space by this reasoning; it remains "barely believable."

The vision we have of space can be compared to the images which we think we perceive in dreams. We have an image of this kind when we think about the infinite space in which the limited spherical Universe is lodged; the space in which the Universe finds a place is, in fact, absolutely empty, for there is nothing outside of what has its place in earth or in heaven. Plato expresses such thoughts in the following terms:

> It is this space which we see as in a dream when we say: It is necessary for universal being to be somewhere, in a certain place, and to occupy a certain space; and, on the other hand, what is neither in earth nor somewhere in heaven is necessarily nothing at all.

Thus Plato poses the problem of the location of the Universe, a problem which haunts most of the philosophers we shall have to deal with.

According to Plato, then, there is, outside of the limited, spherical World, a necessarily unlimited space, where this Universe is located. Since nothing exists in this space, it is empty. In the World itself, Plato, unlike the Atomists, disallows the existence of empty space; he rejects the opinion, professed by the Atomists, according to which the existence of the void is required for the possibility of movement. In a language which anticipates Descartes he states that all motion produced within the completely filled Universe is vortical movement, closed upon itself.[3] He refers to the air we exhale in formulating this doctrine:

> There is no void for any of the moving bodies to enter; therefore when we expel the breath from our chest, it is clear to everyone, in the light of what has just been said, that this breath does not go out into the void, but that it pushes the neighboring air from its place. This air, in turn, always pushes away that which is next to it. By the same necessity, all the air moves in a circle; as soon as the air leaves a place, other air enters as if it were attached to the air that escaped, and it fills that place. This entire effect is produced simultaneously like a wheel turning about its axis. All this comes about because there is no void.

While Plato does not admit the void of the Atomists in his World, neither can one say that he admits what these philosophers call the plenum, that is, the indefinite, but rigid and impenetrable substance, from which they form bodies; in space, in the χώρα, Plato admits no real bodies other than combinations of geometrical figures.

This hybrid reasoning which is geometrical reasoning leads Timaeus to represent specific essences of fire, air, water, and earth, intermediary between being and changing appearances. The theory of regular polyhedra reveals the nature of these essences to him.

Timaeus first describes the three regular polyhedra with triangular faces, namely, the tetrahedron, the octahedron, and the icosahedron; then he defines the cube.[4] He is certainly too much of a geometrician not to know that there is a fifth regular polyhedron, the pentagonal dodecahedron, and that is what he is referring to when he says: "There exists a fifth combination which God used to design the Universe."[5] But only the first four polyhedra represent the specific essences of the elements.

To the earth we shall give the cubic species. Indeed, of the four types of elements earth is the most immobile; among the bodies, it is the most likely to stay in one place. Thus it must have the most stable bases.

The square bases assure the cube a greater stability than the triangular bases provide for the other polyhedra.

To fire we shall attribute the polyhedron which is the most mobile, because its bases are the least numerous, and which is the sharpest, the most likely to divide and to cut – in a word, the tetrahedron. To air and water, which are, because of their decreasing mobility, intermediaries between fire and earth, we shall give the octahedron and the icosahedron.

How are we to understand this correspondence between the four elements and the regular polyhedra? Must one understand the cube, the icosahedron, the octahedron, and the tetrahedron to be merely symbols of the specific essences of earth, water, air and fire? Or must one, like the disciples of Democritus, imagine that the tangible and visible elementary bodies are really combinations of such polyhedral particles? The following passage leaves no doubt that Plato is of the second opinion:

It is therefore likely and fitting to regard the shape of the tetrahedral solid as being the element and seed of fire, the second figure as being the element of air, the third as being the element of water [the cube, finally, as being the element of earth]. These solids must be conceived as so small that it is impossible for us to discern one of them in isolation from others of the same species; but when these solids are found together in great numbers, we see the mass they form.

How can this opinion be accepted without contradiction? Contrary to Leucippus and Democritus, Plato, as we understand him, affirms that there is no void, that all motion is produced in an absolute plenum and therefore takes a vortical form. He explains this with a clarity which Descartes himself will not surpass.

Did he believe then that icosahedra and octahedra could be juxtaposed in such a way as to form continuous masses of air or water without leaving any empty interval? Surely he was too good a geometrician to think that.

We must conclude from this that the different parts of his theory stand in irreducible contradiction to one another. If this seems surprising or scandalous, let us compare the incoherency of Plato with that of Descartes. Descartes also admits that there is no void. He too admits elements each of which is formed of small bodies of a determinate figure. Did he ever wonder how the rigid whorls of his subtle matter could fill the interstices of the spheres which form gross matter without leaving any empty space?

It would seem that Plato (and this is another of the analogies to be pointed out between his thought and Descartes') provided no real and permanent principle for these figures of which matter is formed other than the very fact of their extension. That is why Aristotle quite rightly tells us that in the *Timaeus* Plato identifies the extension of a body, the χώρα, with the principle which remains permanent through all the changes of the body,[6] with what Aristotle calls ὕλη and his Latin commentators call *materia*. "Thus Plato, in the *Timaeus*, says that extension and matter are the same thing.

Aristotle wisely rejects a similar identification between extension [occupied by a body], place, and the principle of permanence which is matter; "matter," he says, "cannot be separated from the real thing; place, on the contrary, can be separated from it." It is because of this, in fact, that local movement is possible. If local movement is to take place, the same matter must leave one place in order to reach another place; therefore matter must be something other than place.

Because Plato makes extension the permanent matter of the elements, which are capable of change, he calls it "the nurse of generation".[7]

This extension receives the diverse forms which constitute fire, air, water and earth; each of these forms (μορφή) is at the same time the source of power (δύρανις); therefore the χώρα loses its homogeneity....

Anyone who sought a rigorous logical order in the theory of space and place put forth in the *Timaeus* would therefore be greatly disappointed. Nevertheless, this theory merits attention, because, in formulating it, Plato was the first to attempt to solve the fundamental problem of place and movement. As Aristotle points out, "Everyone says that place is something; but he alone attempted to say what it was."[8]

NOTES

* From *Le système du monde*, Hermann, Paris, n.d., vol. I, pp. 36–42; transl. by David A. and Mary-Alice Sipfle.

[1] For a comparison of the doctrines of Plato and Democritus, see Albert Rivaud, *Le problème du devenir et la notion de la matière dans la philosophie grecque depuis les origines jusqu'à Théophraste*, Thesis, Paris, 1905, §215, pp. 309–311. Ingeborg Hammer Jensen, *Demokrit und Plato* (*Archiv für Philosophie*, I. *Archiv für Geschichte der Philosophie*, vol. XVI, pp. 92–105 and pp. 211–229; 1910).

[2] On the various attributes of the χώρα in Plato, see Albert Rivaud, *op. cit.*, Bk. III, chs. II, III and IV; pp. 285–315.

[3] Plato, *Timaeus*, 79; Platonis *Opera*, (ed. by Ambroise-Firmin), Didot, Paris, 1846; vol. II, p. 239.

[4] Plato, *Timaeus*, 54–56; *ed. cit.*, pp. 221–222.

[5] According to Jean Philopon, this passage is to be understood as follows: just as the regular dodecahedron has twelve faces, God constructed the world of twelve spheres, nested inside each other, namely earth, water, air, sublunary fire, the seven orbs of the wandering stars, and the sphere of the fixed stars (Ioannes Grammaticus Philoponus Alexandrinus *In Procli Diadochi duodeviginti argumenta de mundi aeternitate...* Ioanne Mahotio Argentenae interprete. Lugduni, 1557. In Procli Diadochi argumentum decimumtertium, p. 244. – Ioannes Philoponus *De aeternitate Mundi contra Proclum*. Lipsiae, MDCCCXCIX. XIII, 18, pp. 536–537). Indeed in the *Phaedrus* and in the fifth book of the *Laws*, Plato constructs the world of twelve concentric spheres, but it does not appear absolutely certain that he ever admitted a sphere of sublunary fire as Aristotle did. It is more likely that the first sphere was, for him as for the Pythagoreans of his time, that of the central fire contained by the earth. On this subject, see Th.-H. Martin, *Études sur le Timée*, Paris, 1841; vol. II, note XXXVII, § 3, pp. 114–119, and note XXXVIII, pp. 141–143.

[6] Aristotle, *Physics*, Bk. IV, ch. II (IV) (Aristotelis *Opera*, ed. Didot, vol. II, pp. 286–287; ed. Bekker, vol. II, p. 209, col. b.).

[7] Plato, *Timaeus*, 52–53; *ed. cit.*, p. 220.

[8] Aristotle, *Physics*, Bk. IV, Ch. II (IV) (Aristotelis *Opera*, ed. Didot, vol. II, p. 286; ed. Bekker, vol. I, p. 209, Col. b.).

SPACE AND THE VOID ACCORDING TO ARISTOTLE*

We have rapidly sketched Aristotle's doctrine on the incorruptible substance of which the heavens are formed. Let us now see what Peripatetic physics teaches us about substances which undergo generation, change and corruption.

The whole of Aristotle's doctrine concerning these substances is dominated by his theory of the heavy and the light. The fundamental idea of this theory is that of natural place. This idea presupposes a concept of place bearing no analogy to the κενόν of the atomists or the χώρα of Plato.

Leucippus, Democritus and Plato had all been greatly influenced by the Pythagorians. Like the Pythagorians, they were above all geometricians, and their whole philosophy was permeated by Geometry. The theories of space which they developed are the work of geometricians who projected their geometrical concepts upon reality.

Aristotle is by no means a geometrician. He is above all an observer; the real, for him, is revealed by experience. This characteristic, essential to all Peripatetic philosophy, is clearly seen in Aristotle's theory of place and movement. As one might expect, therefore, his theory is quite unlike that of Plato.

Aristotle rejects Plato's identification of place with the position occupied in geometrical space.

Place has a kind of power to influence the movement of elementary bodies.[1] Unless it is impeded, each elementary body moves in a determinate direction, upward or downward. These movements of elementary bodies toward their natural places "show not only that place is something [real], but also that it exerts a certain influence, ἔχει τινὰ δύναμιν." Plato also acknowledges this influence when he compares the action of the χώρα on the elements to that of a sieve which separates the heavy bodies from the light ones. Now, how can a power of this sort be attributed to geometrical space? In geometrical space, the six directions, namely, up, down, in front of, behind, right, left, have no real existence; they are determined only by the position we ourselves take within this space. If we turn around or turn over, what was above or to the right becomes below or to the left, and inversely. The χώρα of Plato is like the figures the mathematician uses.

> It is quite clear in the case of mathematical figures that they are not in a place. Yet, they have a right and a left in terms of the position they occupy in relation to us. But it is only in thought that these figures occupy such a position [in relation to us]. By nature they have none of these things [position, right, left, up, down, etc.]

In this passage, Aristotle makes clear the nature of Plato's illusion. Plato unconsciously attributes to the χώρα a certain orientation in relation to himself, so that it is possible to distinguish up and down and to allow some bodies to go in the first direction and others in the second. But, if we assume Plato absent, the nature of the χώρα itself does not allow such distinctions between directions. Plato's error is like that of a geometrician who believes that a cube has a right side and a left side and that it has them of itself, independently of the position which the geometrician attributes to it in thought.

Since it is a fact that, independently of us, heavy bodies move in one direction and light bodies in another, place must be something other than the χώρα, which is essentially undifferentiated as to direction. There must be something of such a nature that the terms up and down have determinate meanings.

A similar conclusion may be drawn from Aristotle's argument against the void of the atomists. Several of the objections by which he intends to prove that local motion would be impossible in the void are drawn from principles proper to his Dynamics. Before examining them, let us first consider objections of more universal application. These objections, as Aristotle himself points out, are quite similar to those he made against the Platonic χώρα. He states them as follows:[2]

> If there exists a place deprived of body which is the void, where will a body placed in this void move to? It cannot, of course, go in all directions at once. The same argument holds against those who take place to be something distinct from bodies in which local movement occurs: How could a body placed in it either move or remain at rest? The argument cited above concerning up and down movements is equally applicable to the void; those who affirm the existence of the void are, in fact, making it a place.... Upon reflection, one sees that those who believe that there must be a void if there is to be motion should come to the opposite conclusion, that is, that nothing can move if there is a void. Some maintain that the Earth is at rest for reasons of symmetry[3]; for the same reason, all bodies would have to be at rest in the void. There is really no place to which

they can move more or less than to another, because the void, insofar as it is void, is undifferentiated.

Thus local movement is possible only in a place where different positions are marked out permitting one to discern that a body is moving in one direction more or less than in another. Being perfectly homogeneous, the void or the χώρα cannot offer such landmarks. Consequently, neither the void of the Atomists nor Plato's χώρα can perform the function of place. Place must be defined in such a way that it provides fixed points in relation to which one is able to discern local motion. This is the fundamental thought which guides Aristotle in his search for a definition of place.

NOTES

* From *Le système du monde*, Hermann, Paris, n.d., vol. I, pp. 189–191; transl. by David A. and Mary-Alice Sipfle. The quotations from Aristotle's *Physics* in this selection are direct translations from Duhem. Although we have consulted the English translation by R. P. Hardie and R. K. Gaye, their interpretation of the Greek often differs from Duhem's.

[1] Aristotle, *Physics*, Bk. IV, ch. I (Aristotelis *Opera*, ed. Didot, vol. II, p. 285; ed. by Bekker, vol. I, p. 208, col. b).

[2] Aristotle, *Physics*, Bk. IV, ch. VIII [IX] (Aristotelis *Opera*, ed. Didot, vol. II, p. 194; ed. by Bekker, vol. I, p. 214, col. b).

[3] On the subject of this argument, see Plato, *Phaedo*, LVIII (Platonis *Opera*, ed. by Didot, vol. I, p. 85); Plato, *Timaeus*, 62–63 (Platonis *Opera*, ed. by Didot, vol. II, p. 227). – Aristotelis *De Caelo*, Bk. II, ch. XIII (Aristotelis *Opera*, ed. Didot, vol. II. p. 406; ed. by Bekker, vol. I, p. 295, col. b, and p. 296, col. a). See also Duhem, *Le système du monde*, vol. I, pp. 88–89.

S. SAMBURSKY

THE STOIC IDEA OF SPACE*

The Stoics, uncompromising as they were in their conception of the physical world as a continuum, supposed that "beyond the cosmos there stretches an infinite, non-physical void". This empty space played an important part in their cosmogony, for the following reason. At the end of every cosmic cycle, when the hot element becomes predominant, the cosmos expands thermically and thus increases in volume. Now, the problem raised by this increase in volume can be solved, if the cosmos is, as assumed, an island in an infinitive void. The pupils of Aristotle, who, like their master, maintained that the cosmos was finite and denied the infinite extension of space, rejected this Stoic theory. The whole question became the subject of a scientific controversy between the two schools which bears a remarkable resemblance to the modern cosmological argument about space and the structure of the matter in it. We are told by a Stoic source:

Aristotle and his school maintained also that if there were a void outside the confines of the cosmos, all the matter would be poured out into the infinite and scattered and dissipated. But we maintain that this could never happen. For matter has a coherence which holds it together and against which the surrounding void is powerless. For the material world preserves itself by an immense force, alternately contracting and expanding into the void following its physical transmutations, at one time consumed by fire, at another beginning again the creation of the cosmos.

The cohesion of the cosmos results from the tension of the pneuma, as we know from Stoic doctrine. This cohesive force, which binds the parts of the cosmos together into a single entity, offsets the dissipating influence of the surrounding infinite void and makes the cosmos a closed universe whose unity is not vitiated by changes in its size. There is an unmistakable analogy here with the theoretical argumentation of modern relativistic cosmologies. Once again we see that the inner logic of scientific patterns of thought has remained unchanged by the passage of centuries and the coming and going of civilizations: the same models and associations recur, only in new forms suited to the more advanced stage reached by physical knowledge.

NOTE

* From *The Physical World of the Greeks* (transl. by Marton Dagut), MacMillan, New York and Routledge, London, 1956, pp. 202–203.

THE CONTINUITY AND INFINITY OF SPACE ACCORDING TO EPICURUS AND LUCRETIUS*

We must turn now to Epicurus' conception of space, for, although from the nature of the case it is not so complicated as that of the atoms, it involves certain difficulties which cannot be disguised. The syllogistic argument by which he inferred the existence of space from the fact of motion has been discussed already:[1] we must now inquire more closely what it was that he meant by space. The mathematical conception of space as extension may be put out of court: it is impossible that Epicurus should have meant that for several reasons; (1) it would have been inconsistent with his whole attitude to the mathematical point of view, (2) it would have clashed with his theory of area as a succession of discrete *minima*, (3) it is sufficiently contradicted by the many synonyms which he employs to describe it, and particularly with its definition as 'intangible existence'. Space is an 'existence' just as much as body, it is not mere measurement or extension: it is a 'thing', but a thing whose sole property is that it cannot touch or be touched, it can offer no sort of resistance to body. Here then is his answer to the difficulties of the earlier Atomists: he does not trouble himself with their subtle discussions as to whether space is 'nothing' (οὐκ ὄν), or 'non-existent' (μὴ ὄν); he simply affirms, with the same meaning but much closer precision, that it is an 'intangible existence'. The conception is not abstract but concrete: it is derived from that of body by a negation of its properties.

Yet considerable difficulty remains. Are we to conceive space as absolutely continuous and universal, coextensive with the universe itself, or as discrete and consisting only of the intervals between bodies? In other words, is there space in a place which is occupied by body, or is there not? does he mean 'place' or 'empty space'? The question is a very difficult one to decide and there seem many indications on either side. If we consider the synonyms which Epicurus uses, we see that two of them, 'place' (τόπος) and 'room', (χώρα) are in favour of the former view, that by space he means occupied as well as empty space – a continuous whole: the same conclusion may be drawn from the definition of space as that "in which things exist and through which they move", and possibly (though I think it need not be interpreted in this sense) the contention that space is infinite in extent: for if there is no space, where bodies are, then there is a limit to space. On the other side, we have the fourth synonym, 'the empty' (κενὸν),[2] which clearly suggests only unoccupied space, and the frequent reference to the void in compound bodies as 'intervals' (διαστήματα) between the component atoms. Most of the ancient commentators too seem to interpret Epicurus in this sense, and among their comments is the express statement of Simplicius that the Epicureans regarded space as "the interval between the boundaries of that which surrounds it". A consideration of the main passages in Epicurus and Lucretius, where space is mentioned, seems to show that they both oscillate between the two conceptions. Are we then to leave this difficulty – so fundamental in the system – as a point which Epicurus never really thought out? I believe that it arises largely from the fact that we are not easily able to approach Epicurus with a sufficiently concrete conception. Giussani, in one of the most interesting of his Essays, has very largely cleared the matter up. He points out that we must think of Epicurus' notion primarily in relation to the ideas which he was combatting. The Parmenideans, for instance, would readily admit the conception of space in the sense of 'extension', but they would maintain that there is matter everywhere, there is no such thing as empty space. In strong opposition to this view Epicurus wished to maintain not merely that there was empty space between portions of matter but that empty space was a necessary presupposition to that of matter: there must be empty space in order that things may exist at all; otherwise there would be "nowhere for them to be and nothing through which they might move". 'Void' then is the fundamental notion always in the mind of Epicurus and his disciple, and therefore he is most often apt to think of it as completely empty space, or the intervals between matter.

We come back then to the original conception of the Universe as atoms moving in space and we must ask finally whether this universe and its two constituents are or are not infinite. Epicurus gives the traditional answer of the atomic school, but once again supports it with argument.

The universe is boundless. For that which is bounded has an extreme point: and the extreme point is seen against something else. So that as it has no extreme point, it has no limit; and as it has no limit it must be boundless and not bounded.

Lucretius puts the proof rather more lucidly:

it is seen that nothing can have an extreme point, unless there is something beyond to bound it, so that there is seen to be a spot farther than which the nature of our sense cannot follow it. As it is, since we must admit that there is nothing outside the whole sum, it has not an extreme point, it lacks therefore bound and limit.

The appeal is then once again to phenomena: the condition of limitation there, the existence of something else beyond, is one which cannot be applied to our mental conception of the universe. Lucretius brings out his point by the famous illustration of the hurling of the spear.

Go, if you can, he challenges the doubter, to the extreme limit of the universe and hurl a spear: either it will be stopped or it will go on: if it is stopped, there will be matter beyond, if it goes on there will be empty space: in either case you did not start from the end of the universe. The same will happen wherever you take your stand.

The universe then cannot have a limit.

Moreover the two constituents are also infinite, though in different senses, the atoms in number and the void in extent. For, as Lucretius argues, in order that the sum total, the universe, may be unlimited, either both or one or other of its constituents must be infinite. Epicurus then deals with the two questions separately.

If the void were boundless, and the bodies limited in number, the bodies could not stay anywhere, but would be carried about and scattered through the infinite void, not having other bodies to support them and keep them in place by means of collisions.

The statement is careful and precise: the condition of the creation of things is the constant collision of atoms and their crowding together in such numbers as to be able to enter into the combined existence of a compound, which in its turn is kept together and held in its place by the external blows of other countless atoms. That this may occur in an infinite universe, it is necessary that there should be an infinite supply of matter: otherwise the comparatively few collisions which would take place would just send individual atoms wandering far out into space, where they would have no chance of meeting their fellows. Lucretius elaborates the idea with a fine imaginative description of the chaos which must ensue.

Similarly, space is infinite in extent: for "if the void were limited, the infinite bodies would have no room wherein to take their place." Lucretius' argument here is rather different and perhaps less satisfactory:

if space were limited, the atoms through the downward motion due to weight would all have sunk to the bottom and there remained in an inert mass: it is because there is no bottom that they are still kept in eternal restlessness.

The limitation of space would in fact preclude the ceaseless motion of the atoms which is an essential part of the atomic conception. The argument is more esoteric and less likely than Epicurus' own to convince a non-Epicurean. It is simpler and more cogent to maintain that unless space were infinite, there would not be room for infinite atoms.

The idea of the infinity of space raises again the question of the conception of space and presents the same difficulty. One is inclined to ask: does not the existence of the infinite number of atoms really preclude the infinity of space: for each atom, inasmuch as it is not itself empty space, is really a limitation to it? This question requires a careful answer. It is tempting to argue that the instantaneous occupation of any 'piece' of space by an atom does not interfere with the conception of space as continuous and infinite 'place', in which the atoms have their momentary station. But there is a good reason against this: if this were the Epicurean conception, then space would itself be coextensive with the universe, whereas Epicurus always speaks of the universe as "body plus void": the sum total of matter, divided though it is into infinite particles in ceaseless motion has to be added to infinite void to make the sum total of the infinite universe. Similarly it is significant that in this section space is spoken of throughout as 'the void' (κενὸν) and not 'place' (τόπος) as before. It would probably be a more correct solution of the difficulty to say that this 'internal' limitation of space, if it may be so described, was not here present to Epicurus' mind. He was thinking rather, as he clearly was when speaking of the infinity of the universe, of an 'external' limit (πέρας). In extension outward space is unlimited (ἄπειρον), even though internally it might be thought of as limited by the presence of atomic matter. Once again the notion of 'empty space' seems to be uppermost and that from which Epicurus started: the kinetic view of matter helps to an understanding of his point of view, but the particular difficulty would not have troubled him.

We have then at last reached the traditional atomic conception of an infinite universe, consisting of atoms infinite in number moving in space infinite in extent. The conception has not varied since Democritus, but in its gradual unfolding in Epicurus it seems almost to have changed its nature. Each step in the argument has now been thought out under the definite rules of the Canonice: the detached notions about the character of the atoms have been correlated into a self-dependent whole: a universe seems not to have been assumed but created in thought. There has been occasion here and there to point out weak points in the argument or possibly hazy and ill-defined conceptions. But the result is one worthy of a great thinker: it is no mere wholesale adoption of the theory of Democritus – in certain places it has been seen to differ conspicuously from it: nor is it the work of a preacher, who hastily patched together some kind of physical theory to act as a basis for his moral teaching. With all its limitations, it is the construction of a master-mind, working on definite lines and with a deep and penetrating interest in his subject for its own sake. Nor will these characteristics be missed, as the further development of the system is traced in detail.

NOTES

* From *Greek Atomists...* pp. 293–297.
[1] To wit: "If no void exists, there is no motion.
    But motion exists.
    Therefore the void must exist."
                              Ed.)
[2] 'Lucretius' 'inane'.

PLACE AND THE VOID ACCORDING TO JOHN PHILOPON*

Place is space with its three dimensions.[1] Space must be entirely separate in thought from the bodies which occupy it. It must be considered as an incorporeal volume, extended in length, width and depth, so that place is identical with the void.

Philopon states this thesis in terms which closely follow the thought of Chrysippus and Cleomedes:

> Place is not the adjacent part of the surrounding body.... It is a given interval, measurable in three directions; it is distinct from the bodies found in it, and is, by its very nature, incorporeal. In other words, it is the dimensions alone, devoid of any body. Indeed, insofar as their matter is concerned, place and the void are essentially the same thing.

That is not to say that the void can ever exist in actuality[2], nor does it mean that there can be a volume which is occupied by no body. Although reason distinguishes it from all bodies and regards it as essentially incorporeal, the void is, nevertheless, always filled by some body. Place and the body which is in it form one of those pairs of things which are indissolubly linked, so that one of these things cannot exist without the other. Pure reason distinguishes place from body, but place can never exist in actuality without body. In the same way, reason distinguishes matter from form, but matter can never exist in actuality unless it is united with a certain form.

This space, distinct from all bodies and void by nature, remains absolutely motionless as a whole and in each of its parts.[3] A given part of space can successively receive different bodies, which, one after another, take their place there. It always remains the same part of space, however; it does not move.

As soon as a moving body leaves a given place, another body comes to occupy this same place, since it must never remain empty.[4] In the same way, as soon as one form becomes corrupt in matter, it is replaced by another form, so that matter never becomes bare and deprived of all form. John the Grammarian thus draws a perfect parallel between local movement and qualitative change. Place and the body which occupies it play the same role in the first kind of motion as matter and form play in the second.

Philopon anticipates that the Peripatetics will raise objections to his theory, and he attempts to disarm them in advance.

The following argument seems compelling:

This three-dimensional space, which is taken to be the place of bodies, is infinite. How can that be, given that it cannot subsist without bodies and that the bodies taken as a whole form a finite mass?

The Grammarian is surprised that one would attribute the least importance to this objection. Just as the intelligence has a conception of three-dimensional space, it can, he believes, conceive of an abstract surface which limits this space in such a way that it has the exact size required to contain the corporeal Universe.

John Philopon is clearly diverging from the orthodox view of the Stoics; since the time of Zeno and Chrysippus they had never ceased to maintain that an infinite void extended beyond the limits of the Universe. The Grammarian, on the other hand, teaches that, beyond the spherical surface which limits the World, there is nothing except a space conceived by reason, but deprived of reality, which he refuses to call the void. In so doing, he abandons the teaching of the Stoa in favor of the Peripatetic tradition.

NOTES

* From *Le système du monde*, Hermann, Paris, n.d., vol. I, pp. 317–319; transl. by David A. and Mary-Alice Sipfle.
[1] John Philopon, *In Aristotelis physicorum libros quinque posteriores commentaria.* Edidit Hieronymus Viteli Berolini, 1888, p. 567.
[2] *Ibid.*, pp. 569 and 579.
[3] *Ibid.*, p. 569.
[4] *Ibid.*, p. 579.

P. DUHEM

ABSOLUTE FRAME OF REFERENCE ACCORDING TO ST. THOMAS*

Let us turn to St. Thomas' general doctrine concerning the nature and the immobility of place.

As we have seen, Aristotle, in treating this question, successively adopted two mutually incompatible definitions of place. First of all, he called the *place of a body* that part of the surrounding matter immediately contiguous to the body. But, thus defined, place is not immobile. In order to assure its immobility, Aristotle then declared the place of the body to be the first immobile surface surrounding this body.

It has been the principal aim of several of the Scholastic commentators to avoid this change of definition, which constitutes a grave logical error. To this end, they have generally distinguished two senses of the word 'place': in the first of these senses, place is movable; in the second it is immobile.

Such a distinction is already indicated, although too briefly to be really clear, in the extremely concise *Summa* which Robert Grosseteste, Bishop of Lincoln, composed on Aristotle's *Physics*.

Robert Grosseteste notes that the location of a body is an accident of that body, in such a manner that it must move with the body.[1] He devotes only a single sentence to the problem this creates: "Materially,[2] the place is mobile; formally, it is immobile." What constitutes *material place* and what constitutes *formal place*, the Bishop of Lincoln does not tell us.

A reading of St. Thomas will explain this.[3]

First of all, the part of the surrounding matter which is immediately contiguous to the body can be called the *place of a body*. This place, insofar as it is formed of such and such matter, is mobile. The body in question was surrounded by this air or that water; a little later, the air or water surrounding it may have changed.

Besides place thus understood, which is mobile, we must consider another place. This place is limited by the same contiguous parts of the surrounding bodies which define the first; but it consists of a certain relationship between these contiguous parts and the whole of the celestial sphere. It determines the order or the position of the body contained by these parts in relation to the whole of the Heavens or in relation to the immobile Universe. This place is *rational place* (*ratio loci*).

The closest part of the container,[4] insofar as it is formed of such and such matter, is not immobile. However, insofar as one considers the position it occupies in the Heavens as a whole, it does not move. Any new body which comes to form this closest part has the same relative position, when compared to the Heavens as a whole, as the body which previously formed it and which has moved out.

The immobile *rational place* is a fixed relation to the Heavens as a whole. This whole itself is determined by the central body and by the poles. As a result one could define *rational place* as the position in relation to the central body and the poles.

The *rational place* of any container whatsoever is determined by the ultimate container, the ultimate abode, that is, the Heavens.

This is an example, suggested by the text of Aristotle himself, showing how all *ratio loci* is, in the last analysis, derived from consideration of the supreme orbit. In the domain of heavy or light elements, lower or higher places are determined by comparison to the center of the World and to the concave surface of the Lunar sphere. Now, we have seen how the immobility of the central body was required by the rotational movement of the supreme sphere.

As for the concave surface which, on our side, marks the limit of all the rotating celestial spheres, it is, indeed, revolving; however, it remains unchanging, in that it always maintains the same distance from us,

that is, from the motionless center.

NOTES

* From *Le Système du monde*, Hermann, Paris, 1956, vol. VII, pp. 176–178; transl. by David A. and Mary-Alice Sipfle.
[1] Divi Roberti Linconensis, *Super octo libris physicorum brevis et utilis summa*; in lib. IV. This summa is found at the end of the following work: *Emptor et lector Aveto. Divi Thome Aquinatis, In libros physicorum Aristotelis interpretatio sum et expositio…* Colophon:… Anno a nativitate Domini quarto supra millesimum quinquiesque centesimum, sexto Idus Aprilis.
[2] The text says: *naturaliter*; it should, I think, read *materialiter*.
[3] Sancti Thomae Aquinatis, *In libros physicorum Aristotelis expositio*; in lib. IV, lect. VI.
[4] To wit: The inner surface of the surrounding body. *Ed.*]

P. DUHEM

THE EMPYREAN AS THE PLACE OF THE UNIVERSE*

Most of the discussions we have just summarized arise because of two Peripatetic propositions which it is difficult to reconcile for all the bodies which constitute the Universe. These propositions are:
 The place of a body surrounds the body;
 The place of a body is at rest.

At this point, Christian theology seemed to offer Aristotelian physics a way out of the difficulty by surrounding the Universe with a sphere exempt from all local movement. Theologians believed that the Scriptures affirmed the existence of this ultimate fixed sphere. Proclus had earlier tried to make such a sphere the place of the Universe and the point of reference for all movement.[1]

A number of theologians, guided by certain Scriptural passages, posited an outermost immobile Heaven beyond the various mobile heavens imagined by the astronomers; among them were Isidore of Sevilla, the Venerable Bede, Rabanus Maurus, the Pseudo-Bede, St. Anselm and Peter Lombard. Supported by this theological opinion, several physicists had sought physical reasons, foremost among them being Michael Scot, William of Auvergne, St Bonaventura and Vincent of Beauvais. Some physicists, perplexed by the 'great question' of the location of the ninth sphere, thought they could find the solution by appealing to this tenth immobile sphere. This Empyrean, this 'aqueous Heaven' surrounding the highest orbit, provided a place for it, and served as the fixed terminus to which the movements of the Heavens could be referred. It assured the fixity of the two poles around which all the other spheres turn.

This theory may already have been current in the time of St. Bonaventura. He seems to allude to the role of the Empyrean as universal place when he speaks of it as an immobile sphere "which is a container and is not contained."[2]

Some passages in St. Thomas would lend themselves to a similar interpretation.

In any case, the doctrine in question is clearly formulated in *The Theory of the Planets*, which Campanus of Novara wrote at the request of Pope Urban IV.[3] This learned astronomer, the pope's chaplain, expressed himself as follows:

> Is there anything else beyond the convex surface of the ninth sphere, for example another sphere? From the standpoint of reason, this conclusion is not necessary. Instructed by faith, however, respectfully bowing to the opinion of the holy doctors of the Church, we shall confess that beyond the ninth Heaven is found the Empyrean, which is the dwelling place of the angels.

Is the Empyrean the tenth Heaven, directly contiguous to the ninth sphere? Or must one place between this orbit and the Empyrean an aqueous Heaven, which would make the supreme Heaven the eleventh in rank? Campanus hesitates between these two alternatives. But he states the following conclusion with assurance:

> Beyond the convex surface of the Empyrean, there is nothing. It is the supreme limit of all corporeal things, the surface farthest from the common center of all the spheres, that is, from the center of the Earth. That is why it is the universal and common place of all things which are contained, for it contains all things, and nothing else contains it.

These last words, "*Omnia continens et a nullo alio contenta*," repeat almost word for word the phrase used by St. Bonaventura.

There is a singularly clear account of Campanus' theory in the *Summa philosophiae* which some manuscripts attribute to Robert Grosseteste, but which, as we have said, is the work of some disciple of Roger Bacon. In this work, we find the following passage on the subject of the Empyrean Heaven:

> The first moving sphere must move in relation to something absolutely at rest. This can be demonstrated both in physics and in mathematics. Now, this immobile thing is not primordially the center of the World, as Aristotle and the other Peripatetics thought; it is the sphere of the Empyrean heaven, which is, in an entirely natural manner, immobile in all its parts. It is in relation to this sphere that the diverse parts of the first sphere and of the other moving spheres are able to move and actually do move. It is the same sphere, as we have said, which enables the World to have a center and this center to be fixed. It is not the existence and the fixity of the center which cause the fixity of the motionless and immobile heaven; quite the opposite is true. Otherwise, that which is by nature the lowest and most vile would cause that which is by nature the highest and most noble; such a thing is impossible.
>
> If it is therefore impossible for the circular motion of the sphere to take place unless one conceives of an absolutely motionless center, and not only a mathematical center but a natural center, about which the sphere moves; if, granting the existence of the empyrean heaven which contains all other corporeal things, which precedes them all, both in time or by nature, it is still necessary to assume a center, a thing which plays

the role of a center (*ratio centralis*), a center about which the moving heavens will necessarily revolve – if all this be so, then it is clear that the immobility of the empyrean heaven will be the universal cause of all change undergone by beings capable of generation and corruption. Just as the first cause is more a cause than are secondary causes, the empyrean heaven will be more a cause of such change than are the first moving sphere and the other inferior spheres.'

NOTES

* From *Le système du monde*, Hermann, Paris, 1956, vol. VII, pp. 197–200; transl. by David A. and Mary-Alice Sipfle.
[1] See *Le système du monde*, pt. I, ch. V, § XVI, vol. I, pp. 341–342.
[2] Celebratissimi Patris Domini Bonaventurae, Doctoris Seraphici, Ordinis Minorum, *In secundum librum Sententiarum disputata*; Secunda pars; libri secundi distinctionis XIV pars quarta; quaest. III: Utrum conveniat alicui orbi moveri absque stellis?
[3] This document bears the title: *Opus Campani de modo adaequandi planetas, sive de quantitatibus motuum caelestium, orbiumque proportionibus, centrorumque distantiis, ipsorumque corporum magnitudinibus*, in ms. no. 7298 in the Latin collection of the Bibliothèque nationale; in no. 7401 of the same collection, it is simply designated by the words: *Theorica planetarum* Campani. The passage which interests us at the moment is found in the second chapter after the *Prooemium*.

A. KOYRÉ

THE INFINITE SPACE IN THE FOURTEENTH CENTURY*

With Thomas Bradwardine[1] we return to the domain of theology, for Bradwardine is above all a theologian. What preoccupies him is not the problem of the world but the conditions of salvation; the cosmological structure of the Universe and even its ontological structure interest him only insofar as his study sheds light on the being of man and of God.[2]

Bradwardine's theological thought is based on a crucial experience:[3] the experience of man's fundamental helplessness, of his inherent inability to accomplish, by himself, an act of positive freedom. Any doctrine which, in the manner of Pelagianism, fails to recognize this fact, is therefore false, pernicious and sinful. Furthermore, by imposing on man a task beyond his capabilities, it leads him to despair, to a disavowal of self. Thus, nothing is more important than to combat this 'Cainite'[4] theology, to confront it with the consoling theology of absolute predestination, the theology of hope in the goodness and infinite power of a God who is infinitely free because He is infinitely perfect.

Thomas Bradwardine's theology is, almost as much as Calvin's, a theology of divine omnipotence, omnipotence which he will not and cannot allow to be diminished or impeded by anything whatsoever: neither by the necessities of an Aristotelian ontology, nor by those of even a Christian psychology. The God of Bradwardine does not hesitate before Aristotelian or psychological absurdities: He is able to act in the void and constrain one to be free.[5]

Nevertheless, one would be wrong to suppose that *nothing* interferes with His omnipotence, and that *nothing* limits His freedom. In fact, there is a limit, and an absolute limit, beyond which it is completely impossible for Him to go: the limit of metaphysical and mathematical coherence.

This is because Bradwardine, though he is a theologian, is not only a theologian. He is also – and at the same time – an able metaphysician and a talented mathematician (a geometrician). Since he is the heir and most orthodox representative of the Anselmian tradition and mentality,[6] it is as a metaphysician that he constructs his notion of God. Once he has posited that God is the absolutely and infinitely perfect being and that, consequently, we must attribute to Him all that which *melius est esse quam non esse*, it is with the imperturbability of a mathematician that Bradwardine follows the innumerable consequences of the definition to their ultimate conclusions.[7]

Nothing is able to limit the divine essence. However, God cannot be without acting, nor act without being, nor even act without being there, that is, without being present. Thus He is present in all creation, in all creatures, the source of their being and their action, being and acting in them much more than they themselves. And, because He is supremely immutable and consequently cannot be 'moved', the only way He can act in a given place is, necessarily, through His active presence. It results from this (exactly as Walter Burleigh had stated) that the creation of the world is possible only if the site, or place, exists prior to the actual existence of the world. Furthermore, it implies the real presence of God in the place or the site where the world now is, a presence prior to creation – this the authors studied by Burleigh had not dared affirm. And, as it is ridiculous for a geometrician to imagine a limited empty space, it implies, *ipso facto*, His active – and eternal – presence in all of infinite space, which extends beyond the limits of the Universe. 'Imaginary' space thereby becomes real. And yet, non-created.[8]

Consequently, God is necessarily, eternally, infinitely in every part of the infinite imaginary site, which is why, in truth, He may be called omnipresent as well as omnipotent. Likewise, for analagous reasons, He can be said to be, in some manner, infinite, infinitely great, or of infinite magnitude.[9] These terms are to be understood in a metaphysical sense, however, and not in the sense of extension properly speaking, since He is both infinitely extended and, at the same time, unextensible and dimensionless. Indeed, he coexists at the same time and in His entirety with magnitude and infinite imaginary extension and with each part of this extension. For this reason, and in the same sense, He can be called immense, for He is neither measured nor measurable by any measure; He can also be called uncircumscribed, because He is neither circumscribed by anything which would embrace Him in His entirety, nor can He be circumscribed by anything. He Himself circumscribes, contains, and embraces everything.[10]

Sextus the Pythagorian, one of the twenty-four philosophers who left us their definitions of God, was right, in Bradwardine's opinion, when he said: You would not find the measure of God, even if you had wings. And another of these philosophers rightly called Him a circle whose center is everywhere and whose circumference is nowhere'.[11]

Thus, it is not as a result of theological preoccupations alone, no more

THE INFINITE SPACE IN 14TH CENTURY 49

than of purely scientific preoccupations, but as a result of the conjuncture, in one mind, of the theological notion of divine infinity and the geometrical notion of spatial infinity, that Bradwardine formulated the paradoxical conception of the reality of imaginary space. Three centuries later, this empty space, a veritable nothingness made real, was to engulf the celestial spheres which held together the beautiful Aristotelian and medieval Cosmos, and they were to vanish. Then, during the next three centuries, the world, which was no longer a Cosmos, appeared to man to be placed in Nothingness, surrounded by Nothingness, and everywhere penetrated by Nothingness.

NOTES

* From 'Le vide et l'espace infini au XIXe siècle', *Études d'Histoire de la Pensée philosophique*, Armand Colin, Paris, 1961, pp. 72–74; 83–84; transl. by Mary-Alice and David A. Sipfle.
[1] On Thomas Bradwardine, *cf.* K. Werner, *Die Scholastik des späteren Mittelalters*, vol. III, Vienna, 1883; – S. Hahn, *Thomas Bradwardinus und seine Lehre von der menschlichen Freiheit* (Beiträge zur Geschichte der Philosophie des Mittelalters, vol. V, 2) Münster, 1905. On Bradwardine as a mathematician, *cf.* M. Cantor, *Vorlesungen über die Geschichte der Mathematik*, vol. II, 2nd ed., Leipzig, 1900, and P. Duhem, *Études sur L. de Vinci*, vol. III, Paris, 1913; – Marshall Clagett, *Giovanni Marliani and Late Medieval Physics*, New York, 1941; – Anneliese Maier, *Die Vorläufer Galilei's im XIV. Jahrhundert*, Rome, 1949; – H. Lamar Crosby, Jr., *Thomas of Bradwardine, His 'Tractatus de proportionibus'*; *Its Significance for the Development of Mathematical Physics*, Madison (Wis.), 1955; Marshall Clagett, *Science of Mechanics in the Middle Ages*, Madison (Wis.), 1959. Also see E. Gilson, *La philosophie au moyen âge*, p. 618ff Bradwardine's principal work, very popular in the seventeenth century, Thomae Bradwardini, *Archiepiscopi Olim Cantuarensis, De Causa Dei contra Pelagium et de Virtute causarum ad suos Mertonenses Libri tres*, London, 1618, in folio, has been edited *opera et studio Dr. Henrici Savilii.* Now, and this is perhaps more than a simple biographical fact, Henry Savile was the founder of the famous Savillian chairs of mathematics and astronomy at Oxford, holders of which at the time of Newton were Seth Ward, the future Bishop of Salisbury, and John Wallis, the noted author of the *Arithmetica Infinitorum.*
[2] Of the three books which make up the *De Causa Dei*, the first is devoted to God, the second to man, the third to the problem of the reconciliation of human freedom with divine omnipotence.
[3] Bradwardine tells us (*De Causa Dei*, Bk. I, ch. I, cor. 17) that he was a Pelagian himself in his youth, like all those around him.
[4] Pelagian theology is 'Cainite' because it convinces the man who is conscious of his sin and his impotence that he is definitively and irremediably damned. Cf. *ibid.*, cor. 18.
[5] *De Causa Dei*, Bk. III, ch. I, p. 637; ch. II, p. 646.
[6] For Bradwardine the great authorities are St. Augustine and St. Anselm; after them come Robert Grosseteste and John Peckham. Among the moderns, Duns Scotus.

[7] The *Causa Dei* begins with these two postulates or axioms, which are set forth in Chapter I. A *Corollarium* in forty parts follows from these two suppositions'.

[8] 'Imaginary' space is there *before* the creation of the world.

[9] Thus God is also infinitely great in the sense of extension. This conception, characteristic of a mathematician's thinking, will be found later, *mutatis mutandi*, in Malebranche and in Newton.

[10] *De Causa Dei*, ch. V, p. 179.

[11] *Ibid.*, p. 180. Bradwardine refers to the *Book of the twenty-four Philosophers* (cf. Clemens Baeumker, *Das Pseudo – Hermetische Buch der XXIV Meister*, Beiträge zur Geschichte der Philosophie des Mittelalters, fasc. XXV, Münster, 1928). In this book the definition quoted by Bradwardine is proposition II. For the history of this definition, cf. D. Mahnke, *Unendliche Sphaere und Allmittelpunkt*, Halle, 1937.

A. KOYRÉ

THE FINITE WORLD OF COPERNICUS*

I need not insist on the overwhelming scientific and philosophical importance of Copernican astronomy, which, by removing the earth from the center of the world and placing it among the planets, undermined the very foundations of the traditional cosmic world-order with its hierarchical structure and qualitative opposition of the celestial realm of immutable being to the terrestrial or sublunar region of change and decay. Compared to the deep criticism of its metaphysical basis by Nicholas of Cusa, the Copernican revolution may appear rather half-hearted and not very radical. It was, on the other hand, much more effective, at least in the long run; for, as we know, the immediate effect of the Copernican revolution was to spread skepticism and bewilderment of which the famous verses of John Donne give such a striking, though somewhat belated, expression, telling us that the

> ... new Philosophy calls all in doubt,
> The Element of fire is quite put out;
> The Sun is lost, and th'earth, and no mans wit
> Can well direct him where to looke for it.
> And freely men confesse that this world's spent,
> When in the Planets, and the Firmament
> They seeke so many new; then see that this
> Is crumbled out againe to his Atomies.
> 'Tis all in peeces, all cohaerence gone;
> All just supply, and all Relation.

To tell the truth, the world of Copernicus is by no means devoid of hierarchical features. Thus, if he asserts that it is not the skies which move, but the earth, it is not only because it seems irrational to move a tremendously big body instead of a relatively small one, "that which contains and locates and not that which is contained and located," but also because "the condition of *being at rest* is considered as nobler and more divine than that of *change* and *inconsistency*; the latter therefore, is more suited to the earth than to the universe." And it is on account of its supreme perfection and value – source of light and of life – that the place it occupies in the world is assigned to the sun: the central place which, following the Pythagorean tradition and thus reversing completely the Aristotelian and mediaeval scale, Copernicus believes to be the best and the most important one.

Thus, though the Copernican world is no more hierarchically structured (at least not fully: it has, so to say, two poles of perfection, the sun and the sphere of the fixed stars, with the planets in between), it is still a well-ordered world. Moreover, it is still a finite one.

This finiteness of the Copernican world may appear illogical. Indeed, the only reason for assuming the existence of the sphere of the fixed stars being their common motion, the negation of that motion should lead immediately to the negation of the very existence of that sphere; moreover, since, in the Copernican world, the fixed stars must be exceedingly big – the smallest being larger than the whole *Orbis magnus* – the sphere of the fixed stars must be rather thick; it seems only reasonable to extend its volume indefinitely 'upwards'.

It is rather natural to interpret Copernicus this way, that is, as an advocate of the infinity of the world, all the more so as he actually raises the question of the possibility of an indefinite spatial extension beyond the stellar sphere, though refusing to treat that problem as not scientific and turning it over to the philosophers. As a matter of fact, it is in this way that the Copernican doctrine was interpreted by Gianbattista Riccioli, by Huygens, and more recently by Mr. McColley.

Though it seems reasonable and natural, I do not believe this interpretation to represent the actual views of Copernicus. Human thought, even that of the greatest geniuses, is never completely consequent and logical. We must not be astonished, therefore, that Copernicus, who believed in the existence of material planetary spheres because he needed them in order to explain the motion of the planets, believed also in that of a sphere of the fixed stars which he no longer needed. Moreover, though its existence did not explain anything, it still had some usefulness: the stellar sphere, which "embraced and contained everything and itself," held the world together and, besides, enabled Copernicus to assign a determined position to the sun.

In any case, Copernicus tells us quite clearly that

THE FINITE WORLD OF COPERNICUS 53

... the universe is spherical; partly because this form, being a complete whole, needing no joints, is the most perfect of all; partly because it constitutes the most spacious form which is thus best suited to contain and retain all things; or also because all discrete parts of the world, I mean the sun, the moon and the planets, appear as spheres.

True, he rejects the Aristotelian doctrine according to which "outside the world there is no body, nor place, nor empty space, in fact that nothing at all exists" because it seems to him "really strange that something could be enclosed by nothing" and believes that, if we admitted that "the heavens were infinite and bounded only by their inner concavity", then we should have better reason to assert "that there is nothing outside the heavens, because everything, whatever its size, is within them," in which case, of course, the heavens would have to be motionless: the infinite, indeed, cannot be moved or traversed.

Yet he never tells us that the *visible world*, the world of the fixed stars, is infinite, but only that it is immeasurable (*immensum*), that it is so large that not only the earth compared to the skies is "as a point" (this, by the way, had already been asserted by Ptolemy), but also the whole orb of the earth's annual circuit around the sun; and that we do not and cannot know the limit, the dimension of the world. Moreover, when dealing with the famous objection of Ptolemy according to which "the earth and all earthly things if set in rotation would be dissolved by the action of nature," that is, by the centrifugal forces produced by the very great speed of its revolution, Copernicus replies that this disruptive effect would be so much stronger upon the heavens as their motion is more rapid than that of the earth, and that, "if this argument were true, the extent of the heavens would become infinite." In which case, of course, they would have to stand still, which, though finite, they do.

Thus we have to admit that, even if outside the world there were not nothing but space and even matter, nevertheless the *world* of Copernicus would remain a finite one, encompassed by a material sphere or orb, the sphere of the fixed stars – a sphere that has a centrum, a centrum occupied by the sun. It seems to me that there is no other way of interpreting the teaching of Copernicus. Does he not tell us that

... the first and the supreme of all [spheres] is the sphere of the fixed stars which contains everything and itself and which, therefore, is at rest. Indeed, it is the place of the world to which are referred the motion and the position of all other stars. Some [astronomers] indeed, have thought that, in a certain manner, this sphere is also subjected to change: but in our deduction of the terrestrial motion we have determined another cause why it appears so. [After the sphere of the fixed stars] comes Saturn, which performs its circuit in thirty years. After him, Jupiter, which moves in a duodecennial revolution. Then Mars which circumgirates in two years. The fourth place in this order is occupied by the annual revolution, which, as we have said, contains the Earth with the orb of the Moon as an epicycle. In the fifth place Venus revolves in nine months. Finally, the sixth place is held by Mercury, which goes around in the space of eighty days.

But in the center of all resides the Sun. Who, indeed, in this most magnificent temple would put the light in another, or in a better place than that one wherefrom it could at the same time illuminate the whole of it? Therefore it is not improperly that some people call it the lamp of the world, others its mind, others its ruler. Trismegistus [calls it] the visible God, Sophocles' Electra, the All-Seeing. Thus, assuredly, as residing in the royal see the Sun governs the surrounding family of the stars.

We have to admit the evidence: the world of Copernicus is finite. Moreover, it seems to be psychologically quite normal that the man who took the first step, that of arresting the motion of the sphere of the fixed stars, hesitated before taking the second, that of dissolving it in boundless space; it was enough for one man to move the earth and to enlarge the world so as to make it immeasurable – *immensum*; to ask him to make it infinite is obviously asking too much.

Great importance has been attributed to the enlargement of the Copernican world as compared to the mediaeval one – its diameter is at least 2000 times greater. Yet, we must not forget, as Professor Lovejoy has already pointed out, that even the Aristotelian or Ptolemaic world was by no means that snug little thing that we see represented on the minatures adorning the manuscripts of the Middle Ages and of which Sir Walter Raleigh gave us such an enchanting description. Though rather small by our astronomical standards, and even by those of Copernicus, it was in itself sufficiently big not to be felt as built to man's measure: about 20000 terrestrial radii, such was the accepted figure, that is, about 125000000 miles.

Let us not forget, moreover, that, by comparison with the infinite, the world of Copernicus is by no means greater than that of mediaeval astronomy; they are both as nothing, because *inter finitum et infinitum non est proportio*. We do not approach the infinite universe by increasing the dimensions of our world. We may make it as large as we want: that does *not* bring us any nearer to it.

Notwithstanding this, it remains clear that it is somewhat easier, psychologically if not logically, to pass from a very large, immeasurable and ever-growing world to an infinite one than to make this jump starting with a rather big, but still determinably limited sphere: the world-bubble has to swell before bursting.

NOTE

* From *From the Closed World to the Infinite Universe*, The Johns Hopkins Press, Baltimore, 1957, pp. 29–35.

H. HÖFFDING

ESTABLISHMENT AND EXTENSION OF THE NEW WORLD SCHEME: GIORDANO BRUNO*

Giordano is one of the first thinkers to realise clearly that the great thoughts are due to a long successive series of experiences. He believed himself to have uttered great thoughts, but, at the same time, he is very well aware of how much he owes to his predecessors, especially to the astronomers, on whose observations he relied. While, in the age of the Renaissance, men were still inclined to look back to antiquity as the source of all truth, just as the Church looked back to the time when the revelation took place, Bruno asserts that the men of the present time are older than "the ancients," since they have a richer experience on which to build than the latter. Eudoxus did not know so much as Hipparchus, nor Hipparchus as Copernicus. And he commends Copernicus, not only for having carried on the work of his predecessors, but more especially for his strong and magnanimous spirit, which raised him above the prejudices of the many, and the illusions of the senses, and enabled him to establish a new world-scheme. In his Latin didactic poem, 'On the Immeasurable and the Countless Worlds', he breaks out into a hymn of praise in honour of Copernicus. He reproaches him, however, for having halted too soon, i.e. before he had deduced all the consequences of his ideas. He therefore needed a commentator, able "to think out all that was involved in his discovery," and this office Bruno claims for himself. He opened men's eyes to the infinitude of the universe, showed that this can no more have absolute limits than there can be fixed 'spheres' separating the different regions of the world from one another, and that a single law and a single force prevail throughout the world, so that wherever we may find ourselves we cannot get away from God who rules throughout the same; moreover we have no need to go beyond ourselves to find Him. Our present task is to assign the grounds on which Bruno builds in his establishment and further development of the Copernican world-scheme. These may be reduced to two main considerations – of which one is epistemological and the other religio-philosophical.

The old world-scheme, with the earth as the central point and the fixed spheres as the outermost limits, has no right to appeal to the evidence of the senses. If we examine the different sense-images which we receive when we move, we see that the horizon continually changes as we change our place. Rightly interpreted, so far from proving to us that there is an absolute centre and an absolute limit to the world, sense-perception shows us the contrary, i.e. the possibility of conceiving any place whatever, wherever we may be, or can convey ourselves in imagination, as the central point, and also the possibility of constantly changing and extending the limits of our world. And in harmony with this testimony of sense-perception is the capacity of our imagination and of our thought to continue unceasingly to add number to number, magnitude to magnitude, form to form; moreover we are impelled to do this by an impulse and striving which stir within us and which are never satisfied with what we have already attained. It would be inconceivable, thinks Bruno, that our imagination and our thought should surpass Nature, and that this continual possibility of taking new views should correspond to no reality in the world. From the subjective impossibility of setting a limit and of affirming an absolute central point he now argues that there is no limit and no central point. In proof of this Bruno relies, as he himself tells us, on the fundamental condition of our knowledge (*la conditione del modo nostro de intendere*). In strict consistency with this view – a consequence, however, which he only incidentally points out – Bruno somewhere remarks that we have, properly speaking, no right to conceive the universe as a totality, if it has no limit.

Since the horizon forms itself anew round every place occupied by the spectator as its central point, every determination of place must be relative. The universe looks different according to whether we conceive it to be regarded from the earth, the moon, Venus, the sun, etc. One and the same place will, according to the different points from which it is regarded (*respectu diversorum*), be centre, pole, zenith, or nadir. Determinations such as 'over' and 'under' do not therefore signify, as the old world-scheme presupposes, any absolute relation. It is only when we assume definite points of view that we invest such expressions with definite significance. And as with the relativity of place so with the relativity of motion. Motion is only conceived in its relation to one fixed point, and all depends on where we suppose this fixed point to be. One and the same motion will present a different appearance according to whether I regard it from the earth or from the sun, and wherever I may place myself in thought, my own standpoint will always appear to me to be immovable. We must not demand, therefore, that absolute certainty shall attend the distinction between that which is at rest and that which is in motion. The old world-scheme takes as given exactly what has to be proved, viz. that the earth is the fixed point from which every motion is to be measured. From the relativity of motion follows the relativity of time. For no absolutely regular motion can be discovered, and we possess no records which can prove to us that all the stars have taken up exactly the same position, with regard to the earth, as those they previously occupied, and that their motions are absolutely regular. We can therefore find no absolute measure of time. Since motion appears different when regarded from different stars, there must, if it is to be taken as the measure of time, be as many times in the universe as there are stars.

Nor have the concepts of heaviness and lightness any more absolute significance than have determinations of place. For, according to Aristotle, heaviness was the tendency to seek out the central point of the world, and, since the earth was the heaviest element, it followed that it was the central point of the world. But the qualities of heaviness and lightness are predicable of the particles of every individual heavenly body in their relation to this body as a whole. When that which is heavy falls, it does so because it seeks to return to the place in which it is at home and where it can best maintain itself. The particles of the sun are heavy in relation to the sun, those of the earth in relation to the earth. With regard to the universe as a totality, the concepts of lightness and heaviness have as little validity as motion and the determinations of place and time. They only receive significance in relation to a particular heavenly body or to a particular system. This theory of weight is identical with that held by Copernicus, only that Bruno lays the chief emphasis on the fact that it is the impulse to self-preservation which causes the parts to seek out their whole. Copernicus, too, relies on the relativity of our determinations, but he pauses half-way. It is Bruno's merit to have carried out this principle, and to have shown what are the consequences following from it. In Bruno, too, we meet for the first time with a decided answer to one of the most weighty objections against Copernicus, i.e. that objects falling on to the earth cannot fall on a spot perpendicularly below the point from which they started, but must fall a little to the west of this. For Bruno shows that a stone thrown from the top of a mast will fall at the foot of the mast because, from the beginning of its fall, it has participated, by means of strength imparted to it (*virtù impressa*), in the motion of the ship. If, on the other hand, the stone had been thrown down from a point outside the ship, it would have fallen a little further back. Bruno here enters on a train of thought of very great significance, and which afterwards led Galilei to the discovery of the law of inertia.

Closely connected, in Bruno's theory, with the principle of relativity is the principle that Nature is everywhere essentially the same (*indifferenza della natura*). From relations as we find them with us, he concludes to relations in other places in the universe. An experience of his childhood led him to adopt this method. From the hill Cicada, near Nola, which lay at his feet covered with forests and vines, he looked at the distant Vesuvius which appeared to him small, as well as bare and unfruitful. But when, on one occasion, he had wandered as far as Vesuvius he perceived that the two hills had exchanged aspects. Now it was Vesuvius which was high and wooded, while the Cicada was low and bare. The same principle which led him to establish and extend Copernicanism through the assumption of the infinity of the universe also led him to assume, as a matter of course, that the same relations exist everywhere, where we have no experience to the contrary. He now conceives the other heavenly bodies as similar to the earth, and the other systems as similar to the solar system, so that the fixed stars become suns surrounded by planets. There is no ground for assuming anything else but that the same force is everywhere in operation. But, in that case, Copernicus was not justified in following the ordinary conception and supposing all the fixed stars to be equally distant from us, and to lie in one and the same sphere. Perhaps it only appears as if they always maintained the same distance from us and from one another. Distant ships appear immovable, and yet they are often moving with no small velocity. Whether this is the case with the fixed stars can certainly only be established by observations extending through many years, and which may even, perhaps, not yet have been begun. But the reason that such observations have not yet been set on foot is precisely this firm conviction that the fixed stars never change their place, either in relation to us or to one another! Thus it is evident that the principle of the 'indifference of Nature' (or, as it is called nowadays, the principle of actuality), no less than the principle of relativity from which it is deduced, will be productive, since it leads to new investigations. Bruno has a much keener sense of the necessity of confirming theoretical and subjective considerations by the method of experience than is generally attributed to him. "What could we think without all the observations that have been collected?" he asks. He is certainly no mere enthusiast for the infinite. He sought to show, by means of a thoughtful and critical examination, what are the presuppositions on which the old world-scheme rests, and how justifiable and natural it is to bring forward other assumptions. And the *onus probandi*, he thinks, lies first of all with those who assert the limitation of the universe; for does not experience show us that wherever we may go the boundaries always change with our progress? And why should the universe extend no further than to eight spheres, as even Copernicus still believed? Why not to a ninth, a tenth, and so on? Because our sense-perception is limited, we have no right to conclude that the universe is limited also. Bruno's greatest merit is the energy with which he thought himself into the new world-conception, and demanded its verification in detail. On this account his teaching is more than an anticipation of genius. The epistemological foundation on which he bases it has lasting significance. Nevertheless it cannot be denied that the passionate consistency with which he proceeded often led him to express himself with greater certainty than he was by rights entitled to. Small wonder if the zeal with which he laboured, and which was necessary to the surmounting of obstacles, carried him beyond the goal.

By means of the relativity of place-determinations Bruno had, as we saw, overthrown the old doctrine of the elements according to which they were characterised by heaviness or lightness as absolute qualities, and to each one was assigned its 'natural place' in the universe. But with this doctrine the distinction between the heavenly and the sublunary world vanished, as also the prejucide that no change could take place in heaven. Bruno was especially anxious to overthrow the belief in the fixed spheres. He shows that this belief is a corollary from the assumption that the earth is the absolute central point. As soon as we have thoroughly grasped the idea that every heavenly body is, so to say, a central point, and can move freely in space, as the earth does, the necessity for believing in fixed spheres disappears. And why should the heavenly bodies require external forces to move them? Each one of them, like every other creature in the world, has an inner impulse to motion which carries it forwards; every heavenly body and every little world has in itself a source of life and motion, and space is the great ethereal medium in which the all-embracing world-soul is active; there is therefore no need for special spirits of the spheres to set particular regions in motion. Bruno found a confirmation of his view in Tycho Brahe's investigations into the nature of comets. He may, perhaps, have composed the Latin didactic poem 'On the Immeasurable and the Countless Worlds' on purpose to show how these investigations confirmed the opinions which he had deduced on other grounds in his Italian dialogue 'On the Infinite Universe and the Worlds.' He here eulogises the Danish investigator as the first astronomer of his day (*Ticho Danus, nobilissimus atque princeps astronomorum nostri temporis*) and as the man who put an end to the fixed spheres which were supposed to enclose our world in layers. For the comets go straight through the 'spheres', whose crystal masses are said to divide the different regions of the world from one another!

These are all Bruno's views that can be brought together under the point of view of an epistemological foundation of the new world-scheme. We will now pass on to what may be called the religio-philosophical foundation. This is taken from the idea of the infinity of the Deity, an idea which Bruno had, from the beginning, regarded as unquestionable; and which he might also safely assume to be shared by his readers and opponents, even though they might not have been clear as to all it involved. To Bruno it seemed a contradiction that no infinite effect should correspond to the infinite cause. If the Deity, which in its original unity embraces all that is unfolded in the universe, is infinite, then the universe which is the unfolded form of God's essence must be infinite. No force limits itself and the infinite force has nothing by which it can be limited. If the Deity is conceived as the principle of good, must we not then assume that it will impart all that it can? Shall we suppose it to be envious or niggardly? The infinite perfection must express itself in infinitely many creatures and worlds. It is not justifiable to attribute to the Deity a force or a possibility which does not become reality. This opposition between possibility and reality is only valid for finite creatures, and must not be transferred to the Deity. Otherwise we shall have two Gods, – one possible, and one real or active, – in opposition to one another; a blasphemous theory contradictory to the unity of God. Jakob Boehme, as we saw above, was not afraid of this blasphemy. His religio-philosophical speculations remind us, in several points, of those of Bruno; and for him too the new conception of the world had significance. But Boehme was concerned with the problem of evil, not that of the interconnection of the world. The religio-philosophical proof on which Bruno relies did not originate with himself. As he himself mentions, it had already been established by Pietro Manzoli of Ferrara, who, under the name of Palingenius, published a Latin didactic poem (*Zodiacus vitae*, Lyons, 1552), in which he taught the infinity of the universe; although he conceives the world-scheme with fixed spheres in the traditional way; in addition to the eight spheres Palingenius introduces an incorporeal and infinite world of light. It does not derogate from the originality of Bruno that he thus made use of former thinkers, e.g. Palingenius here, and, in other passages, Cusanus, Copernicus, and the old Atomists. In call cases he reduces their thoughts to greater coherency, and carries them out with greater consistency and on a basis of richer experience than was possible to them.

It seemed to Bruno as if he had never breathed freely until the limits of the universe had been extended to infinity, and the fixed spheres has disappeared. No longer now was there a limit to the flight of the spirit, no 'so far and no farther'; the narrow prison in which the old beliefs had confined men's spirits had now to open its gates and let in the pure air of a new life. He has expressed these thoughts in some sonnets which precede the dialogue on 'the infinite Universe'. The picture of reality, at the shaping of which his thought had laboured so enthusiastically and untiringly, contained for him a symbolic significance. The outer infinity was for him the symbol of the inner. Not all symbolism rests on so firm a basis.

NOTE

* From *A History of Modern Philosophy*, MacMillan, Londen and N.Y. 1900, Dover, 1955, pp. 123-130.

M. JAMMER

GRADUAL EMANCIPATION FROM ARISTOTLE: FROM CRESCAS TO GILBERT*

Crescas' theory of space solved the problem of the outermost sphere: The infinite vacuum provides this sphere with space, so that its eternal rotation becomes a special kind of local motion and the sphere ceases to be the final limit and boundary of space.

Crescas' solution of the problem was not the only one advanced in the beginning of the fifteenth century. Nicholas of Cusa offered another. In his view, universal motion has no center, since in terms of his principle of the '*coincidentia oppositorum*' the absolute minimum must coincide with the absolute maximum. But God alone may be thought of as the absolute maximum of existence, so that Cusanus comes to the conclusion: "Qui igitur est Centrum mundus? scilicet est Deus benedictus, ille est Centrum terrae, et omnium sphaerarum."[1] However, from the purely physical point of view, the identification of the center of the universe with its circumference is an obvious absurdity. To Cusanus the world has neither a center nor a circumference. "Quia minimum cum maximo coincidere necesse est. Centrum igitur mundi coincideret cum circumferentia. Non habet igitur mundus circumferentiam."[2] So it is clear that the earth is not the center of the universe or of space. "Terra non est centrum mundi."[3] The manner in which Cusanus goes on to derive the motion of the earth, thereby anticipating certain ideas of the Copernican theory, is not part of our subject. But it is important for us to note that the absence of a body absolutely at rest (the earth) does away with the possibility of absolute motion and absolute space. It is this relative character of position and motion that brands Cusanus' theory of space as modern. Another modern feature is its rejection of the idea that a hierarchy of values rules different regions of space. Of Aristotelian origin, the idea is implied in the doctrine of physico-moral parallelism. As is well known, Aristotelian biology assigns to the upper parts of the human body a greater degree of nobility than to its lower parts. In consequence of this conception, as well as that of the parallelism between macrocosm and microcosm, the terms 'high' and 'low', though primarily purely geometric notions of spatial orientation, came in most languages to stand for distinctions of value.[4] The conception of a spatial hierarchy of values found its most perfect expression in Dante's *Divine comedy*, which from this point of view is a spatial metaphor of the gradations of sin and blessedness. How far this anthropomorphic conception became an integral part of medieval natural philosophy can be illustrated by the fact that Nicolaus of Autrecourt had to renounce his untimely thesis: "Quod non potest evidenter ostendi nobilitas unius rei super aliam."[5]

Cusanus, objecting the spatial hierarchy of values, states explicitly: "Neque dici debet, quod quia terra est minor sole et ab eo recipit influentiam, quod propterea sit vilior."[6] To Cusanus the earth is certainly not the smallest celestial body, the moon and Mercury being smaller; nor can any conclusion be drawn from the fact that the earth depends on the sun since the earth as a celestial body influences also in some degree the sun and its region.

The rejection of a spatial hierarchy of values is the logical conclusion of a more general principle which Cusanus advances in his *Docta ignorantia*: wherever in the heavens anyone may be placed, it would seem to him as if he were the center of the universe. This statement is evidently a rudimentary expression of the so-called 'cosmological principle' of modern science as far as the spherical symmetry of space is concerned. The general validity of the principle that the universe presents the same aspect from every point (and according to a modern school of cosmologists also at every time), except for local irregularities, is accepted in modern science as a necessary condition for the repeatability of experiments, since space and time are the only parameters which, at least in principle, are beyond the control of the experimenter and cannot be reproduced at his will. Since this postulate in modern cosmology – not only with respect to the purely geometric aspect of space, but also with regard to its kinematic and dynamic aspects – has gained so much importance recently, it is not without interest to note that in Cusanus' writings we encounter, probably for the first time, an explicit enunciation of its spatial implications. If there were any justification for regarding Nicholas of Cusa as marking the turning point in the history of astronomy, it would be rather because of this enunciation than on account of the insufficient evidence of his astronomical discoveries (the triple motion of the earth).[7] One has, however, to keep in mind that Cusanus' principally mystic-speculative approach to his conclusions is fundamentally different from the scientific method of the Renaissance.

The theories of both Crescas and Cusanus, nevertheless, were far in advance of their time. If the notion of space was to be emancipated from the Aristotelian tradition, it would have to be done, as history proved, more gradually. It was not done until the sixteenth century. Even in Cardan's *De subtilitate*, space is still conceived in accord with Aristotelian tradition as the concave surface of the limiting body. "Est igitur locus ultima corporis superficies, corpus contentum ambicus."[8]

In contrast to Cardan, Scaliger identifies space with the void, which is coextensive with the body occupying it. Under the influence of atomistic thought, Scaliger presupposes the vacuum as a necessary condition of motion. "In natura vacuum dari necesse est."[9] Scaliger's vacuum, however, is not an infinite empty extension beyond all bodies, but merely the receptacle coexistent with matter and penetrated by matter. The terms '*vacuum*', '*locus*', and '*spatium*' are synonymous in Scaliger's doctrine. "Idemque esse vacuum et locum; neque differre, nisi nomine."[10] Although Scaliger's theory represents an important step forward in the demolition of Aristotelian doctrine, it is still not the decisive step. For to Scaliger space, in its logical as well as metaphysical significance, is only secondary to matter. In a word, Scaliger's physics is still dominated by Aristotelian categories. As Ernst Cassirer points out, the real turning point is Bernardino Telesio's and Franciscus Patritius' theories of space.[11]

In his general philosophic outlook Telesio adopted certain materialistic and Stoic conceptions of Antiquity, which led him to ascribe to spiritual functions a certain degree of corporeality. This may account for his tendency to attribute independent reality to space and time, to place them on the same level with concrete matter. Space ceases with Telesio to be a mere quality and assumes an independent existence, parallel to matter or '*moles*', *moles* being a concept that comes very near to the Newtonian notion of mass. Space is the great receptor of all being whatever. If a body leaves its place or is expelled from it, place itself does not leave, nor is it expelled, but remains the same, promptly becoming the receptacle of another body.

Itaque locus entium quorumvis receptor fieri queat et in existentibus entibus recedentibus expulsisve nihil ipse recedat expellaturve, sed idem perpetuo remaneat et succedentia, entia promptissime suscipiat omnia, tantusque assidue ipse sit, quantaquae in ipso locantur sunt entia; perpetio nimirum iis, quae in ea locata sunt, aequalis, at eorum nulli idem sit nec fiat unquam, sed penitus ab omnibus diversus sit.[12]

Thus space, though equal to the things which occupy it, is not the same as any of these things. First of all, space is incorporeal, and being pure aptitude to receive matter ("aptitude ad corpora suscipienda"), it is free of all actions and operations. Space shows no qualitative differentiation; it is completely homogeneous in its structure, so that the existence of 'natural places' is impossible. All parts of space show equal aptitude to receive any kind of matter. The motion of bodies in space is not caused by any qualitative differences inherent in space itself, but is the result of physical forces. Space as a whole is immobile ("universum perpetuo immobile permanet"). It is accessible to sense perception ("ipso comprehensum est sensu"), as experiments with vacua clearly show. Basing himself on physical grounds, Telesio attacks Aristotle's argument against the possibility of empty space, while disdaining to deal with demonstrations of the nonexistence of things whose existence is yet patently observable.

The considerations adduced by Telesio show clearly the new spirit of Italian natural philosophy of the sixteenth century. Nothing less than the formulation of a new physics is at issue. But another obstacle has still to be removed before these ideas could be assimilated and a new mechanics reared on their basis. The traditional substance-accident doctrine, the great bulwark of scholastic thought, had to be set aside. It was not enough to revise the physical foundations of the theory of space: it had to be provided with a new metaphysical foundation as well.

Franciscus Patritius undertook this task.

Quid ergo est? hypostasis, diastema, est, diastasis, ectasis est, extensio est, intervallum est, capedo est, atque intercapedo. Ergo quantitas? Ergo accidens? Ergo accidens ante substantiam? & ante corpus? Archytas uterque, & senior Pythagorae auditor, & iunior Platonis amicus, & quicos secuti sunt scriptores categoriam, hoc spacium non cognovere.[13]

Is space a substance or an accident, is it corporeal or incorporeal, he asks in the chapter called 'De spacio physico' of his comprehensive work. None of these concepts applies to space, since they are only ways of characterizing things in space. Space must be presupposed as a necessary condition of all that exists in it. "Id enim ante omnia necesse est esse, quo posito alia poni possunt omnia; quo ablato alia omnia tollantur."[14] Further, qualities themselves are still dependent on space. It is therefore clear that space does not fit into the substance-accident scheme. "Nulla ergo categoriarum spatium complectitur; ante eas est, extra eas omnes est..." Patritius thus achieves the important result of emancipating the concept of space from the Aristotelian doctrine of categories. But, he asks, has not space magnitude? And is it not therefore subjected to the category of quantity? And this is his answer:

Itaque aliter de eo philosophandum est quam ex categoriis. Spatium ergo extensio est hypostatica per se substans, nulli inhaerens. Non est quantitas. Et si quantitas est, non est illa categoriarum, sed ante eam ejusque fons et origo.[15]

This view of space as being ontologically and epistemologically the primary basis of all existence led Patritius, as Cassirer points out,[16] to reverse the relation between mathematics and physics. The study of space must come before the study of matter. To Patritius, since space conditions not only matter as such, but its qualities as well, the investigation of space is an indispensable prerequisite to all natural science. Space makes not only nature, but the knowledge of nature, possible.

Before we go on to analyze Patritius' influence on the development of seventeenth-century physics, we may pause for a moment to say something about Giordano Bruno's place in the history of the development of the concept of space. As the exponent of the philosophy of infinity, Bruno is obliged to dispose of the idea of the world's finiteness, and he is thus confronted with the Peripatetic physics, in particular, with Aristotle's definition of place. "If the world is finite, and beyond the world there is nothing at all, where then is the world?" asks Bruno. Aristotle's answer that the world is in itself, although it follows logically from the definition of place, does not satisfy Bruno. So without attacking the validity of the logical conclusion Bruno confines himself to the premise itself. It is the definition itself that is wrong, and only a wrong conclusion could follow. To define place as the adjacent boundary of the containing body is to preclude the existence of space for the outermost sphere, and this renders meaningless any question as to what is outside the world. Before stating his own ideas, Bruno, in the manner of Crescas, mentions the arguments of Aristotle: "The convex surface of the primal heaven is universal space, which being the primal container is by naught contained. For position in space is no other than the surfaces and limit of the containing body, so that he who hath no containing body hath no position in space."[17] On the question "Where is the universe?" Aristotle, on the basis of these definitions, can only answer: "It is in itself." Here it is where Bruno's criticism begins. He says: "What then dost thou mean, O Aristotle, by this phrase, that 'space is within itself'? What will be thy conclusion concerning that which is beyond the world? If thou sayest, there is nothing, then the heaven and the world will certainly not be anywhere." After the discussion on the importance of the convex surface of the outermost sphere for spatial relations, Bruno (through the words of Philotheo) confesses: "Thus let this surface be what it will, I must always put the question, what is beyond?"[18] Bruno's restless temperament and his constantly searching disposition of mind did not let him find satisfaction with Peripatetic dialectic. Rejecting the finite categories of Peripatetic thought, he forms an ecstatic vision of an infinite universe in his mind.

> Henceforth I spread confident wings to space;
> I fear no barrier of crystal or of glass;
> I cleave the heavens and soar to the infinite.
> And while I rise from my own globe to others
> And penetrate ever further through the eternal field,
> That which others saw from afar, I leave far behind me.[19]

It is therefore only natural that Bruno expresses a new conception of infinite space on the ground that "Si non superficies sed spatium quoddam locus est, nullum corpus neque ulla corporis illocata erit sive maximum, sive minimum sive finitum sit ipsum, sive infinitum."[20] Bruno's definition of space is contained in Philotheo's answer on Albertino's theses in the fifth dialogue of *On the infinite universe and worlds*. Replying to Albertino's fifth and sixth arguments Philotheo says:

There is a single general space, a single vast immensity which we may freely call VOID; in it are innumerable (*innumerabili et infiniti*) globes like this one on which we live and grow. This space we declare to be infinite, since neither reason, convenience, possibility, senseperception nor nature assign to it a limit... It diffuseth throughout all, penetrateth all and it envelopeth, toucheth and is closely attached to all, leaving nowhere any vacant space; unless, indeed, like many others, thou preferest to give the name of void to this which is the site and position of all motion, the space in which all have their course.[21]

Although this definition, or description, of space is characteristic of the spirit of Italian natural philosophy of the sixteenth century, it is yet the case, as Wolfson points out, that a certain indebtedness of Bruno to

Crescas is likely. Both Crescas and Bruno focus their critique of Aristotle's definition on the problem of the outermost sphere; both attempt to demonstrate the existence of a vacuum on similar grounds; both refute Aristotle's theory of lightness in much the same way.

That two men separated by time and space and language, but studying the same problems with the intention of refuting Aristotle, should happen to hit upon the same arguments is not intrinsically impossible, for all these arguments are based upon inherent weaknesses in the Aristotelian system. But knowing as we do that a countryman of Bruno, Giovanni Francesco Pico della Mirandola, similarly separated from Crescas in time and space and language, obtained knowledge of Crescas through some unknown Jewish intermediary, the possibility of a similar intermediary in the case of Bruno is not to be excluded.[22]

Campanella develops Patritius' theory of space still further, maintaining that space is the immovable basis of all existence: "basin omnis creati, omniaque praecedere esse saltem origine et natura."[23] At another place he calls space "locus, basis existentiae, in quo pulcrum Opificium, hoc est mundus, sedet."[24] In Campanella's view space is homogeneous and undifferentiated, penetrated corporeally and penetrating incorporeally. Its homogeneity excludes such differentiations as 'down' or 'up', which attach to the diversities of bodies, rather than to space. It goes without saying that the existence of 'natural places' is emphatically rejected. God created space as a 'capacity', a receptacle for bodies. "Locum dico substantiam primam incorpoream, immobilem, aptam ad receptandum omne corpus."[25]

The works of Telesio, Patritius, and Campanella show that Italian natural philosophy must be credited with having emancipated the concept of space from the scholastic substance-accident scheme. In the physics of the early seventeenth century space becomes the necessary substratum of all physical processes. It is this emancipated concept, divested of all inherent differentiations or forces. Gilbert in his *Philosophia nova* expresses these ideas in a concise way:

Sed non locus in natura quicquam potest: locus nihil est, non existit, vim non habet; potestas omnis in corporibus ipsis. Non enim Luna movetur, nec Mercurii, aut Veneris stella, propter locum aliquem in mundo, nec stellae fixae quietae manent propter locum.[26]

Place does not affect the nature of things, it has no bearing on their being at rest or being in motion.

NOTES

* From *Concepts of Space*, Harvard Univ. Press, 1957, pp. 80–89.
[1] Nicholas of Cusa, *De docta ignorantia*, II, 11; see A. Petzelt (ed.) *Nicolaus von Cues, Texte seiner philosophischen Schriften, nach der Ausgabe von Paris 1514, sowie nach der Drucklegung von Basel 1565*, Kohlhammer, Stuttgart, 1949, vol. I.
[2] *Ibid.*, 21.
[3] *Ibid.*
[4] The designations 'right' and 'left' ('dextra' 'sinistra') have their origin in a somewhat opposite development: the 'propitious' or 'faithful' (Hebrew: 'yamin') hand became the 'right', the 'sinister', malignant, the 'left'. A reference to the widespread belief that the left side is ill-omened is encountered in the Ebers Papyrus, the famous document on early Egyptian medicine, dating most probably from 3400 B.C.
[5] Denifle-Chatelain, *Chartularium Universitatis Parisiensis*, vol. II, 544.
[6] *De docta ignorantia*; see reference 1, p. 106.
[7] See Lynn Thorndike, *Science and Thought in the Fifteenth Century*, New York 1929, p. 133.
[8] Jerome Cardan, *De subtilitate*, lib. I.
[9] C. Scaliger, *Exotericarum exercitationum liberi ad Hieronymum Cardanum*, Lutet. 1557.
[10] *Ibid.*
[11] E. Cassirer, *Das Erkenntnisproblem in der Philosophie und Wissenschaft der neueren Zeit*, Berlin 1911.
[12] Telesio, *De natura rerum juxta propria principia libri novem*, Naples 1586, vol. I, 25.
[13] Patritius, *Nova de universis philosophia libris quinquaginta comprehensa*, Venice 1593, fol. 65.
[14] *Ibid.*
[15] Patritius, *Pancosmia. De spatio physico*, 65f.
[16] Cassirer, *Das Erkenntnisproblem*, vol. I, p. 232.
[17] Bruno, *On the Infinite Universe and Worlds*, transl. by Dorothea Waley Singer in *Giordano Bruno*, Schuman, New York, 1950, p. 251.
[18] *Ibid.*, p. 254.
[19] *Ibid.*, p. 249.
[20] Bruno, *Acrotismus*, Vitebergae, 1588, vol. I, 1, p. 121.
[21] Bruno, *On the Infinite Universe and Worlds*; see reference 17, pp. 363, 373.
[22] Wolfson, *Crescas' critique of Aristotle*, p. 36.
[23] Thomas Campanella, *De sensu rerum* (1620), I, cap. 12.
[24] Campanella, *Metaphysicarum rerum juxta propria dogmata* (1638), pars I, lib. 2, cap. 13.
[25] Campanella, *Physiologia*, Paris 1637, I, 2.
[26] William Gilbert, *De mundo nostro sublunari philosophia nova*, Amsterdam 1651, lib. II, cap. 8, p. 144.

R. DESCARTES

VIEW OF SPACE AS PLENUM*

PRINCIPLE III

That the perceptions of the senses do not teach us what is really in things, but merely that whereby they are useful or hurtful to man's composite nature.

It will be sufficient for us to observe that the perceptions of the senses are related simply to the intimate union which exists between body and mind, and that while by their means we are made aware of what in external bodies can profit or hurt this union, they do not present them to us as they are in themselves unless occasionally and accidentally. For [after this observation] we shall without difficulty set aside all the prejudices of the senses and in this regard rely upon our understanding alone, by reflecting carefully on the ideas implanted therein by nature.

PRINCIPLE IV

That the nature of body consists not in weight, nor in hardness, nor colour and so on, but in extension alone.

In this way we shall ascertain that the nature of matter or of body in its universal aspect, does not consist in its being hard, or heavy, or coloured, or one that affects our senses in some other way, but solely in the fact that it is a substance extended in length, breadth and depth. For as regards hardness we do not know anything of it by sense, excepting that the portions of the hard bodies resist the motion of our hands when they come in contact with them; but if, whenever we moved our hands in some direction, all the bodies in that part retreated with the same velocity as our hands approached them, we should never feel hardness; and yet we have no reason to believe that the bodies which recede in this way would on this account lose what makes them bodies. It follows from this that the nature of body does not consist in hardness. The same reason shows us that weight, colour, and all the other qualities of the kind that is perceived in corporeal matter, may be taken from it, it remaining meanwhile entire: it thus follows that the nature of body depends on none of these.

PRINCIPLE V

That this truth regarding the nature of body is obscured by prejudices regarding rarefaction and the vacuum.

There still remain two reasons which may cause us to doubt whether the true nature of body consists solely in extension. The first is that prevalent opinion that most bodies are capable of being rarefied and condensed, so that when rarefied they have greater extension than when condensed; and some have even subtilized to such an extent that they desire to distinguish the substance of a body from its quantity, and its quantity from its extension. The second reason is that when we conceive that there is extension in length, breadth and depth only, we are not in the habit of saying that there is a body, but only space and further empty space, which most people persuade themselves is a mere negation.

PRINCIPLE VI

In what way rarefaction takes place.

But as regards rarefaction and condensation, whoever will examine his own thoughts and refuse to admit anything which he does not clearly perceive, will not allow that there is anything in these processes but a change of figure [in the body rarefied or condensed]: that is to say, rare bodies are those between whose parts there are many interstices filled with other bodies; and those are called dense bodies, on the other hand, whose parts, by approaching one another, either render these distances less than they were, or remove them altogether, in which case the body is rendered so dense that it cannot be denser. And yet it does not possess less extension than when the parts occupied a greater space, owing to their being further removed from one another. For we ought not to attribute to a body the extension of the pores or the interstices which its parts do not occupy [when it is rarefied], but to the other bodies which occupy these interstices. Just as when we see a sponge filled with water or some other liquid, we do not suppose that for this reason each part of the sponge is more extended than when it is compressed and dry, but only that its pores are wider, and that it is therefore distributed over a larger space.

PRINCIPLE VII

That rarefaction cannot be intelligibly explained in any other way.

I am indeed unable to say why this rarefaction of bodies has been explained by some as the result of augmentation of quantity rather than by the example of the sponge. For although when air or water are rarefied we do not see any of the pores which are rendered large, nor any new body that is added to occupy them, it is yet less consonant with reason to suppose something that is unintelligible in order to give a merely verbal explanation of how bodies are rarefied, than to conclude in consequence of that rarefaction, that there are pores or interstices which become greater, and which are filled with some new body, although we do not percieve this new body with the senses. For there is no reason which obliges us to believe that we should perceive by our senses all the bodies which exist around us. And we perceive that it is very easy to explain rarefaction in this manner though not in any other. And finally it would be undoubtedly contradictory to suppose that any body should be increased by a fresh quantity or fresh extension, without the addition to it of a new extended substance, i.e. a new body. Because it is impossible to conceive any addition of extension or quantity, without the addition of a substance having quantity or extension, as will be more clearly shown below.

PRINCIPLE VIII

That quantity and number differ only in thought[1] from what has quantity and is numbered.

For quantity differs from extended substance, or number from what is numbered, not in reality but only in our conception. Thus, to take an example, we may consider the whole nature of corporeal substance which is comprised within a space of ten feet, although we do not attend to this measure of ten feet; because it is clear that the thing conceived is the same in any one part of that space as in the whole. And *vice versa*, we can comprehend the number ten, as also a continuous quantity of ten feet without attending to any particular determinate substance, because the conception of the number of ten is plainly the same, whether considered in reference to the measure of ten feet, or to any other ten; and we cannot conceive a continuous quantity of ten feet without thinking of some extended substance of which it is the quantity, but yet we can conceive it without thinking of that determinate substance. In reality it is however impossible that even the least part of such quantity or extension can be taken away without taking away likewise an equal amount of substance; on the other hand, not the least part of the substance can be removed without our diminishing its quantity and extension by the same amount.

PRINCIPLE IX

That corporeal substance, when distinguished from its quantity, is confusedly conceived as something incorporeal.

Although however, some express themselves otherwise on this subject, I cannot think that they regard it otherwise than as I have just said; for when they distinguish substance from extension or quantity, they either mean nothing by the word substance, or they merely form in their minds a confused idea of incorporeal substance which they falsely attribute to corporeal, and leave to extension, which they nevertheless call an accident, that true idea of this corporeal substance, and thus it is easy to see that their words are not in harmony with their thoughts.

PRINCIPLE X

What space or internal place is.

Space or internal place and the corporeal substance which is contained in it, are not different otherwise than in the mode in which they are conceived of by us. For, in truth, the same extension in length, breadth, and depth, which constitutes space, constitutes body; and the difference between them consists only in the fact that in body we consider extension as particular and conceive it to change just as body changes; in space, on the contrary, we attribute to extension a generic unity, so that after having removed from a certain space the body which occupied it, we do not suppose that we have also removed the extension of that space, because it appears to us that the same extension remains so long as it is of the same magnitude and figure, and preserves the same position in relation to certain other bodies, whereby we determine this space

PRINCIPLE XI

In what sense it may be said that space is not different from corporeal substance.

And it will be easy for us to recognise that the same extension which constitutes the nature of body likewise constitutes the nature of space, nor do the two mutually differ, excepting as the nature of the genus or species differs from the nature of the individual, provided that, in order to discern the idea that we have of any body, such as stone, we reject from it all that is not essential to the nature of body. In the first place, then, we may reject hardness, because if the stone were liquefied or reduced to powder, it would no longer possess hardness, and yet would not cease to be a body; let us in the next place reject colour, because we have often seen stones so transparent that they had no colour; again we reject weight, because we see that fire although very light is yet body; and finally we may reject cold, heat, and all the other qualities of the kind either because they are not considered as in the stone, or else because with the change of their qualities the stone is not for that reason considered to have lost its nature as body. After examination we shall find that there is nothing remaining in the idea of body excepting that it is extended in length, breadth, and depth; and this is comprised in our idea of space, not only of that which is full of body, but also of that which is called a vacuum.

PRINCIPLE XII

How space is different from body in our mode of conceiving it.

There is, however, some difference in our mode of conceiving them; for if we remove a stone from the space or place where it was, we conceive that the extension of this stone has also been removed from it, because we consider this to be singular, and inseparable from the stone itself. But meantime we suppose that the same extension of place occupied by the stone remains, though the place which it formerly occupied has been taken up with wood, water, air, and any other bodies, or even has been supposed to be empty, because we now consider extension in general, and it appears to us that the same is common to stones, wood, water, air, and all other bodies, and even to a vacuum, if there be such a thing, provided that it is of the same magnitude and figure as before, and preserves the same situation in regard to the external bodies which determine this space.

PRINCIPLE XIII

What external place is.

The reason of this is that the words place and space signify nothing different from the body which is said to be in a place, and merely designate its magnitude, figure, and situation as regards other bodies. For it is necessary in order to determine this situation to observe certain others which we consider to be immovable; and according as we regard different bodies we may find that the same thing at the same time changes its place, and does not change it. For example, if we consider a man seated at the stern of a vessel when it is carried out to sea, he may be said to be in one place if we regard the parts of the vessel with which he preserves the same situation: and yet he will be found continually to change his position, if regard be paid to the neighbouring shores in relation to which he is constantly receding from one, and approaching another. And further, if we suppose that the earth moves, and that it makes precisely the same way from west to east as the vessel does from east to west, it will again appear to us that he who is seated at the stern does not change his position, because that place is determined by certain immovable points which we imagine to be in the heavens. But if at length we are persuaded that there are no points in the universe that are really immovable, as will presently be shown to be probable, we shall conclude that there is nothing that has a permanent place except in so far as it is fixed by our thought.

PRINCIPLE XIV

Wherein place and space differ.

The terms place and space are however different, because place indicates situation more expressly than magnitude or figure; while, on the contrary, we more often think of the latter when we speak of space. For we frequently say that a thing has succeeded to the place of another, although it does not possess exactly either its magnitude or its figure; but we do not for all that

VIEW OF SPACE AS PLENUM   79

mean that it occupies the same space as the other; and when the situation is changed, we say that the place also is changed, although the same magnitude and figure exist as before. And hence if we say that a thing is in a particular place, we simply mean that it is situated in a certain manner in reference to certain other things; and when we add that it occupies a certain space or place, we likewise mean that it is of a definite magnitude or figure [so as exactly to fill the space].

PRINCIPLE XV

How external place is rightly taken to be the superficies of the surrounding body.
And thus we never distinguish space from extension in length, breadth and depth; but we sometimes consider place as in the thing placed, and sometimes as outside of it. Internal place is indeed in no way distinguished from space; but we sometimes regard external place as the superficies which immediately surrounds the thing placed in it. And it is to be observed that by superficies we do not here mean any portion of the surrounding body, but merely the extremity which is between the surrounding body and that surrounded, which is but a mode; or that we mean the common surface which is a surface that is not a part of one body rather than of the other, and that it is always considered the same, so long as it retains the same magnitude and figure. For although all the surrounding body with its superficies is changed, we should not imagine that the body which was surrounded by it had for all that changed its place, if it meanwhile preserved the same situation in regard to other bodies that are regarded as immovable. Thus if we suppose that a ship is carried along in one direction by the current of a stream, and is impelled by a contrary wind in another direction in an equal degree, so that its situation is not changed with regard to the banks, we are ready to admit that it remains in the same place although we see that the whole surrounding superficies is in a state of change.

PRINCIPLE XVI

That it is contrary to reason to say that there is a vacuum or space in which there is absolutely nothing.
As regards a vacuum in the philosophic sense of the word, i.e. a space in

which there is no substance, it is evident that such cannot exist, because the extension of space or internal place, is not different from that of body. For, from the mere fact that a body is extended in length, breath, or depth, we have reason to conclude that it is a substance, because it is absolutely inconceivable that nothing should possess extension, we ought to conclude also that the same is true of the space which is supposed to be void, i.e. that since there is in it extension, there is necessarily also substance.

PRINCIPLE XVII

That a vacuum, in the ordinary sense, does not exclude all body.

And when we take this word vacuum in its ordinary sense, we do not mean a place or space in which there is absolutely nothing, but only a place in which there are none of those things which we expected to find there. Thus because a pitcher is made to hold water, we say that it is empty when it contains nothing but air; or if there are no fish in a fish-pond, we say that there is nothing in it, even though it be full of water; similarly we say a vessel is empty, when, in place of the merchandise which it was designed to carry, it is loaded only with sand, so that it may resist the impetuous violence of the wind; and finally we say in the same way that a space is empty when it contains nothing sensible, even though it contain created matter and self-existent substance; for we are not wont to consider things excepting those with which our senses succeed in presenting us[2]. And if, in place of keeping in mind what we should comprehend by these words – vacuum and nothing – we afterwards suppose that in the space which is termed vacuum there is not only nothing sensible, but nothing at all, we shall fall into the same error as if, because a pitcher is usually termed empty since it contains nothing but air, we were therefore to judge that the air contained in it is not a substantive thing.

PRINCIPLE XVIII

How the prejudice concerning the absolute vacuum is to be corrected.

We have almost all lapsed into this error from the beginning of our lives, for, seeing that there is no necessary connection between the vessel and the body it contains, we thought that God at least could remove all the body contained in the vessel without its being necessary that any other body should take its place. But in order that we may be able to correct this error, it is necessary to remark that while there is no connection between the vessel and that particular body which it contains, there is an absolutely necessary one between the concave figure of the vessel and the extension considered generally which must be comprised in this cavity; so that there is not more contradiction in conceiving a mountain without a valley, than such a cavity without the extension which it contains, or this extension without the substance which is extended, because nothing, as has already been frequently remarked, cannot have extension. And therefore, if it is asked what would happen if God removed all the body contained in a vessel without permitting its place being occupied by another body, we shall answer that the sides of the vessel will thereby come into immediate contiguity with one another. For two bodies must touch when there is nothing between them, because it is manifestly contradictory for these two bodies to be apart from one another, or that there should be a distance between them, and yet that this distance should be nothing; for distance is a mode of extension, and without extended substance it cannot therefore exist.

PRINCIPLE XIX

That this confirms what was said of rarefaction.

After we have thus remarked that the nature of material substance consists only in its being an extended thing, or that its extension is not different from what has been attributed to space however empty, it is easy to discover that it is impossible that any one of these parts should in any way occupy more space at one time than another, and thus that it may be rarefied otherwise than in the manner explained above; or again it is easy to perceive that there cannot be more matter or corporeal substance in a vessel when it is filled with gold or lead, or any other body that is heavy and hard, than when it only contains air and appears to be empty; for the quantity of the parts of matter does not depend on their weight or hardness, but only on the extension which is always equal in the same vessel.

PRINCIPLE XX

That from this may be demonstrated the non-existence of atoms.

We also know that there cannot be any atoms or parts of matter which are indivisible of their own nature [as certain philosophers have imagined]. For however small the parts are supposed to be, yet because they are necessarily extended we are always able in thought to divide any one of them into two or more parts; and thus we know that they are divisible. For there is nothing which we can divide in thought, which we do not thereby recognise to be divisible; and therefore if we judged it to be indivisible, our judgment would be contrary to the knowledge we have of the matter. And even should we suppose that God had reduced some portion of matter to a smallness so extreme that it could not be divided into smaller, it would not for all that be properly termed indivisible. For though God had rendered the particle so small that it was beyond the power of any creature to divide it, He could not deprive Himself of His power of division, because it is absolutely impossible that He should lessen His own omnipotence as was said before. And therefore, absolutely speaking, its divisibility remains [to the smallest extended particle] because from its nature it is such.

PRINCIPLE XXI

That extension of the world is likewise indefinite.

We likewise recognise that this world, or the totality of corporeal substance, is extended without limit, because wherever we imagine a limit we are not only still able to imagine beyond that limit spaces indefinitely extended, but we perceive these to be in reality such as we imagine them, that is to say that they contain in them corporeal substance indefinitely extended. For, as has been already shown very fully, the idea of extension that we perceive in any space whatever is quite evidently the same as the idea of corporeal substance.

PRINCIPLE XXII

Thus the matter of the heavens and of the earth is one and the same, and there cannot be a plurality of worlds.

It is thus not difficult to infer from all this, that the earth and heavens are formed of the same matter, and that even were there an infinitude of worlds, they would all be formed of this matter; from which it follows that there cannot be a plurality of worlds, because we clearly perceive that the matter whose nature consists in its being an extended substance only, now occupies all the imaginable spaces where these other worlds could alone be, and we cannot find in ourselves the idea of any other matter.

PRINCIPLE XXIII

That all the variety in matter, or all the diversity of its forms, depends on motion.

There is therefore but one matter in the whole universe, and we know this by the simple fact of its being extended. All the properties which we clearly perceive in it may be reduced to the one, viz. that it can be divided, or moved according to its parts, and consequently is capable of all these affections which we perceive can arise from the motion of its parts. For its partition by thought alone makes no difference to it; but all the variation in matter, or diversity in its forms, depends on motion. This the philosophers have doubtless observed, inasmuch as they have said that nature was the principle of motion and rest, and by nature they understood that by which all corporeal things become such as they are experienced to be.

PRINCIPLE XXIV

What motion is in common parlance.

But motion (i.e. local motion, for I can conceive no other kind, and do not consider that we ought to conceive any other in nature), in the vulgar sense, is nothing more than the *action by which any body passes from one place to another*. And just as we have remarked above that the same thing may be said to change and not to change its place at the same time, we can say that it moves and does not move at the same time. For he who is seated in a ship setting sail, thinks he is moving when he looks at the shore he has left, and considers it as fixed, but not if he regards the vessel he is on, because he does not change his position in reference to its parts. Likewise, because we are accustomed to think that there is no motion without action and that in rest there is cessation of action, the person thus seated may more properly be said to be in repose than in motion, since he is not conscious of any action in himself.

PRINCIPLE XXV

What movement properly speaking is.

But if, looking not to popular usage, but to the truth of the matter, let us consider what ought to be understood by motion according to the truth of the thing; we may say, in order to attribute a determinate nature to it, that it is the *transference of one part of matter or one body from the vicinity of those bodies that are in immediate contact with it, and which we regard as in repose, into the vicinity of others.* By *one body* or by a *part of matter* I understand all that which is transported together, although it may be composed of many parts which in themselves have other motions. And I say that it is the *transportation* and not either the force or the action which transports, in order to show that the motion is always in the mobile thing, not in that which moves; for these two do not seem to me to be accurately enough distinguished. Further, I understand that it is a mode of the mobile thing and not a substance, just as figure is a mode of the figured thing, and repose of that which is at rest.

NOTES

* From *Philosophical Works of Descartes*; transl. by E. S. Haldane and G. R. T. Ross, Cambridge University Press, 1931, and Dover, N.Y. 1955, pp. 255–266.
[1] ratione.
[2] "consider bodies near to us excepting in so far as they cause in our organs of sense impressions strong enough to enable us to perceive them." French version.

H. MORE

ON THE DIFFERENCE BETWEEN EXTENSION AND MATTER*

(From his First Letter to René Descartes)

December 11, 1648

But truly, most illustrious Descartes, let me hide nothing. Although I greatly admire the exquisite body and essence of your philosophy, I nevertheless confess that there are a few very minor points in the second part of *The Principles*, which my mind is either too little prepared to accept or too loath to admit.

But the substance of your excellent philosophy is in no danger from these, since they are such that, even if they can deservedly be judged either false or uncertain, they will have no effect on the essence or foundation of your philosophy which can stand quite well without them. If it does not trouble you, I will now explain which these points are.

First, that the definition of Matter or Body is inappropriately too broad. For God seems to be an extended thing, and angels: indeed, anything subsisting in itself would seem to be such; and this in such a way that the extension and the absolute essence of things would seem to be encompassed by the same limits, despite any differences in their essences. Indeed, I judge it evident that God is extended in his own manner, namely, that He is omnipresent and pervades intimately the whole machine of the world and each of its particles. For how could He impart motion to matter, which, you will admit, He at times has done and still does, unless He somehow comes into close contact with the matter of the universe, or at least did so at some time? This He certainly could not have done unless He were present everywhere, and pervaded every region. Therefore, God, in his own manner, is extended and spread out, and is therefore an extended thing.

And yet He is not that body, or matter, which your mind – ingenious artist that it is – has so skillfully turned into globules and striated particles. Hence, the concept of *extended thing* is broader than that of *body*.

I am further minded to add that I disagree with you in this respect, that to confirm this definition of yours, *you use such a twisted and most sophistical argument*. Especially when you claim that a body can be a body without being soft, or hard, or heavy, or light, and so forth, and that these and all other sensible qualities of a material body can be removed while the body itself would remain intact. This is the same as if you had said that a pound of wax could be a pound of wax, although it was deprived of shape, whether spherical or cubical or pyramidal, and so forth, and can, with no shape at all, remain a pound of wax. But that is impossible. For although this or that shape does not necessarily belong to the wax to make other shape impossible, nevertheless, it is of the supreme and inherent necessity that wax always be of some shape. In the same way, although it is not necessary that matter be soft, or hard, or hot, or cold, yet it is absolutely necessary that it be *sensible*, or if you prefer, *tangible*, as Lucretius has well defined it,

"For nothing can touch or be touched except a body."

This notion ought to trouble you very little, since your philosophy together with ancient authors mentioned by Theophrastus, plainly holds that all sense is touch, a doctrine I most easily admit as true. But since it is less satisfactory to define body in relation to our senses, this notion of tangibility should be broader and should signify that mutual contact and power of touch between any sort of body, whether animate or inanimate, that is, the immediate juxtaposition of the surfaces of two or more bodies. This notion also signifies another feature of matter or body, which you could call impenetrability: that is, that body can neither penetrate other bodies nor be penetrated by them. From which the distinction between divine and corporeal nature is quite clear, since divine nature can penetrate the corporeal but corporeal entities cannot penetrate each other. Therefore, it seems to me that Virgil with his Platonists philosophizes better than Descartes himself, when he thus poetically expresses their opinion,

"... Mind, being spread through the joints, moves
the whole mass, and unites itself to the great body."

I will pass over other more important features of the Divine extension since there is no need to mention them here. These few minor points will have sufficed to demonstrate that it would have been much safer to have defined matter as *tangible* or – in the way already explained – *impenetrable* substance. For both tangibility and impenetrability adequately belong to body while your definition fails to observe the law of logic since it is not equivalent to that defined.

Secondly, when you state that it is not possible even by divine power that a vacuum properly speaking exist, and that, if every body were removed from a vessel the sides would necessarily collapse, these views certainly seem to me not only false, but inconsistent with what you said earlier. For if God imparts motion to matter, which you had maintained, could He not press against the sides of the vessel and keep them from coming together? It is a contradiction that the sides of a vessel remain apart and yet nothing come between them. But the writings of Antiquity do not agree with this: Epicurus, Democritus, Lucretius, and others. But to lighten a little the force of that sort of argument, I argue that the divine extension lies between the sides of the vessel, and that your supposition on this point, that only matter of itself is extended, is weak. However, that the sides would come together as mentioned above, is not a necessity of logic, but of nature; and God alone could prevent this. Since the particles, of especially the first and second element, are moved by such a violent motion, they must rush to any place wherever matter yields, and drag those particles contiguous with themselves along.

Therefore, it is unfortunate that you based such a marvelous theorem on Rarefaction and Condensation, which I agree to be true on other grounds, on such a slippery foundation.

NOTE

* Transl. by the editor and Walter Emge, from R. Descartes, *Oeuvres* (ed. by Ch. Adam and P. Tannery), Vrin, Paris, 1956, V, pp. 238–241.

B. PASCAL

THE RELATIVITY OF MAGNITUDE*

"But to show him [i.e., man] another prodigy equally astonishing, let him examine the most delicate things he knows. Let a mite be given him, with its minute body and parts incomparably more minute, limbs with their joints, veins in the limbs, blood in the veins, humors in the blood, drops in the humors, vapors in the drops. Dividing these last things again, let him exhaust his powers of conception, and let the last object at which he can arrive be now that of our discourse. Perhaps he will think that here is the smallest point in nature. I will let him see therein a new abyss. I will paint for him not only the visible universe, but all that he can conceive of nature's immensity in the womb of this abridged atom. Let him see therein the infinity of the universes, each of which has its firmament, its planets, its earth, in the same proportion as in the visible world; in each earth animals, and in the last mites, in which he will find again all that the first had, finding still in these others the same thing without end and without cessation. Let him lose himself in wonders as amazing in their littleness as the others in their vastness. For who will not be astounded at the fact that our body, which a little while ago was imperceptible in the universe, itself imperceptible in the bosom of the whole, is now a colossus, a world, or rather a whole, in respect of the nothingness which we cannot reach?"

NOTE

* From *Pensées*; transl. by W. F. Trotter, Dutton, New York, 1931.

P. GASSENDI

THE REALITY OF INFINITE VOID ACCORDING TO ARISTOTLE*

Thus we must say that Place indeed is a quantity, or some extension, that is the space or volume (interval) of three dimensions, length, width and depth, in which a body is contained or through which it can pass. But likewise we must say that its dimensions are incorporeal and thus that place is an incorporeal volume or space, or in other words, an incorporeal quantity. And from the beginning we must distinguish these two sorts of dimensions, of which one may be called corporeal and the others spatial. For the corporeal, for example, are the length, width and depth of water contained in some vessel; but the spatial we conceive as the length, width and depth which would exist within the sides of this vessel, were the water removed and any other body kept out. Evidently, Aristotle denies that there are other than the corporeal ones, or that there is any volume or διάστηνα as he calls it, other than that of the body contained in the vessel or place. Yet very many of the Ancients thought that there were incorporeal dimensions, namely those of volume, or space, from which we take the term 'spatial'. Indeed, rather than cite Epicurus and the others, I will let the following from Nemesius stand for them all: "Every body is endowed with three dimensions. But not everything endowed with three dimensions is a body. For of this sort are Place and Quality, which are incorporeal entities." Thus, in order to understand that there are also volumes of space or places besides the corporeal ones, let us consider something in which there is no body; yet we would regard it as a container in a more obvious sense than any vessel. Let us consider, if you please, the lunar heaven itself, as the common people understand it, and likewise let us conceive the whole mass of the Aristotelian elements contained within its sphere as destroyed by God and reduced to nothingness, in such a way that nothing whatsoever takes its place. I ask, whether or not, having made this reduction to nothingness, we still conceive, within the concave surface of the lunar sphere, that very same region which had existed, but now devoid of elements and empty of all bodies? That it is possible for God to preserve the lunar sphere itself intact, while reducing the bodies contained within it to nothingness, and further to prevent any other body from taking their place, no one would go so far to deny, on the pains of denying the power of God...

Therefore, nothing prevents us from imagining the entire sublunary region, that is the region contained within the heaven, as being empty. And I think that there is no one who, having made this supposition, can not conceive such a notion easily.

Now I ask whether in this empty region while the spherical lunar heaven still persists, we do not conceive a distance or interval between one point of its concave surface and another, opposite to it? Has not this distance some length, that is, an incorporeal and invisible line, which would be the diameter of that region, and the midpoint of which would be the center of that region and of heaven and which was previously the center of the earth itself? Do we not immediately understand how much of the space around this center was previously filled by earth, by water, by air and by fire? Do we not mentally calculate how much surface would correspond to any given depth for each particular element? Accordingly, do there not still remain there the dimensions of length, width and depth which we are free to imagine? Certainly, wherever it is possible to conceive an interval or some distance, there we may also conceive a dimension, however great be the actual or possible measurement of such an interval or distance. Of this sort, therefore, are those dimensions we call both incorporeal and spatial. Moreover, if we suppose again that the whole machine of the heavens was reduced by God in a similar way to nothingness, then we conceive that that region would be in a similar way empty, and would cohere with the empty region of what was the sublunary sphere and in both regions there would equally exist as many spatial dimensions as had existed corporeal dimensions in the whole world spread throughout them. And since, if there were a larger world, and a larger one yet, on to infinity, God successively reducing each of them equally to nothingness, we understand that the spatial dimensions would always be greater and greater, on to infinity, we likewise conceive that that space with its dimensions would be extended in all directions into infinity. Let us further assume that God reproduces the world exactly as it was before in quantity and quality, then we conceive that it will be made in the way it had been made in the first creation, and from this we seem to understand three things.

The first is, that there were immense spaces before God created the World, that these would continue to exist were He, perchance, to destroy the world; and that of these God has chosen for his own good pleasure this specific region in which to create the World (the rest being left all around and commonly called Imaginary): and that this region can be taken as a whole with respect to the World, just as it coincides with the whole space of the World, so any part of space is equal to any part of the World: and thus, there is no part of the World, whether great or small, to which, in respect of its mass, there does not correspond its own part of the World's space.

Secondly, that these spaces are entirely immobile. For it is not the case that if God were to move the World from its present location, that space would follow accordingly and move along with it. But the World alone would be moved, its space certainly remaining unmoved, and it would go on to some other unmoved space, passing through intermediate, equally unmoved, spaces. In the same way, were anything whatever, or a part of the World, to change its place, the space in which it presently is would not move with it, but remain unmoved while being abandoned. And so this thing or part of the World would occupy another unmoved space to which it goes, and that space through which it passes would remain motionless.

Thirdly, that spatial dimensions, without which these spaces would be endlessly open in length, width and depth, as they are immobile, are thus incorporeal, and so have no resistance, or can be penetrated by bodies, or, as it is even commonly said, can coexist with them; so that wherever there is a body either permanently or transiently, it accordingly occupies an equal part of space. And so whatever corporeal dimensions can be designated, we understand that there are also corresponding incorporeal dimensions. What Empiricus says also leads to this conclusion, when he represents the Epicureans as saying: "A straight line (i.e. of the void), is indeed straight, and not twisted at all, since the void neither in whole nor in part is mobile." And from these considerations we finally see why Aristotle does not persuade us that there is no interval unless corporeal, nor dimensions unless corporeal....

There are some who say that God must be believed to be immense in Himself and that His essence is of entirely illimitable perfection; for to be in place is a wholly external denomination in His case. And so, if anyone were to ask, "Where was God before He created the world?" the best answer in the past was that He was in Himself. That is, that neither the world nor any other thing, but Himself alone, is His suitable and appropriate place. For the perfection of the divine essence is truly unlimited or infinite. But this lack of limitation or infinity is not properly called immensity. For it is an infinity of perfection, which belongs to the very concept, while immensity refers to extension. I go on, however, to note, that intensity and extension are distinct .... But indeed, we conceive in God, beyond an infinity of perfection in His essence, an infinity of duration also, which we call eternity, by which we hold that he shall always exist. And beyond this, we conceive an infinity as if of extension, which we call immensity, by which we hold that he is everywhere. But, I say, *as if* of extension, lest we imagine that the divine substance were extended through space like bodies are. Indeed, although the divine substance is supremely indivisible and whole at any time and at any place, yet doubtless as corporeal substance is said to be extended – that it is not at one point only but is spread out through many parts of space – so there is a kind of divine extension, which does not exist in one place only, but in many, indeed, in all places. But let me add that since it follows from the perfection of the divine essence that it be eternal and immense, all time and all space are therefore connoted, without which neither eternity nor immensity could be understood. Therefore, God, indeed, both exists supremely in Himself and is infinitely perfect, but He also necessarily exists in all time and in every place. And when it is asked, "Where was God before He created the world?" one cannot deny that He was in Himself; but it must be conceded at the same time that He was everywhere, that is, in every place; that is, not only in that place in which the future world would be, but also in an infinity of other places. That God be in space is thought to be a characteristic external to His essence, but not with respect to His immensity, the conception of which necessarily involves the conception of space. In other words, to say that God is his own place is obviously metaphorical; and since the same thing can be said about any other thing namely that it is in itself, this does not prevent God from being, properly speaking, in every place and particular things from being in their particular places.

From all of this it follows, that once admitting spaces beyond the World, in which God exists, it can be clearly understood that God is immobile and can, though unmoved, move everything, and act. One cannot conceive this from opposite premises. For if God were only in the World, or in the space which He created as a place for the world to exist and for Himself... it would follow that if the World were moved from the place in which it exists, God would no longer be here, but would be moved with the World and so would not be immobile. In other words, I insist that if God were limited to the place we now occupy, he would not have occupied space before He created the world, nor would he remain there after he had destroyed it, and would have therefore begun to be where He had not been, would cease to be where He had been; and how, therefore, could He be immobile in space?

NOTE

* From *Syntagma philosophicum*, Physica, Sectio I, Liber 2 'De loco et Duratione Rerum', *Opera omnia*, Florencia 1727, I, pp. 162–163, 170; transl. by the editor and Walter Emge.

I. NEWTON

ON ABSOLUTE SPACE AND ABSOLUTE MOTION*

II. Absolute space, in its own nature, without relation to anything external, remains always similar and immovable. Relative space is some movable dimension or measure of the absolute spaces; which our senses determine by its position to bodies; and which is commonly taken for immovable space; such is the dimension of a subterraneous, an aerial, or celestial space, determined by its position in respect of the earth. Absolute and relative space are the same in figure and magnitude; but they do not remain always numerically the same. For if the earth, for instance, moves, a space of our air, which relatively and in respect of the earth remains always the same, will at one time be one part of the absolute space into which the air passes; at another time it will be another part of the same, and so, absolutely understood, it will be continually changed.

III. Place is a part of space which a body takes up, and is according to the space, either absolute or relative. I say, a part of space; not the situation, nor the external surface of the body. For the places of equal solids are always equal; but their surfaces, by reason of their dissimilar figures, are often unequal. Positions properly have no quantity, nor are they so much the places themselves, as the properties of places. The motion of the whole is the same with the sum of the motions of the parts; that is, the translation of the whole, out of its place, is the same thing with the sum of the translations of the parts out of their places; and therefore the place of the whole is the same as the sum of the places of the parts, and for that reason, it is internal, and in the whole body.

IV. Absolute motion is the translation of a body from one absolute place into another; and relative motion, the translation from one relative place into another. Thus in a ship under sail, the relative place of a body is that part of the ship which the body possesses; or that part of the cavity which the body fills, and which therefore moves together with the ship: and relative rest is the continuance of the body in the same part of the ship, or of its cavity. But real, absolute rest, is the continuance of the body in the same part of that immovable space, in which the ship itself, its cavity, and all that it contains, is moved. Wherefore, if the earth is really at rest, the body, which relatively rests in the ship, will really and absolutely move with the same velocity which the ship has on the earth. But if the earth also moves, the true and absolute motion of the body will arise, partly from the true motion of the earth, in immovable space, partly from the relative motion of the ship on the earth; and if the body moves also relatively in the ship, its true motion will arise, partly from the true motion of the earth, in immovable space, and partly from the relative motions as well of the ship on the earth, as of the body in the ship; and from these relative motions will arise the relative motion of the body on the earth. As if that part of the earth, where the ship is, was truly moved towards the east, with a velocity of 10010 parts; while the ship itself, with a fresh gale, and full sails, is carried towards the west, with a velocity expressed by 10 of those parts; but a sailor walks in the ship towards the east, with 1 part of the said velocity; then the sailor will be moved truly in immovable space towards the east, with a velocity of 10001 parts, and relatively on the earth towards the west, with a velocity of 9 of those parts.

As the order of the parts of time is immutable, so also is the order of the parts of space. Suppose those parts to be moved out of their places, and they will be moved (if the expression may be allowed) out of themselves. For times and spaces are, as it were, the places as well of themselves as of all other things. All things are placed in time as to order of succession; and in space as to order of situation. It is from their essence or nature that they are places; and that the primary places of things should be movable, is absurd. These are therefore the absolute places; and translations out of those places, are the only absolute motions.

But because the parts of space cannot be seen, or distinguished from one another by our senses, therefore in their stead we use sensible measures of them. For from the positions and distances of things from any body considered as immovable, we define all places; and then with respect to such places, we estimate all motions, considering bodies as transferred from some of those places into others. And so, instead of absolute places and motions, we use relative ones; and that without any inconvenience in common affairs; but in philosophical disquisitions, we ought to abstract from our senses, and consider things themselves, distinct from what are only sensible measures of them. For it may be that there is no body really at rest, to which the places and motions of others may be referred.

But we may distinguish rest and motion, absolute and relative, one from the other by their properties, causes, and effects. It is a property of rest, that bodies really at rest do rest in respect to one another. And therefore as it is possible, that in the remote regions of the fixed stars, or perhaps far beyond them, there may be some body absolutely at rest; but impossible to know, from the position of bodies to one another in our regions, whether any of these do keep the same position to that remote body, it follows that absolute rest cannot be determined from the position of bodies in our regions.

It is a property of motion, that the parts, which retain given positions to their wholes, do partake of the motions of those wholes. For all the parts of revolving bodies endeavor to recede from the axis of motion; and the impetus of bodies moving forwards arises from the joint impetus of all the parts. Therefore, if surrounding bodies are moved, those that are relatively at rest within them will partake of their motion. Upon which account, the true and absolute motion of a body cannot be determined by the translation of it from those which only seem to rest; for the external bodies ought not only to appear at rest, but to be really at rest. For otherwise, all included bodies, besides their translation from near the surrounding ones, partake likewise of their true motions; and though that translation were not made, they would not be really at rest, but only seem to be so. For the surrounding bodies stand in the like relation to the surrounded as the exterior part of a whole does to the interior, or as the shell does to the kernel; but if the shell moves, the kernel will also move, as being part of the whole, without any removal from near the shell.

A property, near akin to the preceding, is this, that if a place is moved, whatever is placed therein moves along with it; and therefore a body, which is moved from a place in motion, partakes also of the motion of its place. Upon which account, all motions, from places in motion, are no other than parts of entire and absolute motions; and every entire motion is composed of the motion of the body out of its first place, and the motion of this place out of its place; and so on, until we come to some immovable place, as in the before-mentioned example of the sailor. Wherefore, entire and absolute motions can be no otherwise determined than by immovable places; and for that reason I did before refer those absolute motions to immovable places, but relative ones to movable places. Now no other places are immovable but those that, from infinity to infinity, do all retain the same given position one to another; and upon this account must ever remain unmoved; and do thereby constitute immovable space...

The causes by which true and relative motions are distinguished, one from the other, are the forces impressed upon bodies to generate motion. True motion is neither generated nor altered, but by some force impressed upon the body moved; but relative motion may be generated or altered without any force impressed upon the body. For it is sufficient only to impress some force on other bodies with which the former is compared, that by their giving way, that relation may be changed, in which the relative rest or motion of this other body did consist. Again, true motion suffers always some change from any force impressed upon the moving body; but relative motion does not necessarily undergo any change by such forces. For if the same forces are likewise impressed on those other bodies, with which the comparison is made, that the relative position may be preserved, then that condition will be preserved in which the relative motion consists. And therefore any relative motion may be changed when the true motion remains unaltered, and the relative may be preserved when the true suffers some change. Thus, true motion by no means consists in such relations.

The effects which distinguish absolute from relative motion are, the forces of receding from the axis of circular motion. For there are no such forces in a circular motion purely relative, but in a true and absolute circular motion, they are greater or less, according to the quantity of the motion. If a vessel, hung by a long cord, is so often turned about that the cord is strongly twisted, then filled with water, and held at rest together with the water; thereupon, by the sudden action of another force, it is whirled about the contrary way, and while the cord is untwisting itself, the vessel continues for some time in this motion; the surface of the water will at first be plain, as before the vessel began to move; but after that, the vessel, by gradually communicating its motion to the water, will make it begin sensibly to revolve, and recede by little and little from the middle, and ascend to the sides of the vessel, forming itself into a concave figure (as I have experienced), and the swifter the motion becomes, the higher will the water rise, till at last, performing its revolutions in the same times with the vessel, it becomes relatively at rest in it. This ascent of the water shows its endeavor to recede from the axis of its motion; and the true and absolute circular motion of the water, which is here directly contrary to the relative, becomes known, and may be measured by this endeavor. At first, when the relative motion of the water in the vessel was greatest, it produced no endeavor to recede from the axis; the water showed no tendency to the circumference, nor any ascent towards the sides of the vessel, but remained of a plain surface, and therefore its true circular motion had not yet begun. But afterwards, when the relative motion of the water had decreased, the ascent thereof towards the sides of the vessel proved its endeavor to recede from the axis; and this endeavor showed the real circular motion of the water continually increasing, till it had acquired its greatest quantity, when the water rested relatively in the vessel. And therefore this endeavor does not depend upon any translation of the water in respect of the ambient bodies, nor can true circular motion be defined by such translation. There is only one real circular motion of any one revolving body, corresponding to only one power of endeavoring to recede from its axis of motion, as its proper and adequate effect; but relative motions, in one and the same body, are innumerable, according to the various relations it bears to external bodies, and, like other relations, are altogether destitute of any real effect, any otherwise than they may perhaps partake of that one only true motion. And therefore in their system who suppose that our heavens, revolving below the sphere of the fixed stars, carry the planets along with them; the several parts of those heavens, and the planets, which are indeed relatively at rest in their heavens, do yet really move. For they change their position one to another (which never happens to bodies truly at rest), and being carried together with their heavens, partake of their motions, and as parts of revolving wholes, endeavor to recede from the axis of their motions.

Wherefore relative quantities are not the quantities themselves, whose names they bear, but those sensible measures of them (either accurate or inaccurate), which are commonly used instead of the measured quantities themselves. And if the meaning of words is to be determined by their use, then by the names time, space, place and motion, their [sensible] measures are properly to be understood; and the expression will be unusual, and purely mathematical, if the measured quantities themselves are meant. On this account, those violate the accuracy of language, which ought to be kept precise, who interpret these words for the measured quantities. Nor do those less defile the purity of mathematical and philosophical truth who confound real quantities with their relations and sensible measures.

It is indeed a matter of great difficulty to discover, and effectually to distinguish, the true motions of particular bodies from the apparent; because the parts of that immovable space, in which those motions are performed, do by no means come under the observation of our senses. Yet the thing is not altogether desperate; for we have some arguments to guide us, partly from the apparent motions, which are the differences of the true motions; partly from the forces, which are the causes and effects of the true motions. For instance, if two globes, kept at a given distance one from the other by means of a cord that connects them, were revolved about their common centre of gravity, we might, from the tension of the cord, discover the endeavor of the globes to recede from the axis of their motion, and from thence we might compute the quantity of their circular motions. And then if any equal forces should be impressed at once on the alternate faces of the globes to augment or diminish their circular motions, from the increase or decrease of the tension of the cord, we might infer the increment or decrement of their motions; and thence would be found on what faces those forces ought to be impressed, that the motions of the globes might be most augmented; that is, we might discover their hindmost faces, or those which, in the circular motion, do follow. But the faces which follow being known, and consequently the opposite ones that precede, we should likewise know the determination of their motions. And thus we might find both the quantity and the determination of this circular motion, even in an immense vacuum, where there was nothing external or sensible with which the globes could be compared. But now, if in that space some remote bodies were placed that kept always a given position one to another, as the fixed stars do in our regions, we could not indeed determine from the relative translation of the globes among those bodies, whether the motion did belong to the globes or to the bodies. But if we observed the cord, and found that its tension was that very tension which the motions of the globes required, we might conclude the motion to be in the globes, and the bodies to be at rest; and then, lastly, from the translation of the globes among the bodies, we should find the determination of their motions. But how we are to obtain the true motions from their causes, effects, and apparent differences, and the converse, shall be explained more at large in the following treatise. For to this end it was that I composed it....

This most beautiful system of the sun, planets, and comets, could only proceed from the counsel and dominion of an intelligent and powerful Being. And if the fixed stars are the centres of other like systems, these, being formed by the like wise counsel, must be all subject to the dominion of One; especially since the light of the fixed stars is of the same nature with the light of the sun, and from every system light passes into all the other systems: and lest the systems of the fixed stars should, by their gravity, fall on each other, he hath placed those systems at immense distances from one another.

This Being governs all things, not as the soul of the world, but as Lord over all; and on account of his dominion he is wont to be called *Lord God* παντοκράτωρ, or *Universal Ruler*; for *God* is a relative word, and has a respect to servants; and *Deity* is the dominion of God not over his own body, as those imagine who fancy God to be the soul of the world, but over servants. The Supreme God is a Being eternal, infinite, absolutely perfect; but a being, however perfect, without dominion, cannot be said to be Lord God; for we say, my God, your God, the God of *Israel*, the God of Gods, and Lord of Lords; but we do not say, my Eternal, your Eternal, the Eternal of *Israel*, the Eternal of Gods; we do not say, my Infinite, or my Perfect: these are titles which have no respect to servants. The word God[1] usually signifies *Lord*; but every lord is not a God. It is the dominion of a spiritual being which constitutes a God: a true, supreme, or imaginary dominion makes a true, supreme, or imaginary God. And from his true dominion it follows that the true God is a living, intelligent, and powerful Being; and, from his other perfections, that he is supreme, or most perfect. He is eternal and infinite, omnipotent and omniscient; that is, his duration reaches from eternity to eternity; his presence from infinity to infinity; he governs all things, and knows all things that are or can be done. He is not eternity and infinity, but eternal and infinite; he is not duration or space, but he endures and is present. He endures forever, and is everywhere present; and, by existing always and everywhere, he constitutes duration and space. Since every particle of space is *always*, and every indivisible moment of duration is *everywhere*, certainly the Maker and Lord of all things cannot be *never* and *nowhere*. Every soul that has perception is, though in different times and in different organs of sense and motion, still the same indivisible person. There are given successive parts in duration, coexistent parts in space, but neither the one nor the other in the person of a man, or his thinking principle; and much less can they be found in the thinking substance of God. Every man, so far as he is a thing that has perception, is one and the same man during his whole life, in all and each of his organs of sense. God is the same God, always and everywhere. He is omnipresent not *virtually* only, but also *substantially*; for virtue cannot subsist without substance. In him[2] are all things contained and moved; yet neither affects the other: God suffers nothing from the motion of bodies; bodies find no resistance from the omnipresence of God. It is allowed by all that the Supreme God exists necessarily; and by the same necessity he exists *always* and *everywhere*. Whence also he is all similar, all eye, all ear, all brain, all arm, all power to perceive, to understand, and to act; but in a manner not at all human, in a manner not at all corporeal, in a manner utterly unknown to us. As a blind man has no idea of colors, so have we no idea of the manner by which the all-wise God perceives and understands all things. He is utterly void of all body and bodily figure, and can therefore neither be seen, nor heard, nor touched; nor ought he to be worshiped under the representation of any corporeal thing. We have ideas of his attributes, but what the real substance of anything is we know not. In bodies, we see only their figures and colors, we hear only the sounds, we touch only their outward surfaces, we smell only the smells, and taste the savors; but their inward substances are not to be known either by our senses, or by any reflex act of our minds: much less, then, have we any idea of the substance of God. We know him only by his most wise and excellent contrivances of things, and final causes; we admire him for his perfections; but we reverence and adore him on account of his dominion: for we adore him as his servants; and a god without dominion, providence, and final causes, is nothing else but Fate and Nature. Blind metaphysical necessity, which is certainly the same always and everywhere, could produce no variety of things. All that diversity of natural things which we find suited to different times and places could arise from nothing but the ideas and will of a Being necessarily existing. But, by way of allegory, God is said to see, to speak, to laugh, to love, to hate, to desire, to give, to receive, to rejoice, to be angry, to fight, to frame, to work, to build; for all our notions of God are taken from the ways of mankind by a certain similitude, which, though not perfect, has some likeness, however. And thus much concerning God; to discourse of whom from the appearances of things, does certainly belong to Natural Philosophy.

NOTES

* From *Mathematical Principles of Natural Philosophy*; transl. by André Motte, revised by Florian Cajori, Univ. of California Press, Berkeley and Los Angeles, 1962, I, pp. 6–7, 8–12; II, pp. 544–546.

[1] Dr. Pocock derives the Latin word *Deus* from the Arabic *du* (in the oblique case *di*), which signifies *Lord*. And in this sense princes are called *gods*, *Psal*. xxxii. ver. 6; and *John* x. ver. 35. And *Moses* is called a *god* to his brother *Aaron*, and a *god* to *Pharaoh* (*Exod*. iv. ver. 16; and vii. ver. 1). And in the same sense the souls of dead princes were formerly, by the Heathens, called *gods*, but falsely, because of their want of dominion.

[2] This was the opinion of the Ancients. So *Pythagoras*, in *Cicer. de Nat. Deor.* lib. i. *Thales*, *Anaxagoras*, *Virgil*, Georg. lib. iv. ver. 220; and Aeneid, lib. vi. ver. 721. *Philo Allegor*, at the beginning of lib. i. *Aratus*, in his Phaenom. at the beginning. So also the sacred writers: as *St. Paul*, *Acts* xvii. ver. 27, 28. *St John's Gosp*. chap. xiv. ver. 2. *Moses*, in *Deut*. iv. ver. 39; and x. ver. 14. *David*, *Psal*. cxxxix, ver. 7, 8, 9. *Solomon*, I Kings viii ver. 27. *Job*, xxii. ver. 12, 13, 14. *Jeremiah*, xxiii. ver. 23, 24. The Idolaters supposed the sun, moon, and stars, the souls of men, and other parts of the world, to be parts of the Supreme God, and therefore to be worshiped; but erroneously.

J. LOCKE

ON INFINITE SPACE AND ITS DIFFERENCE FROM MATTER*

11. *Extension and body not the same.* – There are some that would persuade us, that body and extension are the same thing; who either change the signification of words, which I would not suspect them of, they having so severely condemned the philosophy of others, because it hath been too much placed in the uncertain meaning, or deceitful obscurity, of doubtful or insignificant terms. If, therefore, they mean by body and extension, the same that other people do, viz., by body, something that is solid and extended, whose parts are separable and moveable different ways; and by extension, only the space that lies between the extremities of those solid coherent parts, and which is possessed by them, they confound very different ideas one with another. For I appeal to every man's own thoughts, whether the idea of space be not as distinct from that of solidity, as it is from the idea of scarlet colour? It is true, solidity cannot exist without extension, neither can scarlet colour exist without extension; but this hinders not but that they are distinct ideas. Many ideas require others as necessary to their existence or conception, which yet are very distinct ideas. Motion can neither be, nor be conceived, without space; and yet motion is not space, nor space, motion: space can exist without it, and they are very distinct ideas; and so, I think, are those of space and solidity. Solidity is so inseparable an idea from body, that upon that depends its filling of space, its contact, impulse and communication of motion upon impulse. And if it be a reason to prove, that spirit is different from body, because thinking includes not the idea of extension in it; the same reason will be as valid, I suppose, to prove, that space is not body, because it includes not the idea of solidity in it; space and solidity being as distinct ideas, as thinking and extension, and as wholly separable in the mind one from another. Body then, and extension, it is evident, are two distinct ideas. For,

12. *First,* Extension includes no solidity, nor resistance to the motion of body, as body does.

13. *Secondly,* The parts of pure space are inseparable one from the other; so that the continuity cannot be separated, neither really nor mentally. For I demand of any one to remove any part of it from another, with which it is continued, even so much as in thought. To divide and separate actually, is, as I think, by removing the parts one from another, to make two superficies, where before there was a continuity: and to divide mentally, is to make in the mind two superficies, where before there was a continuity; and consider them as removed one from the other; which can only be done in things considered by the mind as capable of being separated; and by separation of acquiring new distinct superficies, which they then have not, but are capable of: but neither of these ways of separation, whether real or mental, is, as I think, compatible to pure space.

It is true, a man may consider so much of such a space as is answerable or commensurate to a foot, without considering the rest, which is, indeed, a partial consideration, but not so much as mental separation or division: since a man can no more mentally divide, without considering two superficies, separate one from the other, than he can actually divide without making two superficies disjoined one from the other: but a partial consideration is not separating. A man may consider light in the sun, without its heat; or mobility in body, without its extension, without thinking of their separation. One is only a partial consideration, terminating in one alone; and the other is a consideration of both, as existing separately.

14. *Thirdly,* the parts of pure space are immoveable, which follows from their inseparability; motion being nothing but change of distance between any two things: but this cannot be between parts that are inseparable; which, therefore, must needs be at perpetual rest one amongst another.

Thus the determined idea of simple space, distinguishes it plainly and sufficiently from body; since its parts are inseparable, immoveable, and without resistance to the motion of body.

21. *A vacuum beyond the utmost bounds of body.* – But to return to our idea of space. If body be not supposed infinite, which I, think, no one will affirm, I would ask, whether, if God placed a man at the extremity of corporeal beings, he could not stretch his hand beyond his body? If he could, then he would put his arm where there was before space without

body; and if there he spread his fingers, there would still be space between them without body. If he could not stretch out his hand, it must be because of some external hindrance (for we suppose him alive, with such a power of moving the parts of his body that he hath now, which is not in itself impossible, if God so pleased to have it; or, at least, it is not impossible for God so to move him); and then I ask, whether that which hinders his hand from moving outwards, be substance or accident, something or nothing? and when they have resolved that, they will be able to resolve themselves what that is, which is or may be between two bodies at a distance, that is not body, and has no solidity. In the mean time, the argument is at least as good, that where nothing hinders (as beyond the utmost bounds of all bodies), a body put in motion may move on, as where there is nothing between, there two bodies must necessarily touch: for pure space between, is sufficient to take away the necessity of mutual contact; but bare space in the way, is not sufficient to stop motion. The truth is, these men must either own, that they think body infinite, though they are loth to speak it out; or else affirm, that space is not body. For I would fain meet with that thinking man, that can, in his thoughts, set any bounds to space, more than he can to duration; or, by thinking, hope to arrive at the end of either: and, therefore, if his idea of eternity be infinite, so is his idea of immensity; they are both finite or infinite alike.

22. *The power of annihilation proves a vacuum.* – Farther, those who assert the impossibility of space existing without matter, must not only make body infinite, but must also deny a power in God to annihilate any part of matter. No one, I suppose, will deny, that God can put an end to all motion that is in matter, and fix all the bodies of the universe in a perfect quiet and rest, and continue them so long as he pleases. Whoever then will allow, that God can, during such a general rest, annihilate either this book, or the body of him that reads it, must necessarily admit the possibility of a vacuum: for it is evident, that the space that was filled by the parts of the annihilated body, will still remain, and be a space without body. For circumambient bodies being in perfect rest, are a wall of adamant, and, in that state, make it a perfect impossibility for any other body to get into that space. And, indeed, the necessary motion of one particle of matter, into the place from whence another particle of matter is removed, is but a consequence from the supposition of plentitude,

which will, therefore, need some better proof than a supposed matter of fact, which experiment can never make out; our own clear and distinct ideas plainly satisfying us, that there is no necessary connexion between space and solidity, since we can conceive the one without the other. And those who dispute for or against a vacuum, do thereby confess they have distinct ideas of vacuum and plenum, i.e. that they have an idea of extension void of solidity, though they deny its existence, or else they dispute about nothing at all. For they who so much alter the signification of words, as to call extension, body, and consequently make the whole essence of body to be nothing but pure extension, without solidity, must talk absurdly whenever they speak of vacuum, since it is impossible for extension to be without extension: for vacuum, whether we affirm or deny its existence, signifies space without body, whose very existence no one can deny to be possible, who will not make matter infinite, and take from God a power to annihilate any particle of it.

23. *Motion proves a vacuum.* – But not to go so far as beyond the utmost bounds of body in the universe, nor appeal to God's Omnipotency to find a vacuum, the motion of bodies that are in our view and neighbourhood, seems to me plainly to evince it. For I desire any one so to divide a solid body of any dimension he pleases, as to make it possible for the solid parts to move up and down freely every way within the bounds of that superficies, if there be not left in it a void space, as big as the least part into which he has divided the said solid body. And if where the least particle of the body divided is as big as a mustard-seed, a void space equal to the bulk of a mustard-seed be requisite to make room for the free motion of the parts of the divided body within the bounds of its superficies, where the particles of matter are 100000000 less than a mustard-seed; there must also be a space void of solid matter, as big as 100000000 part of a mustard-seed: for if it hold good in one, it will hold in the other, and so on in infinitum. And let this void space be as little as it will, it destroys the hypothesis of plentitude. For if there can be a space void of body, equal to the smallest separate particle of matter now existing in nature, it is still space without body, and makes as great a difference between space and body, as if it were μέγα χάσμα, a distance as wide as any in nature. And, therefore, if we suppose not the void space necessary to motion, equal to the least parcel of the divided solid matter,

but to $\frac{1}{10}$ or $\frac{1}{1000}$ of it, the same consequence will always follow of space without matter.

27. *Ideas of space and solidity distinct.* – To conclude: whatever men shall think concerning the existence of vacuum, this is plain to me, that we have as clear an idea of space, distinct from solidity, as we have of solidity, distinct from motion, or motion from space. We have not any two more distinct ideas; and we can as easily conceive space without solidity, as we can conceive body or space without motion, though it be never so certain, that neither body nor motion can exist without space. But whether any one will take space to be only a relation resulting from the existence of other beings at a distance, or whether they will think the words of the most knowing King Solomon, "The heaven, and the heaven of heavens, cannot contain thee"; or those more emphatical ones of the inspired philosopher, St. Paul, "In him we live, move, and have our being," are to be understood in a literal sense, I leave every one to consider; only our idea of space is, I think, such as I have mentioned, and distinct from that of body. For whether we consider, in matter itself, the distance of its coherent solid parts, and call it, in respect of those solid parts, extension; or, whether considering it as lying between the extremities of any body in its several dimensions, we call it length, breadth, and thickness; or else considering it as lying between any two bodies, or positive beings, without any consideration whether there be any matter or no between, we call it distance. However named or considered, it is always the same uniform simple idea of space, taken from objects about which our senses have been conversant, whereof having settled ideas in our minds, we can revive, repeat, and add them one to another, as often as we will, and consider the space or distance so imagined, either as filled with solid parts, so that another body cannot come there without displacing and thrusting out the body that was there before; or else as void of solidity, so that a body of equal dimensions to that empty or pure space, may be placed in it without the removing or expulsion of any thing that was there. But to avoid confusion in discourses concerning this matter, it were possibly to be wished, that the name extension were applied only to matter, or the distance of the extremities of particular bodies; and the term expansion to space in general, with or without solid matter possessing it, so as to say, space is expanded, and body extended. But in this every one has

liberty; I propose it only for the more clear and distinct way of speaking.

NOTE

* From *Essay Concerning Human Understanding*, Book II, Chap. 13, §§11–14, 21–23, 27.

ARGUMENT FOR THE REALITY OF ABSOLUTE SPACE*

Axiom 1

78. Every body, apart from any relation to other bodies, is either at rest or in motion. That is, is either absolutely at rest or absolutely in motion.

Explanation 1

79. Thus far, following the senses, we have not recognized any other motion or rest than that with respect to other bodies, whence we have called both motion and rest relative. But, if we now mentally take away all bodies but one, and if thus the relation by which we have hitherto distinguished its rest and motion is withdrawn, it will first be asked whether or not the conclusion respecting the rest or motion of the remaining body still stands. For, if this conclusion can be drawn only from a comparison of the place of the body in question to that of other bodies, it follows that, when these bodies are gone, the conclusion must go with them. But, albeit we do not know of the rest or motion of a body except from its relation to other bodies, it is nevertheless not to be concluded that these things (rest and motion) are nothing in themselves but mere relations established by the intellect, and that there is nothing inherent in the bodies themselves which corresponds to our ideas of rest and motion. For indeed, we have no way of knowing quantity except by comparison, yet, when the things with which we established comparison are removed, there still remains in the body, so to speak, the foundation of quantity (*fundamentum quantitatis*); for, if the body were made larger or smaller by expansion or contraction, we would have to take it as a true change. Hence, if only one body existed, we should have to say that it was either at rest or in motion, though it could not be simultaneously in both states or in neither. Therefore I conclude that rest and motions are not mere ideal entities, born from comparison alone, so that there would be nothing inherent in the body corresponding to them, but that

we can properly ask even of solitary body, whether it is in motion or at rest; whereby I fear least those philosophers who reduce everything to relations since they themselves attribute so much to motion that they regard moving force as something substantial.

Explanation 2

80.   Therefore, since we can properly ask even of a solitary body, without reference to other bodies, or under the assumption that they are annihilated, whether it is at rest or in motion, it must necessarily be in one of these states. But what this rest or motion will be, in view of the fact that there is no change of position with respect to other bodies, we cannot even conceive without admitting an absolute space in which our body would occupy some place and from which it could pass into other places. For since those philosophers who most emphatically deny the existence of absolute space are most interested in the question whether any body is in motion or at rest, they ought to indicate what is the basis of this distinction once the reference to other bodies is denied. Do they say that the body is truly in motion when it changes its position with respect to its immediate neighbourhood? But the true motion could be in its neighbourhood while the body itself could be at rest. Ought we make a comparison with the more distant bodies? But with which shall we start? Then, why with these rather than those? They will finally answer that we should start with those that are themselves at rest. But then I shall go on to ask, not how we can recognize which of these bodies are at rest, but what it is for the body itself to be at rest? since it is now impossible to speak again of the position with respect to other bodies. Thus they are finally forced to concede that those bodies are themselves at rest which remain in the same place in space and by this admission, once every reference to other bodies is removed, they are led to absolute space itself with respect to which those bodies which are either at rest or in motion are defined as being absolutely either at rest or in motion.

Scholium

81.   Whoever denies absolute space, falls into the gravest difficulties. For, since he must reject absolute motion and rest as empty sounds without meaning, he is forced to reject not only the laws of motion which are based on this principle, but the very existence of any such laws. For, if

the question which has led us to this point, "What would happen in a body deprived of any connection with other bodies?" is itself absurd, then even those effects produced in the body by the action of other would become uncertain and undeterminable, and thus everything will have to be regarded as happening by chance and without reason. Or, if someone wish to escape this consequence, he would have to deny all motion. He could hardly find a comfort in such a view even if he succeeded to refute all arguments to the contrary since he would not be able to explain what would be the meaning of 'rest' in a completely motionless universe. The opposition to such obvious absurdities is the surest basis for our view.

Axiom 2

82. A body which is absolutely at rest will continue at rest perpetually if it is not subject to any external action.

Explanation

83. This is a common axiom about bodies and it seems so obvious in itself as to need no proof. But to make its force more clearly understood, let us consider just a point or an element of a body which, if once it were at rest, must remain at rest perpetually. For, since there is no reason in it why it would begin to move in one direction rather than in all others, and since every external cause of motion is removed, it cannot be conceived to move in any direction. This truth then rests upon the principle of sufficient reason. However, we must still recognize in this very point or corporeal element a cause of its remaining at rest, such that this truth may be held to be necessary. But whatever is proved with regard to any body must hold equally for all bodies taken together as well as for any body whatsoever, for if its individual elements are at rest and remain at rest, then no one can deny that the whole body will be at rest. Yet we can raise a doubt about this sort of body, since perhaps its parts, though at rest, might act upon each other and cause motion. But even if we concede this point it makes no difference with respect to the axiom, as long as we free not only the whole body but also its individual parts from all external actions; and it is enough for us to admit the axiom in this sense, that if all the particles of a body, even the smallest, are at rest, so long as they do not act upon one another, they will continue to be at rest.

Scholium

84. The law which we have here established with respect to absolute rest can by no means be extended to relative rest. For if a body which had been at rest with respect to its particles were suddenly struck, it would no longer remain at rest in this respect. Imagine a sphere upon a board lying on a ship proceeding at a uniform speed, this sphere will remain at rest with respect to the ship. But if the ship runs into a rock, this relative rest will suddenly cease and the sphere will take on motion with respect to the ship, even though it was not subjected to any external cause. Therefore this law is necessarily restricted to absolute rest and since the law is necessary, so also the relation of bodies to whatever place they occupy is necessary. That is, since this law of rest implies remaining at rest in the same place, it may not be interpreted except with respect to absolute place. However, absolute place cannot be defined by a relation of coexistents, since otherwise our law would apply to relative rest.

Axiom 3

85. A body which is absolutely in motion will continue to proceed in the same direction in uniform motion if it is not subjected to any external action.

Explanation 1

86. This axiom must also be understood properly to apply to the smallest particles of bodies, as if points, for it would not hold for bodies endowed with magnitude unless all the particles moved with the same velocity in the same direction. For if they began to move at unequal velocities or in different directions, the individual particles could not conserve this motion without separating from each other and dissolving the structure of the body. But this need not be feared if the velocities of all the particles were the same and of the same direction or if the body were so small that there could be no disparity of place within it. Therefore let us consider a corporeal point of this kind, as if existing alone, and if it moves, it begins to move with a given velocity in a given direction; according to the axiom this point will conserve perpetually both the same velocity and the same direction. If we take this as an axiom it needs no demonstration, though

ARGUMENT FOR THE REALITY OF ABSOLUTE SPACE 117

it is not at all difficult to give a reason for it. For in the first place it will not change direction since there can be no reason why it would deviate from the original direction in one rather than in all other sense. It will, to wit, as certainly conserve the same direction as the point at rest will remain at rest. Furthermore, this pertains as well to velocity, for if it does not remain perpetually the same, it would have to be said to increase or decrease, neither of which can be claimed without absurdity, for were it to increase or decrease this would have to be according to some law. But what sort of law this would be can in no way be conceived since it is not certain that any deserves such a priority over the others. Hence, if anyone were by chance to claim that the velocity decreases in reference to time, he has not yet defined anything; for he would have to go on to determine what part of the velocity vanished at any particular moment of time, and this cannot be admitted since whatever diminution of velocity is assigned to any particular moment of time, there is no reason to support it. And this will be true for any other law. Therefore, we are left with no option but to state that velocity also remains perpetually the same, just as is the case with direction.

Explanation 2

87. To both this axiom and the preceeding one, there is opposed the opinion of those philosophers who hold that all bodies are endowed with a certain hidden power of continually changing their own state of motion or rest. This opinion, which has no rational support, is completely destroyed by the very fact that it contradicts our axiom. Yet this axiom normally appears, at first glance, to be contrary to experience, since in all our experiments we observe that motion gradually slows down and finally ceases altogether, so that on this basis we would deny perpetual motion, though in virtue of our axiom we would have to hold that all motion is perpetual. However, in these very experiments the cause of this slowing down is clearly detected, be it friction or the resistance of the air or other obstacles to motion, which can never be entirely removed. And if we carefully weigh these circumstances we must conclude from these experiments themselves that if all these obstacles were absent, motion would indeed last perpetually. Wherefore, since all obstacles were expressly excluded in the axiom, these experiments are so far from opposing it as to confirm its truth by a sensory evidence. For the rest, we should

take proper care lest this axiom, restricted to absolute motion, be extended also to relative motion.

Definition 10

88. While a body is either at absolute rest or in absolute motion with uniform velocity in the same direction, it is said *to remain in the same state*.

Corollary 1

89. Therefore, taking both axioms together, we can state that bodies, insofar as they are not impeded by others, remain in the same state.

Corollary 2

90. Therefore, if a body which had previously been at rest begins to move, or on in motion undergoes a change either in velocity or in direction, it must be held to have changed its state.

Scholium

91. If it is not incongruous to call permanence at rest or in motion with uniform velocity in a straight line a *state*, since a body is so determined by itself; then for as long as a body is left to itself and is subjected to no external action, it is correct to say that it remains in the same state, if indeed a change in state seems to imply an external action. Therefore, staying in the same state differs greatly from staying in the same place, since that occurs when the body is at rest. The axioms previously established have led us to this idea of state, yet the idea of state, which is of itself arbitrary, could not have led us to know the axioms; rather this idea itself has received its fixed meaning from them.

Definition 11

92. That property of bodies by which they inherently contain the reason for their remaining in the same state is called *inertia*, even sometimes *the force of inertia. (vis inertiae.)*

Corollary 1

93. Therefore, inertia is the proper cause of bodies remaining in the same state. For since the cause must be sought in the body itself, it whithout doubt must be held to be a common property of all bodies.

Corollary 2

94. But if, therefore, it is asked, why a body absolutely at rest continues to be at rest or a body in motion continues in motion at the same velocity and in the same direction, one cannot assign to it any other cause but inertia; nor can one seek any cause of this phenomenon outside the body.

NOTE

* From *Theoria Motus Corporum Solidorum*, Cap. II, §§ 78–84; transl. by the editor and Walter Emge.

J. C. MAXWELL

ON ABSOLUTE SPACE*

ON THE IDEA OF SPACE[1]

We have now gone through most of the things to be attended to with respect to the configuration of a material system. There remain, however, a few points relating to the metaphysics of the subject, which have a very important bearing on physics.

We have described the method of combining several configurations into one system which includes them all. In this way we add to the small region which we can explore by stretching our limbs the more distant regions which we can reach by walking or by being carried. To these we add those of which we learn by the reports of others, and those inaccessible regions whose positions we ascertain only by a process of calculation, till at last we recognise that every place has a definite position with respect to every other place, whether the one place is accessible from the other or not.

Thus from measurements made on the earth's surface we deduce the position of the centre of the earth relative to known objects, and we calculate the number of cubic miles in the earth's volume quite independently of any hypothesis as to what may exist at the centre of the earth, or in any other place beneath that thin layer of the crust of the earth which alone we can directly explore.

ERROR OF DESCARTES

It appears, then, that the distance between one thing and another does not depend on any material thing between them, as Descartes seems to assert when he says (*Princip.* Phil., II. 18) that if that which is in a hollow vessel were taken out of it without anything entering to fill its place, the sides of the vessel, having nothing between them, would be in contact.

This assertion is grounded on the dogma of Descartes, that the extension in length, breadth, and depth which constitute space is the sole essential property of matter. "The nature of matter," he tells us, "or of body considered generally, does not consist in a thing being hard, or heavy, or coloured, but only in its being extended in length, breadth, and depth" (*Princip.*, II. 4). By thus confounding the properties of matter with those of space, he arrives at the logical conclusion that if the matter within a vessel could be entirely removed, the space within the vessel would no longer exist. In fact he assumes that all space must be always full of matter.

I have referred to this opinion of Descartes in order to show the importance of sound views in elementary dynamics. The primary property of matter was indeed distinctly announced by Descartes in what he calls the "First Law of Nature" (*Princip.*, II. 37): "That every individual thing, so far as in it lies, perseveres in the same state, whether of motion or of rest."

We shall see when we come to Newton's laws of motion that in the words 'so far as in it lies,' properly understood, is to be found the true primary definition of matter, and the true measure of its quantity. Descartes, however, never attained to a full understanding of his own words (*quantum in se est*), and so fell back on his original confusion of matter with space – space being, according to him, the only form of substance, and all existing things but affections of space. This error runs through every part of Descartes' great work, and it forms one of the ultimate foundations of the system of Spinoza. I shall not attempt to trace it down to more modern times, but I would advise those who study any system of metaphysics to examine carefully that part of it which deals with physical ideas.

We shall find it more conducive to scientific progress to recognise, with Newton, the ideas of time and space as distinct, at least in thought, from that of the material system whose relations these ideas serve to coordinate.

ABSOLUTE SPACE

Absolute space is conceived as remaining always similar to itself and immovable. The arrangement of the parts of space can no more be altered than the order of the portions of time. To conceive them to move from their places is to conceive a place to move away from itself.

But as there is nothing to distinguish one portion of time from another except the different events which occur in them, so there is nothing to distinguish one part of space from another except its relation to the place of material bodies. We cannot describe the time of an event except by reference to some other event, or the place of a body except by reference to some other body. All our knowledge, both of time and place, is essentially relative[2]. When a man has acquired the habit of putting words together, without troubling himself to form the thoughts which ought to correspond to them, it is easy for him to frame an antithesis between this relative knowledge and a so-called absolute knowledge, and to point out our ignorance of the absolute position of a point as an instance of the limitation of our faculties. Any one, however, who will try to imagine the state of a mind conscious of knowing the absolute position of a point will ever after be content with our relative knowledge.

NOTES

[*] From *Matter and Motion*, Dover, New York, n.d., pp. 9–12.
[1] Following Newton's method of exposition in the *Principia*, a space is assumed and a flux of time is assumed, forming together a framework into which the dynamical explanation of phenomena is set. It is part of the problem of physical astronomy to test this assumption, and to determine this frame with increasing precision. Its philosophical basis can be regarded as a different subject, to which the recent discussions on relativity as regards space and time would be attached.
[2] The position seems to be that our knowledge is relative, but needs definite space and time as a frame for its coherent expression.

C. NEUMANN

ON THE NECESSITY OF THE ABSOLUTE FRAME OF REFERENCE*

The principles of Galilei-Newton's theories consist of two laws: the law of inertia, formulated already by Galilei, and the law of gravity, added later by Newton. And even though we must give up any attempt of *explanation* of these two basic principles, we must so much more strongly insist on having at least a clear statement of their content; and even on this point we shall encounter several difficulties. These difficulties will force us to divide these laws in a greater number of simple basic ideas, in a greater number of fundamental principles.

> A material point, provided no external force acts upon it, that is, provided it is left to itself, moves along a *straight line* and covers in equal times *equal distances* – This is the law of inertia, formulated by Galilei.

It is impossible to regard this proposition in this form as the *cornerstone* of the scientific system, as the *starting point* of mathematical deductions. For it is completely *unintelligible*. We do not know what is meant by "a motion along a straight line"; or rather we know that these words can be interpreted in various ways and thus can acquire infinite number of different meanings. For any motion which, for instance, observed from the earth is *rectilinear*, when observed from the sun will appear *curvilinear*, – while from the standpoint of the observers on Jupiter, Saturn and other celestial bodies will in each case be represented by *another* curve. Briefly: every motion which is *rectilinear* with respect to one celestial body, will appear *curvilinear* with respect to *another* celestial body.

Those words of Galilei that a material point left to itself is moving along a straight line therefore appear to us as devoid of definite meaning; they require a definite context to become intelligible. There must exist one particular body in the cosmic space as that object to which all motions should be referred; only then we shall be able to give a definite meaning to the words above. Now which is that body to which we should assign such a privileged status? Or are there perhaps *many* such bodies? Should perhaps the motions in the neighborhood of the earth be referred to the earth, the motions in the vicinity of the sun to be referred to the sun?

Unfortunately, either in Galilei or in Newton we do not find any definite answer to these questions. But if we attentively scrutinize the theoretical construction which they erected and which was being more and more extended until today, its foundations cannot remain hidden to us. We then easily recognize that all existing and all conceivable motions in the universe must be referred to *one and single body*. *Where* this body is located and which are the grounds for endowing it with such outstanding and even sovereign status – to this we have *no* answer.

> From this it follows that as the following proposition should stand as the first principle of Galilei-Newton's theory: there is at some unknown place in the cosmic space one unknown body, more specifically, an *absolutely rigid* body whose shape and dimensions are forever unchangeable.

Permit me to designate this body briefly as body *Alpha*. It should be then added that by the *motion* of a point we should not understand the change of its position with respect to the earth or to the sun, but that with respect to the body Alpha.

From this point of view Galilei's law acquires a clearly discernible content. It is presented as:

> the *second principle*, that a material point, left to itself, is moving along a straight line, that is along the path which is rectilinear with respect to the body Alpha.

It is true that we usually ignore the body Alpha; we speak of *absolute* space and *absolute* motion. These are merely *other* words for the same thing. We cannot further specify the fact that the very essence of absolute motion consists – as nobody would deny – in the possibility to refer all displacements *to one and the same object*, that is, to the object which is spatially extended and unchangeable. Now it is this object which is designated by me as the unknown *rigid* body, the body Alpha.

But then we face another question, whether this body has the same concrete existence as the earth, the sun and other celestial bodies. It seems to me that to this we could answer that its existence can be posited with an equal justification as the existence of the luminiferous aether or of the electrical fluid...

NECESSITY OF THE ABSOLUTE FRAME OF REFERENCE 127

In the same sense in which the electric fluids in a given substance are quantitatively undetermined and – without any damage to the theory – can be conceived either larger or smaller, a certain indefiniteness is inherent in the body Alpha. For this body can be replaced – without any restriction of Galilei-Newton's theory – by another body Alpha, provided that we assign to the latter a *rectilinear uniform* motion with respect to the former. These conditions are, of course, necessary. For the substitution of another body Alpha which with respect to the first body Alpha would have a *different* motion, for instance a *rotary* motion, is *altogether inadmissible*.

On this point certain reflections come irresistibly to our mind from which it clearly follows how intolerable are the contradictions which occur as soon as motion is regarded not as something absolute, but relative.

Let us assume that there is among the stars one which consists of the fluid mass and which – like our earth – is rotating around its axis. This star by the effect of its rotation and of the resulting centrifugal forces would take on the shape of a flattened ellipsoid. Let us ask now: *which form would the star have if all other celestial bodies were completely annihilated?*

These centrifugal forces depend solely upon the state of the star itself; they are wholly independent of the other celestial bodies. These forces, therefore, as well as the ellipsoidal form, will persist, irrespective of the continued existence or disappearance of the other bodies. But, if motion is defined as something relative – as a relative change of place of two points – the answer is very different. If, on this assumption, we suppose all other celestial bodies to be annihiliated, nothing remains but the material points of which the star in question itself consists. But, then, these points do not change their relative positions, and are therefore at rest. It follows that the star must be at rest at the moment when the annihilation of the other bodies takes place, and therefore must assume the spherical form taken by all bodies in a state of rest. A contradiction so intolerable can be avoided only by abandoning the assumption of the relativity of motion, and conceiving motion as absolute, so that thus we are again led to the principle of the body Alpha.

NOTE

* From *Über die Principien der Galilei-Newtonschen Lehre*, Leipzig 1870, pp. 14–16, 20–22, 27–28; transl. by the editor.

EARLY DEFENSE OF NEWTON'S ABSOLUTE SPACE*

463. In the justly famous scholium to the definitions, Newton has stated, with admirable precision, the doctrine of absolute space, time, and motion. Not being a skilled philosopher, he was unable to give grounds for his views, except an empirical argument derived from actual Dynamics. Leibniz, with an unrivalled philosophical equipment, controverted Newton's position in his letters against Clarke[1]; and the victory, in the opinion of subsequent philosophers, rested wholly with Leibniz. Although it would seem that Kant, in the Transcendental Aesthetic inclines to absolute position in space, yet in the *Metaphysische Anfangsgründe der Naturwissenschaft* he quite definitely adopts the relational view. Not only other philosophers, but also men of science, have been nearly unanimous in rejecting absolute motion, the latter on the ground that it is not capable of being observed, and cannot therefore be a datum in an empirical study.

But a great difficulty has always remained as regards the argument from absolute rotation, adduced by Newton himself. This argument, in spite of a definite assertion that all motion is relative, is accepted and endorsed by Clerk Maxwell[2]. It has been revived and emphasized by Heymans[3], combated by Mach[4], Karl Pearson[5], and many others, and made part of the basis of a general attack on Dynamics in Professor Ward's *Naturalism and Agnosticism*. Let us first state the argument in various forms, and then examine some of the attempts to reply to it. For us, since absolute time and space have been admitted, there is no need to avoid absolute motion, and indeed no possibility of doing so. But if absolute motion is in any case unavoidable, this affords a new argument in favour of the justice of our logic, which, unlike the logic current among philosophers, admits and even urges its possibility.

464. If a bucket containing water is rotated, Newton observes, the water will become concave and mount up the sides of the bucket. But if the bucket be left at rest in a rotating vessel, the water will remain level in spite of the relative rotation. Thus absolute rotation is involved in the

phenomenon in question. Similarly, from Foucault's pendulum and other similar experiments, the rotation of the earth can be demonstrated, and could be demonstrated if there were no heavenly bodies in relation to which the rotation becomes sensible. But this requires us to admit that the earth's rotation is absolute. Simpler instances may be given, such as the case of two gravitating particles. If the motion dealt with in Dynamics were wholly relative, these particles, if they constituted the whole universe, could only move in the line joining them, and would therefore ultimately fall into one another. But Dynamics teaches that, if they have initially a relative velocity not in the line joining them, they will describe conics about their common centre of gravity as focus. And generally, if acceleration be expressed in polars, there are terms in the acceleration which, instead of containing several differentials, contain squares of angular velocities: these terms require absolute angular velocity, and are inexplicable so long as relative motion is adhered to.

If the law of gravitation be regarded as universal, the point may be stated as follows. The laws of motion require to be stated by reference to what have been called *kinetic* axes: these are in reality axes having no absolute acceleration and no absolute rotation. It is asserted, for example, when the third law is combined with the notion of mass that, if m, m' be the masses of two particles between which there is a force, the component accelerations of the two particles due to this force are in the ratio $m_2:m_1$. But this will only be true if the accelerations are measured relatively to axes which themselves have no acceleration. We cannot here introduce the centre of mass, for, according to the principle that dynamical facts must be, or be derived from, observable data, the masses, and therefore the centre of mass, must be obtained from the acceleration, and not *vice versa*. Hence any dynamical motion, if it is to obey the laws of motion, must be referred to axes which are not subject to any forces. But, if the law of gravitation be accepted, no *material* axes will satisfy this condition. Hence we shall have to take *spatial* axes, and motions relative to these are of course absolute motions.

465. In order to avoid this conclusion, C. Neumann[6] assumes as an essential part of the laws of motion the existence, somewhere, of an absolutely rigid 'Body *Alpha*', by reference to which all motions are to be estimated. This suggestion misses the essence of the discussion,

which is (or should be) as to the logical *meaning* of dynamical propositions, not as to the way in which they are discovered. It seems sufficiently evident that, if it is necessary to invent a fixed body, purely hypothetical and serving no purpose except to be fixed, the reason is that what is really relevant is a fixed *place*, and that the body occupying it is irrelevant. It is true that Neumann does not incur the vicious circle which would be involved in saying that the Body *Alpha* is fixed, while all motions are relative to it; he asserts that it is rigid, but rightly avoids any statement as to its rest or motion, which, in his theory, would be wholly unmeaning. Nevertheless, it seems evident that the question whether one body is at rest or in motion must have as good a meaning as the same question concerning any other body; and this seems sufficient to condemn Neumann's suggested escape from absolute motion.

466. A development of Neumann's views is undertaken by Streintz[7], who refers motions to what he calls "fundamental bodies" and "fundamental axes." These are defined as bodies or axes which do not rotate and are independent of all outside influences. Streintz follows Kant's *Anfangsgründe* in regarding it as possible to admit absolute rotation while denying absolute translation. This is a view which I shall discuss shortly, and which, as we shall see, though fatal to what is desired of the relational theory, is yet logically tenable, though Streintz does not show that it is so. But apart from this question, two objections may be made to this theory. (1) If motion *means* motion relative to fundamental bodies (and if not, their introduction is no gain from a logical point of view), then the law of gravitation becomes strictly meaningless if taken to be universal – a view which seems impossible to defend. The theory requires that there should be matter not subject to any forces, and this is denied by the law of gravitation. The point is not so much that universal gravitation must be *true*, as that it must be significant – whether true or false is an irrelevant question. (2) We have already seen that absolute *accelerations* are required even as regards translations, and that the failure to perceive this is due to overlooking the fact that the centre of mass is not a piece of matter, but a spatial point which is only determined by means of accelerations.

467. Somewhat similar remarks apply to Mr. W. H. Macaulay's article on 'Newton's Theory of Kinetics'.[8] Mr. Macaulay asserts that the true

way to state Newton's theory (omitting points irrelevant to the present issue) is as follows:

Axes of reference can be so chosen, and the assignment of masses so arranged, that a certain decomposition of the rates of change of momenta, relative to the axes, of all the particles of the universe is possible, namely one in which the components occur in pairs; the members of each pair belonging to two different particles, and being opposite in direction, in the line joining the particles, and equal in magnitude (p. 368).

Here again, a purely logical point remains. The above statement appears unobjectionable, but it does not show that absolute motion is unnecessary. The axes cannot be material, for all matter is or may be subject to forces, and therefore unsuitable for our purpose; they cannot even be defined by any fixed geometrical relation to matter. Thus our axes will really be spatial; and if there were no absolute space, the suggested axes could not exist. For apart from absolute space, any axes would have to be material or nothing. The axes can, in a sense, be defined by relation to matter, but not by a constant geometrical relation; and when we ask what property is changed by motion relative to such axes, the only possible answer is that the absolute position has changed. Thus absolute space and absolute motion are not avoided by Mr. Macaulay's statement of Newton's laws.

468. If absolute rotation alone were in question, it would be possible, by abandoning all that recommends the relational theory to philosophers and men of science, to keep its logical essence intact. What is aimed at is, to state the principles of Dynamics in terms of sensible entities. Among these we find the metrical properties of space, but not straight lines and planes. Collinearity and coplanarity may be included, but if a set of collinear material points change their straight line, there is no sensible intrinsic change. Hence all advocates of the relational theory, when they are thorough, endeavour, like Leibniz[9], to deduce the straight line from distance. For this there is also the reason that the field of a given distance is all space, whereas the field of the generating relation of a straight line is only that straight line, when the latter, but not the former, makes an intrinsic distinction among the points of space, which the relational theory seeks to avoid. Still, we might regard straight lines as relations between *material* points, and absolute rotation would then appear as change in a relation between material points, which is logically

compatible with a relational theory of space. We should have to admit, however, that the straight line was not a *sensible* property of two particles between which it was a relation; and in any case, the necessity for absolute translational accelerations remains fatal to any relational theory of motion.

469. Mach[10] has a very curious argument by which he attempts to refute the grounds in favour of absolute rotation. He remarks that, in the actual world, the earth rotates relating to the fixed stars, and that the universe is not given twice over in different shapes, but only once, and as we find it. Hence any argument that the rotation of the earth could be inferred *if* there were no heavenly bodies is futile. This argument contains the very essence of empiricism, in a sense in which empiricism is radially opposed to the philosophy advocated in the present work[11]. The logical basis of the argument is that all propositions are essentially concerned with actual existents, not with entities which may or may not exist. For if, as has been held throughout our previous discussions, the whole dynamical world with its laws can be considered without regard to existence, then it can be no part of the *meaning* of these laws to assert that the matter to which they apply exists, and therefore they can be applied to universes which do not exist. Apart from general arguments, it is evident that the laws are so applied throughout rational Dynamics, and that in all exact calculations, the distribution of matter which is assumed is not that of the actual world. It seems impossible to deny significance to such calculations; and yet if they have significance, if they contain propositions at all, whether true or false, then it can be no necessary part of their *meaning* to assert the existence of the matter to which they are applied. This being so, the universe is given, as an entity, not only twice, but as many times as there are possible distributions of matter, and Mach's argument falls to the ground. The point is important, as illustrating a respect in which the philosophy here advocated is to be reckoned with idealism and not with empiricism, in spite of the contention that what exists can only be known empirically.

Thus, to conclude: Absolute motion is essential to Dynamics, and involves absolute space. This fact, which is a difficulty in current philosophies, is for us a powerful confirmation of the logic upon which our discussions have been based.

NOTES

* From *The Principles of Mathematics*, W. W. N. Y., 1964, and Allen and Unwin, London 1903, pp. 489–493.

[1] *Phil. Werke*, ed. by Gerhardt, vol. VII.
[2] *Matter and Motion*, Art. CV. Contrast Art. XXX.
[3] *Die Gesetze und Elemente des wissenschaftlichen Denkens*, Leyden 1890.
[4] *Die Mechanik in ihrer Entwickelung*, Leipzig 1883; transl., London 1902.
[5] *Grammar of Science*, London 1892; 2nd ed. 1900.
[6] *Die Galilei-Newtonsche Theorie*, Leipzig 1870, p. 15.
[7] *Die physikalischen Grundlagen der Mechanik*, Leipzig 1883; see esp. pp. 24, 25.
[8] *Bulletin of the American Math. Soc.*, Vol. III. (1896-7). For a later statement of Mr. Macaulay's views, see Art. *Motion, Laws of*, in the new volumes of the *Encycl. Brit.* (Vol XXXI).
[9] See my article 'Recent Work on Leibniz', in *Mind*, 1903.
[10] *Die Mechanik in ihrer Entwickelung*, 1st ed., p. 216.
[11] Cf. Art. 'Nativism' in the *Dictionary of Philosophy and Psychology*, ed. by Baldwin, Vol. II, 1902.

PART 2

THE CLASSICAL AND ANCIENT CONCEPTS OF TIME

F. M. CORNFORD

THE ELIMINATION OF TIME BY PARMENIDES*

Parmenides means that all men – common men and philosophers alike – are agreed to believe in the reality of the world our senses seem to show us. The premiss they start from is neither the recognition of the One Being only (from which follows the Way of Truth and nothing more) nor the recognition of an original state of sheer nothingness (which would lead to the impassable Way of Not-being). What mortals do in fact accept as real and ultimate is a world of diversity, in which things 'both are and are not', passing from non-existence to existence and back again in becoming and perishing, and from being *this* ('the same') to being *something else* ('not the same') in change. The elements, they think, are modified or transformed on a 'way to and fro', that turns back upon itself'. Becoming, change, and the diversity they presuppose must be assumed in any cosmogony. They will be assumed in the cosmogony of the second part. But Parmenides alone perceives that at this point error begins to go beyond the limits of truth.

Premisses of the Way of Truth

In these passages Parmenides has stated the premisses from which the Way of Truth will deduce the attributes of the real.

(1) *That which is, is, and cannot not-be; that which is not, is not, and cannot be.* The real exists and can never be non-existent. It follows that there is no such thing as coming-to-be out of non-existence or perishing into non-existence. 'Being' has for Parmenides a strict and absolute sense: a thing either is or is not. If it is, it is completely and absolutely; if it is not, it is simply nothing. There are no degrees of being; a thing cannot be partly real and partly unreal. There can never be a state of not-being in which what is could ever be; and there can be no transition from not-being to being or from being to not-being. Nor can there be any change of that which is; for that would mean that it *is not* at one time what it is at another.

(2) *That which is can be thought or known, and uttered or truly named;*

that which is not, cannot. This premiss is concerned with the relation of the real to thought and language. "It is the same thing that can be thought and that can be". 'Thinking and the thought that *'it is'* are one and the same. For you will not find thought apart from that which is, in respect of which thought is uttered.' Thought is uttered in names that are true, i.e., names of what really is. In names that are not true no thought or meaning is expressed. You will not find thought (meaning) apart from something real, which is meant by the utterance of that thought in words. There is nothing else for words to mean. Frag. 8 continues: "For there is and shall be no other thing besides what is, since Destiny has fettered it so as to be whole and immovable." (Since it is 'whole', complete and all-containing, there is no second thing beside it, to be thought or spoken of. And it is 'immovable' or unchangeable; so there will never be a second thing arising out of it. The real cannot cease to be just what it is and become something else). "Therefore all those (names) will be a mere word – all the (names) that mortals have agreed upon, believing that they are true: becoming and perishing, both being and not being, change of place and interchange of bright colour." All these terms are dismissed as empty names which are meaningless since they do not apply to what is, and there is nothing else for them to mean.

Only what is can be thought or truly named; and only what can be thought can be. The real must be the same as the conceivable and logically coherent, what is thinkable by reasoning (λόγος) as opposed to the senses (frag. 7, 5). The real is the same as the rational. And the real is the only thing that can be named or 'uttered'. In a sense Parmenides does not deny that it is possible to believe and say what is false; mortals are accused of doing both. But he appears to hold the view, which was maintained later, that all false statements are meaningless. Plato formulates it as follows: "To think (or say) what is false is to think what is not; but that is to think nothing; and that, again, is not to think at all." In a word, it is impossible to say or think what is false, because there is nothing for a false statement to mean or refer to. So Parmenides holds that false names like 'becoming', 'perishing', are meaningless. Only thought as distinct from belief founded on the senses, has a real object.

(3) *That which is, is one and cannot be many.* This is a third premiss, for which Parmenides gives no proof. Theophrastus supplied it as follows: "What is beside that which is, is not; what is not is nothing; therefore that

which is, is one." Theophrastus was probably following Aristotle: 'Claiming that, besides that which is, that which is not is nothing, he thinks that that which is is of necessity one and there is nothing else"; and Aristotle himself was perhaps expanding Frag. 8, 36, "There is and shall be no other thing besides what is." That the real is ultimately one had been assumed from the outset of philosophy; that may be why Parmenides takes his premiss for granted. What is new is his insistence that what is one cannot also be many, or become many. The unity of the real is affirmed as strictly and absolutely as its being. The real is *unique*; there is no second thing beside it. It is also *indivisible*; it does not contain a plurality of distinct parts, and it can never be divided into parts. There cannot be a plurality of things that are.

THE WAY OF TRUTH

From the premisses stated above we can now turn to the Way of Truth, in which their consequences are deduced. We possess here what appears to be a continuous fragment of 61 lines. It opens, like a geometrical theorem, with a sort of enunciation of the conclusion to be proved.

Frag. 8, 1–6. *Enunciation.*

> There is only one Way left to be spoken of namely that *It is*. And on this way are many marks, that what is is unborn and imperishable; whole and unique, and immovable, and without end (in time); nor was it ever, nor will it be, since it is now all at once, one, continuous.

The several attributes here enumerated are now established by a series of arguments.

Frag. 8, 6–21. *No coming-to-be or perishing.*

First comes the proof that what is is unborn and imperishable.

> For what birth of it will thou look for? In what way and whence did it grow?

Birth and growth both suggest a living creature that grows by feeding on something from without. So Empedocles says of the sum of his four elements: "What could augment this all and whence could it come?"

(17, 32). Plato too declares that the world, though living, does not draw nourishment from outside (*Tim.* 33C). Both deny the Milesian doctrine of a boundless circumambient (περιέχον), from which fresh material could be drawn and into which the world's substance could return when it perished. In the Pythagorean cosmogony, too, the world grew from a first unit or seed and drew in breath from the unlimited, which exists 'outside the Heaven'. Parmenides is rejecting the notion that what is can have been born in this way and have grown to its present dimensions. It must always exist as a whole (οὖλον, l. 4). Nor yet, he continues, could it have come out of sheer nothingness.

> Nor shall I let thee say or think that it came from what is not; for it cannot be said or thought that 'it is not'.

What is can never have been in a state of not-being; for such a state is inconceivable and the assertion is meaningless: there is nothing for the words 'it is not' to refer to. So Melissus: "What was, was always and will always be. For if it had come into being, before it came into being it must have been nothing; and if it was nothing, nothing could ever come out of nothing" (frag. 1).

> And what need could have stirred it up, starting from nothing, to be born later rather than sooner?
> Thus it must either be altogether or not at all.

This is an acute and unanswerable objection to current cosmogonies. They all assumed a process of birth or becoming which started at some moment of time. They could give no reason why it should not have started at any earlier or later moment. The last line rejects any process of becoming during which being was growing to completion and at the end of which it would be all there. 'It is now, all at once.' 'It must be *altogether* or not at all.' He now adds: Granted that it is always there as a whole, nothing further can arise alongside of it and in addition to it. It is 'unique' (μουνογενές 8, 4).

> Nor will the force of belief suffer to arise out of what is not something over and above it (viz. what is).

This further something would have to come out of not-being; but that is impossible. At 8, 36, he repeats: "there is and shall be no other (ἄλλο) besides what is (πάρεξ τοῦ ἐόντος)," with the inference that all becoming and change must be mere meaningless words. The One Being exists always as a whole; nothing more and nothing different can be added. The multiplicity of forms (sensible opposites) and changes of quality which mortals believe in, cannot be real. The conclusion is that there is no way in which anything can come to be out of not-being.

> Wherefore Justice with her fetters does not let it loose or suffer it either to come into being or to perish, but holds it fast.
> The decision concerning these things lies in this: *It is, or it is not.* But the decision has been given, as is necessary: to leave alone the one Way as unthinkable and unnamable – for it is no true Way – and that the other Way is real and true.

This refers to the decision given in frag. 2, where the Way of Not-being was finally dismissed as an 'utterly undiscernible path', because Not-being is unknowable and unutterable (p. 31).

> And how could what is be going to be in the future? And how could it come to be? For if it came into being, it *is* not; nor *is* it, if it is at some time going to be.

Thus becoming is extinguished and perishing is not to be heard of. The statement in the enunciation, "Nor was it ever, nor will it be, since it is now all at once," is here echoed. Only the present '*is*' may be used, for there is no process of becoming starting at one time and ending at another, during which we could say that it is not yet all there, but is going to be all there in the future.

Aristotle summarises the Parmenidean argument, where he remarks that his own account of becoming out of potential existence is the only solution of the problem.

'The first philosophic inquirers into the truth and the nature of things turned aside, as it were, into another way into which they were thrust by lack of experience. They say that nothing that is either comes into being or perishes because what comes to be must do so either from what is or from what is not, and both are impossible, For what is cannot come to be, because it already is; and nothing could have come to be out of what is not, for there must be something present as a substrate. So too they exaggerated the consequence which follows and denied the very existence of a plurality of things, saying that only Being itself is'. (*Phys*, 191*a*,23.)

Parmenides intended his denial of becoming to include all change; for in change something which was not comes to be, and something which is so-and-so comes to be not so-and-so but different and such as it was not before. All this seemed to him irrational.

The universal assumption of previous cosmogonies is thus rejected. No one, indeed, had believed that something could come out of nothing; and the philosophers of the sixth century had regarded their primary Being as a permanent and imperishable substance. But, not content with that, they had professed to derive from this one Being a manifold and changing world, which they had regarded as real. Out of a One, which always is, had come a many, which were not before and will again not be. And this had begun to happen at some moment of time. Parmenides declares all this to be not only inexplicable, but impossible. Their real primary Being admittedly never began and will never cease to exist. But besides real ordered world of things was to be born and grow. Out of what? Not out of the original real Being, for that already was, absolutely and completely; no second being could come out of it. Not out of nothing, for all agreed that nothing could come out of nothing. Therefore a changing world of many real things can never arise.

This first conclusion: 'No becoming or perishing of anything real', was accepted by subsequent thinkers. They agreed that the ultimately real factors – elements, atoms, etc. – could not begin or cease to exist. But they evaded the conclusion that a manifold world could never exist by making their ultimately real things a plurality instead of a unity, and by reducing the 'becoming' of things composed of them to a rearrangement of the ultimately real factors.

NOTE

* From *Plato and Parmenides*; transl. with an Introduction and running commentary by Francis M. Cornford, Routledge & Kegan Paul, London, 1950, pp. 33–39.

C. BAILEY

THE RELATIONAL THEORY OF TIME IN ANCIENT ATOMISM*

Of the atoms it may be said finally that their permanent and only 'accompaniments' are size, shape, and weight: but the 'accompaniments' of compound bodies differ with all the variety of things which they constitute: each one of them is the sum of those 'accompaniments', those properties and qualities, the removal of any one of which would alter the physical constitution of the thing.

The notion of the 'things which happen' (συμπτώματα), which we may roughly translate 'accidents',[1] follows almost directly from that of the 'accompaniments'. They are similarly neither imperceptible, nor incorporeal, nor independent corporeal existences like the whole body:[2] nor yet are they, as are the properties, the invariable and 'permanent' accompaniments of the whole. Rather they are occasional accompaniments, which may come and go, and are perceptible not in all acts of apprehension of the senses, but only in some. We cannot, for instance, have any perception of a man which does not tell us of his size and shape, but we may have many which do not tell us whether he is rich or poor, free or slave, asleep or awake. Lucretius[3] again puts it well for us from the atomic point of view. Accidents are such things "by whose going and coming the nature of things abides untouched", that is, their presence or absence involves no atomic dissolution (though it may involve atomic rearrangement), and does not alter the nature of things as such. The main difficulty in the idea is that the term is extremely wide: Epicurus wishes it to cover not merely what we might call 'secondary qualities', but also states of action and suffering, and the whole field of occurrences. It may be well again to illustrate from the three kinds of existing things: it is an 'accident' of the void that a certain atom is moving in it in a certain direction, it is an 'accident' of an atom to have a certain 'position' or 'place' or to enter into the formation of a certain compound body. These are simple cases, but when we come to compound things the idea is more far-reaching. All kinds of qualities, colour, heat, cold, goodness, badness, hardness, softness, and so on, may be the 'accidents' of bodies, and so is everything that they do and everything that happens to them: 'slavery and liberty, poverty and wealth, war and peace' are Lucretius' examples, and later on he explains that all the events of history are the accidents of the persons concerned in them or of the places where they occur.

Under the head of 'accidents' Epicurus deals with Time. To us who are accustomed to class Time, whether as a condition of perception or in any other category, always with space, this treatment may seem strange. But it must be remembered that Epicurus' conception of space was essentially concrete, and so conceived it had taken its place for him as one of the two ultimate realities. To it Time offered no apparent analogy: it did not, like space, enter into the physical composition of things, nor could it, he held, be at all conceived, as space could, as existing apart from things happening in it. He must then search for an explanation of time in accordance with ordinary experience and in strict conformity with atomic principles. Now all other things, concrete existences and their attributes, are recognized by means of reference to a general concept or 'anticipation': we know a horse because of the general image of 'horse' which is stored in our mind, we know hardness or wetness by similar reference to a concept built up out of a number of individual experiences. But this is not so with time: we do not recognize a 'time 'by reference to a general notion of 'times' resulting from many experiences. On the contrary it is a continuous experience, incapable of general conception, but always with us, and always an immediate 'clear perception'. Moreover it cannot be explained by means of any description, or by any reference to anything analogous to it: it is in our experience *sui generis*. Is it then impossible to analyse further this perception? Epicurus thinks not except in relation to the ordinary associations of everyday life. Time is something that we connect with other things, yet not directly, as attributes and qualities are associated with things: time is not thought of as immediately dependent on things or persons. Rather it is associated with the actions of persons and things, with what happens to them, with their movements or cessations from movement, with the succession of light and darkness which we call day and night. Now all these things are themselves the accidents of things, and time stands to them in the same relation as they themselves occupy to things: it is then, in other words, as Sextus Empiricus tells us that Epicurus himself defined it, "an accident of accidents, accompanying days and nights and seasons, and states of suffering and quiescence and movement and rest." Lucretius again assists us with a clear statement free from technicalities: "Time exists not by itself, but from actual things comes a feeling, what was brought to a close in time past, then what is present now, and further what is going to be hereafter." This is not perhaps a profound investigation into the problems which surround the conception of time – problems more apparent to the modern than the ancient world – but it is at least a more satisfactory account than that of Epicurus' Stoic opponents, who regarded time as an independent concrete entity super-added to things, and it is reached by the consistent analysis of the 'clear intuition' of everyday experience.

NOTES

* From *The Greek Atomists and Epicurus*, Oxford and Clarendon Press, 1928, pp. 304–307.
[1] So *Lucr.* i. 458, explaining his use of the current term, renders: 'haec soliti sumus, ut par est, eventa vocare'.
[2] *Ep.* i, §70.
[3] i. 456,7.

ARISTOTLE

ON TIME, MOTION AND CHANGE*

But what time really is and under what category it falls, is no more revealed by anything that has come down to us from earlier thinkers than it is by the considerations that have just been urged. For (*a*) some have identified time with the revolution of the all-embracing heaven, and (*b*) some with that heavenly sphere itself. But (*a*) a partial revolution is time just as much as a whole one is, but it is not just as much a revolution; for any finite portion of time is a portion of a revolution, but is not a revolution. Moreover, if there were more universes than one, the re-entrant circumlation of each of them would be time, so that several different times would exist at once. And (*b*) as to those who declare the heavenly sphere itself to be time, their only reason was that all things are contained 'in the celestial sphere' and also take place 'in time', which is too childish to be worth reducing to absurdities more obvious than itself.

Now the most obvious thing about time is that it strikes us as some kind of 'passing along' and changing; but if we follow this clue, we find that, when any particular thing changes or moves, the movement or change is in the moving or changing thing itself or takes place only where that thing is; whereas 'the passage of time' is current everywhere alike and is in relation with everything. And further, all changes may be faster or slower, but not so time; for fast and slow are defined by time, 'faster' being more change in less time, and 'slower' less in more. But time cannot measure time thus, as though it were a distance (like the space passed through in motion) or a qualitive modification as in other kinds of change. It is evident, therefore, that time is not identical with movement; nor, in this connection need we distinguish between movement and other kinds of change....

On the other hand, time cannot be disconnected from change; for when we experience no changes of consciousness, or, if we do, are not aware of them, no time seems to have passed, any more than it did to the men in the fable who 'slept with the heroes'[1] in Sardinia, when they awoke; for under such circumstances we fit the former 'now' on to the later, making them one and the same and eliminating the interval between them, because we did not perceive it. So, just as there would be no time if there were no distinction between this 'now' and that 'now', but it were always the same 'now'; in the same way there appears to be no time between two 'nows' when we fail to distinguish between them. Since, then, we are not aware of time when we do not distinguish any change (the mind appearing to abide in a single indivisible and undifferentiated state), whereas if we perceive and distinguish changes, then we say that time has elapsed, it is clear that time cannot be disconnected from motion and change.

Plainly, then, time is neither identical with movement nor capable of being separated from it.

In our attempt to find out what time is, therefore, we must start from the question, in what way it pertains to motion. For when we are aware of motion we are thereby aware of time, since, even if it were dark and we were conscious of no bodily sensations, but something were 'going on' in our minds, we should, from that very experience, recognize the passage of time. And conversely, whenever we recognize that there has been a lapse of time, we by that act recognize that something 'has been going on'. So time must either itself be motion, or if not, must pertain to motion; and since we have seen that it is not identical with motion, it must pertain to it in some way.

Well then, since anything that moves from a 'here' to a 'there',[2] and magnitude as such is continuous, movement is dependent on magnitude; for it is because magnitude is continuous that movement is also, and because movement is continuous so is time; for (excluding differences of velocity) the time occupied is conceived as proportionate to the distance moved over. Now, the primary significance of before-and-afterness is the local one of 'in front of' and 'behind'. There it is applied to order of position; but since there is a before-and-after in magnitude, there must also be a before-and-after in movement in analogy with them. But there is also a before-and-after in time, in virtue of the dependence of time upon motion. Movement, then, is the objective seat of before-and-afterness both in movement and in time; but conceptually the before-and-afterness is distinguishable from movement. Now, when we determine a movement by defining its first and last limit, we also recognize a lapse of time; for it is when we are aware of the measuring of movement by a prior and posterior limit that we may say time has passed. And our determination consists in distinguishing between the initial limit and the final one, and seeing that what lies between them is distinct from both; for when we distinguish between the extremes and what is between them, and the mind pronounces the 'nows' to be two – an initial and a final one – it is then that we say that a certain time has passed; for that which is determined either way by a 'now' seems to be what we mean by time. And let this be accepted and laid down.

Accordingly, when we perceive a 'now' in isolation, that is to say not as one of two, an initial and a final one in the movement, nor yet as being a final 'now' of one period and at the same time the initial 'now' of a succeeding period, then no time seems to have elapsed, for neither has there been any corresponding movement. But when we perceive a distinct before and after, then we speak of time; for this is just what time is, the calculable measure or dimension of motion with respect to before-and-afterness.

Time, then, is not movement, but that by which movement can be numerically estimated. To see this, reflect that we estimate any kind of more-and-lessness by number; so, since we estimate all more-or-lessness on some numerical scale and estimate the more-or-lessness of motion by time, time is a scale on which something (to wit movement) can be numerically estimated. But now, since 'number' has two meanings (for we speak of the 'numbers' that are counted in the thing in question, and also of the 'numbers' by which we count them and in which we calculate), we are to note that time is the countable thing that we are counting, not the numbers we count in – which two things are different.[3]

And[4] as motion is a continuous flux, so is time; but at any given moment time is the same everywhere, for the 'now' itself is identical in its essence, but the relations into which it enters differ in different connexions, and it is the 'now' that marks off time as before and after. But this 'now,' which is identical everywhere, itself retains its identity in one sense, but does not in another; for inasmuch as the point in the flux of time which it marks is changing (and so to mark it is its essential function) the 'now' too differs perpetually, but inasmuch as at every moment it is performing its essential function of dividing the past and future it retains its identity. For there is a dependent sequence, as we have shown, of movement upon magnitude and (we may add) of time upon movement;

and the moving object, by which we become aware of movement and its before-and-afterness, may be regarded as a point;[5] and throughout its course this – whether point or stone or what you like – retains its identity, but its relations alter: as the Sophists distinguish between Koriscos in the Lyceum and Koriscos in the market-place, so this moving object also is different in so far as it is perpetually marking a different position. And as time follows the analogy of movement, so does the 'now' of time follow the analogy of the moving object, since it is by the moving object that we come to know the before-and-after in motion, and it is in virtue of the countableness of its before-and-afters that the 'now' exists;[6] so that the 'now,' wherever found in the before-and-afters, is identical (for it is simply the mark of the before-and-afters in motion), but the before-and-afternesses it marks differ; though the nature of the 'now' depends on the markableness of any before-and-after in general, not on the specific before-and-after marked by it. And it is this specifically related 'now' that is nearest to our apprehension,[7] just as motion-change is apprehended through the changing object, and translation through the translated object, for this object is a concrete thing, which motion is not. There is a sense, then, in which what we mean when we say 'now' is always the same, and a sense in which it is not, just as is the case with anything that is in motion.

It is evident, too, that neither would time be if there were no 'now,' nor would 'now' be if there were no time; for they belong to each other as the moving thing and the motion do, so that whatever ticks off the position of the one ticks off the other. For time is the dimension proper to motion, and the 'now' corresponds to the moving object as the numerical monad.[8]

So, too, time owes its continuity to the 'now,' and yet is divided by reference to it, since in this respect also the analogy with the translation and the object translated holds good; for the movement or translation is one-and-continuous in virtue of the identity of the translated object – not its identity *qua* object (for it would preserve that if it stopped) but its unbroken identity *qua* 'the thing that is being moved'; and it is this that also marks the division between the movement before and the movement after. And there is an analogy also between such a 'body that is being moved' and a point; for it is a point that both constitutes (by its movement) the continuity of the line it traces and also marks the end of the line that is behind and the beginning of the line in front. If, however, one ascribes the latter function to it, regarding the one point in two capacities – as the end of one section of the line and the beginning of another – it must have been arrested, since its identity in this 'statical' relation must be preserved. But the 'now,' as it follows the object in motion, marks a perpetually different position, so that time is not counted as if by one and the same point, – since each point in it so counted is a double point, being end and beginning at once, – but rather as the two extremities of the line,[9] and not as *parts* of it, for the reason already stated (that, if one were to count the dividing point in its two capacities that would involve a pause), and because it is obvious that the 'now' is not a portion of time, just as the division of motion is not part of motion any more than points are of a line; it is the two sections that are *parts* of the one line.[10] The 'now,' therefore, as a limit is not time, but is incidental to time, while as the numerator it is a number; for limits are limits only of the particular thing they limit, whereas the number 10, for instance, pertains equally to the ten horses (say) the sum of which it has defined, and to anything else numerable.

That time, then, is the dimension of movement in its before-and-afterness, and is continuous (because movement is so), is evident.

Whatever line you take for the unit, two is the smallest number of such units, but in magnitude there is no minimum, for any line whatever may itself be divided into smaller lines. So too with time, 'two' is the smallest possible number of time units, but there is no smallest possible time unit itself that may be selected.

Observe too that we do not speak of time itself as 'swift or slow,' but as consisting of 'many or few' of the units in which it is counted, or as 'long and short' when we regard it as a continuum. It would not be swift or slow, even if we supposed it to be the counter that counts, not the dimension that is counted (which it really is); for abstract numbers are in no case swift or slow, though the counting of them may be.

Moreover, though time is identical everywhere simultaneously, it is not identical if taken twice successively;[11] for the change it measures, likewise, is one when considered as present, but not one if considered as partly past and partly future. And time considered numerically is concrete, not abstract; whereby follows that it changes from the former to the latter 'now,' inasmuch as these 'nows' themselves are different; just as the number of a hundred horses is identical with that of a hundred men, but the horses enumerated are different from the men enumerated. Now note further that as there may be movement (of rotation to wit) that covers the same course over and over again, in like manner we mark off time by the year or by spring or autumn.

And not only do we measure the length of uniform movement by time, but also the length of time by uniform movement, since they mutually determine each other; for the time taken determines the length moved over (the time units corresponding to the space units), and the length moved over determines the time taken. And when we call time 'much' or 'little' we are estimating it in units of uniform motion, as we measure the 'number' of anything we count by the units we count it in – the number of horses, for example, by taking one horse as our unit. For when we are told the number of horses, we know how many there are in the troop; and by counting how many there are, horse by horse, we know their number. And so too with time and uniform motion, for we measure them by each other either way. And this is only natural, for movement corresponds to linear magnitude, and time to movement, in being a quantity, in being continuous, and in being divisible; for it is from linear magnitude that motion takes on these qualities, and from motion that time does. That we do measure linear magnitude by movement, and *vice versa*, is evidenced from our saying that it is a great 'way' if it is a great 'walk,' or *vice versa*. So too with time and movement: we speak of a 'long walk' taking a 'long time,' or *vice versa*.

It[12] is by reference to the standard unit of time that we determine the relative velocity of two several movements. For we ask what distance either movement has covered during the lapse of the standard unit of time, and pronounce the movement itself fast or slow in proportion as that distance is great or small. But that same standard unit of time measures the duration of movement. So the way in which movement exists in time is by both itself and its duration being measured by time. For time measures both the movement and its duration by the same act, and its duration being so measured constitutes it as existing in time. But it is obvious that other things as well as movement exist in time, because their existence too is measured by time....

From all this it is clear that things which exist eternally, as such, are not in time; for they are not embraced by time, nor is their duration measured by time. This is indicated by their not suffering anything under the action of time as though they were within its scope.

And since time is the measure of movement, it will also incidentally be the measure of rest; for all rest is in time. For a thing being in movement necessitates that it should be moving, but its being in time does not; for time is not identical with movement, but is that in terms of which movement is counted; and even if a thing is at rest, it may be countable by the same count as motion. For not everything that is unmoved is at rest, but that only which by its nature is capable of moving but now lacks its actual movement, as we have already noted. But a thing existing in number means that it 'has' a number and that its existence is measured by that number; and so too in the case of time. And time will measure that which is in motion and that which is at rest, *as such*; for it is their movement and their rest of which it determines the amount. So that the thing in motion is not measured by time in all respects in its capacity of a quantum, but in so far as its movement is defined in quantity; hence that which is neither in movement nor at rest is not in time, since to be 'in time' means to be measured by time, and it is movement and rest of which time is the measure.

All this being so established, it becomes clear that all changes and everything that moves are conditioned by time. For it is a patent fact that every change may be quicker or slower. And what I mean by one change being quicker than another is that, of two homogeneous change-movements (either both on a periphery, for instance, or both on a straight line, if it be a local movement, and *mutatis mutandis* in other kinds of change), that one is the quicker which reaches a certain determined stage or point in its course 'before' the other reaches the point at the same distance from the starting-point in its course. Now this 'before' means before 'in time,' for both 'before' and 'after' are expressions of an interval between the 'nows' of arrival;[13] and since the 'now' is a boundary between past and future, it follows that the two 'nows' (of the former and latter arrival, namely) being both phenomena of time, so must their 'before' and 'after' be. For, whatever it be that the 'now' pertains to, to that must the interval determined by it pertain. (But note that 'before' has opposite meanings according to whether it refers to past or future time; for in the past we regard the event that is farther from the present as 'before' the other and the nearer event as 'after' it, but in the future the nearer as 'before' and the farther as 'after' the other.) So, inasmuch as 'before' pertains to time, and may be a 'before' of arrival at a point of any kind of change-movement, it follows that every change or movement takes place 'in time.'

The relation of time to consciousness deserves examination, and so does the question why we conceive of time as immanent in everything in earth and sea and sky. As to the latter point, it is because time, being the numerator of motion, pertains to such motion wherever it exists, as an affection or disposition of it (namely, that it is either actually counted in units or potentially countable in such); and all things in the material universe are susceptible of motion (for they all have position which is subject to change), and time and movement run in pairs both potentially and actually.

The question remains, then, whether or not time would exist if there were no consciousness; for if it were impossible for there to be the factor that does the counting, it would be impossible that anything should be counted; so that evidently there could be no number, for a number is either that which has actually been counted or that which can be counted. And if nothing can count except consciousness, and consciousness only as intellect (not as sensation merely), it is impossible that time should exist if consciousness did not; unless as the 'objective thing' which is subjectively time to us, if we may suppose that movement could thus objectively exist without there being any consciousness. For 'before' and 'after' are objectively involved in motion, and these, *qua* capable of numeration, constitute time.

It may be asked further to what kind of motion-change time does pertain. We may answer, 'It does not matter.' For things begin and cease to be, and grow, and change their qualities and their places 'in time'; so far, then, as change can be regarded as movement, so far time must be a numerator of every such kind of movement. We conclude, then, that time is the numeration of continuous movement, without any qualification, not only of some particular kind.

But if we take one kind of change and say 'now' with respect to it, other kinds of change, each of which has a specifically different unit to be counted in, will be at a certain stage of their change at this same 'now.' Can each of them have a different time, and must there be more than one time running concurrently? No; for it is the same lapse of time that is counted by two 'nows,' everywhere at once, whatever the units of movement or change; whereas the one-and-sameness of the units is determined by their kind and not by their 'at-once-ness'; just as if there were dogs and horses, seven of each, the number would be the same, but the units numbered different. So, too, of all movement-changes determined simultaneously the time is the same; one may be quick and another slow, and one a change of place and the other of quality; the time, however, is the same, if the counting has reached the same number and been made simultaneously, whether of the qualitative modification or of the place-movement. So the movements or changes are different and stand apart, but the time is the same everywhere, because the numeration, if made simultaneously and up to the same figure, is one and the same.

And now, keeping local motion and especially rotation in mind, note that everything is counted by some unit of like nature to itself – monads monad by monad, for instance, and horses horse by horse – and so likewise by some finite unit of time. But as we have said, movement and time mutually determine each other quantitively; and that because the standard of time established by the movement we select is the quantitive measure both of that movement and of time. If, then, the standard once fixed measures all dimensionality of its own order,[14] a uniform rotation will be the best standard, since it is easiest to count.[15]

Neither qualitative modification nor growth nor genesis has the kind of uniformity that rotation has; and so time is regarded as the rotation of the sphere, inasmuch as all other orders of motion are measured by it, and time itself is standardized by reference to it. And this is the reason of our habitual way of speaking; for we say that human affairs and those of all other things that have natural movement and become and perish seem to be in a way circular, because all these things come to pass in time and have their beginning and end as it were 'periodically'[16]; for time itself is conceived as 'coming round'; and this again because time and such a standard rotation mutually determine each other. Hence, to call the happenings of a thing a circle is saying that there is a sort of circle of time; and that is because it is measured by a complete revolution, and the whole measurement of a thing is nought else but a defined number of the units of its measurements.[17]

NOTES

* From *Physics*; transl. by G. P. Wicksteed and F. M. Cornford, II, 10, 12, 14, Harvard Univ. Press, Cambridge, 1927.
[1] [Sons of Herakles and of the daughters of Thespius, who were said to have colonized Sardinia (see Frazer on Apollodorus ii. 7.6). – C.]

² From here to there implies an interval between them, and this implies some kind of magnitude.

³ The contrast is between the *numeri numerati* and the *numeri numerantes* and between the 'concrete' and 'abstract' of recent arithmetical terminology. In counting the successive 'nows,' we are counting sections of continuous time; but we are counting them in abstract numbers. Time, then, is a concrete numerable, not an abstract numerator.

⁴ The following paragraph points out "that as movement is recognized by observing a single moving body successively at different points, the passage of time is recognized by noting that the single character of 'nowness' has been attached to more than one experienced event" (Ross, *Aristotle*, p. 90). – C.]

⁵ 'As a point,' that is, as a mere indicator of motion, in abstraction from all its other properties.

⁶ [Or, "the 'now' is the before and after, qua countable." Themist. 150. 23.]

⁷ I understand this to be a reference to the general principle of the more easy apprehensibility of the concrete and the greater intellectual luminosity of the abstract. [Themistius, 150. 27 explains: The 'now,' as a sort of particular existent, is more *cognizable* than time (γνωριμώτερον τοῦ χρόνου), just as the moving body is more so than its motion. – C.]

⁸ The moving object (whatever it may be) is apprehended through the senses. It is related to the flux of Motion (apprehended through the intelligence) as 'now' is related to the flux of Time. It is no more a *part* of Motion than 'now' is a *part* of Time. It divides perpetually different past and future Motion just as 'now' divides perpetually different past and future Time.

Time, Distance and Motion are all divisible into innumerable parts or units. The 'numbers' by which Time units are counted are static 'nows,' the 'numbers' which count Distance units are static points, and the 'numbers' which count Motion units are static objects. None of these numerators are *parts* of the numerables which they count.

⁹ [The 'now' is analogous to the dividing point in a bisected line, in so far as either can be regarded as the end-point of what comes before and also the starting-point of what comes after. But the 'now' is not stationary; so when time is *counted*, we do not take the same 'nows' twice as we took the dividing point), but different 'nows,' like the two extremeties of the line – two different points. – C.]

¹⁰]If the line AB is divided at the point C, the two lines AC, CB are *parts* of AB; the point C is *not* a part of AB. – C.]

¹¹ What time measures is not the nature of the change, but its before-and-afterness. Therefore it is not differentiated by the specific nature of the several contemporaneous changes, but it is by its very definition differentiated by the before-and-afterness that it counts.

¹² [*Literally*, "Since time is a measure of movement and of the movement's actually taking place, and this measurement of movement by time is effected by determining a certain unit of movement which shall serve to measure off the whole movement (as the cubit serves to measure length by being fixed upon as a unit of magnitude which will serve to measure off the whole length), and for movement to be 'in time' means that both the movement itself and its existence (or taking place) are measured by time – for in measuring the movement time does also measure the movement's existence and this fact that its existence is measured is what it means for a movement to be 'in time' – evidently for all other things also to be 'in time' means that their existence is measured by time." – C.]

[13] [*Literally*, "we use the terms 'earlier' and 'later' with respect to the distance (from the event so described) to the present moment." – C.]

[14] The unit fixed is the unit of time fixed in reference to some particular motion, and this becomes the measure of all time-lapses, whether of movement, modification, change of size, etc.; for the lapses of all these different manifestations of time are akin to each other with respect to before and after.

[15] A circumference, being a re-entrant curve, any part of which will fit any other part, is uniform all over; and motion upon it (supposing its velocity to be uniform) has a natural spatial unit in each completed circle. It is therefore easy to count, and since its natural unit is easy to divide or multiply into convenient secondary units, it furnishes a perfect standard by which to determine time. Aristotle does not say what particular circular motion we are to take as our standard, but commentators are probably right in supposing that for practical purposes he accepted the solar day, though on all theoretical grounds he should have taken the stellar day, as registering (in his astronomy) the prime heavenly movement and one that by the science of the day (which recognized differences in the length of the solar day, measured from one southing of the sun to another, at different seasons of the year) would seem to be demonstrably uniform.

[16] 'Period' is a Greek word which means 'passage round'.

[17] Here in the time of Themistius (fourth century?) the chapter and book seem to have ended, except perhaps the final summary. The interpolated passage is worthless. [Themistius, however, (at 162. 23) does mention the 'ten sheep' of this paragraph. It recurs to the statement (223 b 10) that one universal time is the measure or 'number' of movements which are distinct from one another and of different sorts (qualitative modification, growth, etc.), although these movements are not 'uniform,' whereas the rotation which measures time is uniform (b 12–21). A technical explanation is now given of how it is that *the same* number' can serve to measure things of different kinds. – C.]

S. SAMBURSKY

THE STOIC VIEWS OF TIME*

At both levels of scientific attainment, the modern as the classical, there has always been the cognizance of the formidable difficulties which are rooted in the fact that time as a continuous extension presupposes the existence of an extensionless instant, a dividing mark within the continuum, whereas immediate awareness of perpetual change associated with time renders that concept of a pointlike 'now' null and void. In Greek antiquity it was again the Stoics who, by virtue of their dynamic notion of the continuum, succeeded more than anyone else during the whole period to develop a satisfactory theory of the structure of time and to present a lucid analysis of the nature of its ultimate elements. The significance of this theory of the Stoics will become more evident when seen against the background of the attainments of their predecessors.

Many definitions of Time, such as those attributed to the Pythagorean Archytas, and those of Aristotle, and of the Stoics, Zeno and Chrysippos, contain, with certain variations, statements about its texture manifesting itself in the ordered succession of events, as well as about the mode of its measurement which is the result of its serial character. Aristotle's definition – "time is number of motion in respect of 'before' and 'after'" – expresses both the association of time with change and the possibility of enumerating this change. It is also evident from his analysis that he realized that the prerequisite for time measurement is a clock, i.e. a periodic mechanism, and that the revolution of the celestial sphere, being a regular circular motion, is the best measure of time "because the number of it is the best known".

Aristotle's definition was criticized from several angles by his pupil Strato, whose interesting arguments, some of which would perhaps not stand up to closer examination, probably also influenced the Stoic view on Time. Strato objected to the use of the term 'number' in connection with time, as number is a discrete quantity whereas time is continuous. Zeno and Chrysippos put 'interval' in place of 'number', a term which fits the idea of continuity better and which also expresses a certain kinship between the spatial and the temporal dimensions, keeping in mind that time elapsed can be measured by an arc of a circle. It should perhaps be mentioned that, according to Simplicios, 'interval' was used already in Archytas' definition, but it seems possible that this quotation is from a spurious source.

Strato, by defining time as "a quantity which exists in all actions", eliminated the word 'motion' from his definition in order to avoid a confusion of 'actions', i.e. of all kinematical aspects of physical phenomena including rest, with the uniform, constant flux of time by which those aspects are supposed to be measured. An action is slow, if little happens during a long stretch of time, and vice versa. This was already indicated by Aristotle, but the point is that Strato's remarks obviously induced the Stoics to introduce expressly into their definition the function of time as the measure of swiftness and slowness. Thus Zeno speaks of Time as "the interval of movement which holds the measure and standard of swiftness and slowness", and Chrysippos, amplifying on it, defines Time as "interval of movement in the sense in which it is sometimes called measure of swiftness and slowness, or the interval proper to the movement of the cosmos, and it is in Time that everything moves and exists".

It is worth while to compare Chrysippos' definition with that of Newton:

> Absolute, true and mathematical time, of itself, and from its own nature, flows equably without relation to anything external, and by another name is called duration; relative, apparent, and common time, is some sensible and external measure of duration by the means of motion, which is commonly used instead of true time; such as an hour, a day, a month, a year.

What Chrysippos had in mind was apparently identical with Newton's relative time, but it is most improbable that he had the conception of a flux of time existing independently of physical occurrences, i.e., of absolute time in the Newtonian sense. For him, as for Plato, "time was created with the heavens". Whether there was in Greek antiquity any notion of time flowing "without relation to anything external" seems extremely doubtful. Strato came perhaps nearest to this notion when he emphasized that "day and night and year are not Time nor part of Time but they are respectively light and darkness and the revolution of the moon and sun; Time, however, is the quantity in which these exist". Galen, too, seems to have been of the same opinion, according to a tenth-century source (Ibn Abi Said) which declares: "Galen states that motion does not produce time for us; it only produces for us days, months, and years. Time, on the other hand, exists *per se*, and is not an accident consequent upon motion."

The assumption that for the Stoics time was inseparably bound up with events is supported by the continuation of the passage in Stobaios quoting Chrysippos' definition. There it says:

> It seems that Time is to be taken in two senses, just like the earth and the sea and the void, namely in the sense of the Whole and its parts. In the same way as the void is all infinite everywhere, so time is all infinite in both directions; indeed, past and future are both infinite.

Reference to the void is made here mainly as a parallel to the infinite extension of time, as the non-empty cosmos was regarded by the Stoics as a finite island within the infinite void. However, the analogy with earth and sea has to be interpreted as follows: in the same way as every part of the earth is earth and every part of the sea is sea, so every part of Time is time. An element of time, the shortest duration, exhibits the same character of an 'interval of movement' as does a macroscopic stretch of time. It is not to be compared to the mathematical continuum of a line which can be whittled down to extensionless points, but is composed of substantial elements of events similar to the earth and the sea, each of whose elements are again of an earthlike and sealike character.

The last quotation bears a strong similarity to Kant's passage on intensive quantities in the *Critique of Pure Reason*, in the section on anticipation of perception:

> Space and Time are *quanta continua*, because there is no part of them that is not enclosed between limits (points and moments) such that this part itself is again a space or a time. Space, therefore, consists of spaces only, time of times.

The idea underlying this last sentence corresponds exactly to that of Chrysippos in his simile of the earth and the sea, and nobody before the Stoics had expressed it with such clarity. The fact that points and moments are only limits and that moments cannot be synthesized into time, as Kant emphasized further, was already clear to Aristotle, who said "in so far as the 'now' is a boundary, it is not time, but an attribute of it".

Here, however, we have the crux of the problem which was first attacked successfully by Chrysippos. On the one hand the 'now' is supposed to be a limit of time only, a mathematical boundary which itself is not time. On the other hand this 'now' coincides with the present moment, i.e. with the only event lived by me, and which, in contradistinction to the moment that has passed and the moment to come, is coupled with the immediate awareness of reality. Greek science has through all periods been vexed by this dilemma, an echo of which is to be found in the frustrating and not very helpful analysis of the Sceptics.

A plausible solution seems to offer itself in the application of Xenocrates' hypothesis of atomic lengths to the temporal dimension and the assumption of the existence of indivisible atoms of time. Such an assumption would avoid the reduction of the 'now' to a shadow, a mere mathematical point. This solution was apparently suggested, either by Xenocrates himself or by someone else, but it was of course rejected by the Stoics because it is contrary to the very idea of continuity that no part of a quantity can be the smallest possible one, i.e. indivisible.

> The Stoics do not admit the existence of a shortest element of time, nor do they concede that the 'now' is indivisible, but that which someone might assume and think of as present is according to them partly future and partly past. Thus nothing remains of the Now, nor is there left any part of the present, but what is said to exist now is partly spread over the future and partly over the past.

And by the same source we are told a little further on that Chrysippos "in the third, fourth and fifth book *On Parts* declares that part of the present is future and part past".

At first sight it would seem as if the Stoic refutation of Xenocrates' atomic time elements is a mere playing with words. Is not the assertion that the present moment consists of a small stretch of time spread over past and future substantially the same as the Xenocratic idea of the indivisible atom of time which was suggested for the very sake of avoiding the point-like Now? In fact, we have here two completely different conceptions whose disparity again stems from the difference between the customary static notion of the continuum and the dynamic one of the Stoics. This can be elucidated by further sentences from the passage quoting Chrysippos' view on Time: "He states most clearly that no time is entirely present. For the division of continua goes on indefinitely, and by this distinction time, too, is infinitely divisible; thus no time is strictly present but is defined only loosely." The 'loose' definition of the present obviously results from the limiting process of infinite convergence by which it is 'caught' in an operation similar to that which defined a body according to the Stoic conception. There the surface of the body was intercepted by two infinite sequences of surfaces of inscribed and circumscribed bodies. In the case of time, the limiting process consists in an infinite approach to the mathematical Now both from the direction of the past and from the future. In this sense, "no time is entirely present" and the present is "partly spread over the future and partly over the past", because the present is given by an infinite sequence of nested time intervals shrinking towards the mathematical 'now', whereby the 'lower' boundaries of each interval are points of the past and the 'upper' ones points of the future. In strict conformity with the dynamic conception of continua – spatial as well as temporal – the present *qua* limit of time is not sharp but forms a fringe covering the immediate past and future. In contradistinction to the static concept of an 'atom of time' we have thus to regard the Stoic present as a shrinking duration of only indistinctly defined boundaries. The physical significance of such a duration is that it still represents an eventlike structure, it is an elementary event, and macroscopic time is composed of the succession of such events in the same sense as every part of the earth is earth and every part of the sea is sea. The mathematical 'now' towards which the shrinking intervals of duration converge has no physical significance. This 'now' of Aristotle and Strato is 'fleeting and next to nothing', it is "unreal and exists only in pure thought".

One is tempted to atrribute to Plato an anticipation of the Stoic conception of Time, when he says in the *Timaios:* "Moreover, when we say that what has become has become and what is becoming is becoming, and that what will become will become, and that what is not is not – all these are inaccurate modes of expression." However, this remark is too aphoristic to allow for a conclusive interpretation. On the other hand, there is a remarkable similarity of the Stoic doctrine to the ideas on time of some modern philosophers, and especially noticeable in this respect is the theory of Whitehead[1] as some quotations from his works will prove: "A moment has no temporal extension, and is in this respect to be contrasted with a duration which has such extension. ..." "A moment is a limit to which we approach as we confine attention to durations of minimum extension."

There is no such thing [as the instantaneous present] to be found in nature. As an ultimate fact it is a nonentity. What is immediate for sense-awareness is a duration. Now a duration has within itself a past and a future; and the temporal breadths of the immediate durations of sense-awareness are very indeterminate and dependent on the individual percipient. ... The passage of nature leaves nothing between the past and the future. What we perceive as present is the vivid fringe of memory tinged with anticipation. ... The past and the future meet and mingle in the ill-defined present.

The striking close kinship of Whitehead's doctrine with that of the Stoics could be proved by many more quotations. It lends support to the assumption that, like Whitehead, the Stoics made nature prior to time and identified the flux of time with the passage of events of nature. The conception that the world is made up of events or of 'drops of experience' is inconsistent with that of an 'absolute time', and no doubt the Stoics would have agreed with Whitehead's statement that "there is time because there are happenings and apart from happenings there is nothing". More specifically, the character of these happenings and therefore that of time on a macroscopic scale reveals itself as essentially cyclic and periodic. In this sense one has to interpret the recurrent allusion to the 'movement of the cosmos' in Stoic definitions of time. In addition to the daily and yearly cycles of this movement there is the cosmic period of the Great Year. Thus the Stoic Apollodoros of Seleuca (end of second century B.C.) said: "Time is the interval of movement of the cosmos ... and the whole time is passing just as we say that the year passes, on a larger circuit." This 'larger circuit' is obviously nothing else but the greatest of the cosmic periods, the Great Year whose length the Stoics reckoned from one state of the universe to the next identical one recurring after the world has passed through the stage of *ekpyrosis*.[2] The idea of the Great Year has oriental precursors and is found also in Heracleitos and Plato. The Stoics, however, were the first to identify it with their hypothetical full cycle of cosmic transmutation of matter which consists of the preponderance of fire in the beginning, the differentiation of the other three elements in the intermediate stage, and the ascendancy of fire at the end, after which the cycle of transmutation begins anew.

The essential point in our context is that for the Stoics the length of the cycle was determined by the return of the actual material state of the Whole, whereby each state, given by the sum of all hexeis, follows from the former through a continuous transition. We have seen already an indication of this conception of the dynamic evolution of a cycle in the instance given for the comparative proposition 'rather day than night', where different states of the daily period were regarded as 'mixed cases'. The same approach was found in the description of the yearly cycle, where the different seasons and their transitions were defined in terms of different thermal states of the air, which are dependent on the position of the sun. In cosmic dimensions it is the continuous change in the mixture of the elements on a universal scale which fixes the cosmic period, and, in Stoic terminology, this change is given by the variation of the composition of the pneuma, taken over the cosmos as a whole. We have seen that *psyche* is pneuma in its driest and hottest state, whereas *physis* and the inorganic state are characterized by an increasing admixture of the humid and the cold. We have to apply this differentiation to the cosmos as an entity regarded by the Stoics as a living rational being endowed with *psyche* and *nous*. The cosmic cycle thus means that the cosmos, although subject to continuous metabolism, never dies and that its immortality is only another expression of the infinite extension of time, of the never-ceasing succession of events.

Luckily, Plutarch has preserved for us two fragments from Chrysippos' book *On Providence* which give a good illustration of the Stoic conception of the eternal cosmic metabolism. "As death is the separation of soul from body and as the soul of the cosmos does not leave it but is growing continuously until it has consumed all matter, one cannot say that the cosmos is mortal." And further: "When the cosmos is completely in the fiery state, so at the same time are its soul and its *hegemonikon*. But if that what is left over of the soul is changing into the humid state, the cosmos is in a certain way transformed into body and soul, and thus, composed of both, it exhibits another order." Plutarch, in continuing this passage, says that when *ekpyrosis*, i.e. the thermal part of the cycle, has reached its maximum, the cosmos is wholly soul-like, whereas in the cooling down part of the cycle, i.e. the early part of the recurring creation, the cosmic soul slackens (because of the reduced tension of the pneuma) and becomes more humid, which means that the cosmos becomes body-like. Generally, therefore, the cosmos is in a mixed state, being both body and soul, and its eternal life is characterized by periods exhibiting different ratios of this mixture and changing continuously in a slow process of dynamic transformation. However, we shall not enter here into a more detailed discussion of the doctrine of ekpyrosis as we are concerned solely with the analysis of the Stoic theory of time.

NOTES

* From *Physics of the Stoics*, Routledge & Kegan Paul, London, 1959, pp. 100–108.
[1] A. N. Whitehead, *The Concept of Nature*, C.U.P., 1920, ch. III; *Science and the Modern World*, Macmillan, 1925, ch. VII; *Process and Reality*, Macmillan, 1929, ch. II.
[2] The usual translation of ekpyrosis as 'conflagration' is misleading, because it suggests a sudden catastrophe. In fact, ekpyrosis originally denoted that period of the cosmic cycle where the preponderance of the fiery element reaches its maximum.

S. SAMBURSKY

THE STOIC DOCTRINE OF ETERNAL RECURRENCE*

The conception of the formation and decay of the cosmos was given a different twist in Stoic doctrine. The views of the Stoic School were considerably influenced by the philosophy of Heracleitus (beginning of the fifth century B.C.), who, like Empedocles after him, regarded the harmony prevailing in the universe as the result of a dynamic equilibrium of opposite forces. For Heracleitus, this dynamics was built up round fire: "There is an exchange: all things for Fire and Fire for all things, like goods for gold and gold for goods." The sun and stars were created from, and are still fed by, the evaporation of the water on the surface of the earth; and this evaporation is brought about by the heat reaching the earth from the heavenly bodies. This double motion, upwards and downwards, is characteristic of the harmony of opposites whereby there is a simultaneous process of coming into being and decay within the existing cosmos. This was Heracleitus' first theory, as confirmed by the extant fragments of his works, and we should understand the following passage from Aristotle accordingly:

All thinkers agree that it has a beginning, but some maintain that having begun it is everlasting, others that it is perishable like any other formation of nature, and others again that it alternates, being at one time as it is now, and at another time changing and perishing, and that this process continues unremittingly.

The notion of the eternal order of the cosmos as maintained by a simultaneous process of creation and decay was, in the generations after Heracleitus, mixed up with the idea of creation and destruction occurring one after the other, and with cyclical processes in which fire still played the essential part: the cosmos develops out of fire and eventually returns to fire, and so on in endless cycle. This idea appears in ancient mythologies. But the Stoics incorporated it into their scientific doctrine as a modification of Heracleitus' theory and set him up as the authority for their conception of the final conflagration of the cosmos. It is, therefore, hardly surprising that later commentators and compilers completed the confusion by identifying the theory of the Stoics with that of Heracleitus:

"Heracleitus too says that the cosmos goes up in fire and is once again formed out of fire at certain periods of time in which, according to him, it is 'kindled in measure and quenched in measure'. This opinion was later arrived at by the Stoics too."

We have seen that in Stoic physics fire occupied a special position amongst the elements by virtue of its active character. The Stoics were the first to grasp the special significance of thermodynamic processes in inorganic nature, as well as in biology, where their great importance had already been recognized before. It is thus easy to understand why the Stoic cosmogony was also built upon the concept of fire as the symbol of thermic phenomena: "Zeno said that fire is the essence of what exists. ... At periods of time allocated by fate all the cosmos is conflagrated and after that it returns to its first order." This doctrine of the founder of the school was carried on by his followers: "Zeno, Cleanthes and Chrysippus hold that matter undergoes transmutation, as e.g. fire turns into seed, from which once again is restored the same world order, that was before." The process of the emergence of the cosmos from the primary state pictured as primeval fire is essentially the differentiation of the other three elements from fire: "Chrysippus, in his first book *On Nature*, says: The transmutation of fire is as follows. It turns first into air, then into water; and from the water at the bottom of which earth settles, air rises. As the air becomes rarefied, the aether is spread around in a circle; and the stars and the sun are kindled from the sea." This is the first stage of the creation in which the nature of the cosmic continuum is defined principally by the hot and dry elements. The emergence of air from fire as a second element is pictured on the analogy of the ascent of smoke and hot air from flames. This air (including all kinds of vapour) then condenses into water. From here onwards the development proceeds in two different directions: downwards for the sedimentation of earth from water, and upwards for the evaporation of vapours which, through rarefication, become fire, and this fire then agglomerates into the heavenly bodies. This completes the creation of the physical universe, which continues to maintain itself in being by the dynamic equilibrium of Heracleitus. However, it appeared that this equilibrium between the water in the seas and the fire in the sun is only a first approximation to reality: on a cosmic scale of time fire holds the ascendancy. Evidence for this was found in geological indications (already remarked on by pre-Stoic natural philosophers), such as the presence of shells on dry land showing that the sea has retreated, and in slow meteorological changes over the course of centuries:

> Swamps and wet places become habitable through dryness, while places that were previously habitable become uninhabitable through the increase of dryness. This is explained by the change and decay of the universe. From these signs there are some, like Heracleitus and his followers, together with the Stoics, who believe that eventually there will be a conflagration of the universe.

It follows from this that we are not to regard the conflagration of the cosmos as an actual fire, a kind of holocaust which in a moment destroys the whole cosmos: the process, in fact, is an extremely slow one, like all cosmic processes, the rate of decay being equal to that of coming into being. In the end, the original situation will be restored and the active element will dominate all the vast expanses of the cosmos which will then be reborn from the primeval fire. With the 'thermic' cosmogony of the Stoics before us, we can hardly fail to be reminded of the 'thermic death' of the universe. This idea formed the subject of many controversies amongst nineteenth-century physicists, when the second law of thermodynamics was applied to the universe as a whole. Every physical process ultimately leads to an increase in thermic energy. Thus, after the equalization of temperatures, the universe will reach a state similar to that of the final state of the Stoic cosmos.

Since the human mind is not unreasonably appalled by this prospect of an absolute, irrevocable end, various physical solutions out of the dilemma have been proposed to enable the universe to pass safely over the dead point of its final doom. Statistical mechanics explains the thermal phenomena as kinetic energies of the atoms and molecules. In the final state of the universe the movement of all these elementary particles will be one of ideal disorder and their velocities will be equal, on the statistical average. This is where statistics step in and save the situation: through a great fluctuation, a deviation from the statistical mean is likely to occur at some time, as the law of large numbers tells us. A great number of particles with a velocity considerably above the average will then agglomerate in one place and the resulting difference in potentials will bring about a renewal of the life of the cosmos.

Thus statistical mechanics offered a possibility of endless cycles of cosmic creation and destruction in a way that is closely parallel to the Stoic conception. As a corollary of this statistical picture of the cosmos, the old problem of the 'return of the identical' came to life again in the philosophical debates of the last century. If the state of the cosmos is defined by a certain combination of its ultimate particles, and if every new combination necessarily results, by the laws of causality, from its predecessor, it follows that eventually, after all the permutations have been exhausted, the whole cycle of previous combinations will be repeated. Hence the idea of cosmic cycles, when comprehended in terms of number, involves the identical repetition of the present, a fact which was seized on by philosophers like Nietzsche for the purpose of their own doctrine. It is most interesting to observe that the very same inner logic which in our times has linked together thermodynamics, statistical mechanics and the idea of the 'return of the identical', was also the driving force in Greek thought. Then, too, within the limits of the scientific comprehension and terminology of the time, views on the cosmological significance of thermal processes were closely connected with the theories about the formation and destruction of the cosmos and about the identical return of a situation. Of the Stoics we are told that "in their view, after the conflagration of the cosmos everything will again come to be in numerical order, until every specific quality too will return to its original state, just as it was before and came to be in that cosmos. This is what Chrysippus says in his book *On the Cosmos*." An even clearer statement is found in another source which quotes Chrysippus' own words from his book *On Providence*: "Clearly, it is not impossible that, after our death, when long aeons of time have passed, we shall return to the form we have at present." From Eudemus of Rhodes, a pupil of Aristotle's and the first to write a history of astronomy in the Ancient World, we learn that Pythagoras had anticipated the Stoics' conception of the return of the identical. In the following passage, Eudemus addresses his pupils during his lecture:

It may be asked if the same time will return, as some say, or not. ... If you believe the Pythagoreans everything will eventually return in the selfsame numerical order and I shall converse with you staff in hand and you will sit as you are sitting now, and so it will be in everything else. It is reasonable to assume that time too will be the same. For movement is one and the same, and likewise the sequence of many things that repeat themselves is one and the same, and this applies also to their number. Hence everything will be identical, including time.

We see, then, that the Pythagoreans, like the Stoics after them, also tried to put the theory of cosmic cycles on a scientific basis, in their case in terms of their theory of number. Even though they had no knowledge of the theory of permutations nor any conception of the law of large numbers, the Pythagoreans realized that number by its very nature includes a certain order determined by law. The periodicity of cyclical motion symbolizes the periodicity of every cosmic event and therefore that of time itself. We have already seen in another Stoic fragment that the Stoics also conceived of this Pythagorean idea of repetition 'in numerical order'; they completed the physical picture of their cosmogony by the use of mathematical terminology.

NOTE

* From *The Physical World of the Greeks*, Macmillan, New York, and Routledge and Kegan Paul, London 1956; transl. by Menton Dagut, pp. 198–202.

CRITICISM OF THE RELATIONAL THEORIES OF TIME*

Perhaps we can, in the first instance, make a threefold division of the accounts of time which have been given, for either time is movement, as it is called, or one might say that it is what is moved, or something belonging to movement,[1] for to say that it is rest, or what is at rest, or something belonging to rest, would be quite remote from our interior awareness of time, which is never in any way the same. Now of those who say it is movement, some seem to mean that it is all movement,[2] others the movement of the universe; those who say that it is what is moved seem to mean that it is the sphere of the universe; those who say that it is something belonging to movement, that it is the distance covered by the movement[3] or (others of them) the measure,[4] or (others again) that it is in a general way a consequence of movement;[5] and either of all movement or only of ordered movement.[6]

8. It is not possible for it to be movement, whether one takes all movements together and makes a kind of single movement out of them, or whether one takes it as ordered movement, for what we call movement, of either kind, is in time; but if someone says that it is not in time, then it would be still further from being time, since that in which movement is, is something different from movement itself. And, though other arguments can be brought, and have been brought, against this position, this one is enough, and also that movement can stop altogether or be interrupted, but time cannot. But, if someone says that the movement of the universe is not interrupted, this, too (if he means the circuit of the heavens), is in a period of time; and it would go round to the same point not in the time in which half its course was finished, and one would be half, the other double time; each movement would be movement of the universe, one going from the same place to the same place again, and the other reaching the half-way point. And the statement that the movement of the outermost sphere is the most vigorous and quickest is evidence for our argument that its movement is something different from time. For

it is, obviously, the quickest of all the spheres because it covers a greater distance than the others, in fact, the greatest distance, in less time; the others are slower because they cover only a part of the distance [covered by the outermost sphere] in a longer time. If, then, time is not the movement of the sphere, it can hardly be the sphere itself, which was supposed to be time because it is in motion.

Is it, then, something belonging to movement? If it is the distance covered by the movement, first, this is not the same for all movement, not even uniform movement, for movement is quicker and slower, even movement in space. And both these distances covered [by the quicker and the slower movement] would be measured by some one other thing, which would more correctly be called time. Well then, of which of the two of them is the distance covered time, or rather of which of all the movements, which are infinite in number? But if it is the distance covered by the ordered movement, then not by all ordered movement, or by one particular kind of ordered movement, for there are many of these; so that there will be many times at once. But if it is the distance covered by the movement of the universe, if the distance in the movement itself is meant, what would this be other than the movement? The movement, certainly is quantitatively determined; but this definite quantity will either be measured by the space, because the space which it has traversed is a certain amount of space, and this will be the distance covered; but this is not time but space; or the movement itself, by its continuity and the fact that it does not stop at once but keeps on for ever, will contain the distance. But this would be the multiplicity of movement; and if one, looking at movement, shows that it is multiple (as if one were to say there was a great deal of heat), time will not appear or come into one's mind but movement which keeps on coming again and again, just like water flowing which keeps on coming again and again, and the distance observed in it. And the 'again and again' will be a number, like the number two or three, but distance belongs to magnitude. So the amplitude of movement will be like the number ten or the distance from end to end which appears on what you might call the bulk of the movement, and this does not contain our idea of time, but this definite quantity will be something which came to be in time; otherwise time will not be everywhere but in movement as its substrate, and we are back again at the statement that time is movement, for the distance covered is not outside movement but is movement which does not happen all at once; but the comparison of movement which does not happen all at once with what is all at once [the instantaneous] can only be made in time. In what way will the non-instantaneous differ from the instantaneous? By being in time, so that movement which extends over a distance and the distance covered by it are not the actual thing, time, but are in time. But if someone were to say that the distance of movement is time, not in the sense of the distance of movement itself, but that in relation to which the movement has its extension, as if it was running along with it, what this is has not been stated. For it is obvious that time is that in which the movement has occurred. But this was what our discussion was trying to find from the beginning, what time essentially is; since this is like, in fact, the same as, an answer to the question "What is time?" which says that it is distance of movement in time. What, then, is this distance which you call time and put outside the proper distance of the movement? Then, again, on the other side, the person who puts the distance in the movement itself, will be hopelessly perplexed about where to put the interval of rest. For something else could rest for the same space as something was moved, and you would say that the time in each case was the same, as being, obviously, different from both. What, then, is this distance, and what is its nature? For it cannot be spatial, since this also lies outside movement.

9. We must now enquire in what sense it is number of movement or measure[7] – for it is better to call it measure of movement, since movement is continuous. First of all, then, a doubt must arise here, too, about its being the measure of all movement alike, just as its being the measure of all movement alike, just as it did with the distance of movement, if there was said to be a number or measure of all movement. For how could one number disordered and irregular movement? What would its number or measure be, or what its scale of measurement? But if one uses the same measure for both kinds of movement [regular and irregular] and in general for all movement, quick and slow, the number and measure will be like the ten which counts both horses and cows, or like the same measure for liquids and solids. Now, if it is a measure of this kind, then it has been said what time is a measure of, that it is a measure of movements, but we have not yet been told what it is itself. But if, just as when one takes the ten even without the horses it

is possible to think of the number, and the measure is a measure, with
a certain nature, even if it is not yet measuring, so time, too, must have
its own nature since it is a measure, and if it is a thing of this kind on its
own like number, how can it differ from this number we were considering
in the case of the ten, or from any other number made up of abstract
units? But if it is a continuous measure, then it will be a measure because
it is of a certain size, like a length of one cubit. It will be a magnitude, then,
like a line which will obviously run along with movement. But how will
this line running along measure that with which it runs? Why should one
of them measure the other rather than the other the one? And it is better
and more plausible to assume that it is not the measure of all movement
but of the movement it runs along with. But this must be something continuous, or the line which runs with it will stop. But one ought not to
take what measures as something coming from outside or separate but
to consider the measured movement as a whole. And what will the measurer be? Movement will be measured, and the measurer will be magnitude. And which of them will be time? The measured movement or the
measuring magnitude? For either the movement which is measured by the
magnitude will be time, or the magnitude which measures, or what uses
the magnitude, as one uses the cubit to measure how much the movement is. But in all these cases one must assume (which we said was more
plausible), uniform movement, for unless there is uniformity, and, besides
that, the movement is single, and a movement of the whole thing,[8] the
way of proof becomes still more obstructed for whoever holds that time
is in any sense a measure. But now, if time is a measured movement, and
one measured by quantity; just as the movement, if it had to be measured,
could not be measured by itself but by something else, so it is necessary,
if the movement is to have another measure besides itself, and this was
the reason why we needed the continuous measure for measuring it – in
the same way there is need of a measure for the magnitude itself, in order
that the movement, by the fixing at a certain length of that by which it
is measured as being a certain length, may itself be measured. And the
number of the magnitude which accompanies the movement, but not the
magnitude which runs along with the movement, will be that time which
we were looking for. But what could this be except number made up of
abstract units? And here the problem must arise of how this abstract
number is going to measure. Then, even if one does discover how it can,

one will not discover time measuring but a certain length of time; and this is not the same thing as time. It is one thing to say 'time' and another to say 'a certain length of time'; for before saying 'a certain length of time' one ought to say what it is that is of a certain length. But perhaps the number which measures the movement from outside the movement is time, like the ten which counted the horses taken apart from the horses. Well, then, in this version it has not been said what this number is which is what it is before it begins to measure, like the ten.[9] Perhaps it is the number which runs beside the movement and measures it by the sequence of 'before' and 'after'[10] But it is not yet clear what this number which measures by the sequence of 'before' and 'after' is. And then, too, anyone who measures by 'before' and 'after, either with a point or with anything else, will in any case be measuring according to time. So, then, this time of theirs which measures movement by 'before' and 'after' is bound to time and in contact with time in order to measure. For one either takes 'before and 'after' in a spatial sense, like 'the beginning of the race-track, or else one must take them in a temporal sense. For in general, 'before' and 'after' mean, 'before', the time which stops at the 'now,' and 'after,' the time which begins from the 'now'. Time, then, is something different from the number which measures by 'before' and 'after' not only any kind of movement but even ordered movement. Then, why, when number is added to movement, either on the measured or the measuring side – for there is the possibility that the same number could be both measured and measuring – why should time result from its presence, though when movement exists and, certainly, has a 'before' and 'after' belonging to it, there will be no time? This is like saying that a magnitude would not be the size it is unless someone understood that it was that size. But again, since time is, and is said to be, unbounded, how could it have a number? Unless, of course, someone took off a piece of it and measured it, but time would be in the piece before it was measured, too. But why can time not exist before the soul which measures it? Unless perhaps one is going to say that it originated from soul. But this is not in any way necessary because of measuring it, for it exists in its full length, even if no one measures it. One might say that the soul is what uses magnitude to measure time; but how could this help us to form the concept of time?

10. As for calling it an accompaniment of movement, this does not ex-

plain at all what it is, nor has the statement any content before it is said what this accompanying thing is, for perhaps just this might turn out to be time. But we must consider whether this accompaniment comes after movement, or at the same time as it, or before it – if there is any kind of accompaniment which comes before, for whichever may be said, it is said to be in time. If this is so, time will be an accompaniment of movement in time.

NOTES

* From *Enneads* with an English translation by A. H. Armstrong, Harvard University Press, Cambridge, Mass., 1967, III, pp. 319–335.

[1] That it was something belonging to movement was held in different senses by some Academics, Aristotle, Stoics and Epicureans: see notes below.

[2] Some Stoics: cp. *Stoic. Vet. Fr.* II. 514.

[3] Stoics (Zeno and Chrysippus): cp. *Stoic. Vet. Fr.* II. 509–510.

[4] An Academic view taken up and developed by Aristotle: cp. 'Ὅροι l.c. Aristotle, *Physics* Δ 10 ff.

[5] Epicureans: cp. *Stobaeus Ecl.* I. 8 [I] 103. 6.

[6] Cp. *Stoic. Vet. Fragm.* II. 509–510. It is only among Stoics that the distinction between all movement and ordered movement (the movement of the universe) appears.

[7] Aristotle uses both terms (ἀριθμὸς κινήσεως, *Physics* Δ 11. 219b2; μέτρον κινήσεως, 12. 221a1) without distinction.

[8] Aristotle points out that only a uniform movement can be considered a single movement in *Physics* E4. 228b15 ff.; but for him time is the measure of absolutely any kind of movement (*Physics* Δ 14, 223a20 ff.); though the most uniform movement, the circular movement of the heavens, is the standard by which in fact we measure other movements and time itself (223b).

[9] Plotinus assumes here his own view that number has a separate substantial existence prior to the things which it numbers: see *Enneads* VI, 6 [34] 5.

[10] Aristotle defines time as ἀριθμὸς κινήσεως κατὰ τὸ πρότερον καὶ ὕστερον (*Physics* Δ 4. 219b2–3).

ST. AUGUSTINE

VIEWS ON TIME

A. ON THE BEGINNING OF TIME[1]

That we ought not to seek to comprehend the infinite ages of time before the world, nor the infinite realms of space

Next, we must see what reply can be made to those who agree that God is the Creator of the world, but have difficulties about the time of its creation, and what reply, also, they can make to difficulties we might raise about the place of its creation. For, as they demand why the world was created then and no sooner, we may ask why it was created just here where it is, and not elsewhere. For if they imagine infinite spaces of time before the world, during which God could not have been idle, in like manner they may conceive outside the world infinite realms of space, in which, if any one says that the Omnipotent cannot hold His hand from working, will it not follow that they must adopt Epicurus' dream of innumerable worlds? with this difference only, that he asserts that they are formed and destroyed by the fortuitous movements of atoms, while they will hold that they are made by God's hand, if they maintain that, throughout the boundless immensity of space, stretching interminably in every direction round the world, God cannot rest, and that the worlds which they suppose Him to make cannot be destroyed. For here the question is with those who, with ourselves, believe that God is spiritual, and the Creator of all existences but Himself. As for others, it is a condescension to dispute with them on a religious question, for they have acquired a reputation only among men who pay divine honours to a number of gods, and have become conspicuous among the other philosophers for no other reason than that, though they are still far from the truth, they are near it in comparison with the rest. While these, then, neither confine in any place, nor limit, nor distribute the divine substance, but, as is worthy of God, own it to be wholly though spiritually present everywhere, will they perchance say that this substance is absent from such immense spaces outside the world, and is occupied in one only, (and that a very little one compared with the infinity beyond,) the one, namely, in which is the world? I think they will not proceed to this absurdity. Since they maintain that there is but one world, of vast material bulk, indeed, yet finite, and in its own determinate position, and that this was made by the working of God, let them give the same account of God's resting in the infinite times before the world as they give of His resting in the infinite spaces outside of it. And as it does not follow that God set the world in the very spot it occupies and no other by accident rather than by divine reason, although no human reason can comprehend why it was so set, and though there was no merit in the spot chosen to give it the precedence of infinite others, so neither does it follow that we should suppose that God was guided by chance when He created the world in that and no earlier time, although previous times had been running by during an infinite past, and though there was no difference by which one time could be chosen in preference to another. But if they say that the thoughts of men are idle when they conceive infinite places, since there is no place beside the world, we reply that, by the same showing, it is vain to conceive of the past times of God's rest, since there is no time before the world.

That the world and time had both one beginning, and the one did not anticipate the other

For if eternity and time are rightly distinguished by this, that time does not exist without some movement and transition, while in eternity there is no change, who does not see that there could have been no time had not some creature been made, which by some motion could give birth to change, – the various parts of which motion and change, as they cannot be simultaneous, succeed one another, – and thus, in these shorter or longer intervals of duration, time would begin? Since then, God, in whose eternity is no change at all, is the Creator and Ordainer of time, I do not see how He can be said to have created the world after spaces of time had elapsed, unless it be said that prior to the world there was some creature by whose movement time could pass. And if the sacred and infallible Scriptures say that in the beginning God created the heavens and the earth, in order that it may be understood that He had made nothing previously, – for if He had made anything before the rest, this thing would rather be said to have been made 'in the beginning', – then assuredly the world was made, not in time, but simultaneously with time. For that which is made in time is made both after and before some time, – after that which is past, before that which is future. But none could then be past, for there was no creature by whose movements its duration could be measured. But simultaneously with time the world was made, if in the world's creation change and motion were created, as seems evident from the order of the first six or seven days. For in these days the morning and evening are counted, until, on the sixth day, all things which God then made were finished, and on the seventh the rest of God was mysteriously and sublimely signalized. What kind of days these were it is extremely difficult, or perhaps impossible for us to conceive, and how much more to say!

B. TIME IS NOT THE MOTION OF BODIES[2]

He clears this Question, what Time is

I heard a learned man once deliver it, that the motions of the sun, moon, and stars, were the very true times; and I did not agree. For why should not the motions of all bodies in general rather be times? Or if the lights of heaven should cease, and the potter's wheel run round; should there be no time by which we might measure those whirlings about, and might pronounce of it, that either it moved with equal pauses: or, if it turned sometimes slower, and other whiles quicker, that some rounds took up longer time, and others shorter? Or even whilst we were a saying this should we not also speak in time? Or should there in our words be any syllables short, and others long, but for this reason only, that those took up a shorter time in sounding, and these a longer? Grant unto us men the skill, O God, in a little thing to descry those notions as be common to things both great and small. The stars and lights of heaven, 'tis true, be appointed for signs, and for seasons, and for years, and for days. They be indeed: yet should I never, (on the one side) affirm, the whirling about of that little wooden wheel to be the day; nor should he affirm, (on the other side) that therefore there were no time at all.

I for my part, desire to understand the force and nature of time, by which we measure the motions of bodies; and say, (for example) this motion to be twice longer than that. For I demand: seeing this is it which is called the day, not the stay only of the sun upon the earth, (according to which account the day is one thing, and the night another;) but its whole circuit that it runs from east to east again; (according to which account we say, There are so many days past): – for the days being reckoned with their nights, are usually called so many days, and the nights are not out of the reckoning: – seeing therefore that a day is made complete by the motion of the sun, and by his circuit from east to east again, I thereupon demand, whether the motion itself makes the day; or the stay in which that motion is finished; or both? For if the first be the day; then should we have a day of it, although the sun should finish that course of his in so small a space of time as one hour comes to. If the second, then should not that make a day, if between one sunrise and another, there were but so short a stay as one hour comes to, but the sun must go four and twenty times about for the making of one day. If both, then could not this neither be called a day, if the sun should run his whole round in the space of one hour; no, nor that, if while the sun stood still, so much time should overpass, as the sun usually makes his whole course in, from morning to morning. I will not therefore demand now what that should be which is called day: but, what time should be, by which we measuring the circuit of the sun, should say, that he had then finished it in half the time he was wont to do, if so be he had gone it over in so small a space as twelve hours come to: and when upon comparing of both times together, we should say, that this is but a single time, and that a double time, notwithstanding that the sun should run his round from east to east sometimes in that single time, and sometimes in that double time. Let no man therefore say unto me hereafter, that the motions of the celestial bodies be the times; because that when at the prayer of a certain man, the sun had stood still, till he could achieve his victorious battle, the sun stood indeed, but the time went on: for in a certain space of time of his own, (enough to serve his turn) was that battle strucken and gotten. I perceive time therefore to be a certain stretching. But do I perceive it, or do I seem to perceive it? Thou, O Light and Truth, shalt show it.

Time it is, by which we measure the Motion of Bodies

Dost thou command me to allow of it, if any man should define time to be the motion of a body? No, thou dost not bid me. For there is no body that I hear of, moved, but in time; this thou sayest: but that the motion of a body should be time, I never did hear: nor dost thou say it. For when a body is moved, I by time then measure how long it may have moved, from the instant it first began to move, until it left moving. And if so be I did not see the instant it began; and if it continues to move so long as I cannot see when it ends; I am not then able to measure it, but only perchance from that instant I first saw it begin, until I myself leave measuring. And if I look long upon it, I can only signify it to be a long time, but not how long: because when we pronounce how long, we must do it by comparison: as for example: This is as long as that; or This twice so long as that, or the like. But were we able to make observation of the distances of those places, whence and whither a body or his parts go, which moveth; (as if, suppose it were moved in a lathe) then can we say, how much time the motion of that body or his part, from this place unto that, was finished in. Seeing therefore the motion of a body is one thing, and that by which we measure how long it is, another thing; who cannot now judge which of the two is rather to be called time? For and if a body be sometimes moved uncertainly, and stands still other sometimes; then do we measure, not his motions only, but his standing still too: and we say, It stood still as much as it moved; or It stood still twice or thrice so long as it moved; or any other space which our measuring hath either perfectly taken, or guessed at, more or less, as we use to say. Time therefore is not the motion of a body.

NOTES

[1] From *The City of God;* transl. and ed. by Marcus Dods, D.D., Hafner Publishing Company, New York, 1948, T. and T. Clark, Edinburgh, 1872, pp. 441-443.

[2] From St. Augustine, *Confessions;* transl. by W. Watts, MacMillan, Harvard University Press, Cambridge, Mass., and Heinemann Ltd., London, 1912, in The Loeb Classical Library, pp. 259-269.

THE PROBLEM OF THE ABSOLUTE CLOCK*

All local movement presupposes a fixed reference point to which the positions of the moving body may be successively related. Just what is this immovable reference point? For Aristotle it is the Earth, because it is contradictory to suppose that the Earth, the center of the circular celestial movements, could be set in motion. This is an error, according to Stephen Tempier and the doctors of Paris; if He wished, God could cause the whole universe, and the Earth at its center, to undergo a movement of translation. William of Ockham then declared that the fixed point to which all local movements are referred is neither the Earth nor any body which actually exists in nature, for all natural bodies are or can be in motion; this reference point is simply an imaginary body. It is a simple concept, added Bonet, a geometrical concept which exists only as *esse cognitum* within the mind of the mathematician. Thus the whole peripatetic theory of place and motion was overthrown.

All change takes place in time. In order to determine this time, there must be a privileged, absolutely uniform movement which marks the duration of all other changes. Where is this primary clock to be found? For Aristotle the absolute clock is the diurnal movement of the first celestial sphere. Since the first sphere is perfect and divine, its movement is necessarily an absolutely uniform rotation. Moreover, since there can be only one World, this unique clock records the same time for all the movements which take place in Heaven and on Earth. This is an error, according to Stephen Tempier and his counsellors, for, if He wished, God could impose a movement of translation upon the Heaven. It is also an error, because God could, if He wished, create several Worlds. Then Nicolas Bonnet and Gradazei [of Ascoli] declared that this perfectly uniform movement, this precisely regulated clock which indicates the duration of all changes, does not exist in nature; it is a pure concept which resides in the mind of the mathematician. It matters little whether the highest Heaven actually accelerates or retards in its course; the duration of the abstract sidereal day conceived by the astronomer remains none the less invariable. Thus the peripatetic theory of time was destroyed.

Nicolas Bonet pushed this point further: it is obvious that what he has said about place and time can be readily extended to geometrical measurements. In order to measure lengths there must be a fixed length. Where are we to find the standard which maintains this invariable length? Does the wooden ruler we call a foot have the same length today that it had yesterday? How can we be certain? The immutable length, replied Bonet, does not exist in any concrete entity such as a wooden or stone rod. It exists as a figure, abstracted from all matter, of which the geometre conceives and about which he reasons.

Thus the unit or standard, for any kind of magnitude, is not a thing which really exists outside of our mind. It is an abstraction having only conceptual existence within our mind.

> All philosophers, ancient as well as modern, agree in declaring that the unit is indivisible, for to be one is to be individual. But note carefully here that this is understood in a mathematical sense, insofar as the unit is abstracted from all sensible matter. Otherwise it would not be true that the unit was absolutely indivisible; indeed, the unit which makes a piece of wood one, which has its basis in the wood (*quae subjective est in ligno*), can be divided as a result of the division of the wood, just like the other accidents of the wood. In the same way that the piece of wood allows itself to be divided into pieces of wood, the unit of the piece of wood allows itself to be subdivided into other units. However, if one considers it from the mathematical[1] point of view, the unit is indivisible, and, being one, it is indiviudal.[2]

The unit, therefore, is really a unit only for the mathematician who conceives of it apart from all sensible matter. It cannot be made real without ceasing to be a unit.

And the same holds true for the whole number. If the number ten always remains the one and identical number ten, whether it be realized in a herd of ten horses or a pack of ten dogs, it is because it is an abstract number. As soon as one attempts to change it from purely conceptual existence to real existence, from *esse cognitum* to *esse realis existentiae*, the number ceases to exist.

NOTES

* From *Le système du monde*, Hermann, Paris, 1956, VII, pp. 439–441; transl. by David A. and Mary-Alice Sipfle.
[1] The manuscript, obviously in error, reads *metaphysice* instead of *mathematice*.
[2] Nicolai, Boneti, *Tractatus de praedicamentis*, libellus de quantitate, ch. XIII; ms. no. 16.132, fol. 159, col. d.

B. TELESIO

INDEPENDENCE OF TIME FROM MOTION*

Aristotle was quite correct in teaching that our knowledge of time comes from our knowledge of motion and change of things and that we can have absolutely no sense of time without sensing some motion; that is, unless we perceive or remember some motion. But it appears that from these premises he incorrectly inferred that time does not exist without motion. But why should time depend on the motion of things and why in the absence of motion and change should it likewise cease to exist and to flow, when it exists by itself when no motion and no change are present? Likewise it was not correct for him to declare that, since we always perceive time and motion together – never time without motion nor motion without time – time is a kind of state or affection of motion. For even had motion no affinity or relationship whatever with time and time none with it, being both completely distinct from one another; since every motion and change takes place in time, and time manifests to us its brevity or length, and no time elapses in which we would not perceive or remember some motion, it is necessarily the case that wherever we sense any motion – since we also sense its duration – we are likewise aware of time in which it takes place. Therefore the fact that we always perceive them together is no reason for claiming that one of them is the essence (quid) of the other; but only, what seems to be the case, namely that every motion occurs in its own time and that no motion can take place without time. This is the only sense in which their mutual affinity and relationship should be and is understood.

There was even less reason to impose on time continuity and succession because continuity and succession inhere in quantity and through quantity also in motion. For, as it has been said, time exists by itself and in no way depends on motion; whatever characteristics it has, it has them all from itself and none from motion. Furthermore, it is astonishing that 'before' and 'after' belong, according to Aristotle, to motion on the ground that they belong also to magnitude under which motion is subsumed. Indeed no part of the celestial sphere – of whose motion in particular time was the number and measure, according to Aristotle – is 'before' and 'after'. But even granted all this, we certainly cannot concede to Aristotle that time is the number or measure of motion or some state or effect. For how can a thing, which for Aristotle is without beginning and end, and which is always homogeneous and never interrupted by any intermediate thing of a different kind, be the number or measure of the thing which here and there ends and changes, and to which we can always assign some limit or some difference? Or which thing could be an effect or a state of that thing in which it does not inhere and with which it has nothing common, being altogether different from it? Thus if time and motion have any conjunction or connection or affinity, it should be attributed to them in the following way: time is the interval, duration and extent, not over which or through which, but *in* which all motion and change occur. And motion is the measure of time: though not every motion, but only the motion of the celestial sphere which is truly one, continuous and uniform.

NOTE

* From *De rerum natura*, I, 29; transl. by the editor and Walter Emge.

G. BRUNO

HESITATIONS BETWEEN ABSOLUTE AND RELATIONAL THEORY OF TIME*

ARTICLE XXXVIII

Time which is the measure of motion is not in the heaven, but in the stars, and the first motion which we conceive is nowhere except subjectively on the earth.

If we hold with Aristotle that time is the measure of motion, it does not follow from this that it is placed in the sky to which that motion by which all times and movements are measured does not belong; this motion does exist, however, in any star. If that motion by which all things seem to be driven with an extreme velocity around the earth is in truth only subjectively observed on the earth, there are as many times in the universe as there are the stars. Nor is it possible that there be such single motion in the universe which would be the measure of all motions. For if we were on another star, the shortest motion would be quite obviously different from that one; as on the moon it is well known that there is a different diurnal motion since it requires for its rotation twenty-eight days while this star, the earth, only twenty-four hours. Hence, when we imagine time as something flowing, we should conceive it as flowing continuously and with a perfect uniformity; but how then can time be measured by that diurnal motion when the motion along the [celestial] equator is neither diurnal nor uniform? If this circular motion does not equally deviate from two regular motions, one of which supposedly belongs to the eighth, and the other to the ninth sphere, why is it at one time regular, at another time irregular? And also which two parts of that motion will you regard as equal? Where would you find two motions which are proportional? ...

Furthermore, all the circles which are seen and verified by our astronomers on the sky are only inappropriately called so and cannot be truly regarded as such according to the definition of circle. Therefore the law of diurnal motion, whether it be taken from the earth alone, or from the motion of the sun only, or from both, whether from these or from other circular motions, is not and cannot be truly geometrical. If it is impossible to divide uniformly a helical line and if the uniform motion – in the strict sense of the word – along this line is not natural; how can time and motion measure each other? For where is the measure of time? Where is that self-identical unit by means of which the equality and inequality of other intervals will be judged? Since Aristotle held the first motion as the most regular of all, and since according to him the motion of the eighth sphere was precisely this first regular motion, the reckoning of time and the measure of the duration of everything was based on that very same motion. But what would Aristotle say now if he were aware of other motions and saw that the measure of the daily motion is disturbed by thousands of irregularities? You already see into how many and how great errors the ignorance of one thing and also the presumption led that man! See how in this way the six alleged reasons by which time was regarded as the motion of the heaven fall to the ground: since supposedly (1) by this motion successive duration is measured; (2) since this motion is most reknown; (3) since it is common to all; (4) because it is invariable; (5) because it is prior to all; (6) because it is minimal in the sense that it is fastest. ...

ARTICLE XXXIX

Motion is the measure of time rather than time being the measure of motion: we truly came to know duration through motion rather than conversely. Although they both measure each other, no time is the measure of motion unless previously some motion was the measure of time.

Argument

Though time is some duration which can indeed be separately conceived and defined by mind, nevertheless it is not found separately from things because it is predicated in relation to duration and to something which endures. Therefore just as the place, one and infinite, must be the common infinite space for the infinite universe, so ought there be one common time, one duration, without end and without beginning. Also just as the infinitely many places of particular things and the proper places in which they are individually located are conceived to be in one, infinite, continuous, common space, so the different durations and times belong to one single duration, common to all of them. Yet, space and duration differ since the former is believed to remain everywhere immobile both in general and in a particular sense while time is understood as flowing most rapidly in those things which move very fast, at a slower rate in those which change more slowly, and with a minimum speed in those which are not subject to any change; and if these can be located in time, it is because we have a single concept of time which applies not only to those things which move, but also to all those which simply are and, consequently, are understood to endure. As we call the universal place the immense space, so we call the universal time eternity. Of the things existing in one duration some are called eternal, some simply temporal, and some of them of a greater, some of a lesser age; and in these mind conceives time sometimes as the measure, sometimes as the thing meassured while we alternately inquire about the quantity of duration by means of motion and about the quantity of motion by means of duration. Yet, time is never the measure of motion unless it was previously defined by motion. Indeed, time is always the amount of some revolving motion. Therefore motion, by itself, is more properly the measure of time while time is less properly and only by accident the measure of motion. Thus, finally, it is more accurate to say not that motion is measured by time, but that motion is measured by motion. Yet, we take a certain duration of motion as a measure through which we evaluate the durations of other motions. From this it is clear that we are aware of time by means of motion rather than *vice versa*. And indeed when Aristotle sees in the example of those seven who slept among the heroes at Sardes that there is no perception of time without some perception of motion, he should have or could have concluded that motion is the measure of time rather than time is the measure of motion.

ARTICLE XL

Nevertheless we say that time exists even if all things were at rest. Therefore Aristotle should have related not time itself, but the knowledge of time to motion.

Argument

It is certain that if there were no motion nor change, nothing could be called temporal; but there would still be the same one time for everything and one and the same duration, called eternity. Indeed, time in the sense of the age of any particular thing would not exist. Therefore the existence of time in its particular kinds depends on motion.

Now if it happened that all things were at rest, would this mean that they would not endure? Indeed, they would endure, they would all endure by one and the same duration. But without any existing motion there will will be no measure of their duration. Therefore motion will be the measure of time or of duration, especially in those things which admit a certain and definite duration. But Aristotle certainly guessed this for the most part, although he preferred to regard time as the measure of motion; he wanted even to equate it with the diurnal motion. Thus he did not recognize the duration of the absolute kind, but only that related to some kind of motion. Thus if there were no movement, there would be no different kinds of duration, but only one nameless eternity ... But if we assume that everything is at rest, I cannot see clearly why it should be impossible to dissociate the number of time from the number of motion – unless I accept as the principle that the meaning of time is restricted by the Aristotelian definition instead of being taken (as it should be) sometimes absolutely, sometimes relatively, now in one sense, now in another. But from this it follows that time should be understood not only as numbering and measuring motion, but also as being measured and numbered by motion. As far as the rest is concerned, we affirm that it also is measured by motion and time; for if all things were at rest, it would not follow that time would cease to measure duration since there would be one duration, one rest for all things. Hence, just as we judge the motion of one thing by that of another when all things move, so time will measure the rest of one thing by the rest of all things and vice versa ...

All this we accept as a part of the definition of time, conceived not as a species of duration, but as duration itself. Therefore perpetual time is eternity, finite time is a period of which there are innumerable species. If one prefers to regard time one-sidedly as duration of things in motion, we shall not object as long as this will not prevent him from considering other meanings which we now and then have mentioned and formulated. From this it is known how well and felicitously the knowledge of time is united to motion. It is enough for us to have pointed out here that: time which is the uniform and universal measure of motion cannot be perceived and imagined except through motion, either naturally, by the circular motion of the sun or the moon or any other star; or artificially, by the flow of water or sand or by the reversion of the openings [of clepsydra]. Each of these motions we conceive as being imperceptibly different from each other and each of them varying in its rate. Also we regard as sufficiently established that there is nowhere a perfect correspondence between the geometrical properties of motion and the physically real motion as far as its quantity, its shape and mass is concerned.

NOTE

* From *Camoeracensis Acrotismus*, Art. XXXVIII–XL, *Opera latine conscripta*, F. Fromann, Stuttgart, 1962, I, pp. 143–150; Transl. by the editor and Walter Emge.

P. GASSENDI

REALITY OF ABSOLUTE TIME*

It seems that the Stoics have been more perceptive than Epicurus when they held that time is something incorporeal which is understood to exist *by itself* and not as something which accidentally pertains to things in the sense that there would be no time without things which endure in it or at least without our mind imagining them to endure. We comprehend that even before there were any things time flowed; and from this we acknowledge that they could have been created by God earlier than they had been created – that is either a short time or long time or even eternal time earlier. Even now, while they exist, we understand that time flows in the same tenor as it flowed before; and if God reduced the whole universe to nothing, we comprehend that time will still flow; and we also understand that if God would wish to recreate the universe, time still would flow in the interval between its destruction and recreation. There is a difficulty for us here, since, although we apparently speak appropriately and accurately and we seem to comprehend what we are speaking of when we say that 'time flows', 'time passes', 'time follows', 'time came near', 'time will come', etc., yet we hesitate, and in attempting to clarify the meaning of 'flow', 'passage', 'coming' etc. of time, we notice that this was a mere metaphorical way of speaking; we cannot point out with a finger the flow of time in the way we point out the flow of water. In truth, since we cannot speak of incorporeal things except by analogy with corporeal things, it ought to suffice that if we understand the flow of water whose parts pass in successive order, one after another, so we should understand the flow of time whose parts pass in the same way in succession, one after another, or prior and posterior. And yet it is more accurate to compare time with the flame of a lamp which also consists in flux in such a way that in every moment it is always different and what has been before is never more again and nothing is yet present what will be in the future. Truly in the same way the nature of time consists in flowing so that whatever is past, now no longer exists nor anything yet exists of whatever will pass. From this it follows that just as the flame does not cease as a whole to be something corporeal and continuous, even though each of its parts is momentary, so time as a whole does not cease to be something incorporeal and continuous, though any of its parts is momentary, or rather it is the very moment itself which is called 'now' or 'instant' or 'present'. For as much as each present bit of flame is connected with the immediately preceding one and with the immediately subsequent one and for the same reason it is a continuation of the whole; so any moment of time is connected with the immediately preceding and the immediately subsequent moment, and since this argument applies to the whole, a continuous succession of the whole is being created.

Thus, when someone objects that time is nothing because, since it supposedly consists of the past, present and future, yet the past is gone, the future is not yet, the present is altogether vanishing, we can answer that it would be the same as to object that the flame is nothing because whatever of its parts was before, does not exist any longer, whatever part will be, does not yet exist, and whatever is present, is vanishing.

But there is clearly a paralogism here since the things of different kind are treated as the things of the same kind, that is, the things which are successive are regarded as if they were permanent, although they are altogether different or, as we say, worlds apart. Those who apply the categories valid for one kind of thing to the other, act as if they were measuring a straight line by a curve, a pound by a yard, a length by a pound. They are seeking in successive things that which is not in their nature for, if it were there, they would not be successive. Let their parts be standing still, let them not flow, let them remain the same; then they would not be successive, but permanent; but is it true that nothing except permanent can exist? Obviously, nothing can exist permanently except what is permanent; but what is successive, can exist in its own way, that is, successively. For just as much as the nature of the former consists in the fact that its parts are simultaneous, and about the whole we can repeat the words 'it is', 'it is', 'it is', so the nature of the latter consists in the fact that its parts are not simultaneous, and about the whole we can say 'it was', 'it is', 'it will be'. Undoubtedly, we lack a simple word to express the total existence which, while not being simultaneous, is contained not only in the present, but also in the past and in the future. But we ought not dwell any longer on this point, especially since it appears to be a terminological dispute. We should only observe that Possidonius was right when, according to Stobaeus, he expressed the opinion – doubtless in order not to be entangled in subtleties – that time should not be conceived as a point-like present in the strict mathematical sense, but more broadly, as the perceptible minimum of time in which the future and the past are being joined.

Likewise, Aristotle conceded that it is said: "It happened now because it happened today, and now it will happen since it will happen today." We certainly use ordinary language, but not inappropriately, when we speak of the present day. And again Apollodorus with whom we speak in ordinary language – and also correctly – of 'the present year'; neither would it be wrong, were we to say 'the present century' and so forth.

Indeed already Epicurus could not say that a day or night is long or short in virtue of that time which our thought assigns to it, but rather it is long or short in virtue of that time which flows whether we think or not. For it is not true that a day which in the state of hope appears long and in the state of fear short is either extended or contracted by the effect of such thoughts. Nor could Aristotle say that time is the number of motion which would not exist without the numbering mind; for whatever time is, whether it is numbered or not, if flows and is successive in its nature. Rather it is man who by designating, distinguishing and measuring time perceives, adjusts and counts the parts of some motion, especially of the celestial motion. But this does not make time dependent either on motion itself nor on its parts whether they are numbered or not; and most of all, since it existed before the celestial motions, we ought to understand very clearly that the fact that there are many celestial motions is no reason for the existence of multiple time; nor should there be plural times if God produced many worlds and many moving heavens. I have already covered this before when I suggested that Aristotle was not sufficiently careful when to those who defined time as the motion of the heaven, he objected that if there were many worlds and many heavens, there would be also many times since there would be many motions. He was not careful, I say, since the same argument could be turned against him when he goes on to define time as the number of this motion. For if there were many worlds and many motions of *prima mobilia*, could not one infer that there would be also many "numbers of motion according to before and after"? ... Perhaps you will distinguish, as it is customary, intrinsic from external time; for as you say, that the particular motions of inferior things have times which are intrinsic and proper to them and beyond these there is external and universal time, namely that of the first moving sphere, so you would say to any motion first belongs a particular time, while the general time belongs to all. But you will not be able to determine this general time since there is no general motion having its before and after. Nor do your particular times exist unless you would also admit that ten hours elapse when ten bodies or ten spheres move during one hour and one hour would elapse twice as fast as another when one of two motions were twice as fast as another.

Thus it seems that Aristotle in his objection correctly guessed the true nature of time, but he missed it when he defined time as the number of motion. For if time is a kind of flow, as we said already several times, it is independent of motion no less than of rest; and with it not only many, but countless and most diverse motions can co-exist. Also it is equally false to say that time is the measure of the celestial motion; but rather the celestial motion itself is the measure of time for the same reason the measure ought to be better known than the thing to be measured. This is correctly suspected by those who discern and admit the so-called imaginary time. For they admit that time flowed even before the creation of the heavens and furthermore, they concede that the world could have been created before it was created; and that time flows when the world exists and will flow when the world will cease to exist. But since they are prejudiced, they immediately retreat and claim that besides that imaginary time there is still another time which is truly real in the sense defined by Aristotle; this time began with the celestial motion, it would stop with the interruption of this motion and would cease entirely in its absence. They are prejudiced, I say; for if you look at this matter seriously, there is clearly no other time except that which is called imaginary and which is necessary and which, as they admit, continued to flow alone when the heavens were standing still as long as Joshua was fighting the kings of Amorites.

But to make this point clearer, we have to consider again the already mentioned comparison and kind of parallellism between this time or imaginary duration and place or imaginary space; indeed from the insight into the nature of the latter, the nature of the former can be considerably clarified. That there is a certain affinity between these two things was not only recognized by Aristotle when he explained in the same book together place, void and time; but even much more by Plato when, as we did before, distinguished place and time as two different genera from all other things. Thus he, after dividing the properly existing things into five genera, added the sixth genus, in the sense of the passage of Seneca, quoted above: "There is the sixth genus of the things existing in a certain fashion such as the void and time." It also seems that Chrisippus had someinkling of it when, according to Stobaeus, compared these two things, especially with regard to their infinity. This seems to be true also of Philo when he said: "Place is understood to pertain to the things which are at rest while time to the things which are moving"; that is, place is conceived as immobile, time as flowing. It seems to us that, instead of looking for more illustrations, the matter should be explained more explicitly in the following way.

As Place as a whole is unlimited, so Time as a whole has neither the beginning nor the end, and, as any moment of Time is the same in all places, so any part of Place is in all times. Likewise as Place persists immobile whether anything exists in it or not, so Time flows with equal tenor whether anything endures in it or not, whether anything is at rest or in motion, whether it moves faster or slower. And as the Place cannot be dislocated by any power, but remains unchangingly continuous and always the same; so Time cannot be stopped or suspended by any power, but, irresistibly advancing, it always flows uniformly. Again as from Place or immense space is taken portion in which the world is located, so from infinite Time a part was selected in which the world exists. Moreover, as every body (or, more generally, every thing), in so far as it is here or there, requires for itself a certain part of the cosmic space, so also every thing, whether it exists now or then, requires for itself a special part of the cosmic duration. Furthermore, as in speaking of Place we say *everywhere* or *somewhere*; so in speaking of time we say *always* or *sometimes*. For this reason Plotinus scolded the Peripatetics for regarding those two categories which they call 'where' and 'when' as distinct from Place and Time. Thus, as it is proper for the created things to be only *somewhere* with respect to place and *sometime* with respect to time, so it is proper for the Creator to be everywhere with regard to place and always with regard to time; and so these distinguished attributes belong to him: immensity by which he is present in every place, and eternity through which he persists in every time. Finally, as Place has constant dimensions which correspond to the length, breadth and depth of the bodies: Time has successive dimensions corresponding to the motions of the bodies. From which it finally follows that just as we measure the length of Place by a length, for example, of a yard, so we measure the flow of time by the movement of a clock. And since there is no other motion more universal, more constant and more known than that of the sun; so we accept the motion of this kind as a sort of clock by which the flow of time is measured. Not that, if the sun were to move faster or slower, Time would accordingly flow more quickly or more slowly; nor that we would accept any kind of solar motion for measuring time. For instance, if this motion were twice as fast, Time would not, on that account, be twice as rapid, but the interval of two days would be equivalent to that of one which we have now. Just as if the motion of the sun were twice as slow, one day would be equivalent to two. It is in this sense, I believe, that we must understand the view which Plutarch attributed to Empedocles that at the beginning of the world the days were much longer than they are now, and in particular, that day on which men were first created was the equivalent of two months now. This means that we have to believe that the motion of the sun since that time became sixty times faster.

Finally, from these considerations it does not appear that time is something dependent upon motion or subsequent to it, but something indicated by it as the measured thing is indicated by its measure. For since it cannot be known by any other way how much time we spent while we were doing something or not doing anything, so we behold the celestial motion and from its amount we evaluate the passage of time. And since the observation of this kind of motion seems to be difficult for the ordinary man, the motions of other things, familiar to the ordinary man, such as water, sand, wheels and sundials, are adjusted to the celestial motion, so that by quickly glancing at them he could count both that motion and Time itself. This is why I said immediately above that the heaven is a kind of general clock; indeed it is, as our clocks imitate it as much as they can and are used as substitutes for that which is less easily seen. Hence also why shortly before, to prove that time is independent of the celestial motion and that it does not need succession in order to exist, I stressed that time could flow even if the heavens were at rest, and it flows while they are in motion. As an example, I pointed to the story of Joshua in Scripture. No one really believes that while Joshua was fighting with

Amorites, and the heaven stood still, that no time passed and that the number of hours almost equivalent to the full day did not elapse. As the Scripture testifies: "Never before, nor after there was such a long day." But this length cannot be understood in any other way than as a passage of time. Supposing now that heaven stood still (doubtless it can be stopped by God), do you not see that time would flow in the same way as when the heaven was in motion? You might ask: how could there be the hours if the motion of the sun would not mark them off? They would exist, not because they were marked off by the motion of the sun, but because they *could be marked off* by this motion which then could exist. (Nay, they could be marked off by a waterclock or some other time measuring instrument.) Thus we say that the world could have been created a thousand years before its actual creation, not because there were years marked off by the repeated revolutions of the sun, but because Time flowed of which the appropriate measures, the revolving motions of the sun, in the way they exist now, *could* have existed then. And do not say that all these times were imaginary; certainly we cannot conceive them in any other way than Time which flows while the heaven moves.

Therefore, it is worth wondering why Aristotle did not want time to exist without change, so that when in the sleep we do not perceive change, we usually connect the time when we fall asleep with the time of awakening. "As if," he says, "it happened to those who slept with the heroes in Sardinia." [sic] (Simplicius adds the example of those who being drunk lie down in a cave and sleep continuously for two days and after waking up the second night and seeing the stars, they lie down again; on the dawn of the third day they believe that they slept through one night.) This, I say, is really worth wondering about. For no matter how much some change is necessary for the perception of the flow of time, it does not mean that it is necessary for the very existence of this flow.

NOTE

* From *Syntagma philosophicum, Physicae Sectio I, Liber II;* transl. by the editor and Walter Emge.

I. BARROW

ABSOLUTE TIME

LECTURE I

Now pray tell me what Time is? You know the very trite Saying of *St. Augustin, If no one asks me, I know; but if any Person should require me to tell him, I cannot.* But because *Mathematicians* frequently make use of Time, they ought to have a distinct Idea of the meaning of that Word, otherwise they are Quacks. My Auditors may therefore, on this Occasion, very justly require an Answer from me, which I shall now give, and that in the plainest and least ambiguous Expressions, avoiding as much as possible all trifling and empty Words. *Time*, (to speak abstractedly) is the continuance of any Thing in its own Being. But some Things continue longer in their Beings than others; those *were* when these were not, and *are* when these are not; they enter'd first into Being, and cease to be after these; nor is there any Person but perceives, that some Things enter into Being, and cease to be at the same Time; keeping an equal Pace, as it were, from the beginning to the end of their Duration. Time absolutely therefore is Quantity, as admitting in some Manner the chief Affections of Quantity, Equality, Inequality, and Proportion; nor do I believe there is any One but allows that those Things existed equal Times, which rose and perished together; and that those Things had unequal Durations, when the one was in Being before the other had existance, and continue in its Being, after the other had ceased to be. But a longer and shorter Time is common in every Body's Mouth; and there is no Man but seems to understand the Meaning of these Words. Common Sense, therefore allows Time to partake of Quantity, as the Measure of the Continuance of Things in their Being. But perhaps you may ask, whether Time was not before the World was created? And if Time does not flow in the Extramundane Space, where nothing is: A mere Vacuum? I answer, that since there was Space before the World was created, and that there now is an Extramundane, infinite Space, (where God is present;) inasmuch as there might have been of old, and now may be, such, and so many Bodies, which then were not, and now are not; consequently Time existed before the World began, and does exist together with the World in the Extramundane Space, because 'tis possible that some Thing might have existed long before the World was made; and there may now be something in the Extramundane Space, capable of such a Continuance: Some *Sun* might have given Light long before; and at present this, or some other like it, may diffuse Light thro' Imaginary Spaces. Time therefore does not imply an actual Existence, but only the Capacity or Possibility of the Continuance of Existence; just as Space expresses the Capacity of a Magnitude contain'd in it. But you may perhaps wonder why I explain Time without Motion, and will say, does not Time imply Motion? I answer no, as to its absolute and intrinsic Nature; any more than it does Rest. The Quantity of Time, in itself, depends not on either of them; for whether Things move on, or stand still; whether we sleep or wake, Time flows perpetually with an equal Tenor. If you suppose all the fixed Stars to have stood still from their Beginning; not the least Portion of Time wou'd be lost by this; for so long as that Rest continues, so long has this Motion flowed. There may be what we may call *first* and *last*, beginning and ending together, (with Regard to the first Appearance and Disappearance of Things) even in that State of Tranquillity, which some Mind more perfect than ours may possibly comprehend. But as Magnitudes themselves are absolute *Quantums* Independent on all Kinds of Measure, tho' indeed we cannot tell what their Quantity is, unless we measure them; so Time is likewise a *Quantum* in itself, tho' in Order to find the Quantity of it, we are obliged to call in Motion to our Assistance, as a Measure whereby we may esteem and compare the Quantity of it; and thus Time as measurable signifies Motion; for if all Things were to continue at Rest, it would be impossible to find out by any Method whatsoever how much Time has elaps'd; and the several Ages wou'd roll on imperceptibly and undistinguish'd. Do I say we shou'd not perceive how Time flows? No indeed, nor any Thing else, but remain like Stocks or Stones in a continual Insensibility. We perceive nothing, unless so far as we may be instigated by some Change affecting the Senses, or that our Souls are mov'd and excited by the internal Operation of the Mind. We esteem the Quantities and different Degrees of Things according to the Extension or Intension of Motions striking upon us either interiorly or exteriorly: So that the Quantity of Time so far as we can observe, depends upon the Extension of Motion. 'Tis not improperly observ'd by *Lucretius*.

Nec per se quenquam tempus sentire satendum est.
Semotum ab rerum motu placidâque quiete.
No Thought can think on Time; that's still confess'd:
But thinks on Things in Motion or at Rest.

Nor by the *Philosopher himself. When we wake we cannot perceive or tell how much Time has passed during our Sleep*: which is certainly true: But it cannot be justly inferr'd from thence. *We do not perceive the Thing, therefore there is no such Thing*, that is a false Illusion, a deceitful Dream, that wou'd cause us to join together two remote Instants of Time. But nevertheless this is very True. That is, *for as much Motion as there was, so much Time seems to have been elapsed*; nor, when we mention such a Quantity of Time, do we merely mean any Thing else, than the Performance of so much Motion, to the continued successive Extension of which we imagine the Permanency of Things is coextended. Moreover, because we conceive Time to flow always in an equal Channel, not by Starts, some times slower, and faster at others. (For were such Disparity allow'd, it wou'd admit of no Manner of Computation or Dimension.) Every Motion must therefore not be consider'd as equally fit to determine and shew the Quantity of Time, but that chiefly which being the most simple and uniform, goes on always with an equal Pace; the Thing moved every where retaining an equal Force, and being carried along in a uniform Medium. Such a moveable Body must therefore be pitched upon for the Determination of Time, as at least constantly keeps an equal Impetus, with Regard to the Periods of its Motion, and runs thro' an equal Space: and indeed, for common Use, the most remarkable Motion possible ought to be taken; such a one as is immediately obvious to all, and strikes the Senses of every one; of this Kind is the Motion of the Stars, and particularly of the *Sun* and *Moon*, which observe a wonderful Regularity one to the other, in all Things, and are conspicuous to every Part of the Earth; those Bodies being deputed for that Purpose, not only by the common Consent of Mankind, but adapted to it by the divine Will of the Creator, who pronounc'd as follows: *Let there be Lights in the Firmament of the Heaven, to divide the Day from the Night: and let them be for Signs, and for Seasons, and for Days, and for Years.* But you may ask, how we *can know that the Sun moves with an equal Motion, that the Time of one Day or Year, for Supposition Sake, is exactly equal to that of another?* I answer, we cannot know this any otherwise (except what we gather from the divine Testimony) but by comparing the Motion of the Sun with other equal Motions. If, for Example, the Sun's Motion on a Sun-Dial (which almost to a Certainty, and in an exquisite Manner, shews the Quantities of the Spaces run through by the Sun in Circles parallel to the Equator) be found to agree with the Motions of any artificial Time-Keeper accurately made, the Sun's Motion must consequently be equal. For since the very Construction of the Machine is such, as to equally move in every successive Repetition of its Motion, as, for Instance, an Hour-Glass, destined to measure an Hour; and because the Water or Sand contain'd in it remain entirely the same as to Quantity, Figure, and Force of descending, and the Vessel that contains them, as likewise the little Hole they run thro' don't undergo any Kind of Mutation, at least in a short Space of Time, and the State of the Air much the same; there is no Manner of Reason for us not to allow the Times of every running out of the Water or Sand to be equal; If therefore the solar Motions, either as to whole Periods, or to the proportional Parts of them, are found to correspond entirely with the repeated Motions of such an Instrument, we thence may very justly conclude them to be entirely equal and uniform. Whence it seems to follow (which possibly wou'd appear wonderful to some,) that the celestial Bodies are not essentially, and properly speaking, the primary and original Measures of Time; but rather those Motions which are near us, that strike upon our Senses, and fall under our Experience, since by their Means, we discover the Regularity of the celestial Motions. Nor is even the Sun itself a proper Judge or Witness in this Affair, any farther than as its Veracity is shown, by the Attestation of an Horary Machine. Neither indeed can we know by any Methods, whether the Periods of the Stars, many Ages since, were altogether equal to their Revolutions in our Age; no one, for Instance, can pretend to assert as a certainty, that the Age of *Methusalem* himself, who lived a thousand Years wanting one, was really longer than that of a Man, who now dies before he arrives at an Hundred. Why might not the Sun, being then younger and more vigorous, have performed his Periods ten Times sooner than at this Time? Perhaps the Air was then purer, and by that Means the Gravity of Bodies becoming more powerful, organical Instruments themselves might have had swifter Motions, and so, if compared with the Instruments of our Time, would very much disagree. *Empedocles* indeed, as we find in *Plutarch*, is said to have been of Opinion, that the Sun in his Infancy caused the Days to be much longer than in his Time. But this does not seem agreeable to Reason, because such circular Motions are won't to Decay continually, rather than Increase. But to return from this scarce serious digression; the Quantity of Time, (or Continuance of Things in their Being, State, or Motion) is found out, as was before observ'd, first by any well known equal Motion, (or as to the Parts used for this Purpose, constantly equal and similar to themselves) and afterwards by any other Motions which upon Trial, proportionably answer to it, and chiefly by the celestial Motions, especially of the Sun and Moon: So that those Times are equal, when the same Hour-Glass is run out once, twice, or any equal Number of Times, or, in which the same Stars perform the same Period, or equal Parts of the same Period; but those unequal, according to any Proportion, in which unequal Periods are similarly or proportionably compleated. Nor shou'd any one object to Time's being commonly taken for the Measure of Motion and consequently in bringing to this the Differences of Motion, swifter, slower, accelerated, or retarded. Time must be taken as foreknown; [yet the Quantity of Time is not therefore to be determined by Motion, but the Quantity of Motion by Time;] for there is no Reason why Time and Motion may not mutually perform these Offices. For, in like Manner as we first of all measure a Space by some Magnitude, and declare it is so much; and afterwards by Means of this Space, compute other Magnitudes correspondent with it: So we first assume Time from some Motion, and afterwards judge thence of other Motions, which in Reality is no more than comparing some Motions with others, by the Assistance of Time; just as we investigate the Ratios of Magnitudes by the help of some Space. For Example, he who computes the Proportion of Motion, by the Proportion of Time, does no more than get the said Ratio of Motions from Clocks, Dials; or from the Proportion of solar Motions performed in the same Time. This *Aristotle* doubtless knew, and has plainly taught: *We not only measure Motion by Time, but also Time by Motion, because they determine each other.* Again, because Time, as has been shewn, is a Quantum uniformly extended, all whose Parts correspond, either Proportionably to the respective Parts of an equal Motion, or to the Parts of Spaces moved through with an equal

Motion; it may therefore be very aptly represented to our Minds, by any Magnitude alike in all its Parts; and especially by the most simple ones, such as a strait or circular Line; between which and Time there happens to be much Likeness and Analogy. For as Time consists of Parts altogether similar, it is reasonable to consider it as a Quantity endowed with one Dimension only; for we imagine it to be made up, as it were, either of the simple Addition of rising Moments, or of the continual Flux of one Moment, and for that Reason ascribe only Length to it, and determine it's Quantity by the length of the Line passed over: As a Line, I say, is looked upon to be the Trace of a Point moving forward, being in some sort divisible by a Point, and may be divided by Motion one Way, *viz.* as to Length; so Time may be conceiv'd as the Trace of a Moment continually flowing, having some Kind of Divisibility from an Instant, and from a successive Flux, inasmuch as it can be divided some how or other. And like as the Quantity of a Line consists of but one Length following the Motion; so the Quantity of Time pursues but one Succession stretched out as it were in Length, which the Length of the Space moved over shews and determines. We therefore shall always express Time by a right Line; first, indeed, taken or laid down at Pleasure, but whose Parts will exactly answer to the proportionable Parts of Time, as its Points do to the respective Instants of Time, and will aptly serve to represent them. Thus much for Time.

NOTE

* From *Lectiones geometricae*, pp. 4–15.

I. NEWTON

ON TIME*

Hitherto I have laid down the definitions of such words as are less known, and explained the sense in which I would have them to be understood in the following discourse. I do not define time, space, place, and motion, as being well known to all. Only I must observe, that the common people conceive those quantities under no other notions but from the relation they bear to sensible objects. And thence arise certain prejudices, for the removing of which it will be convenient to distinguish them into absolute and relative, true and apparent, mathematical and common.

I. Absolute, true, and mathematical time, of itself, and from its own nature, flows equably without relation to anything external, and by another name is called duration: relative, apparent, and common time, is some sensible and external (whether accurate or unequable) measure of duration by the means of motion, which is commonly used instead of true time; such as an hour, a day, a month, a year.

Absolute time, in astronomy, is distinguished from relative, by the equation or correction of the apparent time. For the natural days are truly unequal, though they are commonly considered as equal, and used for a measure of time; astronomers correct this inequality that they may measure the celestial motions by a more accurate time. It may be, that there is no such thing as an equable motion, whereby time may be accurately measured. All motions may be accelerated and retarded, but the flowing of absolute time is not liable to any change. The duration or perseverance of the existence of things remains the same, whether the motions are swift or slow, or none at all: and therefore this duration ought to be distinguished from what are only sensible measures thereof; and from which we deduce it, by means of the astronomical equation. The necessity of this equation, for determining the times of a phenomenon, is evinced as well from the experiments of the pendulum clock, as by eclipses of the satellites of Jupiter.

NOTE

* From *The Mathematical Principles of Natural Philosophy,* A. Motte's translation, revised by Florian Cajori, Univ. of California Press, Berkeley and Los Angeles, 1962, I, pp. 6, 7–8.

J. LOCKE

ON SUCCESSION AND DURATION*

CHAPTER XIV: OF DURATION, AND ITS SIMPLE MODES

§ 1. *Duration is fleeting extension.* – There is another sort of distance, or length, the idea whereof we get, not from the permanent parts of space, but from the fleeting and perpetually perishing parts of succession. This we call duration, the simple modes whereof are any different lengths of it, whereof we have distinct ideas, as hours, days, years, &c., time and eternity.

§ 2. *Its ideas from reflection on the train of our ideas.* – The answer of a great man, to one who asked what time was, *Si non rogas intelligo* (which amounts to this; the more I set myself to think of it, the less I understand it), might, perhaps, persuade one, that time, which reveals all other things, is itself not to be discovered. Duration, time, and eternity, are not, without reason, thought to have something very abstruse in their nature. But however remote these may seem from our comprehension, yet if we trace them right to their originals, I doubt not but one of those sources of all our knowledge, viz., sensation and reflection, will be able to furnish us with these ideas, as clear and distinct as many others which are thought much less obscure; and we shall find, that the idea of eternity itself, is derived from the same common original with the rest of our ideas.

§ 3. To understand time and eternity aright, we ought, with attention, to consider what idea it is we have of duration, and how we came by it. It is evident to any one who will but observe what passes in his own mind, that there is a train of ideas which constantly succeed one another in his understanding, as long as he is awake. Reflection on these appearances of several ideas, one after another, in our minds, is that which furnishes us with the idea of succession; and the distance between any parts of that succession, or between the appearance of any two ideas in our minds, is that we call duration. For whilst we are thinking, or whilst we receive successively several ideas in our minds, we know that we do exist; and so we call the existence, or the continuation of the existence of ourselves, or any thing else, commensurate to the succession of any ideas in our minds, the duration of ourselves, or any such other thing co-existent with our thinking.

§ 4. That we have our notion of succession and duration, from this original, viz., from reflection on the train of ideas which we find to appear, one after another, in our own minds, seems plain to me, in that we have no perception of duration, but by considering the train of ideas that take their turns in our understandings. When that succession of ideas ceases, our perception of duration ceases with it; which every one clearly experiments in himself, whilst he sleeps soundly, whether an hour or a day, a month or a year; of which duration of things, while he sleeps, or thinks not, he has no perception at all, but it is quite lost to him; and the moment wherein he leaves off to think, until the moment he begins to think again, seems to him to have no distance. And so I doubt not but it would be to a waking man, if it were possible for him to keep only one idea in his mind, without variation, and the succession of others; and we see, that one who fixes his thoughts very intently on one thing, so as to take but little notice of the succession of ideas that pass in his mind, whilst he is taken up with that earnest contemplation, lets slip out of his account a good part of that duration, and thinks that time shorter than it is. But if sleep commonly unites the distant parts of duration, it is because, during that time, we have no succession of ideas in our minds. For, if a man, during his sleep, dreams, and variety of ideas make themselves perceptible in his mind one after another, he hath, then, during such a dreaming, a sense of duration, and of the length of it. By which it is to me very clear, that men derive their ideas of duration from their reflections on the train of the ideas they observe to succeed one another in their own understandings; without which observation, they can have no notion of duration, whatever may happen in the world.

§ 5. *The idea of duration applicable to things whilst we sleep.* – Indeed, a man having, from reflecting on the succession and number of his own thoughts, got the notion or idea of duration, he can apply that notion to things which exist while he does not think; as he that has got the idea of extension from bodies by his sight or touch, can apply it to distances, where no body is seen or felt. And, therefore, though a man has no perception of the length of duration, which passed whilst he slept or thought not, yet having observed the revolution of days and nights, and found the length of their duration to be, in appearance, regular and constant, he can, upon the supposition that that revolution has proceeded, after the same manner, whilst he was asleep, or thought not, as it used to do at other times; he can, I say, imagine and make allowance for the length of duration, whilst he slept. But if Adam and Eve (when they were alone in the world) instead of their ordinary night's sleep, had passed the whole twenty-four hours in one continued sleep, the duration of that twenty-four hours had been irrecoverably lost to them, and been for ever left out of their account of time.

§ 6. *The idea of succession not from motion.* – Thus by reflecting on the appearing of various ideas one after another in our understandings, we get the notion of succession; which if any one would think we did rather get from our observation of motion by our senses, he will, perhaps, be of my mind, when he considers, that even motion produces in his mind an idea of succession no otherwise than as it produces there a continued train of distinguishable ideas. For a man looking upon a body really moving, perceives yet no motion at all, unless that motion produces a constant train of successive ideas, v. g. a man becalmed at sea, out of sight of land, in a fair day, may look on the sun, or sea, or ship, a whole hour together, and perceive no motion at all in either; though it be certain that two, and perhaps all of them, have moved, during that time, a great way; but as soon as he perceives either of them to have changed distance with some other body, as soon as this motion produces any new idea in him, then he perceives that there has been motion. But wherever a man is, with all things at rest about him, wihout perceiving any motion at all; if during this hour of quiet he has been thinking, he well perceive the various ideas of his own thoughts, in his own mind, appearing one after another, and thereby observe and find succession, where he could observe no motion.

§ 7. And this, I think, is the reason why motions very slow, though they are constant, are not perceived by us; because, in their remove from one sensible part towards another, their change of distance is so slow, that it causes no new ideas in us, but a good while one after another; and so not causing a constant train of new ideas to follow one another immediately in our minds, we have no perception of motion, which consisting in a constant succession, we cannot perceive that succession, without a constant succession of varying ideas arising from it.

§ 8. On the contrary, things that move so swift, as not to affect the senses distinctly with several distinguishable distances of their motion, and so cause not any train of ideas in the mind, are not also perceived to move. For any thing that moves round about in a circle, in less time than our ideas are wont to succeed one another in our minds, is not perceived to move; but seems to be a perfect entire circle of that matter or colour, and not a part of a circle in motion.

§ 9. *The train of ideas has a certain degree of quickness.* – Hence I leave it to others to judge, whether it be not probable, that our ideas do, whilst we are awake, succeed one another in our minds at certain distances, not much unlike the images in the inside of a lanthorn, turned round by the heat of a candle. This appearance of theirs in train, though, perhaps, it may be sometimes faster, and sometimes slower; yet, I guess, varies not very much in a waking man: there seem to be certain bounds to the quickness and slowness of the succession of those ideas one to another in our minds, beyond which they can neither delay nor hasten.

§ 16. *Ideas, however made, include no sense of motion.* – Whether these several ideas in a man's mind be made by certain motions, I will not here dispute; but this I am sure, that they include no idea of motion in their appearance; and if a man had not the idea of motion otherwise, I think he would have none at all, which is enough to my present purpose, and sufficiently shows, that the notice we take of the ideas of our minds appearing there one after another, is that which gives us the idea of succession and duration, without which, we should have no such ideas at all. It is not then motion, but the constant train of ideas in our minds whilst we are waking, that furnishes us with the idea of duration, whereof motion no otherwise gives us any perception, than as it causes in our minds a constant succession of ideas, as I have before shown: and we have as clear an idea of succession and duration, by the train of other ideas succeeding one another in our minds, without the idea of any motion, as by the train of ideas caused by the uninterrupted sensible change of distance between two bodies, which we have from motion; and, therefore, we should as well have the idea of duration, were there no sense of motion at all.

§ 17. *Time is duration set out by measures.* – Having thus got the idea of duration, the next thing natural for the mind to do, is, to get, some measure of this common duration, whereby it might judge of its different lengths, and consider the distinct order wherein several things exist, without which, a great part of our knowledge would be confused, and a great part of history be rendered very useless. This consideration of duration, as set out by certain periods, and marked by certain measures or epochs, is that, I think, which most properly we call time.

§ 18. *A good measure of time must divide its whole duration into equal periods.* – In the measuring of extension, there is nothing more required but the application of the standard or measure we make use of, to the thing of whose extension we would be informed. But in the measuring of duration, this cannot be done, because no two different parts of succession can be put together to measure one another; and nothing being a measure of duration but duration, as nothing is of extension but extension, we cannot keep by us any standing unvarying measure of duration, which consists in a constant fleeting succession, as we can of certain lengths of extensions, as inches, feet, yards, &c., marked out in permanent parcels of matter. Nothing then could serve well for a convenient measure of time, but what has divided the whole length of its duration into apparently equal portions, by constantly repeated periods. What portions of duration are not distinguished, or considered as distinguished and measured by such periods, come not so properly under the notion of time, as appears by such phrases as these, viz. 'Before all time,' and 'when time shall be no more'.

§ 19. *The revolutions of the sun and moon the properest measures of time.* – The diurnal and annual revolutions of the sun, as having been, from the beginning of nature, constant, regular, and universally observable by all mankind, and supposed equal to one another, have been with reason made use of for the measure of duration. But the distinction of days and years, having depended on the motion of the sun, it has brought this mistake with it, that it has been thought that motion and duration were the measure one of another: for men, in the measuring of the length of time, having been accustomed to the ideas of minutes, hours, days, months, years, &c. which they found themselves, upon any mention of time or duration, presently to think on, all which portions of time were measured out by the motion of those heavenly bodies: they were apt to confound time and motion, or at least to think that they had a necessary connection one with another: whereas any constant periodical appearance or alteration of ideas in seemingly equidistant spaces of duration, if constantly and universally observable, would have as well distinguished the intervals of time, as those that have been made use of. For, supposing the sun, which some have taken to be a fire, had been lighted up at the same distance of time that it now every day comes about to the same meridian, and then gone out again about twelve hours after, and that, in the space of an annual revolution, it had sensibly increased in brightness and heat, and so decreased again; would not such regular appearances serve to measure out the distances of duration to all that could observe it, as well without, as with, motion? for if the appearances were constant, universally observable, and in equidistant periods, they would serve mankind for measure of time as well, were the motion away.

§ 20. *But not by their motion, but periodical appearances.* – For the freezing of water, or the blowing of a plant, returning at equidistant periods in all parts of the earth, would as well serve men to reckon their years by, as the motions of the sun. And, in effect, we see that some people in America counted their years by the coming of certain birds amongst them at their certain seasons, and leaving them at others. For a fit of an ague, the sense of hunger or thirst, a smell, or a taste, or any other idea, returning constantly at equidistant periods, and making itself universally be taken notice of, would not fail to measure out the course of succession, and distinguish the distances of time. Thus we see, that men, born blind, count time well enough by years, whose revolutions yet they cannot distinguish by motions that they perceive not. And I ask, whether a blind man, who distinguished his years either by heat of summer, or cold of winter; by the smell of any flower of the spring, or taste of any fruit of the autumn, would not have a better measure of time than the Romans had before the reformation of their Calendar by Julius Cæsar; or many other people, whose years, notwithstanding the motion of the sun, which they pretend to make use of, are very irregular? And it adds no small difficulty to chronology, that the exact regular lengths of the years that several nations counted by, are hard to be known, they differing very much one from another, and I think I may say all of them from the precise motion of the sun. And if the sun moved from the creation to the flood, constantly in the equator, and so equally dispersed its light and heat to all the habitable parts of the earth, in days all of the same length, without its annual variations to the tropics, as a late ingenious author supposes, I do not think it very easy to imagine, that (notwithstanding the motion of the sun) men should, in the antediluvian world, from the beginning, count by years, or measure their time by periods, that had no sensible marks very obvious to distinguish them by.

§ 21. *No two parts of duration can be certainly known to be equal.* – But perhaps it will be said, without a regular motion, such as of the sun, or some other, how could it ever be known that such periods were equal? To which I answer: The equality of any other returning appearances might be known by the same way that that of days was known, or presumed to be so at first; which was only by judging of them by the train of ideas which had passed in men's minds in the intervals, by which train of ideas discovering inequality in the natural days, but none in the artificial days, the artificial days, or νυχθήμερα, were guessed to be equal, which was sufficient to make them serve for a measure: though exacter search has since discovered inequality in the diurnal revolutions of the sun, and we know not whether the annual also be not unequal; these yet, by their presumed and apparent equality, serve as well to reckon time by (though not to measure the parts of duration exactly), as if they could be proved to be exactly equal. We must, therefore, carefully distinguish betwixt duration itself, and the measures we make use of to judge of its length. Duration in itself, is to be considered as going on in one constant, equal, uniform course: but none of the measures of it, which we make use of, can be known to do so; nor can we be assured, that their assigned parts or periods are equal in duration one to another; for two successive lengths of duration, however measured, can never be demonstrated to be equal.

The motion of the sun, which the world used so long, and so confidently, for an exact measure of duration, has, as I said, been found in its several parts unequal: and though men have of late made use of a pendulum, as a more steady and regular motion than that of the sun, or (to speak more truly) of the earth; yet if any one should be asked how he certainly knows that the two successive swings of a pendulum are equal, it would be very hard to satisfy himself, that they are infallibly so. Since we cannot be sure that the cause of that motion, which is unknown to us, shall always operate equally; and we are sure that the medium in which the pendulum moves, is not constantly the same: either of which varying, may alter the equality of such periods, and thereby destroy the certainty and exactness of the measure by motion, as well as any other periods of other appearances; the notion of duration still remaining clear, though our measures of it cannot any of them be demonstrated to be exact. Since, then, no two portions of succession can be brought together, it is impossible ever certainly to know their equality. All that we can do for a measure of time, is to take such as have continual successive appearances at seeming equidistant periods; of which seeming equality, we have no other measure, but such as the train of our own ideas have lodged in our memories, with the concurrence of other probable reasons, to persuade us of their equality.

§ 22. *Time not the measure of motion.* – One thing seems strange to me, that whilst all men manifestly measured time by the motion of the great and visible bodies of the world, time yet should be defined to be the measure of motion: whereas it is obvious to every one who reflects ever so little on it, that to measure motion, space is as necessary to be considered as time; and those who look a little farther, will find also the bulk of the thing moved, necessary to be taken into the computation by any one who will estimate or measure motion, so as to judge right of it. Nor, indeed, does motion any otherwise conduce to the measuring of duration, than as it constantly brings about the return of certain sensible ideas, in seeming equidistant periods. For if the motion of the sun were as unequal as of a ship driven by unsteady winds, sometimes very slow, and at others, irregularly very swift; or if being equally swift, it yet was not circular, and produced not the same appearances, it would not at all help us to measure time, any more than the seeming unequal motion of a comet does.

§ 23. *Minutes, hours, days and years, not necessary measures of duration.* – Minutes, hours, days, and years, are then no more necessary to time or duration, than inches, feet, yards, and miles, marked out in any matter, are to extension. For though we, in this part of the universe, by the constant use of them, as of periods set out by the revolutions of the sun, or as known parts of such periods, have fixed the ideas of such lengths of duration in our minds, which we apply to all parts of time, whose lengths we should consider; yet there may be other parts of the universe, where they no more use these measures of ours, than in Japan they do our inches, feet, or miles. But yet something analogous to them, there must be; for without some regular periodical returns, we could not measure ourselves, or signify to others the length of any duration, though, at the same time, the world were as full of motion as it is now, but no part of it disposed into regular and apparently equidistant revolutions. But the different measures that may be made use of for the account of time, do not at all alter the notion of duration, which is the thing to be measured, no more than the different standards of a foot and a cubit, alter the notion of extension to those who make use of those different measures.

§ 24. *One measure of time applicable to duration before time.* – The mind having once got such a measure of time, as the annual revolution of the sun, can apply that measure to duration, wherein that measure itself did not exist, and with which, in the reality of its being, it had nothing to do: for should one say, that Abraham was born in the 2712 year of the Julian period, it is altogether as intelligible, as reckoning from the beginning of the world, though there were so far back no motion of the sun, nor any motion at all. For though the Julian period be supposed to begin several hundred years before there were really either days, nights, or years, marked out by any revolutions of the sun, yet we reckon as right, and thereby measure durations as well, as if really at that time the sun had existed, and kept the same ordinary motion it doth now. The idea of duration equal to an annual revolution of the sun, is as easily applicable in our thoughts to duration, where no sun nor motion was, as the idea of a foot or yard taken from bodies here, can be applied in our thoughts to distances beyond the confines of the world, where are no bodies at all.

§ 25. For supposing it were 5639 miles, or millions of miles, from this place to the remotest body of the universe (for being finite, it must be at a certain distance), as we suppose it to be 5639 years from this time to the first existence of any body in the beginning of the world, we can, in our thoughts, apply this measure of a year to duration before the creation, or beyond the duration of bodies or motion, as we can this measure of a mile to space beyond the utmost bodies; and by the one, measure duration, where there was no motion; as well as by the other, measure space in our thoughts, where there is no body.

§ 31. And thus I think it is plain, that from those two fountains of all knowledge before-mentioned, viz., reflection and sensation, we get the ideas of duration, and the measures of it.

For, *First*, By observing what passes in our minds, how our ideas there in train constantly some vanish, and others begin to appear, we come by the idea of succession.

Secondly, By observing a distance in the parts of this succession, we get the idea of duration.

Thirdly, By sensation, observing certain appearances at certain regular and seeming equidistant periods, we get the ideas of certain lengths or measures of duration, as minutes, hours, days, years, &c.

Fourthly, By being able to repeat those measures of time, or ideas of stated length of duration in our minds, as often as we will, we can come to imagine duration, where nothing does really endure or exist; and thus we imagine to-morrow, next year, or seven years hence.

Fifthly, By being able to repeat ideas of any length of time, as of a minute, a year, or an age, as often as we will in our own thoughts, and adding them one to another, without ever coming to the end of such addition, any nearer than we can to the end of number, to which we can always add, we come by the idea of eternity, as the future eternal duration of our souls, as well as the eternity of that infinite being, which must necessarily have always existed.

Sixthly, By considering any part of infinite duration, as set out by periodical measures, we come by the idea of what we call time in general.

CHAPTER XV

§ 2. *Expansion not bounded by matter.* – The mind, having got the idea of the length of any part of expansion, let it be a span, or a pace, or what length you will, can, as has been said, repeat that idea; and so adding it to the former, enlarge its idea of length, and make it equal to two spans, or two paces, and so, as often as it will, till it equals the distance of any parts of the earth one from another, and increase thus, until it amounts to the distance of the sun, or remotest star. By such a progression as this, setting out from the place where it is, or any other place, it can proceed and pass beyond all those lengths, and find nothing to stop it going on, either in or without body. It is true, we can easily, in our thoughts, come to the end of solid extension; the extremity and bounds of all body, we have no difficulty to arrive at; but when the mind is there, it finds nothing to hinder its progress into this endless expansion; of that it can neither find nor conceive any end. Nor let any one say, that beyond the bounds of body there is nothing at all, unless he will confine God within the limits of matter. Solomon, whose understanding was filled and enlarged with wisdom, seems to have other thoughts, when he says, "Heaven, and the heaven of heavens, cannot contain thee;" and he, I think, very much magnifies to himself the capacity of his own understanding, who persuades himself, that he can extend his thoughts farther than God exists, or imagine any expansion where he is not.

§ 3. *Nor duration by motion.* – Just so is it in duration; the mind having got the idea of any length of duration, can double, multiply, and enlarge it, not only beyond its own, but beyond the existence of all corporeal beings, and all the measures of time taken from the great bodies of the world, and their motions. But yet every one easily admits, that though we make duration boundless, as certainly it is, we cannot yet extend it beyond all being. God, every one easily allows, fills eternity; and it is hard to find a reason, why any one should doubt that he likewise fills immensity. His infinite being is certainly as boundless one way as another; and methinks it ascribes a little too much to matter, to say, where there is no body, there is nothing.

§ 5. *Time to duration, is as place to expansion.* – Time in general is to duration, as place to expansion. They are so much of those boundless oceans of eternity and immensity, as is set out and distinguished from the rest, as it were, by land-marks; and so are made use of, to denote the position of finite real beings, in respect one to another, in those uniform infinite oceans of duration and space. These rightly considered, are only ideas of determinate distances from certain known points fixed in distinguishable sensible things, and supposed to keep the same distance one from another. From such points, fixed in sensible beings, we reckon, and from them we measure our portions of those infinite quantities; which so considered, are that which we call time and place. For duration and space being in themselves uniform and boundless, the order and position of things, without such known settled points, would be lost in them; and all things would lie jumbled in an incurable confusion.

§ 9. *All the parts of extension, are extension; and all the parts of duration; are duration.* – There is one thing more, wherein space and duration have a great conformity, and that is; though they are justly reckoned amongst our simple ideas; yet none of the distinct ideas we have of either, is without all manner of composition; it is the very nature of both of them to consist of parts: but their parts being all of the same kind, and without the mixture of any other idea, hinder them not from having a place amongst simple ideas. Could the mind, as in number, come to so small a part of extension or duration, as excluded divisibility, that would be, as it were, the indivisible unit, or idea; by repetition of which, it would make its more enlarged ideas of extension and duration. But since the mind is not able to frame an idea of any space without parts, instead thereof it makes use of the common measures, which, by familiar use, in each country, have imprinted themselves on the memory (as inches and feet; or cubits and parasangs; and so seconds, minutes, hours, days, and years in duration): the mind makes use, I say, of such ideas as these, as simple ones; and these are the component parts of larger ideas, which the mind, upon occasion, makes by the addition of such known lengths, which it is acquainted with. On the other side, the ordinary smallest measure we have of either, is looked on as an unit in number, when the mind, by division, would reduce them into less fractions. Though on both sides, both in addition and division, either space or duration, when the idea under consideration becomes very big, or very small, its precise bulk becomes very obscure and confused; and it is the number of its repeated additions, or divisions, that alone remains clear and distinct, as will easily appear to any one, who will let his thoughts loose in the vast expansion of space, or divisibility of matter. Every part of duration, is duration too; and every part of extension, is extension, both of them capable of addition or division *in infinitum*. But the least portions of either of them, whereof we have clear and distinct ideas, may perhaps be fittest to be considered by us, as the simple ideas of that kind, out of which our complex modes of space, extension, and duration, are made up, and into which they can again be distinctly resolved. Such a small part of duration, may be called a moment, and is the time of one idea in our minds, in the train of their ordinary succession there. The other, wanting a proper name, I know not whether I may be allowed to call a sensible point, meaning thereby the least particle of matter or space we can discern, which is ordinarily about a minute, and to the sharpest eyes, seldom less than thirty seconds of a circle, whereof the eye is the centre.

§ 10. *Their parts inseparable.* – Expansion and duration have this farther agreement, that though they are both considered by us as having parts, yet their parts are not separable one from another, no not even in thought; though the parts of bodies, from whence we take our measure of the one, and the parts of motion, or rather a succession of ideas in our minds, from whence we take the measure of the other, may be interrupted and separated; as the one is often by rest, and the other is by sleep, which we call rest too.

§ 11. *Duration is as a line, expansion as a solid.* – But yet there is this manifest difference between them, that the ideas of length, which we have of expansion, are turned every way, and so make figure, and breadth, and thickness; but duration is but as it were the length of one straight line, extended *in infinitum*, not capable of multiplicity, variation, or figure; but is one common measure of all existence whatsoever, wherein all things, whilst they exist, equally partake. For this present moment is common to all things that are now in being, and equally comprehends that part of their existence, as much as if they were all but one single being; and we may truly say, they all exist in the same moment of time.

§ 12. *Duration has never two parts together, expansion altogether.* – Duration, and time, which is a part of it, is the idea we have of perishing distance, of which no two parts exist together, but follow each other in succession; as expansion is the idea of lasting distance, all whose parts exist together, and are not capable of succession. And, therefore, though we cannot conceive any duration without succession, nor can put it together in our thoughts, that any being does now exist to-morrow, or possess at once more than the present moment of duration; yet we can conceive the eternal duration of the Almighty far different from that of man, or any other finite being. Because man comprehends not in his knowledge or power, all past and future things; his thoughts are but of yesterday, and he knows not what to-morrow will bring forth. What is once passed, he can never recall; and what is yet to come, he cannot make present. What I say of man, I say of all finite beings, who, though they may far exceed man in knowledge and power, yet are no more than the meanest creature, in comparison with God himself. Finite, of any magnitude, holds not any proportion to infinite. God's infinite duration being accompanied with infinite knowledge and infinite power, he sees all things past and to come; and they are no more distant from his knowledge, no farther removed from his sight, than the present; they all lie under the same view; and there is nothing which he cannot make exist each moment he pleases. For the existence of all things depending upon his good pleasure, all things exist every moment that he thinks fit to have them exist. To conclude: expansion and duration do mutually embrace and comprehend each other; every part of space being in every part of duration; and every part of duration in every part of expansion. Such a combination of two distinct ideas, is, I suppose, scarce to be found in all that great variety we do or can conceive, and may afford matter to farther speculation.

NOTE

* From *An Essay Concerning Human Understanding,* Book II, Ch. 14–15.

R. J. BOSCOVICH

ON THE RELATIVITY OF TEMPORAL INTERVALS*

It is conceivable that in some small grain of sand which we can hardly perceive there is hidden a whole world in which there is an immense number of living beings so small that they escape not only our perception, but also the perception of those tiny living beings which we hardly observe under a microscope. Is it not possible that there be a long series of such worlds, which, with respect to one another have the same relation as our single grain of sand has to the whole world? And I frequently meditate on this and think about those large cakes of cheese inside of which there frequently are very tiny insects. But there could be also others, much smaller, which escape the power of our microscopes; very little spheres, below all our sensory perceptions, are for them what the earth is for us; yet, there they have their own provinces and kingdoms. Their astronomers, looking through their telescopes, observe other little globes of the same material around them which they will regard as being very far away, their distance measuring an immense number of their feet and cubits. Their philosophers are proud of their knowledge of their little grains, although no one of them has ever penetrated to the crust nor can ever acquire any knowledge of it....

Whatever the truth of the matter, it really seems beyond doubt that what is for us a vanishing instant seems to be a very long time to those very tiny living beings. In this respect it occurs in those little animals something similar to what we observe in the pendulums of a shorter length whose number of oscillations is in a given time so much larger, the shorter is their length. So these very tiny living beings, embedded in a cake of cheese, if they pass through three or four generations in a day, regard this day as a century, and the two months which elapsed since the formation of the cheese would seem like thousand years from the beginning of their world. The few days duration of the same family would be for them a mobility of many centuries, and a lasting fame would be that which endured for two or three months; an affair of the greatest importance and of momentous concern would be the preservation of the rule of the living beings which occupy a part of the surface of this very little globe. What is their most ancient history? What is the glory of the war made by their grandfathers and great-grandfathers? What hope for enduring glory lasting for many centuries?...

NOTE

* From his commentary to *Philosophiae recentioribus versibus tradita a Benedicto Stay libri decem*, Vol. III, Romae 1755, pp. 421f; transl. by the editor and Walter Emge.

A. SCHOPENHAUER

ON THE NECESSARY ATTRIBUTES OF TIME AND SPACE*

PRAEDICABILIA 'A PRIORI'

Of Time	Of Space
(1) There is only *one* Time, and all different times are parts of it.	(1) There is only *one* Space, and all different spaces are parts of it.
(2) Different times are not simultaneous but successive.	(2) Different spaces are not successive but simultaneous.
(3) Time cannot be thought away, but everything can be thought away from it.	(3) Space cannot be thought away, but everything can be thought away from it.
(4) Time has three divisions, the past, the present, and the future, which constitute two directions and a centre of indifference.	(4) Space has three dimensions – height, breadth, and length.
(5) Time is infinitely divisible.	(5) Space is infinitely divisible.
(6) Time is homogeneous and a *Continuum*, *i.e.*, no one of its parts is different from the rest, nor separated from it by anything that is not time.	(6) Space is homogeneous and a *Continuum*, *i.e.*, no one of its parts is different from the rest, nor separated from it by anything that is not space.
(7) Time has no beginning and no end, but all beginning and end is in it.	(7) Space has no limits, but all limits are in it.

PRAEDICABILIA 'A PRIORI'

Of Time	Of Space
(8) By reason of time we count.	(8) By reason of space we measure.
(9) Rhythm is only in time.	(9) Symmetry is only in space.
(10) We know the laws of time *a priori*.	(10) We know the laws of space *a priori*.
(11) Time can be perceived *a priori*, although only in the form of a line.	(11) Space is immediately perceptible *a priori*.
(12) Time has no permanence, but passes away as soon as it is there.	(12) Space can never pass away, but endures through all time.
(13) Time never rests.	(13) Space is immovable.
(14) Everything that exists in time has duration.	(14) Everything that exists in space has a position.
(15) Time has no duration, but all duration is in it, and is the persistence of what is permanent in contrast with its restless course.	(15) Space has no motion, but all motion is in it, and it is the change of position of what is moved, in contrast with its unbroken rest.
(16) All motion is only possible in time.	(16) All motion is only possible in space.
(17) Velocity is, in equal spaces, in inverse proportion to the time.	(17) Velocity is, in equal times, in direct proportion to the space.

NECESSARY ATTRIBUTES OF TIME AND SPACE

PRAEDICABILIA 'A PRIORI'

Of time	Of Space
(18) Time is not measurable directly through itself, but only indirectly through motion, which is in space and time together: thus the motion of the sun and of the clock measure time.	(18) Space is measurable directly through itself, and indirectly through motion, which is in time and space together: hence, for example, an hour's journey, and the distance of the fixed stars expressed as the travelling of light for so many years.
(19) Time is omnipresent. Every part of time is everywhere, *i.e.*, in all space, at once.	(19) Space is eternal. Every part of it exists always.
(20) In time taken by itself everything would be in succession.	(20) In space taken by itself everything would be simultaneous.
(21) Time makes the change of accidents possible.	(21) Space makes the permanence of substance possible.
(22) Every part of time contains all parts of matter.	(22) No part of space contains the same matter as another.
(23) Time is the *principium individuationis*.	(23) Space is the *principium individuationis*.
(24) The now has no duration.	(24) The point has no extension.
(25) Time in itself is empty and without properties.	(25) Space in itself is empty and without properties.

PRAEDECABILIA 'A PRIORI'

Of Time	Of Space
(26) Every moment is conditioned by the preceding moment, and is only because the latter has ceased to be. (Principle of sufficient reason of existence in time. – See my essay on the principle of sufficient reason.)	(26) By the position of every limit in space with reference to any other limit, its position with reference to every possible limit is precisely determined. (Principle of sufficient reason of existence in space.)
(27) Time makes arithmetic possible.	(27) Space makes geometry possible.
(28) The simple element in arithmetic is unity.	(28) The simple element in geometry is the point.

NOTE

* From *The World as Will and Idea*, Part II (transl. by R. B. Haldane and J. Kemp), Routledge and Kegan Paul, London, pp. 221–223.

J. C. MAXWELL

ABSOLUTE TIME AND THE ORDER OF NATURE*

17. ON THE IDEA OF TIME

The idea of Time in its most primitive form is probably the recognition of an order of sequence in our states of consciousness. If my memory were perfect, I might be able to refer every event whithin my own experience to its proper place in a chronological series. But it would be difficult, if not impossible, for me to compare the interval between one pair of events and that between another pair – to ascertain, for instance, whether the time during which I can work without feeling tired is greater or less now than when I first began to study. By our intercourse with other persons, and by our experience of natural processes which go on in a uniform or a rhythmical manner, we come to recognise the possibility of arranging a system of chronology in which all events whatever, whether relating to ourselves or to others, must find their places. Of any two events, say the actual disturbance at the star in Corona Borealis, which caused the luminous effects examined spectroscopically by Mr Huggins on the 16th May, 1866, and the mental suggestion which first led Professor Adams or M. Leverrier to begin the researches which led to the discovery, by Dr Galle, on the 23rd September, 1846, of the planet Neptune, the first named must have occurred either before or after the other, or else at the same time.

Absolute, true, and mathematical Time is conceived by Newton as flowing at a constant rate, unaffected by the speed or slowness of the motions of material things. It is also called Duration. Relative, apparent, and common time is duration as estimated by the motion of bodies, as by days, months, and years. These measures of time may be regarded as provisional, for the progress of astronomy has taught us to measure the inequality in the lengths of days, months, and years, and thereby to reduce the apparent time to a more uniform scale, called Mean Solar Time.

19. STATEMENT OF THE GENERAL MAXIM OF PHYSICAL SCIENCE

There is a maxim which is often quoted, that "The same causes will always produce the same effects."

To make this maxim intelligible we must define what we mean by the same causes and the same effects, since it is manifest that no event ever happens more than once, so that the causes and effects cannot be the same in *all* respects. What is really meant is that if the causes differ only as regards the absolute time or the absolute place at which the event occurs, so likewise will the effects.

The following statement, which is equivalent to the above maxim, appears to be more definite, more explicitly connected with the ideas of space and time, and more capable of application to particular cases:

The difference between one event and another does not depend on the mere difference of the times or the places at which they occur, but only on differences in the nature, configuration, or motion of the bodies concerned.

It follows from this, that if an event has occurred at a given time and place it is possible for an event exactly similar to occur at any other time and place.

NOTE

* From *Matter and Motion*, §§ 17, 19.

C. NEUMANN

ON THE DEFINITION OF THE EQUALITY OF SUCCESSIVE INTERVALS OF TIME*

So far we have considered *the first part* of Galilei's law. This law furthermore states that a material point left to itself moves not only along a straight line, but also that it moves with a *constant velocity*, that is, in equal intervals of time it covers the same distances. To understand these words, we must first know what 'equal intervals of time' means. In other words, we must know how to evaluate and measure the temporal segments.

We are used to regard the rotation time of our terrestrial globe as our unit of time; we hardly know any other way how to measure time except by choosing as its unit the interval separating two successive culminations of a star. This time unit, the so-called stellar day, we subdivide into hours, the hour into minutes, the minute into seconds. In such a way we regulate the astronomical clocks; and on these clocks our ordinary clocks depend.

By the successive rotations of the earth there arises in forward moving time a scale whose larger segments are the stellar days while its smaller segments are hours, minutes and seconds. Ought we now really to regard such scale as *completely exact*, ought we to view its two corresponding segments - for instance, two stellar days - as *exactly* equal? Should we really view this time-scale, which is imposed on us by our planet, as valid when we consider the whole universe! Do not other celestial bodies lay a similar claim? Or should we assume that the rotational motions of all the celestial bodies are harmonized so that they are giving us the concordant temporal scales in such a way that the equal segments on one scale correspond to the equal segments on another?

There is no doubt now how these questions should be answered. And the last trace of hesitation disappears when we recall that several astronomers arrived at the conclusion that the rotation of the earth is gradually becoming slower and slower. They have found that in every millenium the *last* stellar day is by one thousandth second longer than the *first* one.

It is, of course, doubtful that the computations by which the astronomers arrived at those results are sufficiently certain and whether the empirical data underlying these computations have the desirable reliability. But if we remember that the movements of tides as well as the variations of temperature must influence the rotation of the earth, we cannot doubt for a moment that theoretical astronomy once will obtain such results with a complete certainty and that it will be able to determine exactly how much the rotation time of the earth increased or decreased during a thousand years.

It would be, therefore, absurd to claim that two given intervals of time are equally long as they coincide with the same number of stellar days or stellar seconds. In this way we face with respect to the law of Galilei a peculiar difficulty.

One material point left to itself covers in two equal intervals of time equal distances. And it is impossible to associate with these words a definite meaning as long as we do not know what we mean by the words 'equal intervals of time".

But this difficulty is only apparent. For when we get rid of that stumbling stone, if we eliminate that irrational concept of 'equal intervals of time', there still remains the definite residue of the law which is as follows:

> *Two* material particles when they are left to themselves, move in such a way that the equal trajectories of one correspond always to equal trajectories of another.

In this form and under these limitations the statement above represents the *third* principle of Galileo-Newton's theory, the principle which is as clear as the two already mentioned... In agreement with the spirit of Galilei and Newton, in agreement with the whole development of the theory which they had founded, we can now - when the third principle is stated in the way indicated above – define *equal intervals of time* as those during which a material point, left to itself, moves through equal distances. From this point we can elucidate the proper meaning of what astronomers affirm that after every thousand years the last days will be somehow longer than the first one: its meaning consists in that a material point, left to itself, would on that last day move over a larger distance than on the first day.

NOTE

* From *Über die Principien der Galilei-Newtonschen Lehre*, Leipzig, 1870, pp. 16–19; transl. by the editor.

ON ZENO'S PARADOXES*

325. The word *continuity* has borne among philosophers, especially since the time of Hegel, a meaning totally unlike that given to it by Cantor. Thus Hegel says[1]:

> Quantity, as we saw, has two sources: the exclusive unit, and the identification or equalization of these units. When we look, therefore, at its immediate relation to self, or at the characteristic of selfsameness made explicit by abstraction, quantity is *Continuous* magnitude; but when we look at the other characteristic, the One implied in it, it is *Discrete* magnitude.

When we remember that quantity and magnitude, in Hegel, both mean 'cardinal number', we may conjecture that this assertion amounts to the following:

> Many terms, considered as having a cardinal number, must all be members of one class; in so far as they are each merely an instance of the class-concept, they are indistinguishable one from another, and in this aspect the whole which they compose is called *continuous*; but in order to their maniness, they must be *different* instances of the class-concept, and in this aspect the whole which they compose is called *discrete*.

Now I am far from denying – indeed I strongly hold – that this opposition of identity and diversity in a collection constitutes a fundamental problem of Logic – perhaps even *the* fundamental problem of philosophy. And being fundamental, it is certainly relevant to the study of the mathematical continuum as to everything else. But beyond this general connection, it has no special relation to the mathematical meaning of continuity, as may be seen at once from the fact that it has no reference whatever to order. In this chapter, it is the mathematical meaning that is to be discussed. I have quoted the philosophic meaning only in order to state definitely that this is *not* here in question; and since disputes about words are futile, I must ask philosophers to divest themselves, for the time, of their habitual associations with the word, and allow it no signification but that obtained from Cantor's definition.

326. In confining ourselves to the arithmetical continuum, we conflict

in another way with common preconceptions. Of the arithmetical continuum, M. Poincaré justly remarks[2]:

> The continuum thus conceived is nothing but a collection of individuals arranged in a certain order, infinite in number, it is true, but external to each other. This is not the ordinary conception, in which there is supposed to be, between the elements of the continuum, a sort of intimate bond which makes a whole of them, in which the point is not prior to the line, but the line to the point. Of the famous formula, the continuum is unity in multiplicity, the multiplicity alone subsists, the unity has disappeared.

It has always been held to be an open question whether the continuum is composed of elements; and even when it has been allowed to contain elements, it has been often alleged to be not *composed* of these. This latter view was maintained even by so stout a supporter of elements in everything as Leibniz[3]. But all these views are only possible in regard to such continua as those of space and time. The arithmetical continuum is an object selected by definition, consisting of elements in virtue of the definition, and known to be embodied in at least one instance, namely the segments of the rational numbers. The chief reason for the elaborate and paradoxical theories of space and time and their continuity, which have been constructed by philosophers, has been the supposed contradictions in a continuum composed of elements. The thesis of the present chapter is, that Cantor's continuum is free from contradictions. This thesis, as is evident, must be firmly established, before we can allow the possibility that spatio-temporal continuity may be of Cantor's kind. In this argument, I shall assume as proved the thesis of the preceding chapter, that the continuity to be discussed does not involve the admission of actual infinitesimals.

327. In this capricious world, nothing is more capricious than posthumous fame. One of the most notable victims of posterity's lack of judgment is the Eleatic Zeno. Having invented four arguments, all immeasurably subtle and profound, the grossness of subsequent philosophers pronounced him to be a mere ingenious juggler, and his arguments to be one and all sophisms. After two thousand years of continual refutation, these sophisms were reinstated, and made the foundation of a mathematical renaissance, by a German professor, who probably never dreamed of any connection between himself and Zeno. Weierstrass, by strictly banishing all infinitesimals, has at last shown that we live in an unchang-

ing world, and that the arrow, at every moment of its flight, is truly at rest. The only point where Zeno probably erred was in inferring (if he did infer) that, because there is no change, therefore the world must be in the same state at one time as at another. This consequence by no means follows, and in this point the German professor is more constructive than the ingenious Greek. Weierstrass, being able to embody his opinions in mathematics, where familiarity with truth eliminates the vulgar prejudices of common sense, has been able to give to his propositions the respectable air of platitudes; and if the result is less delightful to the lover of reason than Zeno's bold defiance, it is at any rate more calculated to appease the mass of academic mankind.

Zeno's arguments are specially concerned with motion, and are not therefore, as they stand, relevant to our present purpose. But it is instructive to translate them, so far as possible, into arithmetical language [4].

328. The first argument, that of dichotomy, asserts: "There is no motion, for what moves must reach the middle of its course before it reaches the end." That is to say, whatever motion we assume to have taken place, this presupposes another motion, and this in turn another, and so on *ad infinitum*. Hence there is an endless regress in the mere idea of any assigned motion. This argument can be put into an arithmetical form, but it appears then far less plausible. Consider a variable x which is capable of all real (or rational) values between two assigned limits, say 0 and 1. The class of its values is an infinite whole, whose parts are logically prior to it: for it has parts, and it cannot subsist if any of the parts are lacking. Thus the numbers from 0 to 1 presuppose those from 0 to 1/2, these presuppose the numbers from 0 to 1/4, and so on. Hence, it would seem, there is an infinite regress in the notion of any infinite whole; but without such infinite wholes, real numbers cannot be defined, and arithmetical continuity, which applies to an infinite series, breaks down.

This argument may be met in two ways, either of which, at first sight, might seem sufficient, but both of which are really necessary. First, we may distinguish two kinds of infinite regresses, of which one is harmless. Secondly, we may distinguish two kinds of whole, the collective and the distributive, and assert that, in the latter kind, parts of equal complexity with the whole are not logically prior to it. These two points must be separately explained.

329. An infinite regress may be of two kinds. In the objectionable kind, two or more propositions join to constitute the *meaning* of some proposition; of these constituents, there is one at least whose meaning is similarly compounded; and so on *ad infinitum*. This form of regress commonly results from circular definitions. Such definitions may be expanded in a manner analogous to that in which continued fractions are developed from quadratic equations. But at every stage the term to be defined will reappear, and no definition will result. Take for example the following: "Two people are said to have the *same* idea when they have ideas which are similar; and ideas are similar when they contain an identical part." If an idea may have a part which is not an idea, such a definition is not logically objectionable; but if part of an idea is an idea, then, in the second place where identity of ideas occurs, the definition must be substituted; and so on. Thus wherever the *meaning* of a proposition is in question, an infinite regress is objectionable, since we never reach a proposition which has a definite meaning. But many infinite regresses are not of this form. If A be a proposition whose meaning is perfectly definite, and A implies B, B implies C, and so on, we have an infinite regress of a quite unobjectionable kind. This depends upon the fact that implication is a synthetic relation, and that, although, if A be an aggregate of propositions, A implies any proposition which is part of A, it by no means follows that any proposition which A implies is part of A. Thus there is no logical necessity, as there was in the previous case, to complete the infinite regress before A acquires a meaning. If, then, it can be shown that the implication of the parts in the whole, when the whole is an infinite class of numbers, is of this latter kind, the regress suggested by Zeno's argument of dichotomy will have lost its sting.

330. In order to show that this is the case, we must distinguish wholes which are defined extensionally, *i.e.* by enumerating their terms, from such as are defined intensionally, *i.e.* as the class of terms having some given relation to some given term, or, more simply, as a class of terms. (For a class of terms, when it forms a whole, is merely all terms having the class-relation to a class-concept.) Now an extensional whole – at least so far as human powers extend – is necessarily finite: we cannot enumerate more than a finite number of parts belonging to a whole, and if the number of parts be infinite, this must be known otherwise than by enumeration. But

this is precisely what a class-concept effects: a whole whose parts are the terms of a class is completely defined when the class-concept is specified; and any definite individual either belongs, or does not belong, to the class in question. An individual of the class is part of the whole extension of the class, and is logically prior to this extension taken collectively; but the extension itself is definable without any reference to any specified individual, and subsists as a genuine entity even when the class contains no terms. And to say, of such a class, that it is infinite, is to say that, though it has terms, the number of these terms is not any finite number – a proposition which, again, may be established without the impossible process of enumerating *all* finite numbers. And this is precisely the case of the real numbers between 0 and 1. They form a definite class whose meaning is known as soon as we know what is meant by *real number*, 0, 1, and *between*. The particular members of the class, and the smaller classes contained in it, are not logically prior to the class. Thus the infinite regress consists merely in the fact that every segment of real or rational numbers has parts which are again segments; but these parts are not logically prior to it, and the infinite regress is perfectly harmless. Thus the solution of the difficulty lies in the theory of denoting and the intensional definition of a class. With this an answer is made to Zeno's first argument as it appears in Arithmetic.

331. The second of Zeno's arguments is the most famous: it is the one which concerns Achilles and the tortoise. "The slower", it says, "will never be overtaken by the swifter, for the pursuer must first reach the point whence the fugitive is departed, so that the slower must always necessarily remain ahead." When this argument is translated into arithmetical language, it is seen to be concerned with the one-one correlation of two infinite classes. If Achilles were to overtake the tortoise, then the course of the tortoise would be part of that of Achilles; but, since each is at each moment at some point of his course, simultaneity establishes a one-one correlation between the positions of Achilles and those of the tortoise. Now it follows from this that the tortoise, in any given time, visits just as many places as Achilles does; hence – so it is hoped we shall conclude – it is impossible that the tortoise's path should be part of that of Achilles. This point is purely ordinal, and may be illustrated by Arithmetic. Consider, for example, $1+2x$ and $2+x$, and let x lie between 0

and 1, both inclusive. For each value of $1+2x$ there is one and only one value of $2+x$, and *vice versa*. Hence as x grows from 0 to 1, the number of values assumed by $1+2x$ will be the same as the number assumed by $2+x$. But $1+2x$ started from 1 and ends at 3, while $2+x$ started from 2 and ends at 3. Thus there should be half as many values of $2+x$ as of $1+2x$. This very serious difficulty has been resolved, as we have seen, by Cantor; but as it belongs rather to the philosophy of the infinite than to that of the continuum, I leave its further discussion to the next chapter.

332. The third argument is concerned with the arrow. "If everything is in rest or in motion in a space equal to itself, and if what moves is always in the instant, the arrow in its flight is immovable." This has usually been thought so monstrous a paradox as scarcely to deserve serious discussion. To my mind, I must confess, it seems a very plain statement of a very elementary fact, and its neglect has, I think, caused the quagmire in which the philosophy of change has long been immersed. In Part VII[5], I shall set forth a theory of change which may be called *static*, since it allows the justice of Zeno's remark. For the present, I wish to divest the remark of all reference to change. We shall then find that it is a very important and very widely applicable platitude, namely: "Every possible value of a variable is a constant." If x be a variable which can take all values from 0 to 1, all the values it can take are definite numbers, such as 1/2 or 1/3, which are all absolute constants. And here a few words may be inserted concerning variables. A variable is a fundamental concept of logic, as of daily life. Though it is always connected with some class, it is not the class, nor a particular member of the class, nor yet the whole class, but *any* member of the class. On the other hand, it is not the *concept* 'any member of the class', but it is that (or those) which this concept denotes. On the logical difficulties of this conception, I need not now enlarge. The usual x in Algebra, for example, does not stand for a particular number, nor for all numbers, nor yet for the class *number*. This may be easily seen by considering some identity, say

$$(x+1)^2 = x^2 + 2x + 1.$$

This certainly does not mean what it would become if, say, 391 were substituted for x, though it implies that the result of such a substitution would be a true proposition. Nor does it mean what results from substituting for x the class-concept *number*, for we cannot add 1 to this concept. For the same reason, x does not denote the concept *any number*: to this, too, 1 cannot be added. It denotes the disjunction formed by the various numbers; or at least this view may be taken as roughly correct. The values of x are then the terms of the disjunction; and each of these is a constant. This simple logical fact seems to constitute the essence of Zeno's contention that the arrow is always at rest.

333. But Zeno's argument contains an element which is specially applicable to continua. In the case of motion, it denies that there is such a thing as a *state* of motion. In the general case of a continuous variable, it may be taken as denying actual infinitesimals. For infinitesimals are an attempt to extend to the *values* of a variable the variability which belongs to it alone. When once it is firmly realized that all the values of a variable are constants, it becomes easy to see, by taking *any* two such values, that their difference is always finite, and hence that there are no infinitesimal differences. If x be a variable which may take all real values from 0 to 1, then, taking any two of these values, we see that their difference is finite, although x is a continuous variable. It is true the difference might have been less than the one we chose; but if it had been, it would still have been finite. The lower limit to possible differences is zero, but all possible differences are finite; and in this there is no shadow of contradiction. This static theory of the variable is due to the mathematicians, and its absence in Zeno's day led him to suppose that continuous change was impossible without a state of change, which involves infinitesimals and the contradiction of a body's being where it is not.

334. The last of Zeno's arguments is that of the measure. This is closely analogous to one which I employed against those who regard dx and dy as distances of consecutive terms. It is only applicable, as M. Noël points out (*loc. cit.* p. 116), against those who hold to indivisibles among stretches, the previous arguments being held to have sufficiently refuted the partisans of infinite divisibility. We are now to suppose a set of discrete moments and discrete places, motion consisting in the fact that at one moment a body is in one of these discrete places, in another at another.

Imagine three parallel lines composed of the points $a, b, c, d; a', b', c', d';$

a'', b'', c'', d'' respectively. Suppose the second line, in one instant, to move

```
a   b   c   d           a   b   c   d
.   .   .   .           .   .   .   .
a'  b'  c'  d'          a'  b'  c'  d'
.   .   .   .           .   .   .   .
a'' b'' c'' d''         a'' b'' c'' d''
.   .   .   .           .   .   .   .
```

all its points to the left by one place, while the third moves them all one place to the right. Then although the instant is indivisible, c', which was over c'', and is now over a'', must have passed b'' during the instant; hence the instant is divisible, *contra hyp*. This argument is virtually that by which I proved, in the preceding chapter, that, if there are consecutive terms, then $dy/dx = \pm 1$ always; or rather, it is this argument together with an instance in which $dy/dx = 2$. It may be put thus: Let y, z be two functions of x, and let $dy/dx = 1$, $dz/dx = -1$. Then $(d/dx)(y-z) = 2$, which contradicts the principle that the value of every derivative must be ± 1. To the argument in Zeno's form, M. Evellin, who is an advocate of indivisible stretches, replies that a'' and b' do not cross each other at all[6]. For if instants are indivisible – and this is the hypothesis – all we can say is, that at one instant a' is over a'', in the next, c' is over a''. Nothing has happened between the instants, and to suppose that a'' and b' have crossed is to beg the question by a covert appeal to the continuity of motion. This reply is valid, I think, in the case of motion; both time and space may, without positive contradiction, be held to be discrete, by adhering strictly to distances in addition to stretches. Geometry, Kinematics, and Dynamics become false; but there is no very good reason to think them true. In the case of Arithmetic, the matter is otherwise, since no empirical question of existence is involved. And in this case, as we see from the above argument concerning derivatives, Zeno's argument is absolutely sound. Numbers are entities whose nature can be established beyond question; and among numbers, the various forms of continuity which occur cannot be denied without positive contradiction. For this reason the problem of continuity is better discussed in connection with numbers than in connection with space, time, or motion.

335. We have now seen that Zeno's arguments, though they prove a very

great deal, do not prove that the continuum, as we have become acquainted with it, contains any contradictions whatever. Since his day the attacks on the continuum have not, so far as I know, been conducted with any new or more powerful weapons.

NOTES

* From *The Principles of Mathematics*, Allen & Unwin, London, 1903, pp. 346–353.
¹ *Smaller Logic*, § 100, Wallace's translation, p. 188.
² *Revue de Métaphysique et de Morale* **1**, 26.
³ See *A Critical Exposition of The Philosophy of Leibniz*, by the present author, Chap. IX (London, 1900).
⁴ Not being a Greek scholar, I pretend to no first-hand authority as to what Zeno really did say or mean. The form of his four arguments which I shall employ is derived from the interesting article of M. Noël, 'Le mouvement et les arguments de Zénon d'Elée', *Revue de Métaphysique et de Morale* **1**, 107–125. These arguments are in any case well worthy of consideration, and as they are, to me, merely a text for discussion, their historical correctness is of little importance.
⁵ This is included in this anthology, pp. 251–254.
⁶ *Revue de Métaphysique et de Morale*, 386.

H. BERGSON

ON ZENO'S PARADOXES*

We discover here, at its outset, the illusion which accompanies and masks the perception of real movement. Movement visibly consists in passing from one point to another, and consequently in traversing space. Now the space which is traversed is infinitely divisible; and as the movement is, so to speak, applied to the line along which it passes, it appears to be one with this line and, like it, divisible. Has not the movement itself drawn the line? Has it not traversed in turn the successive and juxtaposed points of that line? Yes, no doubt, but these points have no reality except in a line drawn, that is to say motionless; and by the very fact that you represent the movement to yourself successively in these different points, you necessarily arrest it in each of them; your successive positions are, at bottom, only so many imaginary halts. You substitute the path for the journey, and because the journey is subtended by the path you think that the two coincide. But how should a *progress* coincide with a *thing*, a movement with an immobility?

What facilitates this illusion is that we distinguish moments in the course of duration, like halts in the passage of the moving body. Even if we grant that the movement from one point to another forms an undivided whole, this movement nevertheless takes a certain time; so that if we carve out of this duration an indivisible instant, it seems that the moving body must occupy, at that precise moment, a certain position, which thus stands out from the whole. The indivisibility of motion implies, then, the impossibility of real instants; and indeed, a very brief analysis of the idea of duration will show us both why we attribute instants to duration and why it cannot have any. Suppose a simple movement like that of my hand when it goes from A to B. This passage is given to my consciousness as an undivided whole. No doubt it endures; but this duration, which in fact coincides with the aspect which the movement has inwardly for my consciousness, is, like it, whole and undivided, Now, while it presents itself, *qua* movement, as a simple fact, it describes in space a trajectory which I may consider, for purposes of simplification, as a geometrical line; and the extremities of this line, considered as abstract limits, are no longer lines, but indivisible points. Now, if the line, which the moving body has described, measures for me the duration of its movement, must not the point, where the line ends, symbolize for me a terminus of this duration? And if this point is an indivisible of length, how shall we avoid terminating the duration of the movement by an indivisible of duration? If the total line represents the total duration, the parts of the line must, it seems, correspond to parts of the duration, and the points of the line to moments of time. The indivisibles of duration, or moments of time, are born, then, of the need of symmetry; we come to them naturally as soon as we demand from space an integral presentment of duration. – But herein, precisely, lies the error. While the line AB symbolizes the duration already lapsed of the movement from A to B already accomplished, it cannot, motionless, represent the movement in its accomplishment nor duration in its flow. And from the fact that this line is divisible into parts and that it ends in points, we cannot conclude either that the corresponding duration is composed of separate parts or that it is limited by instants.

The arguments of Zeno of Elea have no other origin than this illusion. *Zeno transfers to the moving body the properties of its trajectory: hence all the difficulties and contradictions*
They all consist in making time and movement coincide with the line which underlies them, in attributing to them the same subdivisions as to the line, in short in treating them like that line. In this confusion Zeno was encouraged by common sense, which usually carries over to the movement the properties of its trajectory, and also by language, which always translates movement and duration in terms of space. But common sense and language have a right to do so and are even bound to do so, for, since they always regard the *becoming* as a *thing* to be made use of, they have no more concern with the interior organization of movement than a workman has with the molecular structure of his tools. In holding movement to be divisible, as its trajectory is, common sense merely expresses the two facts which alone are of importance in practical life: first, that every movement describes a space; second, that at every point of this space the moving body *might* stop. But the philosopher who reasons upon the inner nature of movement is bound to restore to it the mobility which is its essence, and this is what Zeno omits to do. By the

first argument (the Dichotomy) he supposes the moving body to be at rest, and then considers nothing but the stages, infinite in number, that are along the line to be traversed: we cannot imagine, he says, how the body could ever get through the interval between them. But in this way he merely proves that it is impossible to construct, *à priori*, movement with immobilities, a thing no man ever doubted. The sole question is whether, movement being posited as a fact, there is a sort of retrospective absurdity in assuming that an infinite number of points has been passed through. But at this we need not wonder, since movement is an undivided fact, or a series of undivided facts, whereas the trajectory is infinitely divisible. In the second argument (the Achilles) movement is indeed given, it is even attributed to two moving bodies, but always by the same error, there is an assumption that their movement coincides with their path, and that we may divide it, like the path itself, in any way we please. Then instead, of recognizing that the tortoise has the pace of a tortoise and Achilles the pace of Achilles, so that after a certain number of these indivisible acts or bounds Achilles will have outrun the tortoise, the contention is that we may disarticulate as we will the movement of Achilles and, as we will also, the movement of the tortoise: thus reconstructing both in an arbitrary way, according to a law of our own which may be incompatible with the real conditions of mobility. The same fallacy appears, yet more evident, in the third argument (the Arrow) which consists in the conclusion that, because it is possible to distinguish points on the path of a moving body, we have the right to distinguish indivisible moments in the duration of its movement. But the most instructive of Zeno's arguments is perhaps the fourth (the Stadium) which has, we believe, been unjustly disdained, and of which the absurdity is more manifest only because the postulate masked in the three others is here frankly displayed.[1] Without entering on a discussion which would here be out of place, we will content ourselves with observing that motion, as given to spontaneous perception, is a fact which is quite clear, and that the difficulties and contradictions pointed out by the Eleatic school concern far less the living movement itself than a dead and artificial reorganization of movement by the mind....

Philosophy perceived this as soon as it opened its eyes. The arguments of Zeno of Elea, although formulated with a very different intention, have no other meaning.

Take the flying arrow. At every moment, says Zeno, it is motionless, for it cannot have time to move, that is, to occupy at least two successive positions, unless at least two moments are allowed it. At a given moment, therefore, it is at rest at a given point. Motionless in each point of its course, it is motionless during all the time that it is moving.

Yes, if we suppose that arrow can ever *be* in a point of its course. Yes again, if the arrow, which is moving, ever coincides with a position, which is motionless. But the arrow never *is* in any point of its course. The most we can say is that it might be there, in this sense, that it passes there and might stop there. It is true that if it did stop there, it would be at rest there, and at this point it is no longer movement that we should have to do with. The truth is that if the arrow leaves the point A to fall down at the point B, its movement AB is as simple, as indecomposable, in so far as it is movement, as the tension of the bow that shoots it. As the shrapnel, bursting before it falls to the ground, covers the explosive zone with an indivisible danger, so the arrow which goes from A to B displays with a single stroke, although over a certain extent of duration, its indivisible mobility. Suppose an elastic stretched from A to B, could you divide its extension? The course of the arrow is this very extension; it is equally simple and equally undivided. It is a single and unique bound. You fix a point C in the interval passed, and say that at a certain moment the arrow was in C. If it had been there, it would have been stopped there, and you would no longer have had a flight from A to B, but *two* flights, one from A to C and the other from C to B, with an interval of rest. A single movement is entirely, by the hypothesis, a movement between two stops; if there are intermediate stops, it is no longer a single movement. At bottom, the illusion arises from this, that the movement, *once effected*, has laid along its course a motionless trajectory on which we can count as many immobilities as we will. From this we conclude that the movement, *whilst being effected*, lays at each instant beneath it a position with which it coincides. We do not see that the trajectory is created in one stroke, although a certain time is required for it; and that though we can divide at will the trajectory once created, we cannot divide its creation, which is an act in progress and not a thing. To suppose that the moving body *is* at a point of its course is to cut the course in two by a snip of the scissors at this point, and to substitute two trajectories for the single trajectory which we were first considering. It is to distinguish two successive acts where, by the hypothesis, there is only one. In short, it is to attribute to the course itself of the arrow everything that can be said of the interval that the arrow has traversed, that is to say, to admit *a priori* the absurdity that movement coincides with immobility.

We shall not dwell here on the three other arguments of Zeno. We have examined them elsewhere. It is enough to point out that they all consist in applying the movement to the line traversed, and supposing that what is true of the line is true of the movement. The line, for example, may be divided into as many parts as we wish, of any length that we wish, and it is always the same line. From this we conclude that we have the right to suppose the movement articulated as we wish, and that it is always the same movement. We thus obtain a series of absurdities that all express the same fundamental absurdity. But the possibility of applying the movement *to* the line traversed exists only for an observer who, keeping outside the movement and seeing at every instant the possibility of a stop, tries to reconstruct the real movement with these possible immobilities. The absurdity vanishes as soon as we adopt by thought the continuity of the real movement, a continuity of which every one of us is conscious whenever he lifts an arm or advances a step. We feel then indeed that the line passed over between two stops is described with a single indivisible stroke, and that we seek in vain to practice on the movement, which traces the line, divisions corresponding, each to each, with the divisions arbitrarily chosen of the line once it has been traced. The line traversed by the moving body lends itself to any kind of division, because it has no internal organization. But all movement is articulated inwardly. It is either an indivisible bound (which may occupy, nevertheless, a very long duration) or a series of indivisible bounds. Take the articulations of this movement into account, or give up speculating on its nature.

When Achilles pursues the tortoise, each of his steps must be treated as indivisible, and so must each step of the tortoise. After a certain number of steps, Achilles will have overtaken the tortoise. There is nothing more simple. If you insist on dividing the two motions further, distinguish both on the one side and on the other, in the course of Achilles and in that of the tortoise, the *submultiples* of the steps of each of them; but respect the natural articulations of the two courses. As long as you respect them, no difficulty will arise, because you will follow the indications of experience. But Zeno's device is to reconstruct the movement of Achilles according to a law arbitrarily chosen. Achilles with a first step is supposed to arrive at the point where the tortoise was, with a second step at the point which it has moved to while he was making the first, and so on. In this case, Achilles would always have a new step to take. But obviously, to overtake the tortoise, he goes about it in quite another way. The movement considered by Zeno would only be the equivalent of the movement of Achilles if we could treat the movement as we treat the interval passed through, decomposable and recomposable at will. Once you subscribe to this first absurdity, all the others follow.[2]

NOTES

* From *Matter and Memory*, Allen & Unwin, London, 1911, pp. 248–253; and from *Creative Evolution*, Holt, New York, 1911, pp. 308–311.

[1] We may here briefly recall this argument. Let there be a moving body which is displaced with a certain velocity, and which passes simultaneously before two bodies, one at rest and the other moving towards it with the same velocity as its own. During the same time that it passes a certain length of the first body, it naturally passes double that length of the other. Whence Zeno concludes that 'a duration is the double of itself.' A childish argument, it is said, because Zeno takes no account of the fact that the velocity is in the one case double that which it is in the other. – Certainly, but how, I ask, could he be aware of this? That, in the same time, a moving body passes different lengths of two bodies, of which one is at rest and the other in motion, is clear for him who makes of duration a kind of absolute, and places it either in consciousness or in something which partakes of consciousness. For while a *determined* portion of this absolute or conscious duration elapses, the same moving body will traverse, as it passes the two bodies, two spaces of which the one is the double of the other, without our being able to conclude from this that a duration is double itself, since duration remains independent of both spaces. But Zeno's error, in all his reasoning, is due to just this fact, that he leaves real duration on one side, and considers only its objective track in space. How then should the two lines traced by the same moving body not merit an equal consideration, *qua* measures of duration? And how should they not represent the same duration, even though the one is twice the other? In concluding from this that 'a duration is the double of itself.' Zeno was true to the logic of his hypothesis; and his fourth argument is worth exactly as much as the three others.

[2] That is, we do not consider the sophism of Zeno refuted by the fact that the geometrical progression $a(1 + 1/n + 1/n^2 + 1/n^3 + \cdots,$ etc.) – in which a designates the initial distance between Achilles and the tortoise, and n the relation of their respective velocities – has a finite sum if n is greater than 1. On this point we may refer to the arguments of F. Evellin, which we regard as conclusive (see Evellin, *Infini et quantité*, Paris, 1880, pp. 63–97; cf. *Revue philosophique*, l. **11** (1881) 564–568. The truth is that mathematics, as we have tried to show in a former work, deals and can deal only with lengths. It has therefore had to seek devices, first, to transfer to the movement, which is not a length, the divisibility of the line passed over, and then to reconcile with experience the idea (contrary to experience and full of absurdities) of a movement that is a length, that is, of a movement *placed upon* its trajectory and arbitrarily decomposable like it.

ON CHANGE, TIME AND MOTION*

443. The notion of change has been much obscured by the doctrine of substance, by the distinction between a thing's nature and its external relations, and by the pre-eminence of subject-predicate propositions. It has been supposed that a thing could, in some way, be different and yet the same: that though predicates define a thing, yet it may have different predicates at different times. Hence the distinction of the essential and the accidental, and a number of other useless distinctions, which were (I hope) employed precisely and consciously by the scholastics, but are used vaguely and unconsciously by the moderns. Change, in this metaphysical sense, I do not at all admit. The so-called predicates of a term are mostly derived from relations to other terms; change is due, ultimately, to the fact that many terms have relations to some parts of time which they do not have to others. But every term is eternal, timeless, and immutable; the relations it may have to parts of time are equally immutable. It is merely the fact that different terms are related to different times that makes the difference between what exists at one time and what exists at another. And though a term may cease to exist, it cannot cease to be; it is still an entity, which can be counted as *one*, and concerning which some propositions are true and others false.

444. Thus the important point is the relation of terms to the times they occupy, and to existence. Can a term occupy a time without existing? At first sight, one is tempted to say that it can. It is hard to deny that Waverley's adventures occupied the time of the '45, or that the stories in the 1001 Nights occupy the period of Harun al Raschid. I should not say, with Mr Bradley, that these times are not parts of real time; on the contrary, I should give them a definite position in the Christian Era. But I should say that the *events* are not real, in the sense that they never existed. Nevertheless, when a term exists at a time, there is an ultimate triangular relation, not reducible to a combination of separate relations to existence and the time respectively. This may be shown as follows. If

'*A* exists now' can be analyzed into '*A* is now' and '*A* exists,' where *exists* is used without any tense, we shall have to hold that '*A* is then' is logically possible even if *A* did not exist then; for if occupation of a time be separable from existence, a term may occupy a time at which it does not exist, even if there are other times when it does exist.

445. Before applying these remarks to motion, we must examine the difficult idea of occupying a place at a time. Here again we seem to have an irreducible triangular relation. If there is to be motion, we must not analyze the relation into occupation of a place and occupation of a time. For a moving particle occupies many places, and the essence of motion lies in the fact that they are occupied at different times. If '*A* is here now' were analyzable into '*A* is here' and '*A* is now,' it would follow that '*A* is there then' is analyzable into '*A* is there' and '*A* is then.' If all these propositions were independent, we could combine them differently: we could, from '*A* is now' and '*A* is there', infer '*A* is there now,' which we know to be false, if *A* is a material point. The suggested analysis is therefore inadmissible. If we are determined to avoid a relation of three terms, we may reduce '*A* is here now' to '*A*'s occupation of this place is now.' Thus we have a relation between *this time* and a complex concept, *A*'s occupation of this place. But this seems merely to substitute another equivalent proposition for the one which it professes to explain. But mathematically, the whole requisite conclusion is that, in relation to a given term which occupies a place, there is a correlation between a place and a time.

446. We can now consider the nature of motion, which need not, I think, cause any great difficulty. A simple unit of matter, we agreed, can only occupy one place at one time. Thus if *A* be a material point, '*A* is here now' excludes '*A* is there now,' but not '*A* is here then.' Thus any given moment has a unique relation, not direct, but via *A*, to a single place, whose occupation by *A* is at the given moment: but there need not be a unique relation of a given place to a given time, since the occupation of the place may fill several times. A moment such that an interval containing the given moment otherwise than as an end-point can be assigned, at any moment within which interval *A* is in the first place, is a moment when *A* is at rest. A moment when this cannot be done is a moment

when A is in motion, provided A occupies *some* place at neighbouring moments on either side. A moment when there are such intervals, but all have the said moment as an end-term, is one of transition from rest to motion or *vice versa*. Motion consists in the fact that, by the occupation of a place at a time, a correlation is established between places and times; when different times, throughout any period however short, are correlated with different places, there is motion; when different times, throughout some period however short, are all correlated with the same place, there is rest.

We may now proceed to state our doctrine of motion in abstract logical terms, remembering that material particles are replaced by many-one relations of all times to some places, or of all terms of a continuous one-dimensional series t to some terms of a continuous three-dimensional series s. Motion consists broadly in the correlation of different terms of t with different terms of s. A relation R which has a single term of s for its converse domain corresponds to a material particle which is at rest throughout all time. A relation R which correlates all the terms of t in a certain interval with a single term of s corresponds to a material particle which is at rest throughout the interval, with the possible exclusion of its end-terms (if any), which may be terms of transition between rest and motion. A time of momentary rest is given by any term for which the differential coefficient of the motion is zero. The motion is continuous if the correlating relation R defines a continuous function. It is to be taken as part of the definition of motion that it is continuous, and that further it has first and second differential coefficients. This is an entirely new assumption, having no kind of necessity, but serving merely the purpose of giving a subject akin to rational Dynamics.

447. It is to be observed that, in consequence of the denial of the infinitesimal, and in consequence of the allied purely technical view of the derivative of a function, we must entirely reject the notion of a *state* of motion. Motion consists *merely* in the occupation of different places at different times, subject to continuity as explained in Part V. There is no transition from place to place, no consecutive moment or consecutive position, no such thing as velocity except in the sense of a real number which is the limit of a certain set of quotients. The rejection of velocity and acceleration as physical facts (*i.e.* as properties belonging

at each instant to a moving point, and not merely real numbers expressing limits of certain ratios) involves, as we shall see, some difficulties in the statement of the laws of motion; but the reform introduced by Weierstrass in the infinitesimal calculus has rendered this rejection imperative.

NOTE

* From *The Principles of Mathematics*, §§ 443, 445–447.

E. MEYERSON

THE ELIMINATION OF TIME
IN CLASSICAL SCIENCE*

The perfect identity between cause and effect, as the causal tendency postulates it, would imply from all evidence their equivalence – that is, the possibility of reversing the phenomenon, of arriving at the antecedent by starting from the consequent. On the other hand, this 'reversibility,' as we say in physics, does not imply identity. I can exchange a tenfranc gold piece for two five-franc silver pieces, or vice versa, from which it follows that these things have the same value, are equivalent, but not identical.

The postulate of reversibility was clearly formulated by Leibniz. Let us recall here the most trenchant of the sentences quoted before upon the relations between causes and effects:

"The integral effect may reproduce the entire cause or its equal."

Is this really so? We have only to consult our inner consciousness to reply to this question. We notice that we have the absolute feeling that nature follows an immutable course in time. We know that to-day is not like yesterday, that between the two something irreparable has happened: *fugit irreparabile tempus.* We feel ourselves growing old. We cannot invert the course of time. An able novelist has recently tried to render tangible the contrary supposition; his stories are instructive precisely because of the strange effect which emanates from them. And yet Wells, in supplying his hero with a machine which permits him to displace himself in time as we displace ourselves in space, takes the precaution to have him travel chiefly in the future, which appears to us necessarily as undetermined because of our ignorance. But let us suppose a displacement in the past. Well's hero, on the eve of the Battle of Hastings, will warn Harold of the subterfuge meditated by William, and the Normans will be beaten. Thus the entire course of history will be modified; but it will be altered, in fact, without these bold suppositions, for any individual who is added necessarily modifies the state of the universe at a given moment, and from that very fact it becomes impossible that the result should be what it has been. To reascend to the past is to modify the past, and that seems contradictory to us.

We feel, moreover, that not only the entire universe, but also each particular phenomenon which we observe, follows a determined course in time, has a beginning and an end, and that it is impossible for us to represent it in reverse order. We have only to think of events in organic nature: the birth of beings, thei maturation, their decline and death. Who can imagine fruits preceding flowers, the fowl being transformed into an egg? But it is the same also in phenomena where the fact of evolution seems less marked to us; everything unfolds in a certain direction, and if, by chance, we saw something produced in the opposite direction we should be immediately struck by it as by something contrary to the course of nature. We need not, as a matter of fact, have here recourse to imaginations such as those of Wells; we can *see* this reversed world. To do this we have only to provide ourselves with a cinematograph, to insert within it a reel of pictures representing moving phenomena – such as the jumping or the falling of a horse, a drop of water dripping into a pond, the descent of a mass of stones or of sand – and to turn the crank in the opposite direction. It is impossible to describe the strangeness of the impression which comes from the aspect of these pictures. It is not sorcery; but is something more or less like it: it is a world manifestly absurd, which presents no analogy with the one we know.

Doubtless if we look in this way at a machine the impression will not be the same: a locomotive will simply have the air of 'going backward,' and the phenomenon thus reversed in no way shocks our sensation of reality. But if we observe the fireplace we shall see that the smoke, instead of leaving it and of being dissipated, is formed in the distance, approaches, thickens, and finally rushes into the fireplace, a phenomenon which certainly will appear impossible to us. We can, moreover, explain that even the resemblance between one part of the reversed phenomenon and the going backward is only apparent and is due mainly to the imperfection of our senses. The different parts of the machine, however well oiled they may be, rub against each other just as the wheels rub against the rails, all of which become heated, and the heat is dissipated in the air. Everyone knows, moreover, that the friction of the moving wheels on the rail constitutes one of the essential conditions of the phenomenon, the locomotive *skids* when this friction is insufficient. But could we *see* these waves as we see the smoke, the reversed picture of the moving locomotive would shock us as much as what occurs in the fireplace.

It is clear that the observation just formulated is quite general. The stellar movements seem to be an exception; we are, indeed, forced to suppose that the medium in which they act offers no resistance; therefore it seems that all could turn backward and that, endowed with equal velocity and an opposite direction, the planets would pass again in the reverse direction through the same series of perihelions and eclipses. But this is probably only an appearance; where we can study the phenomenon more closely the illusion of reversibility is dissipated. Thus, in respect to the earth, the flow of the tides plays, as we know, the rôle of a brake; it tends to oppose itself to its rotation and converts a part of the kinetic energy of this motion into heat which is dissipated afterwards. On the earth, at any rate, there can be no mechanism absolutely deprived of friction, and therefore there exists none that is really reversible.

It is quite different in rational mechanics: there all motions are reversible. At the outset, with the help of a postulate tacitly assumed, the very essence of the concept of time is completely altered. Mechanical time no longer always flows uniformly in the same direction; on the contrary, one can move in it freely in the desired direction, as we do in space. It is doubtless this which Lagrange felt in affirming that time could be considered as a fourth dimension of space. This is a strikingly strange statement; we have, indeed, the immediate sensation that it is not so, and this sensation, we have seen, is justified, because there is no real parallelism between our concepts of time and space. But in rational mechanics time is indeed something analogous to space. There the effect may really 'reproduce the cause or its like,' according to the postulate of Leibniz.

H. Poincaré puts forth the ingenious supposition that the form of our mechanics is due to the influence of celestial mechanics, a science which was completed before it, and which impressed minds by its beautiful orderliness. The motions of celestial bodies appearing to us, we have just seen, as reversible, one might explain why rational mechanics was based on the same hypothesis. Whithout denying this influence (for it has certainly strenghtened conviction and has favoured the ignoring of real conditions), we believe, however, that the cause has been more profound. We see in it an evident manifestation of the principle of causality, of the general tendency which is in us and which leads us to postulate the equality between antecedent and consequent.

Having at the outset transformed the intimate nature of time with the help of an audacious postulate, rational mechanics next uses all of its efforts to eliminate it altogether from its statements. At the beginning of the development one is often obliged to regard modifications as functions of the time, but the permanent, although often unconscious, concern of the scientist is to eliminate that variable later, to reduce what is variable in time to what is constant.

Nowhere, perhaps, does this tendency manifest itself so clearly as in the development of chemistry. Here is a substance which was called in the eighteenth century *calx of mercury*, or *mercury precipitate per se*. We heat it and we perceive that drops of a liquid and metallic substance are formed in the neck of the vessel. Chemists tell us that the phlogiston uniting with the calx has formed a matter which we call metallic mercury and which is the calx of mercury phlogisticated. The fact that the reaction is found to be accompanied by the appearance of a gas does not essentially modify the explanation; the 'phlogisticians,' and amongst them the discoverers of the gas, Priestley and Scheele, considered this phenomenon as secondary, and formulated on this subject auxiliary hypotheses.

In what did the 'explanation' in question consist? You wished to know, we were told, why this red and powdery substance becomes metallic. It is because the phlogiston which has been added to it has the power of giving to the substance these very metallic qualities. Doubtless the phlogiston does not always manifest these qualities; that is because it is sometimes in a peculiar condition, but as soon as it is united to certain substances these qualities appear. The phlogiston which pre-existed has simply changed place in going from the fire to the calx of mercury. We can sum up this explanation by presenting it in the form of an equation:

Calx of mercury + phlogiston = metallic mercury.

To be sure, the phlogistician did not state this equation, the use of chemical equations coming a little later. But it none the less translates their thought, and modern chemists have often had recourse to this translation in trying to understand and to state precisely the conceptions of their predecessors. The general sense of the explanation was thus really the following: by the reaction nothing was created, nothing was lost; the red mercury and the phlogiston which existed before subsist in the metallic mercury, and this profound change was really only a displacement.

To overthrow these conceptions Lavoisier remarks that weight is a permanent property of the substance, a property which is never obliterated;

we shall, therefore, be able to recognize by an infallible sign whether something has really been added to it. Lavoisier states, making use of the balance, that the calx of mercury weighed more than the metallic mercury obtained after the operation, and that the difference is just about equal to the weight of the gas produced. Hence the logical conclusion that we must reverse the terms of the relation established by the theory of phlogiston, that it is the metallic mercury which is the simple substance, the element, and that the calx of mercury is a compound of metallic mercury and of the gas which Lavoisier called oxygen. This interpretation is thus written in the equation:

$$HgO = Hg + O$$

which has in comparison with the preceding one the inestimable advantage of taking into account weights, of being quantitative, as we say. Indeed, following a formal convention the symbols which we have just used indicate not only the substances, but also the definite weights of these substances, and the equation signifies that 216 grammes of oxide of mercury yield 200 grammes of metallic mercury and 16 grammes of oxygen. Thus the identity between the antecedent and the consequent is still more precise; not only the metallic mercury and the oxygen pre-existed in the oxide, since they are the 'components' of it, but they even existed in determined quantities, equal to those we have just obtained after the separation.

Later it was perceived that the decomposition of oxide of mercury is accompanied by another phenomenon – to wit, the absorption of a certain quantity of heat. It was observed that it was an almost general phenomenon, that, save for certain quite explainable exceptions, the substances in combining liberated heat and that, on the contrary, in dissociating they absorbed it. This quantity of heat (expressed, as we know, in calories) was successfully measured and it was seen that it really constituted a characteristic of the reaction. (we can, therefore, thus complete the above equation:

$$HgO + X \text{ cal.} = Hg + O.$$

We have written out three different equations, yet they represent one and the same phenomenon. Doubtless we might completely set aside the first, declaring it to be false, since we no longer believe in the existence of phlogiston. But this would be wrong, for the theory in question furnished, we have seen, a very acceptable interpretation of the phenomenon as it was observed at the time. It has even been noticed, and rightly, that there is a certain analogy between the most recent conceptions and those of the chemists of phlogiston. These latter has a feeling that a 'principle' must be added to the calx of mercury to produce metallic mercury, and that this was a general condition; the fact that, in replacing the term phlogiston by that of energy, we obtain, in many cases, statements which are nearly true, is not a pure coincidence.

These three equations close up on the phenomenon more and more. But all three are *equations* – that is, they tend to establish a relation of equality between terms representing prior and subsequent states of the phenomenon. As the explanation advances, the identification becomes more and more perfect. In the beginning it treats only of the qualitative side of the fact, and is lacking in numerical precision; then come the considerations of quantity, and finally the thermal energy changes are 'explained' in their turn – that is, they enter into the equation between the antecedent and the consequent.

Moreover, in support of this opinion we may cite the authority of the man from whom is derived this whole manner of thinking in modern chemistry. Lavoisier, in his *Traité élémentaire de chimie*, wrote what may be called the first true chemical equation, and with his usual clearness explained the sense which he attributed to this formula.

After having affirmed that "nothing is created, either in the operations of art, or in those of nature," and that "it may be stated as a principle that in every operation there is an equal quantity of matter before and after the operation," he adds: "It is upon this principle that is based the entire art of making experiments in chemistry. We are obliged to assume in all experiments a true equality or equation between the principles of the substance which we are examining and those which we take from it by analysis. Thus, since grape-must yields carbonic acid gas and alcohol, I can say that grape-must = carbonic acid + alcohol."

Therefore, to Lavoisier the equation is really the expression of complete equality, of identity between the antecedent and the consequent in a chemical reaction.

Poinsot says: "In perfect knowledge we know but one law – that of constancy and uniformity. To this simple idea we try to reduce all others and it is only in this reduction that we believe science to consist." No truer or more penetrating word; the purely empirical law seems external at once to things and to our mind, impenetrable, opaque. Only the laws which affirm identity, which flow from it or lead to it, appear to us as adequate either to the essence of things and to our understanding; these, alone, we know 'in perfect knowledge.' And we see also that science thus defined by the great mathematician is quite other than a group of empirical rules. It is the effort of the mind toward comprehension, of the intellect toward the rationalization of things, an effort which is the normal function of the intellect and which cannot be accomplished except with the aid of the principle of identity in time.

Thus these chemical equations are the expression of the tendency to identify things in time; one can also say 'to eliminate' time. Let us suppose that the process of identification continues and that we really succeed in expressing the entire phenomenon in an equation, in identifying completely the antecedent and the consequent; everything has been preserved, everything has remained as it was – that is, time has exercised no influence. To be sure, we know in advance that that complete identification is impossible. But, in part, we can verify with the aid of our equations that it is really so. Elementary matter which existed before the phenomenon has subsisted afterwards; from that point of view there has been no change. Weight has also remained the same; there, too, nothing has been modified. Finally, energy also has been conserved. On the whole, as far as our explanation reaches, *nothing has happened*. And since phenomenon is only change, it is clear that according as we have explained it we have made it disappear. Every explained part of a phenomenon is a part denied; to explain is to explain away.

It is true that there remain the phenomena of displacement. They are, we have seen, privileged by their very nature in that we can, according as it suits us, consider them sometimes as implying a change and sometimes as conserving identity. It is upon this particularity that all possibility of causal identification rests, and in our equations we likewise have done nothing else than use this means, in assuming that elementary masses and energy displace themselves even while remaining self-identical. Moreover, the nature of this displacement has remained undetermined; an indetermination which evidently is quite provisional, for if we wished to penetrate more deeply into the explanation of the phenomenon, to scrutinize its mechanism, we should be obliged to determine exactly the molecular motions – that is, to commit ourselves upon the mode of displacement.

We have avoided it by the simple fact that the explanation of chemical phenomena is still too little extended; chemistry is not yet advanced enough to admit of true mechanical explanations. But we cannot doubt that universal mechanism demands explanations of this kind. Yet this provisional indetermination has served us in our demonstration, in this sense that it has permitted us to set forth better the identity between antecedent and consequent.

It seems, indeed, more difficult if we turn to the science of motion itself, mechanics. Here we have the fundamental phenomenon, that to which explicative science tends to reduce all others. Shall we say that science tends to disavow it?

And yet if the analysis we have just made is correct, if explanation by mechanism is not an end, but a means, and if the explicative value of mechanical theories rests upon the fact that they satisfy our causal tendency, it is evident that our manner of treating bodies in motion must feel the effects of it. And we can foresee from the double aspect of motion, which is at once conservation and change, that rational science will apply itself to emphasizing the first.

Let us come back to the three principles of conservation. That of mass has nothing to do, strictly speaking, with mechanics, since it is understood by definition that the phenomena with which that science is concerned exclude any transformation whatever of matter. But the principle of conservation of energy is, in part, mechanical, and that of inertia is entirely so. These two propositions completely dominate that part of science, and it is especially under this aspect that we are prone to deal with these phenomena. A stone was attached to something. It has become detached and is in the act of falling – that is, because its energy, which was originally potential, has become kinetic; but the energy was there, and still is there. I threw a stone into the air; it moved at first with a certain velocity, constantly diminishing until it stopped completely. The stone began to come down again, falling faster and faster; but, I am told that during all of the time when this remarkably changing phenomenon was taking place, something supposed to represent its most essential aspect, energy, has, on the contrary, remained immutable, for at every moment the sum of the kinetic and potential energies has remained the same. A cannon-ball passes, hurled with great velocity. Here, it seems, is continual change, clearly characterized; doubtless so, science assures me, if we consider this motion as change of position with respect to surrounding bodies. But if I imagine a being placed upon this cannon-ball, he will believe himself to be at rest just as we have the same sensation on the earth; for him this motion, provided it be uniform and rectilinear, will not exist. It is generally said that the principle of inertia makes the notion of rest disappear, and we ourselves have considered it especially from this point of view; but the truth is that between motion and rest there is a reciprocal assimilation, and we can affirm with as much reason that the motion is suppressed, since immediately, and one might say almost instinctively, we henceforth apply to rectilinear motion all the norms our mind establishes for rest. "Inertia," says Hermann Cohen justly, "does not include motion; it is rather even supposed to exclude it in a certain sense."

That motion, considered in itself, should, like every other change, in reality by inconceivable, is clearly shown by the reasonings or, if you will, the paradoxes of the Eleatics, and more particularly the paradoxes designated under the name of Achilles and the Arrow. It is incomprehensible that Achilles should ever overtake the tortoise, and likewise that the arrow occupying at a given moment a determined spot should leave it. It is generally affirmed that the origin of these paradoxes lies in the fact that we cannot conceive of the actual infinite, and consequently of the continuous, that our reason can only grasp the discrete. Without wishing to probe this question, which is a digression from our subject, let us observe that our understanding does not seem to revolt against the concept of the continuous, so long as the consideration of motion is not in question. In synthetic geometry the body appears to us as really continuous, and also the surface in so far as it is the limit of the body, the line in so far as it is the limit of the surface. The difficulties of the continuous appear only with motion. It may, however, be that these difficulties are not so great for us as they were for the Greeks; thus we need to make a certain effort to grasp the sense of the argument about the arrow which seems, at the very moment when it is at a certain spot, to have nevertheless conserved *velocity*. Our habit of mind has been created in us by the concept of inertia and is doubtless strengthened by the infinitesimal calculus. It has been said that the infinitesimal calculus is an effort to grasp the concept of continuity with the help of the discrete. This observation is quite correct, but it must be added that in this calculus continuity always appears in the act of being generated by motion. It is this motion that we try to grasp by making it discontinuous, by decomposing it into small indivisible phases which are so many small rests. There is, therefore, an analogy between the methods of the mathematician and those of the physicist, in the sense that both reduce motion to immobility.

In a word, science in its effort to become "rational" tends more and more to suppress variation in time. And it is clearly seen that empiricism has nothing to do with it. Indeed, the instinct of preservation demands prevision; it is, therefore, evolution in time which especially interests us, and it seems that the essential form of the law, of the empirical rule, must be that expressing a modification as a function of the time. But it is in no wise so. If many statements expressing functions of the time are found in the sciences of organic nature it is because they are still in the beginning of their evolution. But these statements are the more rare as the science becomes more rational.

Let us suppose for a moment that science can really make the causal postulate prevail; antecedent and consequent, cause and effect, are confused and become indiscernible, simultaneous. And time itself, whose course no longer implies change, is indiscernible, unimaginable, non-existent. It is the confusion of past, present, and future – a universe eternally immutable. The progress of the world is stopped. And, of course, simultaneously, or rather previously, the cause has vanished. For as soon as cause is confused with effect, when there is identity between antecedent and consequent, when nothing happens, there is no more cause. The principle of causality, as Renouvier has justly remarked, is the elimination of the cause.

In appearance it is a paradoxical result. At a glance, however, we can take in the ground covered, and verify the fact that we have not lost ourselves on the way, that the terminus had really been demanded by the starting-point. We have searched for the causes of phenomena and we have searched for them with the help of a principle which, we know, is only the principle of identity applied to the existence of objects in time. The ultimate source of all causes can thus only be identical to itself. It is the universe immutable in space and time, the sphere of Parmenides, imperishable and without change.

NOTE

* From *Identity and Reality*; transl. by Kate Loewenberg, Allen and Unwin, London, 1930, pp. 216–219, 222–224, 226–231.

PART 3

MODERN VIEWS OF SPACE AND TIME AND THEIR ANTICIPATIONS

CRITICISM OF NEWTON*

52. The Peripatetics used to distinguish various kinds of motion corresponding to the variety of changes which a thing could undergo. To-day those who discuss motion understand by the term only local motion. But local motion cannot be understood without understanding the meaning of *locus*. Now *locus* is defined by moderns as 'the part of space which a body occupies,' whence it is divided into relative and absolute corresponding to space. For they distinguish between absolute or true space and relative or apparent space. That is they postulate space on all sides measureless, immoveable, insensible, permeating and containing all bodies, which they call absolute space. But space comprehended or defined by bodies, and therefore an object of sense, is called relative, apparent, vulgar space.

53. And so let us suppose that all bodies were destroyed and brought to nothing. What is left they call absolute space, all relation arising from the situation and distances of bodies being removed together with the bodies. Again, that space is infinite, immoveable, indivisible, insensible, without relation and without distinction. That is, all its attributes are privative or negative. It seems therefore to be mere nothing. The only slight difficulty arising is that it is extended, and extension is a positive quality. But what sort of extension, I ask, is that which cannot be divided nor measured, no part of which can be perceived by sense or pictured by the imagination? For nothing enters the imagination which from the nature of the thing cannot be perceived by sense, since indeed the imagination is nothing else than the faculty which represents sensible things either actually existing or at least possible. Pure intellect, too, knows nothing of absolute space. That faculty is concerned only with spiritual and inextended things, such as our minds, their states, passions, virtues, and such like. From absolute space then let us take away now the words of the name, and nothing will remain in sense, imagination, or intellect. Nothing else then is denoted by those words than pure privation or negation, *i.e.* mere nothing.

54. It must be admitted that in this matter we are in the grip of serious prejudices, and to win free we must exert the whole force of our minds. For many, so far from regarding absolute space as nothing, regard it as the only thing (God excepted) which cannot be annihilated; and they lay down that it necessarily exists of its own nature, that it is eternal and uncreate, and is actually a participant in the divine attributes. But in very truth since it is most certain that all things which we designate by names are known by qualities or relations, at least in part (for it would be stupid, to use words to which nothing known, no notion, idea or concept, were attached), let us diligently inquire whether it is possible to form any idea of that pure, real, and absolute space continuing to exist after the annihilation of all bodies. Such an idea, moreover, when I watch it somewhat more intently, I find to be the purest idea of notihng, if indeed it can be called an idea. This I myself have found on giving the matter my closest attention; this, I think, others will find on doing likewise.

55. We are sometimes deceived by the fact that when we imagine the removal of all other bodies, yet we suppose our own body to remain. On this supposition we imagine the movement of our limbs fully free on every side; but motion without space cannot be conceived. None the less if we consider the matter again we shall find, 1st, relative space conceived defined by the parts of our body; 2nd, a fully free power of moving our limbs obstructed by no obstacle; and besides these two things nothing. It is false to believe that some third thing really exists, *viz.* immense space which confers on us the free power of moving our body; for this purpose the absence of other bodies is sufficient. And we must admit that this absence or privation of bodies is nothing positive.[1]

56. But unless a man has examined these points with a free and keen mind, words and terms avail little. To one who meditates, however, and reflects, it will be manifest, I think, that predications about pure and absolute space can all be predicated about nothing. By this argument the human mind is easily freed from great difficulties, and at the same time from the absurdity of attributing necessary existence to any being except to the good and great God alone.

57. It would be easy to confirm our opinion by arguments drawn, as they say *a posteriori,* by proposing questions about absolute space, *e.g.* Is it substance or accidents? Is it created or uncreated? and showing the absurdities which follow from either answer. But I must be brief. I must

not omit, however, to state that Democritus of old supported this opinion with his vote. Aristotle is our authority for the statement, *Phys.* Bk. I, where he has these words, "Democritus lays down as principles the solid and the void, of which the one, he says, is as what is, the other as what is not." That the distinction between absolute and relative space has been used by philosophers of great name, and that on it as on a foundation many fine theorems have been built, may make us scruple to accept the argument, but those are empty scruples as will appear from what follows.

58. From the foregoing it is clear that we ought not to define the true place of the body as the part of absolute space which the body occupies, and true or absolute motion as the change of true or absolute place; for all place is relative just as all motion is relative. But to make this appear more clearly we must point out that no motion can be understood without some determination or direction, which in turn cannot be understood unless besides the body in motion our own body also, or some other body, be understood to exist at the same time. For *up*, *down*, *left*, and *right* and all places and regions are founded in some relation, and necessarily connote and suppose a body different from the body moved. So that if we suppose the other bodies were annihilated and, for example, a globe were to exist alone, no motion could be conceived in it; so necessary is it that another body should be given by whose situation the motion should be understood to be determined. The truth of this opinion will be very clearly seen if we shall have carried out thoroughly the supposed annihilation of all bodies, our own and that of others, except that solitary globe.

59. Then let two globes be conceived to exist and nothing corporeal besides them. Let forces then be conceived to be applied in some way; whatever we may understand by the application of forces, a circular motion of the two globes round a common centre cannot be conceived by the imagination. Then let us suppose that the sky of the fixed stars is created; suddenly from the conception of the approach of the globes to different parts of that sky the motion will be conceived. That is to say that since motion is relative in its own nature, it could not be conceived before the correlated bodies were given. Similarly no other relation can be conceived without correlates.

60. As regards circular motion many think that, as motion truly circular increases, the body necessarily tends ever more and more away from its

axis. This belief arises from the fact that circular motion can be seen taking its origin, as it were, at every moment from two directions, one along the radius and the other along the tangent, and if in this latter direction only the impetus be increased, then the body in motion will retire from the centre, and its orbit will cease to be circular. But if the forces be increased equally in both directions the motion will remain circular though accelerated – which will not argue an increase in the forces of retirement from the axis, any more than in the forces of approach to it. Therefore we must say that the water forced round in the bucket rises to the sides of the vessel, because when new forces are applied in the direction of the tangent to any particle of water, in the same instant new equal centripetal forces are not applied. From which experiment it in no way follows that absolute circular motion is necessarily recognized by the forces of retirement from the axis of motion. Again, how those terms *corporeal forces* and *conation* are to be understood is more than sufficiently shown in the foregoing discussion.

61. A curve can be considered as consisting of an infinite number of straight lines, though in fact it does not consist of them. That hypothesis is useful in geometry; and just so circular motion can be regarded as arising from an infinite number of rectilinear directions – which supposition is useful in mechanics. Not, however, on that account must it be affirmed that it is impossible that the centre of gravity of each body should exist successively in single points of the circular periphery, no account being taken of any rectilineal direction in the tangent or the radius.

62. We must not omit to point out that the motion of a stone in a sling or of water in a whirled bucket cannot be called truly circular motion as that term is conceived by those who define the true places of bodies by the parts of absolute space, since it is strangely compounded of the motions, not alone of bucket or sling, but also of the daily motion of the earth round her own axis, of her monthly motion round the common centre of gravity of earth and moon, and of her annual motion round the sun. And on that account each particle of the stone or the water describes a line far removed from circular. Nor in fact does that supposed axifugal conation exist, since it is not concerned with some one axis in relation to absolute space, supposing that such a space exists; accordingly I do not see how that can be called a single conation to which a truly circular motion corresponds as to its proper and adequate effect.

63. No motion can be recognized or measured, unless through sensible things. Since then absolute space in no way affects the senses, it must necessarily be quite useless for the distinguishing of motions. Besides, determination or direction is essential to motion; but that consists in relation. Therefore it is impossible that absolute motion should be conceived.

64. Further, since the motion of the same body may vary with the diversity of relative place, nay actually since a thing can be said in one respect to be in motion and in another respect to be at rest, to determine true motion and true rest, for the removal of ambiguity and for the furtherance of the mechanics of these philosophers who take the wider view of the system of things, it would be enough to bring in, instead of absolute space, relative space as confined to the heavens of the fixed stars, considered as at rest. But motion and rest marked out by such relative space can conveniently be substituted in place of the absolutes, which cannot be distinguished from them by any mark. For however forces may be impressed, whatever conations there are, let us grant that motion is distinguished by actions exerted on bodies; never, however, will it follow that that space, absolute place, exists, and that change in it is true place.

65. The laws of motions and the effects, and theorems containing the proportions and calculations of the same for the different configurations of the paths, likewise for accelerations and different directions, and for mediums resisting in greater or less degree, all these hold without bringing absolute motion into account. As is plain from this that since according to the principles of those who introduce absolute motion we cannot know by any indication whether the whole frame of things is at rest, or is moved uniformly in a direction, clearly we cannot know the absolute motion of any body.

NOTES

* From *De motu*, §§ 52–65.
[1] See the arguments against absolute space in my book on *The Principles of Human Knowledge* in the English tongue published ten years ago [1710].

G. W. LEIBNIZ AND S. CLARKE

DISCUSSION ON THE NATURE OF SPACE AND TIME

MR. LEIBNIZ'S THIRD PAPER*[1]

1. According to the usual way of speaking, mathematical principles concern only mere mathematics, viz. numbers, figures, arithmetic, geometry. But metaphysical principles concern more general notions, such as are cause and effect.

2. The author grants me this important principle; that nothing happens without a sufficient reason, why it should be so, rather than otherwise. But he grants it only in words, and in reality denies it. Which shows that he does not fully perceive the strength of it. And therefore he makes use of an instance, which exactly falls in with one of my demonstrations against real absolute space, which is an idol of some modern Englishmen. I call it an idol, not in a theological sense, but in a philosophical one; as Chancellor Bacon says, that there are *idola tribus, idola specus*.[2]

3. These gentlemen maintain therefore, that space is a real absolute being. But this involves them in great difficulties; for such a being must needs be eternal and infinite. Hence some have believed it to be God himself, or, one of his attributes, his immensity. But since space consists of parts, it is not a thing which can belong to God.

4. As for my own opinion, I have said more than once, that I hold space to be something merely[3] relative, as time is; that I hold it to be an order of coexistences, as time is an order of successions. For space denotes, in terms of possibility, an order of things which exist at the same time, considered as existing together; without enquiring into their manner of existing. And when many things are seen together, one perceives that order of things among themselves.

5. I have many demonstrations, to confute the fancy of those who take space to be a substance, or at least an absolute being. But I shall only use, at the present, one demonstration, which the author here gives me occasion to insist upon. I say then, that if space was an absolute being, there would something happen for which it would be impossible there should be a

sufficient reason. Which is against my axiom. And I prove it thus. Space is something absolutely uniform; and, without the things placed in it, one point of space does not absolutely differ in any respect whatsoever from another point of space. Now from hence it follows, (supposing space to be something in itself, besides the order of bodies among themselves,) that 'tis impossible there should be a reason, why God, preserving the same situations of bodies among themselves, should have placed them in space after one certain particular manner, and not otherwise; why every thing was not placed the quite contrary way, for instance, by changing East into West. But if space is nothing else, but that order or relation; and is nothing at all without bodies, but the possibility of placing them; then those two states, the one such as it now is, the other supposed to be the quite contrary way, would not at all differ from one another. Their difference therefore is only to be found in our chimerical supposition of the reality of space in itself. But in truth the one would exactly be the same thing as the other, they being absolutely indiscernible; and consequently there is no room to enquire after a reason of the preference of the one to the other.

6. The case is the same with respect to time. Supposing any one should ask, why God did not create every thing a year sooner; and the same person should infer from thence, that God has done something, concerning which 'tis not possible there should be a reason, why he did it so, and not otherwise: the answer is, that his inference would be right, if time was any thing distinct from things existing in time. For it would be impossible there should be any reason, why things should be applied to such particular instants, rather than to others, their succession continuing the same. But then the same argument proves, that instants, consider'd without the things, are nothing at all; and that they consist only in the successive order of things: which order remaining the same, one of the two states, viz. that of a supposed anticipation, would not at all differ, nor could be discerned from, the other which now is.

7. It appears from what I have said, that my axiom has not been well understood; and that the author denies it, tho' he seems to grant it. 'Tis true, says he, that there is nothing without a sufficient reason why it is, and why it is thus, rather than otherwise: but he adds, that this sufficient reason, is often the simple or mere will of God: as, when it is asked why matter was not placed otherwhere in space; the same situations of bodies

among themselves being preserved. But this is plainly maintaining, that God wills something, without any sufficient reason for his will: against the axiom, or the general rule of whatever happens. This is falling back into the loose indifference, which I have confuted at large, and showed to be absolutely chimerical even in creatures, and contrary to the wisdom of God, as if he could operate without acting by reason.

DR. CLARKE'S THIRD REPLY[4]

1. This relates only to the signification of words. The definitions here given, may well be allowed; and yet mathematical reasonings may be applied to physical and metaphysical subjects.

2. Undoubtedly nothing is, without a sufficient reason why it is, rather than not; and why it is thus, rather than otherwise. But in things in their own nature indifferent; mere will, without any thing external to influence it, is alone that sufficient reason. As in the instance of God's creating or placing any particle of matter in one place rather than in another, when all places are originally alike. And the case is the same, even though space were nothing real, but only the mere order of bodies: for still it would be absolutely indifferent, and there could be no other reason but mere will why three equal particles should be placed or ranged in the order *a*, *b*, *c*, rather than in the contrary order. And therefore no argument can be drawn from this indifferency of all places, to prove that no space is real. For different spaces are really different or distinct one from another, though they be perfectly alike. And there is this evident absurdity in supposing space not to be real, but to be merely the order of bodies; that, according to that notion, if the earth and sun and moon had been placed where the remotest fixed stars now are, (provided they were placed in the same order and distance they now are with regard one to another,) it would not only have been, (as this learned author rightly says,) *la même chose*, the same thing in effect; which is very true: but it would also follow, that they would then have been in the same place too, as they are now: which is an express contradiction.

The ancients[5] did not call all space which is void of bodies, but only extramundane space, by the name of imaginary space. The meaning of which, is not, that such space is not real;[6] but only that we are wholly ignorant what kinds of things are in that space. Those writers, who by the word, *imaginary*, meant at any time to affirm that space was not real; did not thereby prove, that it was not real.

3. Space is not a being, an eternal and infinite being, but a property, or a consequence of the existence of a being infinite and eternal. Infinite space, is immensity: but immensity is not God: and therefore infinite space, is not God. Nor is there any difficulty in what is here alleged about space having parts. For infinite space is one, absolutely and essentially indivisible: and to suppose it parted, is a contradiction in terms; because there must be space in the partition itself; which is to suppose it parted, and yet not parted at the same time. [7] The immensity or omnipresence of God, is no more a dividing of his substance into parts, than his duration; or continuance of existing, is a dividing of his existence into parts. There is no difficulty here, but what arises from the figurative abuse of the word, *parts*.

4. If space was nothing but the order of things coexisting; it would follow, that if God should remove in a straight line the whole material world entire, with any swiftness whatsoever; yet it would still always continue in the same place: and that nothing would receive any shock upon the most sudden stopping of that motion. And if time was nothing but the order of succession of created things; it would follow, that if God had created the world millions of ages sooner than he did, yet it would not have been created at all the sooner. Further: space and time are quantities; which situation and order are not.

5. The argument in this paragraph, is; that because space is uniform or alike, and one part does not differ from another; therefore the bodies created in one place, if they had been created in another place, (supposing them to keep the same situation with regard to each other,) would still have been created in the same place as before: which is a manifest contradiction. The uniformity of space, does indeed prove, that there could be no (external) reason why God should create things in one place rather than in another: but does that hinder his own will, from being to itself a sufficient reason of acting in any place, when all places are indifferent or alike, and there be good reason to act in some place?

6. The same reasoning takes place here, as in the foregoing.

7 and 8. Where there is any difference in the nature of things, there the consideration of that difference always determines an intelligent and perfectly wise agent. But when two ways of acting are equally and alike

good, (as in the instances before mentioned) to affirm in such case, that God cannot act at all, or that 'tis no perfection in him to be able to act, because he can have no external reason to move him to act one way rather than the other, seems to be a denying God to have in himself any original principle or power of beginning to act, but that he must needs (as it were mechanically) be always determined by things extrinsic.

FROM LEIBNIZ' FIFTH LETTER

33. Since space in itself is an ideal thing, like time; space out of the world must needs be imaginary, as the schoolmen themselves have acknowledged. The case is the same with empty space within the world; which I take also to be imaginary, for the reasons before alleged.

34. The author objects against me the vacuum discovered by Mr. Guericke[8] of Magdeburg, which is made by pumping the air out of a receiver; and he pretends that there is truly a perfect vacuum, or a space without matter, (at least in part), in that receiver. The Aristotelians and Cartesians, who do not admit a true vacuum, have said in answer to that experiment of Mr. Guericke, as well as to that of Torricellius[9] of Florence, (who emptied the air out of a glass-tube by the help of quicksilver,) that there is no vacuum at all in the tube or in the receiver; since glass has small pores, which the beams of light, the effluvia of the load-stone, and other very thin fluids may go through. I am of their opinion: and I think the receiver may be compared to a box full of holes in the water, having fish or other gross bodies shut up in it; which being taken out, their place would nevertheless be filled up with water. There is only this difference; that though water be fluid and more yielding than those gross bodies, yet it is as heavy and massive, if not more, than they: whereas the matter which gets into the receiver in the room of the air is much more subtile. The new sticklers for a vacuum allege in answer to this instance, that it is not the grossness of matter, but its mere quantity, that makes resistance; and consequently that there is of necessity more vacuum, where there is less resistance. They add, that the subtleness of matter has nothing to do here; and that the particles of quicksilver are as subtle and fine as those of water; and yet that quicksilver resists about ten times more. To this I reply, that it is not so much the quantity of matter, as its difficulty of giving place, that makes resistance. For instance, float-

ing timber contains less of heavy matter, than an equal bulk of water does; and yet it makes more resistance to a boat, than the water does.

35. And as for quicksilver; 'tis true, it contains about fourteen times more of heavy matter, than an equal bulk of water does; but it does not follow, that it contains fourteen times more matter absolutely. On the contrary, water contains as much matter; if we include both its own matter, which is heavy; and the extraneous matter void of heaviness, which passes through its pores. For, both quicksilver and water, are masses of heavy matter, full of pores, through which there passes a great deal of matter void of heaviness [and which makes no sensible resistance]; such as is probably that of the rays of light, and other insensible fluids; and especially that which is itself the cause of the gravity of gross bodies, by receding from the centre towards which it drives those bodies. For, it is a strange imagination to make all matter gravitate, and that towards all other matter, as if each body did equally attract every other body according to their masses and distances; and this by an attraction properly so called, which is not derived from an occult impulse of bodies; whereas the gravity of sensible bodies towards the centre of the earth, ought to be produced by the motion of some fluid. And the case must be the same with other gravities, such as is that of the planets towards the sun, or towards each other. [A body is never moved naturally, except by another body which touches it and pushes it; after that it continues until it is prevented by another body which touches it. Any other kind of operation on bodies is either miraculous or imaginary.]

36. I objected, that space taken for something real and absolute without bodies, would be a thing eternal, impassible, and independent upon God. The author endeavours to elude this difficulty, by saying that space is a property of God. In answer to this, I have said, in my foregoing paper, that the property of God is immensity; but that space (which is often commensurate with bodies,) and God's immensity, are not the same thing.

37. I objected further, that if space be a property, and infinite space be the immensity of God; finite space will be the extension or mensurability of something finite. And therefore, the space taken up by a body, will be the extension of that body. Which is an absurdity; since a body can change space but cannot leave its extension.

38. I asked also; if space is a property, what thing will an empty limited space (such as that which my adversary imagines in an exhausted receiver),

be the property of? It does not appear reasonable to say, that this empty space, either round or square, is a property of God. Will it be then perhaps the property of some immaterial, extended, imaginary substances which the author seems to fancy in the imaginary spaces?

39. If space is the property or affection of the substance, which is in space; the same space will be sometimes the affection of one body, sometimes of another body, sometimes of an immaterial substance, and sometimes perhaps of God himself, when it is void of all other substance material or immaterial. But this is a strange property or affection, which passes from one subject to another. Thus subjects will leave off their accidents, like clothes; that other subjects may put them on. At this rate, how shall we distinguish accidents and substances?

40. And if limited spaces are the affections of limited substances, which are in them; and infinite space be a property of God; a property of God must (which is very strange,) be made up of the affections of creatures; for all finite spaces, taken together, make up infinite space.

41. But if the author denies, that limited space is an affection of limited things; it will not be reasonable neither, that infinite space should be the affection or property of an infinite thing. I have suggested all these difficulties in my foregoing paper; but it does not appear that the author has endeavoured to answer them.

42. I have still other reasons against this strange imagination, that space is a property of God. If it be so, space belongs to the essence of God. But space has parts: therefore there would be parts in the essence of God. *Spectatum admissi.*[10]

43. Moreover, spaces are sometimes empty, and sometimes filled up. Therefore there will be in the essence of God, parts sometimes empty, and sometimes full, and consequently liable to a perpetual change. Bodies, filling up space, would fill up part of God's essence, and would be commensurate with it; and in the supposition of a vacuum, part of God's essence will be within the receiver. Such a God having parts, will very much resemble the Stoics' God, which was the whole universe considered as a divine animal.

44. If infinite space is God's immensity, infinite time will be God's eternity; and therefore we must say, that what is in space, is in God's immensity, and consequently in his essence; and that what is in time, [is in the eternity of God and] is also in the essence of God. Strange

expressions; which plainly show, that the author makes a wrong use of terms.

45. I shall give another instance of this. God's immensity makes him actually present in all spaces. But now if God is in space, how can it be said that space is in God, or that it is a property of God? We have often heard that a property is in its subject; but we never heard, that a subject is in its property. In like manner, God exists in all time. How then can time be in God; and how can it be a property of God? These are perpetual *alloglossies*.[11]

46. It appears that the author confounds immensity, or the extension of things, with the space according to which that extension is taken. Infinite space, is not the immensity of God; finite space, is not the extension of bodies: as time is not their duration. Things keep their extension; but they do not always keep their space. Every thing has its own extension, its own duration; but it has not its own time, and does not keep its own space.

47. I will here show, how men come to form to themselves the notion of space. They consider that many things exist at once and they observe in them a certain order of co-existence, according to which the relation of one thing to another is more or less simple. This order, is their *situation* or distance. When it happens that one of those co-existent things changes its relation to a multitude of others, which do not change their relation among themselves; and that another thing, newly come, acquires the same relation to the others, as the former had; we then say, it is come into the place of the former; and this change, we call a motion in that body, wherein is the immediate cause of the change. And though many, or even all the co-existent things, should change according to certain known rules of direction and swiftness; yet one may always determine the relation of situation, which every co-existent acquires with respect to every other co-existent; and even that relation which any other co-existent would have to this, or which this would have to any other, if it had not changed, or if it had changed any otherwise. And supposing, or feigning, that among those co-existents, there is a sufficient number of them, which have undergone no change; then we may say, that those which have such a relation to those fixed existents, as others had to them before, have now the *same place* which those others had. And that which comprehends all those places, is called *space*. Which shows, that in order to have an idea of place, and consequently of space, it is sufficient to consider these relations,

and the rules of their changes, without needing to fancy any absolute reality out of the things whose situation we consider. And, to give a kind of a definition: *place* is that, which we say is the same to A and, to B, when the relation of the co-existence of B, with C, E, F, G, etc. agrees perfectly with the relation of the co-existence, which A had with the same C, E, F, G, etc. supposing there has been no cause of change in C, E, F, G, etc. It may be said also, without entering into any further particularity, that *place* is that, which is the same in different moments to different existent things, when their relations of co-existence with certain other existents, which are supposed to continue fixed from one of those moments to the other, agree entirely together. And *fixed existents* are those, in which there has been no cause of any change of the order of their co-existence with others; or (which is the same thing,) in which there has been no motion. Lastly, *space* is that, which results from places taken together. And here it may not be amiss to consider the difference between place, and the relation of situation, which is in the body that fills up the place. For, the place of A and B, is the same; whereas the relation of A to fixed bodies, is not precisely and individually the same, as the relation which B (that comes into its place) will have to the same fixed bodies; but these relations agree only. For, two different subjects, as A and B, cannot have precisely the same individual affection; it being impossible, that the same individual accident should be in two subjects, or pass from one subject to another. But the mind not contented with an agreement, looks for an identity, for something that should be truly the same; and conceives it as being extrinsic to the subjects: and this is what we call *place* and *space*. But this can only be an ideal thing; containing a certain order, wherein the mind conceives the application of relations. In like manner, as the mind can fancy to itself an order made up of genealogical lines, whose bigness would consist only in the number of generations, wherein every person would have his place: and if to this one should add the fiction of a *metempsychosis*, and bring in the same human souls again; the persons in those lines might change place; he who was a father, or a grandfather, might become a son, or a grandson, etc. And yet those genealogical places, lines, and spaces, though they should express real truth, would only be ideal things. I shall allege another example, to show how the mind uses, upon occasion of accidents which are in subjects, to fancy to itself something answerable to those accidents, out of the subjects. The ratio or proportion between two lines L and M, may be conceived three several ways; as a ratio of the greater L, to the lesser M; as a ratio of the lesser M, to the greater L; and lastly, as something abstracted from both, that is, as the ratio between L and M, without considering which is the antecedent, or which the consequent; which the subject, and which the object. And thus it is, that proportions are considered in music. In the first way of considering them, L the greater; in the second, M the lesser, is the subject of that accident, which philosophers call relation. But, which of them will be the subject, in the third way of considering them? It cannot be said that both of them, L and M together, are the subject of such an accident; for if so, we should have an accident in two subjects, with one leg in one, and the other in the other; which is contrary to the notion of accidents. Therefore we must say, that this relation, in this third way of considering it, is indeed out of the subjects; but being neither a substance, nor an accident, it must be a mere ideal thing, the consideration of which is nevertheless useful. To conclude:[12] I have here done much like Euclid, who not being able to make his readers well understand what *ratio* is absolutely in the sense of geometricians; defines what are the *same ratios*. Thus, in like manner, in order to explain what *place* is, I have been content to define what is the *same place*. Lastly; I observe, that the traces of moveable bodies, which they leave sometimes upon the immoveable ones on which they are moved; have given men occasion to form in their imagination such an idea, as if some trace did still remain, even when there is nothing unmoved. But this is a mere ideal thing, and imports only, that if there was any unmoved thing there, the trace might be marked out upon it. And 'tis this analogy, which makes men fancy places, traces and spaces; though those things consist only in the truth of relations, and not at all in any absolute reality.

48. To conclude.[13] If the space (which the author fancies) void of all bodies, is not altogether empty; what is it then full of? Is it full of extended spirits perhaps, or immaterial substances, capable of extending and contracting themselves; which move therein, and penetrate each other without any inconveniency, as the shadows of two bodies penetrate one another upon the surface of a wall? Methinks I see the revival of the odd imaginations of Dr. Henry More (otherwise a learned and well-meaning man), and of some others who fancied that those spirits can make themselves impenetrable whenever they please. Nay, some have fancied, that man,

in the state of innocency, had also the gift of penetration; and that he became solid, opaque, and impenetrable by his fall. Is it not overthrowing our notions of things, to make God have parts, to make spirits have extension? The principle of the want of a sufficient reason does alone drive away all these spectres of imagination. Men easily run into fictions, for want of making a right use of that great principle.

To Paragraph 10

49. It cannot be said that [a certain] duration is eternal but [it can be said] that the things which continue always are eternal, [gaining always a new duration.] Whatever exists of time and of duration, [being successive] perishes continually: and how can a thing exist eternally, which (to speak exactly,) does never exist at all? For, how can a thing exist, whereof no part does ever exist? Nothing of time does ever exist, but instants; and an instant is not even itself a part of time. Whoever considers these observations, will easily apprehend that time can only be an ideal thing. And the analogy between time and space, will easily make it appear, that the one is as merely ideal as the other. [But, if in saying that the duration of a thing is eternal, it is only meant that the thing endures eternally, I have nothing to say against it.]

50. If the reality of space and time, is necessary to the immensity and eternity of God; if God must be in space; if being in space, is a property of God; he will in some measure, depend upon time and space, and stand in need of them. For I have already prevented that subterfuge, that space and time are [in God and like] properties of God. [Could one maintain the opinion that bodies move in the parts of the divine essence?]

To Paragraph 11, and 12

51. I objected that space cannot be in God, because it has parts. Hereupon the author seeks another subterfuge, by departing from the received sense of words; maintaining that space has no parts, because its parts are not separable, and cannot be removed from one another by discerption. But 'tis sufficient that space has parts, whether those parts be separable or not; and they may be assigned in space, either by the bodies that are in it, or by lines and surfaces that may be drawn and described in it.

To Paragraph 13

52. In order to prove that space, without bodies, is an absolute reality; the author objected, that a finite material universe might move forward in space. I answered, it does not appear reasonable that the material universe should be finite; and, though we should suppose it to be finite; yet 'tis unreasonable it should have motion any otherwise, than as its parts change their situation among themselves; because such a motion would produce no change that could be observed, and would be without design. 'Tis another thing, when its parts change their situation among themselves; for then there is a motion in space; but it consists in the order of relations which are changed. The author replies now, that the reality of motion does not depend upon being observed; and that a ship may go forward, and yet a man, who is in the ship, may not perceive it. I answer, motion does not indeed depend upon being observed; but it does depend upon being possible to be observed. There is no motion, when there is no change that can be observed. And when there is no change that can be observed, there is no change at all. The contrary opinion is grounded upon the supposition of a real absolute space, which I have demonstratively confuted by the principle of the want of a sufficient reason of things.

53. I find nothing in the Eighth Definition of the *Mathematical Principles of Nature*, nor in the Scholium belonging to it, that proves, or can prove, the reality of space in itself. However, I grant there is a difference between an absolute true motion of a body, and a mere relative change of its situation with respect to another body. For when the immediate cause of the change is in the body, that body is truly in motion; and then the situation of other bodies, with respect to it, will be changed consequently, though the cause of that change be not in them. 'Tis true that, exactly speaking, there is not any one body, that is perfectly and entirely at rest; but we frame an abstract notion or rest, by considering the thing mathematically. Thus have I left nothing unanswered, of what has been alleged for the absolute reality of space. And I have demonstrated the falsehood of that reality, by a fundamental principle, one of the most certain both in reason and experience; against which, no exception or instance can be alleged.

CLARKE'S FIFTH REPLY

33–35. To the argument drawn against a plenum of matter, from the want of resistance in certain spaces; this learned author answers, that those spaces are filled with a matter which has no (§ 35) gravity. But the argument was not drawn from gravity, but from resistance; which must be proportionable to the quantity of matter, whether the matter had any gravity or no.[14]

To obviate this reply, he alleges that (§ 34) resistance does not arise so much from the quantity of matter, as from its difficulty of giving place. But this allegation is wholly wide of the purpose; because the question related only to such fluid bodies which have little or no tenacity, as water and quicksilver, whose parts have no other difficulty of giving place, but what arises from the quantity of the matter they contain. The instance of a (*ibid.*) floating piece of wood, containing less of heavy matter than an equal bulk of water, and yet making greater resistance; is wonderfully unphilosophical: for an equal bulk of water shut up in a vessel, or frozen into ice, and floating, makes a greater resistance than the floating wood; the resistance then arising from the whole bulk of the water: but when the water is loose and at liberty in its state of fluidity, the resistance is then not made by the whole, but by part only, of the equal bulk of water; and then it is no wonder that it seems to make less resistance than the wood.

36–48. These paragraphs do not seem to contain serious arguments, but only represent in an ill light the notion of the immensity or omnipresence of God; who is not a mere *intelligentia supramundana*, (semota a nostris rebus sejunctaque longe;) *is not far from every one of us; for in him we* (and all things) *live and move and have our being*.[15]

The space occupied by a body, is not the (§ 36–7) extension of the body; but the extended body exists in that space.

There is no such thing in reality, as (§ 38) bounded space; but only we in our imagination fix our attention upon what part or quantity we please, of that which itself is always and necessarily unbounded.

Space is not an (§ 39) affection of one body, or of another body, or of any finite being; nor passes from subject to subject; but is always invariably the immensity of one only and always the same *immensum*.

Finite spaces are not at all the (§ 40) affections of finite substances; but they are only those parts of infinite space, in which finite substances exist.

If matter was infinite, yet infinite space would no more be an (§41) affection of that infinite body, than finite spaces are the affections of finite bodies; but, in that case, the infinite matter would be, as finite bodies now are, in the infinite space.

Immensity, as well as eternity, is (§42) essential to God. The parts of immensity[16] (being totally of a different kind from corporeal, partable, separable, divisible, moveable parts which are the ground of corruptibility;) do no more hinder immensity from being essentially one, than the parts of duration hinder eternity from being essentially one.

God himself suffers no (§43) change at all, by the variety and changeableness of things which live and move and have their being in him.

This (§44) strange doctrine, is the express assertion of St. Paul,[17] as well as the plain voice of nature and reason.

God does not exist (§45) in space, and in time; but his existence[18] causes space and time. And when, according to the analogy of vulgar speech, we say that he exists in all space and in all time; the words mean only that he is omnipresent and eternal, that is, that boundless space and time are necessary consequences of his existence; and not, that space and time are beings distinct from him, and IN which he exists.

(§46) How finite space is not the extension of bodies, I have shown just above, on Paragraph 40. And the two following paragraphs also, (Paragraphs 47 and 48), need only to be compared with what hath been already said.

49–51. These seem to me, to be only a quibbling upon words. Concerning the question about space having parts, see above; Reply 3, Paragraph 3; and Reply 4, Paragraph 11.

52, and 53. My argument here, for the notion of space being really independent upon body, is founded on the possibility of the material universe being finite and moveable: 'tis not enough therefore for this learned writer to reply, that he thinks it would not have been wise and reasonable for God to have made the material universe finite and moveable. He must either affirm, that 'twas impossible for God to make the material world finite and moveable; or else he must of necessity allow the strength of my argument, drawn from the possibility of the world's being finite and moveable. Neither is it sufficient barely to repeat his assertion, that the motion of a finite material universe would be nothing, and (for want of other bodies to compare it with) would (§52) produce no discoverable change: unless he could disprove the instance which I gave of a very great change that would happen; viz. that the parts would be sensibly shocked by a sudden acceleration, or stopping of the motion of the whole: to which instance, he has not attempted to give any answer.

53. Whether this learned author's being forced here to acknowledge the difference between absolute real motion and relative motion, does not necessarily infer that space is really a quite different thing from the situation or order of bodies; I leave to the judgment of those who shall be pleased to compare what this learned writer here alleges, with what Sir Isaac Newton has said in his *Principia*, Lib. I, Defin. 8.

NOTES

* From *Leibniz-Clarke Correspondence,* (ed. by H. G. Alexander), Manchester Univ. Press, pp. 25–27; 30–33, 64–74, 102–105. See also *Gottfried Wilhelm Leibniz: Philosophical Papers and Letters* (ed. by Leroy E. Loemker) D. Reidel Publ. Co., Dordrecht, Holland.

[1] Despatched 25th Feb. 1716 (p. 193).
[2] 'idols of the tribe, idols of the cave', *Novum Organum* I, Aphor. 38ff.
[3] 'purement'.
[4] Transmitted 15th May 1716, delayed (p. 194).
[5] This was occasioned by a passage in the private letter wherein Mr. Leibniz's third paper came inclosed. [Gerhardt says that there is no trace among the Leibniz papers of this letter. Klopp's edition of the Leibniz-Caroline correspondence also contains nothing relevant.]
[6] Of nothing, there are no dimensions, no magnitudes, no quantity, no properties.
[7] See §4 of my Second Reply.
[8] Guericke (1602–86). Inventor of the air pump. He is said to have performed an experiment before the Emperor Ferdinand III in which he took two hollow copper hemispheres, exhausted the air from them with his pump, and then showed that thirty horses, fifteen pulling on each hemisphere, could not separate them. Leibniz corresponded with Guericke about the air-pump in 1671–2 (G.I. 193).
[9] Torricelli (1608–47). Pupil of Galileo, and inventor of the barometer. In his most famous experiment he took a long tube closed at one end, filled it with mercury and closing the open end with his finger, inverted it in a basin of mercury. When he removed his finger, the level of mercury in the tube fell to thirty inches above the surface, leaving an apparent vacuum at the top of the tube.
[10] Spectatum admissi risum teneatis amici, i.e. If you saw such a thing, friends, could you restrain your laughter? Horace, *De Arte Poetica*, I.
[11] 'alloglossies', ἀλλογλωσσία: use of a strange tongue (Liddell and Scott).
[12] 'au reste'.
[13] 'au reste'.

[14] Otherwise, what makes the body of the earth more difficult to be moved (even the same way that its gravity tends) than the smallest ball?
[15] Acts xvii. 27, 28.
[16] See above in my Third Reply, §3; and Fourth Reply, §11.
[17] Acts xvii. 27. 28.
[18] See the note on my Fourth Reply, §10.

R. J. BOSCOVICH

CRITICISM OF NEWTON'S ALLEGED PROOF OF ABSOLUTE MOTION*

After referring to Newton's alleged proof of absolute space by the rotating bucket experiment and after quoting from Newton's *Principia* the description of the experiment with two spheres tied by a string and revolving around the common center of gravity, Boscovich made the following comment:

On this Newton's method I remarked that it appears suspect to me and not appropriate to achieve its purpose. I pass over the fact that even here the existence of the inertial forces as well as the way by which the motions are produced by them must be assumed; about it below. The very tension of the string can be defined only by the distance of the spheres, but we cannot measure this distance with certainty unless we assume that it remains unaffected by the translatory motion, the assumption which in that immense void is risky or at least doubtful. But what above all makes everything useless, this method as well as the previous one and any other one, is the fact that if the parallel and equal motions in the same plane are communicated to us and to those bodies, all the motions mentioned above would remain altogether the same according to the accepted principles of Mechanics; also the application of forces by us would produce altogether the same motions described above. Indeed, if already some common motion in a certain plane were present, the absolute motion, composed of this common motion and the motion mentioned above which we detect in that experiment, would be different from that mentioned above. Also the part [of the revolving motion] which we regard as moving backwards, could really move forwards and vice versa. If this common motion is truly faster than the respective circular motion and if it takes place in the same plane, then these two bodies will not be in absolute motion in that plane as the experiments indicate. If we consider two points of the diameter perpendicular to the direction of the common motion where in one of these points the circular motion is added to the common motion, in other point it is opposed to it, while still being overcome by it, the direction of absolute motion thus in both points would agree with the direction of the common motion. *From this it seems to me absolutely evident that absolute motion cannot be differentiated in any way from the relative one.*

Explanatory diagram by the editor:

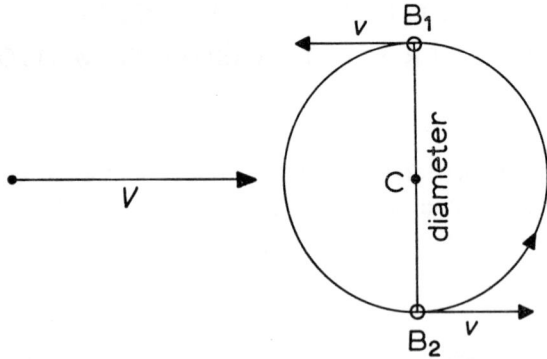

B_1, B_2 the bodies revolving around the common center of gravity (which in the case of two equal masses coincides with the center C of the circular motion.) – v = tangential velocity of the revolving motion, and V = the common motion of translation in the same plane.

NOTE

* From explanatory notes to the work *Philosophiae recentioribus versibus tradita a Benedito Stay Libri Decem*, Romae 1755; Transl. by the editor and Walter Emge.

W. K. CLIFFORD

ON THE BENDING OF SPACE*

As a result, then, of our consideration of one- and two-dimensioned space we find that, if these spaces be not same (*à fortiori* not homaloidal), we should by reason of their curvature have a means of determining absolute position. But we see also that a being existing in these dimensions would most probably attribute the effects of curvature to changes in its own physical condition in nowise connected with the geometrical character of its space.

What lesson may we learn by analogy for the three-dimensioned space in which we ourselves exist? To begin with, we assume that all our space is perfectly *same*, or that solid figures do not change their shape in passing from one position in it to another. We base this postulate of sameness upon the results of observation in that somewhat limited portion of space of which we are cognizant.[1] Supposing our observations to be correct, it by no means follows that because the portion of space of which we are cognizant is for practical purposes same, that therefore *all* space is same.[2] Such an assumption is a mere dogmatic extension to the unknown of a postulate, which may perhaps be true for the space upon which we can experiment. To make such dogmatic assertions with regard to the unknown is rather characteristic of the mediaeval theologian than of the modern scientist. On the like basis with this postulate as to the sameness of our space stands the further assumption that it is homaloidal. When we assert that our space is everywhere same, we suppose it of constant curvature (like the circle as one- and the sphere as two-dimensioned space); when we suppose it homaloidal we assume that this curvature is zero (like the line as one- and the plane as two-dimensioned space). This assumption appears in our geometry under the form that two parallel planes, or two parallel lines in the same plane – that is, planes, or lines in the same plane, which however far produced will never meet – have a *real* existence in our space. This real existence, of which it is clearly impossible for us to be cognizant, we postulate as a result built upon our experience of what happens in a limited portion of space. We may postulate that the portion of space of which we are cognizant is practically homaloidal, but we have clearly no right to dogmatically extend this postulate to *all* space. A constant curvature, imperceptible for that portion of space upon which we can experiment, or even a curvature which may vary in an almost imperceptible manner with the time, would seem to satisfy all that experience has taught us to be true of the space in which we dwell.

But we may press our analogy a step further, and ask, since our hypothetical worm and fish might very readily attribute the effects of changes in the bending of their spaces to changes in their own physical condition, whether we may not in like fashion be treating merely as physical variations effects which are really due to changes in the curvature of our space; whether, in fact, some or all of those causes which we term physical may not be due to the geometrical construction of our space. There are three kinds of variation in the curvature of our space which we ought to consider as within the range of possibility.

(i) Our space is perhaps really possessed of a curvature varying from point to point, which we fail to appreciate because we are acquainted with only a small portion of space, or because we disguise its small variations under changes in our physical condition which we do not connect with our change of position. The mind that could recognize this varying curvature might be assumed to know the absolute position of a point. For such a mind the postulate of the relativity of position would cease to have a meaning. It does not seem so hard to conceive such a state of mind as the late Professor Clerk-Maxwell would have had us believe. It would be one capable of distinguishing those so-called physical changes which are really geometrical or due to a change of position in space.

(ii) Our space may be really same (of equal curvature), but its degree of curvature may change as a whole with the time. In this way our geometry based on the sameness of space would still hold good for all parts of space, but the change of curvature might produce in space a succession of apparent physical changes.

(iii) We may conceive our space to have everywhere a nearly uniform curvature, but that slight variations of the curvature may occur from point to point, and themselves vary with the time. These variations of the curvature with the time may produce effects which we not unnaturally attribute to physical causes independent of the geometry of our space. We might even go so far as to assign to this variation of the curvature of space "what really happens in that phenomenon which we term the motion of matter."[3]

We have introduced these considerations as to the nature of our space to bring home to the reader the character of the postulates we make in the exact sciences. These postulates are *not*, as too often assumed, necessary and universal truths; they are merely axioms based on our experience of a certain limited region. Just as in any branch of physical inquiry we start by making experiments, and basing on our experiments a set of axioms which form the foundation of an exact science, so in geometry our axioms are really, although less obviously, the result of experience. On this ground geometry has been properly termed at the commencement of Chapter II a *physical* science. The danger of asserting dogmatically that an axiom based on the experience of a limited region holds universally will now be to some extent apparent to the reader. It may lead us to entirely overlook, or when suggested at once reject, a possible explanation of phenomena. The hypotheses that space is not homaloidal, and again, that its geometrical character may change with the time, may or may not be destined to play a great part in the physics of the future; yet we cannot refuse to consider them as possible explanation of physical phenomena, because they may be opposed to the popular dogmatic belief in the universality of certain geometrical axioms – a belief which has arisen from centuries of discriminating worship of the genius of Euclid.

NOTES

* From *The Common Sense of the Exact Sciences*, Knopf, New York, 1946, pp. 200–204. [The notes are written by Karl Pearson, editor of the 1st posthumous edition of 1885].
[1] It may be held by some that the postulate of the sameness of our space is based upon the fact that no one has hitherto been able to form any geometrical conception of space-curvature. Apart from the fact that mankind habitually assumes many things of which it can form no geometrical conception (mathematicians the circular points at infinity, theologians transubstantiation), I may remark that we cannot expect any being to form a geometrical conception of the curvature of his space till he views it from space of a higher dimension, that is, practically, never.
[2] Yet it must be noted that, because a solid figure *appears* to us to retain the same shape when it is moved about in that portion of space with which we are acquainted, if does not follow that the figure *really* does retain its shape. The changes of shape may be either imperceptible for those distances through which we are able to move the figure, or if they do take place we may attribute them to 'physical causes' – to heat, light, or magnetism – which may possibly be mere names for variations in the curvature of our space.

[3] This remarkable *possibility* seems first to have been suggested by Professor Clifford in a paper presented to the Cambridge Philosophical Society in 1870 (*Mathematical Papers*, p. 21). I may add the following remarks: The most notable physical quantities which vary with position and time are heat, light, and electro-magnetism. It is these that we ought peculiarly to consider when seeking for any physical changes, which may be due to changes in the curvature of space. If we suppose the boundary of any arbitrary figure in space to be distorted by the variation of space-curvature, there would, by analogy from one and two dimensions, be no change in the volume of the figure arising from such distortion. Further, if we *assume* as an axiom that space resists curvature with a resistance proportional to the change, we find that waves of 'space-displacement' are precisely similar to those of the elastic medium which we suppose to propagate light and heat. We also find that 'space-twist' is a quantity exactly corresponding to magnetic induction, and satisfying relations similar to those which hold for the magnetic field. It is a question whether physicists might not find it simpler to assume that space is capable of a varying curvature, and of a resistance to that variation, than to suppose the existence of a subtle medium pervading an invariable homaloidal space.

W. K. CLIFFORD

ON THE SPACE-THEORY OF MATTER*¹

ABSTRACT. Riemann has shown that as there are different kinds of lines and surfaces, so there are different kinds of space of three dimensions; and that we can only find out by experience to which of these kinds the space in which we live belongs. In particular, the axioms of plane geometry are true within the limits of experiment on the surface of a sheet of paper, and yet we know that the sheet is really covered with a number of small ridges and furrows, upon which (the total curvature not being zero) these axioms are not true. Similarly, he says although the axioms of solid geometry are true within the limits of experiment for finite portions of our space, yet we have no reason to conclude that they are true for very small portions; and if any help can be got thereby for the explanation of physical phenomena, we may have reason to conclude that they are not true for very small portions of space.

I wish here to indicate a manner in which these speculations may be applied to the investigation of physical phenomena. I hold in fact

(1) That small portions of space *are* in fact of a nature analogous to little hills on a surface which is on the average flat; namely, that the ordinary laws of geometry are not valid in them.

(2) That this property of being curved or distorted is continually being passed on from one portion of space to another after the manner of a wave.

(3) That this variation of the curvature of space is what really happens in that phenomenon which we call the *motion of matter*, whether ponderable or etherial.

(4) That in the physical world nothing else takes place but this variation, subject (possibly) to the law of continuity.

I am endeavouring in a general way to explain the laws of double refraction on this hypothesis, but have not yet arrived at any results sufficiently decisive to be communicated.

NOTES

* [From *Proceedings Cambridge Philosophical Society*, II. 1876. Read Feb. 21, 1870, pp. 157, 158.]
[1] Reprinted in *Common Sense of the Exact Sciences*, Knopf, New York, 1946, pp. 21–22.

A. CALINON

GEOMETRICAL SPACES*

1. So-called non-Euclidean geometry has a particular interest for philosophy. We are going to try to point out here, briefly and without the apparatus of formulae, some of the consequences of this geometry with regard to our conception of space.

Although the order in which the different propositions of geometry are presented obviously cannot be arbitrary, nevertheless this order is not absolute. Thus, while it is customary to study plane figures first, one could just as well begin with the study of spherical figures.[1] Once an order has been established, the difinition of a geometrical line must fulfill the following conditions:

(1) This definition must apply to all lines of the kind which one wishes to define, and only these lines; in other words, a line is to be defined in terms of a characteristic property.

(2) The definition must make use only of relationships of the new line to other lines previously studied.

(3) It must be clear on the basis of previously acquired geometrical knowledge that the definition does not give rise to contradiction and that it is quite compatible with the existence of a line.

The same conditions hold for the definition of surfaces.

These general rules of a good definition become inapplicable, however, when the straight line being defined is the first line of the geometry. Our last two rules, at least, are meaningless in this case. We are thus led to ask the following question: how can one recognize that a definition of a straight line is legitimate? Actually, apart from scientific training, we all have a kind of empirical notion of a straight line. For teaching purposes this is enough to serve as a basis for geometry, but let us ask ourselves instead whether the legitimacy of the definition of a straight line can be established in a strictly geometrical way.

Suppose, in a general way, that we are given *a priori* the definition of the first geometrical line, whatever this first line may be. Let us try to apply the geometrical method, that is to say pure reason, to this definition.

Either we will arrive at contradictions and conclude from this that the definition must be rejected; or, on the contrary, it will be possible to continue the deduction as far as one wishes without ever arriving at contradictions, in which case we will consider the definition to be justified and the geometry which results from it to be legitimate.

The geometry so defined, for this is truly a definition, no longer has any experimental basis. It consists simply in the application of the so-called geometrical method to a group of forms (lines or surfaces), the first of which is subject to the single condition of permitting the application of this method.

As we shall see below, geometry thus understood is a more general science than the geometry of the ancients.

2. Euclid seems to define the straight line as follows:

> (a) A line such that only one may pass through two given points, with the result that if two straight lines have two points in common they coincide and are not distinct.

When he arrives at the theory of parallels, Euclid introduces his famous postulate, equivalent to the following proposition:

> (b) Through an [external] point only one line can be drawn parallel to a given straight line.

The geometry with which we are concerned has been called non-Euclidean as opposed to the earlier geometry of Euclid; we prefer to give it the name of general geometry, since, far from being the negation of Euclidean geometry, it includes the latter as a particular case. In the same way we will call the line defined by the property *a* the general straight line, and the line defined by the combination of properties *a* and *b* the Euclidean straight line.

Now, when the formulae of general geometry are compared with those of Euclidean geometry, it is seen that the former all contain the same general parameter which, for a particular value, yields Euclidean geometry.[2]

3. A difficulty arises, however, when one tries to explain the meaning of the parameter which characterizes our general geometry.

In fact, in this general geometry, if one seeks the straight line passing through two points, this line, as one might expect, depends on the general parameter in such a way that there would be an infinite number of straight lines passing through the two points, namely one corresponding to each value of the parameter. This contradicts definition a of the straight line.

This apparent contradiction is not peculiar to the case just cited; it extends to the whole of general geometry. Another example follows:

In Euclidean geometry the volume of a sphere depends only on its radius. In general geometry the formula which expresses this volume as a function of the radius also contains the general parameter, so that our sphere of a given radius would have a particular volume corresponding to each value of the parameter. This would be absurd.

If our general geometry implied such an absurdity in all its formulae and propositions, it would cease to be legitimate and would have to be rejected.

But, as we shall see, the contradiction is only apparent.

Let us call one-, two- and three-dimensional space a line, a surface and a volume, as is customary.

In these terms, two surfaces of different shapes, that is which cannot without deformation be merged into one, are two different two-dimensional spaces; an example would be two spherical surfaces of unequal radius. For the moment let us consider only the two-dimensional spaces of Euclidean geometry. In two dimensional space a line belonging to this space such that only one can be drawn through any two points of the space is called a geodesic: if, for example, the two-dimensional space is a spherical surface, the geodesic must lie on the surface and be defined by any two of its points.

It follows that the geodesics of two different spherical surfaces, for example, have an identical definition but are nevertheless different lines. In a general way, the geodesics of different two-dimensional spaces, although different, have a common definition, common properties, and satisfy one and the same geometry.

Now let us generalize this idea of different two-dimensional spaces, extending it to three-dimensional spaces. One understands, of course, that in this case we are dealing with a purely geometrical generalization which need not have a material realization in order to be conceived of. We shall see that this conception of different three-dimensional spaces is very well suited to our general geometry and furnishes us with an explanation of the parameter which characterizes this geometry.

What characterizes two different two-dimensional spaces in Euclidean geometry is the impossibility of making them coincide without deformation or of transporting a figure unchanged from one into another; thus a spherical triangle cannot be transported from one sphere onto another with a different radius.

We shall likewise say that two three-dimensional spaces are different when the figures of one of them cannot be transported unchanged into the other, since this passage can be effected, as in the case of the spherical triangle, only by modifying the shape and the metrical properties of the figures.

Consequently, in order to eliminate the apparent contradiction to which general geometry gives rise, it is sufficient to consider the parameter of this geometry a spatial parameter, each particular value of which corresponds to a particular three-dimensional space; under these conditions, the definition of the general straight line characterized by property a can take a more precise form which immediately removes all difficulty. A line situated in a two-dimensional space such that only one can be drawn through any two points of that space, we have called a geodesic of that space. Similarly, we shall say that the straight line of a three-dimensional space is a line situated in that space such that only one may be drawn through any two points of that space.

Therefore, to any two particular values of the parameter of general geometry there correspond two different three-dimensional spaces and, in these two spaces, two different straight lines, since a figure which belongs to one of these spaces cannot be transported into the other. In a word, any figure which we define in general geometry is particularized in each particular space, that is for each particular value of the parameter, so that no figure can belong to several spaces at the same time.

In sum, the general geometry based on definition a is the synthetic geometry of an infinite number of absolutely distinct geometrical spaces, and Euclidean geometry is the geometry of only one of these spaces.

4. Let us briefly mention the principal geometric peculiarities which distinguish Euclidean space from all the others. First of all, the straight line of this space possesses property b by definition.

Furthermore, Legendre has shown that property b of the straight line is equivalent to the following proposition: The sum of the [internal] angles of a triangle is equal to two right angles.

It can even be demonstrated that if this is true for a single triangle, it is also true for all other triangles in the same space. Two consequences follow from this:

If, in a space, the sum of the angles of a single triangle is equal to two right angles, the space is Euclidean.

In any non-Euclidean space the sum of the angles of a triangle is always different from two right angles.

Finally, one of the fundamental properties of Euclidean space is that it is the only homogeneous space. Let us explain what this means. Take a triangle with sides which are respectively 7, 8, and 13 m long. As we know, a second triangle can be constructed with the same angles but with sides measuring 7, 8, and 13 mm. This is what is meant by saying that a figure can be constructed on different scales; there is only a simple change in the unit of length (for example, the millimeter instead of the meter), but the angles and the numbers which measure the sides remain the same.

Algebraically this is the same as saying that the metrical relations of figures are always expressed in formulae of the same kind. Only Euclidean space is homogeneous, for the possibility of constructing a figure on different scales is bound to the theory of parallels and to Euclid's postulate. In non-Euclidean spaces it is impossible to reduce the dimensions of a figure, by half for example, while preserving the same angles. This idea is often expressed by saying that in Euclidean space the extension of a figure is relative, while in the other spaces figures have absolute dimensions.

5. Since general geometry is thus constituted without any experimental basis, it is necessary to ask oneself which particular geometry is realized in the material world. The different geometrical spaces for which we have found the general laws cannot all exist at once, because they cannot accommodate the same forms. In order to know which one of these spaces contains the bodies we see around us, we must necessarily look to experience. Several geometricians, to be sure, have allowed that we have a certain number of *a priori* intuitions of our space before any geometrical knowledge, and even before any experience. More particularly, they have allowed that the relativity of the dimensions of bodies, that is the homogeneity of space,

is, for us, a sort of intuitive notion, bound up in the form of our mind itself. We shall return to this point later. Regardless of the status of this notion, however, it is much more consistent with scientific rigor first to establish, as we have suggested, the general geometry of the different spaces apart from its experimental basis and any preconceived idea, and then to seek the geometry peculiar to our space, by means of observation.

One is thus led to the conclusion that, within the limits of the precision of our instruments and our methods of observation, our space does not differ from Euclidean or homogeneous space. We have pointed out that the existence of a single triangle having the sum of its angles equal to two right angles is sufficient to define Euclidean space. This is the simplest mode of verification and the one used in choosing one of the largest triangles astronomical science can observe and measure. Indeed, it is of great importance to choose a triangle with very long sides, for in non-Euclidean space the sum of the angles of triangles most nearly approximates two right angles when the triangles are smallest.

Given this, it would not be surprising if experience confirmed our prior notion of the homogeneity of space, In fact, this notion is like many others of the same kind: since man lives in the midst of the universe, it seems quite impossible to us that the geometry of this universe would have had no influence on the formation of his notion of space, even prior to any scientific knowledge. As we have pointed out, the homogeneity of space is reduced to the relativity of the dimensions of bodies or to the possibility of constructing a figure on different scales without changing the essential elements of its form. Now certainly from the earliest times man has tried to reproduce in miniature the form of the objects which surrounded him, whether by shaping modelling clay or by working a piece of wood with a sharp instrument. This art has been perfected from age to age by more and more precise comparison between the reproduction and the model. Doesn't this make it obvious that an acquired notion of the homogeneity of space had to result from this comparison itself? In sum, far from being an *a priori* intuition of the nature of space, this notion is rather the result of very long experience.

That is, after all, why this notion concerns only the particular space we live in, while it could not lead us to the conception of the different geometrical spaces revealed to us by our general geometry.

One should not consider these different geometrical spaces to be absolutely excluded from the universe, however. Actually the Euclidean form of our space results from observations of only limited accuracy. All that can legitimately be concluded is that the differences which might exist between Euclidean geometry and that realized by the universe are due to experimental error. There remains, then, a doubt which cannot be avoided, but within the limitations of this doubt three different hypotheses concerning our space can be framed. We will limit ourselves to a statement of these, as follows:

(1) Our space is and remains strictly Euclidean.

(2) Our space realizes a geometrical space which differs very little from Euclidean space, but is always the same;

(3) Our space realizes different geometric spaces successively in time; in other words, our spatial parameter varies with time, either by deviating more or less from the Euclidean parameter, or by oscillating around a given parameter very near the Euclidean parameter.

In this last hypothesis, which is the most general possible hypothesis, the shapes of the bodies surrounding us are slowly being modified before our eyes at the same time as our space, since different spaces cannot contain the same shapes.

NOTES

* From his article 'Les espaces géometriques', *Revue philosophique de la France et de l'etranger* **27** (1889); transl. by Mary-Alice and David A. Sipfle.

[1] On this subject see our *Étude sur la sphère, la ligne droite et le plan* (Berger-Levrault, 1888.

[2] This parameter does not appear in Lobachevski's formulae, since he chooses a particular unit. It does appear, however, as soon as any other unit is adopted.

J. B. STALLO

CRITICISM OF NEWTON, EULER, KANT AND NEUMANN*

Euler most strenuously insists on the necessity of postulating an absolute, immovable space.

Whoever denies absolute space,

he says,

falls into the gravest perplexities. Since he is constrained to reject absolute rest and motion as empty sounds without sense, he is not only constrained also to reject the laws of motion, but to affirm that there are no laws of motion. For, if the question which has brought us to this point, What will be the condition of a solitary body detached from its connection with other bodies? is absurd, then those things also which are induced in this body by the action of others become uncertain and indeterminable, and thus everything will have to be taken as happening fortuitously and without any reason.[1]

That the basis of all this reasoning is purely ontological is plain. And, when the thinkers of the eighteenth century became alive to the fallacies of ontological speculation, the unsoundness of Euler's 'axiom', that rest and motion are substantial attributive entities independent of all relation, could hardly escape their notice. Nevertheless, they were unable to emancipate themselves wholly from Euler's ontological prepossessions. They did not at once avoid his dilemma by repudiating it as unfounded – by denying that motion and rest can not be real without being absolute – but they attempted to reconcile the absolute reality of rest and motion with their phenomenal relativity by postulating an absolutely quiescent point or center in space to which the positions of all bodies could be referred. Foremost among those who made this attempt was Kant.[2] In the seventh chapter of his *Natural History of the Heavens* – the same work in which, nearly fifty years before Laplace, he gave the first outlines of the Nebular Hypothesis – he sought to show that in the universe there is somewhere a great central body whose center of gravity is the cardinal point of reference for the motions of all bodies whatever.

If in the immeasurable space,

he says,

wherein all the suns of the milky way have been formed, a point is assumed round which from whatever cause, the first formative action of nature had its play, then at that point a body of the largest mass and of the greatest attractions, must have been formed. This body must have become able to compel all systems which were in process of formation in the enormous surrounding sphere to gravitate toward it as their center, so as to constitute an entire system which was evolved on a small scale out of elementary matter.

– A suggestion similar to that of Kant has recently been made by Professor C. Neumann, who enforces the necessity of assuming the existence, at a definite and permanent point in space, of an absolutely rigid body, to whose center of figure or attraction all motions are to be referred, by physical considerations. The drift of his reasoning appears in the following extracts from his inaugural lecture, 'On the Principles of the Galileo-Newtonian Theory'... [There follows the quotation of the text which is enclosed in Part I.]

After thus showing, or attempting to show, that the reality of motion necessitates its reference to a rigid body unchangeable in its position in space, Neumann seeks to verify this assumption by asking himself the question, what consequences would ensue, on the hypothesis of the mere relativity of motion, if all bodies but one were annihilated.

Let us suppose,

he says,

that among the stars there is one which consists of fluid matter, and which, like our earth, is in rotatory motion round an axis passing through its center. In consequence of this motion, by virtue of the centrifugal forces developed by it, this star will have the form of an ellipsoid. What form, now, I ask, will this star assume if suddenly all other celestial bodies are annihilated?

These centrifugal forces depend solely upon the state of the star itself; they are wholly independent of the other celestial bodies. These forces, therefore, as well as the ellipsoidal form, will persist, irrespective of the continued existence or disappearance of the other bodies. But, if motion is defined as something relative – as a relative change of place of two points – the answer is very different. If, on this assumption, we suppose all other celestial bodies to be annihilated, nothing remains but the material points of which the star in question itself consists. But, then, these points do not change their relative positions, and are therefore at rest. It follows that the star must be at rest at the moment when the annihilation of the other bodies takes place, and therefore must assume the spherical form taken by all bodies in a state of rest. A contradiction so intolerable can be avoided only by abandoning the assumption of the relativity of motion and conceiving motion as absolute, so that thus we are again led to the principle of the body Alpha.

Now, what answer can be made to this reasoning of Professor Neumann? None, if we grant the admissibility of the hypothesis of the annihilation of all bodies in space but one, and the admissibility of the further assumption that an absolutely rigid body with an absolutely fixed place in the universe is possible. But such a concession is forbidden by the universal principle of relativity. In the first place, the annihilation of all bodies but one would not only destroy the *motion* of this one remaining body and bring it to rest, as Professor Neumann sees, but would also destroy its very *existence* and bring it to naught, as he does not see. A body can not survive the system of relations in which alone it has its being; its *presence* or *position* in space is no more possible without reference to other bodies than its *change of position or presence* is possible without such reference. As has been abundantly shown, all properties of a body which constitute the elements of its distinguishable presence in space are in their nature relations and imply terms beyond the body itself.

In the second place the absolute fixity in space attributed to the body Alpha is impossible under the known conditions of reality. The fixity of a point in space involves the permanence of its distances from at least four other fixed points not in the same plane. But the fixity of these several points again depends on the constancy of their distances from other fixed points, and so on *ad infinitum*. In short, the fixity of position of any body in space is possible only on the supposition of the absolute finitude of the universe; and this leads to the theory of the essential curvature of space, and the other theories of modern trancendental geometry, which will be discussed hereafter.

NOTES

* From *The Concepts and Theories of Modern Physics*, 1881; republished by Harvard Univ. Press, 1960, with the introduction by P. W. Bridgman.
[1] *Theoria Motus, etc.*, p. 32.
[2] It is remarkable how many of the scientific discoveries, speculations and fancies of the present day are anticipated or at least foreshadowed in the writings of Kant. Some of them are enumerated by Zoellner (*Natur der Kometen*, p. 455f) – among them the constitution and motion of the system of fixed stars; the nebular origin of planetary and stellar systems; the origin, constitution and rotation of Saturn's rings and the conditions of their stability; the non-coincidence of the moon's center of gravity with her center of figure; the physical constitution of the comets; the retarding effect of the tides upon the rotation of the earth; the theory of the winds, and Dove's law.

E. MACH

CRITICISM OF NEWTON'S CONCEPT OF ABSOLUTE SPACE*

When quite modern authors let themselves be led astray by the Newtonian arguments which are derived from the bucket of water, to distinguish between relative and absolute motion, they do not reflect that the system of the world is only given *once* to us, and the Ptolemaic or Copernician view is *our* interpretation, but both are equally actual. Try to fix Newton's bucket and rotate the heaven of fixed stars and then prove the absence of centrifugal forces.

4. It is scarcely necessary to remark that in the reflections here presented Newton has again acted contrary to his expressed intention only to investigate *actual facts*. No one is competent to predicate things about absolute space and absolute motion; they are pure things of thought, pure mental constructs, that cannot be produced in experience. All our principles of mechanics are, as we have shown in detail, experimental knowledge concerning the relative positions and motions of bodies. Even in the provinces in which they are now recognized as valid, they could not, and were not, admitted without previously being subjected to experimental tests. No one is warranted in extending these principles beyond the boundaries of experience. In fact, such an extension is meaningless, as no one possesses the requisite knowledge to make use of it.

We must suppose that the change in the point of view from which the system of the world is regarded which was initiated by Copernicus, left deep traces in the thought of Galileo and Newton. But while Galileo, in his theory of the tides, quite naïvely chose the sphere of the fixed stars as the basis of a new system of coördinates, we see doubts expressed by Newton as to whether a given fixed star is at rest only apparently or really (*Principia*, 1687, p. 11). This appeared to him to cause the difficulty of distinguishing between true (absolute) and apparent (relative) motion. By this he was also impelled to set up the conception of *absolute space*. By further investigations in this direction – the discussion of the experiment of the rotating spheres which are connected together by a cord and that of the rotating water-bucket (pp. 9, 11) – he believed that he could prove an absolute rotation, though he could not prove any absolute translation. By absolute rotation he understood a rotation relative to the fixed stars, and here centrifugal forces can always be found. "But how we are to collect," says Newton in the Scholium at the end of the Definitions, "the true motions from their causes, effects, and apparent differences, and *vice versa*; how from the motions, either true or apparent, we may come to the knowledge of their causes and effects, shall be explained more at large in the following Tract." The resting sphere of fixed stars seems to have made a certain impression on Newton as well. The natural system of reference is for him that which has any uniform motion or translation without rotation (relatively to the sphere of fixed stars).[1] But do not the words quoted in inverted commas give the impression that Newton was glad to be able now to pass over to less precarious questions that could be tested by experience?

Let us look at the matter in detail. When we say that a body K alters its direction and velocity solely through the influence of another body K', we have asserted a conception that it is impossible to come at unless other bodies A, B, C...are present with reference to which the motion of the body K has been estimated. In reality, therefore, we are simply cognizant of a relation of the body K to A, B, C.... If now we suddenly neglect A, B, C...and attempt to speak of the deportment of the body K in absolute space, we implicate ourselves in a twofold error. In the first place, we cannot know how K would act in the absence of A, B, C...; and in the second place, every means would be wanting of forming a judgement of the behavior of K and of putting to the test what we had predicated – which latter therefore would be bereft of all scientific significance.

Two bodies K and K', which gravitate toward each other, impart to each other in the direction of their line of junction accelerations inversely proportional to their masses m, m'. In this proposition is contained, not only a relation of the bodies K and K' to one another, but also a relation of them to other bodies. For the proposition asserts, not only that K and K' suffer with respect to one another the acceleration designated by $\varkappa(\overline{m+m'}/r^2)$, but also that K experiences the acceleration $-\varkappa m'/r^2$ and K' the acceleration $+\varkappa m/r^2$ in the direction of the line of junction; facts which can be ascertained only by the presence of other bodies.

The motion of a body K can only be estimated by reference to other bodies A, B, C.... But since we always have at our disposal a sufficient

CRITICISM OF NEWTON'S CONCEPT OF ABSOLUTE SPACE   311

number of bodies, that are as respects each other relatively fixed, or only slowly change their positions, we are, in such reference, restricted to no one *definite* body and can alternately leave out of account now this one and now that one. In this way the conviction arose that these bodies are indifferent generally.

It might be, indeed, that the isolated bodies A, B, C...play merely a collateral rôle in the determination of the motion of the body K, and that this motion is determined by a *medium* in which K exists. In such a case we should have to substitute this medium for Newton's absolute space. Newton certainly did not entertain this idea. Moreover, it is easily demonstrable that the atmosphere is not this motion-determinative medium. We should, therefore, have to picture to ourselves some other medium, filling, say, all space, with respect to the constitution of which and its kinetic relations to the bodies placed in it we have at present no adequate knowledge. In itself such a state of things would not belong to the impossibilities. It is known, from recent hydrodynamical investigations, that a rigid body experiences resistance in a frictionless fluid only when its velocity *changes*. True, this result is derived theoretically from the notion of inertia; but it might, conversely, also be regarded as the primitive fact from which we have to start. Although, practically, and at present, nothing is to be accomplished with this conception, we might still hope to learn more in the future concerning this hypothetical medium; and from the point of view of science it would be in every respect a more valuable acquisition than the forlorn idea of absolute space. When we reflect that we cannot abolish the isolated bodies A, B, C..., that is, cannot determine by experiment whether the part they play is fundamental or collateral, that hitherto they have been the sole and only competent means of the orientation of motions and of the description of mechanical facts, it will be found expedient provisionally to regard all motions as determined by these bodies.

5. Let us now examine the point on which Newton, apparently with sound reasons, rests his distinction of absolute and relative motion. If the earth is affected with an *absolute* rotation about its axis, centrifugal forces are set up in the earth: it assumes an oblate form, the acceleration of gravity is diminished at the equator, the plane of Foucault's pendulum rotates, and so on. All these phenomena disappear if the earth is at rest and the other heavenly bodies are affected with absolute motion round it, such that the same *relative* rotation is produced. This is, indeed, the

case, if we start *ab initio* from the idea of absolute space. But if we take our stand on the basis of facts, we shall find we have knowledge only of *relative* spaces and motions. *Relatively*, not considering the unknown and neglected medium of space, the motions of the universe are the same whether we adopt the Ptolemaic or the Copernican mode of view. Both views are, indeed, equally *correct*; only the latter is more simple and more *practical*. The universe is not *twice* given, with an earth at rest and an earth in motion; but only *once*, with its *relative* motions, alone determinable. It is, accordingly, not permitted us to say how things would be if the earth did not rotate. We may interpret the one case that is given us, in different ways. If, however, we so interpret it that we come into conflict with experience, our interpretation is simply wrong. The principles of mechanics can, indeed, be so conceived, that even for relative rotations centrifugal forces arise.

Newton's experiment with the rotating vessel of water simply informs us, that the relative rotation of the water with respect to the sides of the vessel produces *no* noticeable centrifugal forces, but that such forces *are* produced by its relative rotation with respect to the mass of the earth and the other celestial bodies. No one is competent to say how the experiment would turn out if the sides of the vessel increased in thickness and mass till they were ultimately several leagues thick. The one experiment only lies before us, and our business is, to bring it into accord with the other facts known to us, and not with the arbitrary fictions of our imagination.

6. When Newton examined the principles of mechanics discovered by Galileo, the great value of the simple and precise law of inertia for deductive derivations could not possibly escape him. He could not think of renouncing its help. But the law of inertia, referred in such a naïve way to the earth supposed to be at rest, could not be accepted by him. For, in Newton's case, the rotation of the earth was not a debatable point; it rotated without the least doubt. Galileo's happy discovery could only hold approximately for small times and spaces, during which the rotation did not come into question. Instead of that, Newton's conclusions about planetary motion, referred as they were to the fixed stars, appeared to conform to the law of inertia. Now, in order to have a generally valid system of reference, Newton ventured the fifth corollary of the *Principia* (p. 19 of the first edition). He imagined a momentary terrestrial system of

coördinates, for which the law of inertia is valid, held fast in space without any rotation relatively to the fixed stars. Indeed he could, without interfering with its usability, impart to this system any initial position and any uniform translation relatively to the above momentary terrestrial system. The Newtonian laws of force are not altered thereby; only the initial positions and initial velocities – the constants of integration – may alter. By this view Newton gave the *exact* meaning of his hypothetical extension of Galileo's law of inertia. We see that the reduction to absolute space was by no means necessary, for the system of reference is just as relatively determined as in every other case. In spite of his metaphysical liking for the absolute, Newton was correctly led by the *tact of the natural investigator*. This is particularly to be noticed, since, in former editions of this book, it was not sufficiently emphasized. How far and how accurately the conjecture will hold good in future is of course undecided.

The comportment of terrestrial bodies with respect to the earth is reducible to the comportment of the earth with respect to the remote heavenly bodies. If we were to assert that we knew more of moving objects than this their last-mentioned, experimentally-given comportment with respect to the celestial bodies, we should render ourselves culpable of a falsity. When, accordingly, we say, that a body preserves unchanged its direction and velocity *in space*, our assertion is nothing more or less than an abbreviated reference to *the entire universe*. The use of such an abbreviated expression is permitted the original author of the principle, because he knows, that as things are no difficulties stand in the way of carrying out its implied directions. But no remedy lies in his power, if difficulties of the kind mentioned present themselves; if, for example, the requisite, relatively fixed bodies are wanting.

7. Instead, now, of referring a moving body K to space, that is to say to a system of coördinates, let us view directly its relation to the bodies of the universe, by which alone such a system of coördinates can be determined. Bodies very remote from each other, moving with constant direction and velocity with respect to other distant fixed bodies, change their mutual distances proportionately to the time. We may also say, all very remote bodies – all mutual or other forces neglected – alter their mutual distances proportionately to those distances. Two bodies, which, situated at a short distance from one another, move with constant direction and velocity with respect to other fixed bodies, exhibit more complicated relations.

If we should regard the two bodies as dependent on one another, and call r the distance, t the time, and a a constant dependent on the directions and velocities, the formula would be obtained: $d^2r/dt^2 = (1/r)[a^2 - (dr/dt)^2]$. It is manifestly much *simpler* and *clearer* to regard the two bodies as independent of each other and to consider the constancy of their direction and velocity with respect to other bodies.

Instead of saying, the direction and velocity of a mass μ in space remain constant, we may also employ the expression, the mean acceleration of the mass μ with respect to the masses $m, m', m''\ldots$ at the distances $r, r', r''\ldots$ is $=0$, or $d^2(\sum mr/\sum m)/dt^2 = 0$. The latter expression is equivalent to the former, as soon as we take into consideration a sufficient number of sufficiently distant and sufficiently large masses. The mutual influence of more proximate small masses, which are apparently not concerned about each other, is eliminated of itself. That the constancy of direction and velocity is given by the condition adduced, will be seen at once if we construct through μ as vertex cones that cut out different portions of space, and set up the condition with respect to the masses of these separate portions. We may put, indeed, for the *entire* space encompassing μ, $d^2(\sum mr/\sum m)/dt^2 = 0$. But the equation in this case asserts nothing with respect to the motion of μ, since it holds good for all species of motion where μ is uniformly surrounded by an infinite number of masses. If two masses μ_1, μ_2 exert on each other a force which is dependent on their distance r, then $d^2r/dt^2 = (\mu_1 + \mu_2)f(r)$. But, at the same time, the acceleration of the center of gravity of the two masses or the mean acceleration of the mass-system with respect to the masses of the universe (by the principle of reaction) remains $=0$; that is to say,

$$\frac{d^2}{dt^2}\left[\mu_1 \frac{\sum mr_1}{\sum m} + \mu_2 \frac{\sum mr_2}{\sum m}\right] = 0.$$

When we reflect that the time-factor that enters into the acceleration is nothing more than a quantity that is the measure of the distances (or angles of rotation) of the bodies of the universe, we see that even in the simplest case, in which apparently we deal with the mutual action of only *two* masses, the neglecting of the rest of the world is *impossible*. Nature does not begin with elements, as we are obliged to begin with them. It is certainly fortunate for us, that we can, from time to time, turn aside our eyes from the overpowering unity of the All, and allow them to rest on individual details. But we should not omit, ultimately to complete and correct our views by a thorough consideration of the things which for the time being we left out of account.

8. The considerations just presented show, that it is not necessary to refer the law of inertia to a special absolute space. On the contrary, it is perceived that the masses that in the common phraseology exert forces on each other as well as those that exert none, stand with respect to acceleration in quite similar relations. We may, indeed, regard *all* masses as related to each other.

NOTE

* From *Science of Mechanics*, The Open Court, Chicago, 1942, pp. 280–288.
[1] *Principia*, p. 19, Coroll. V: "The motions of bodies included in a given space are the same among themselves, whether that space is at rest or moves uniformly forwards in a right line without any circular motion."

H. POINCARÉ

THE MEASURE OF TIME*

1. So long as we do not go outside the domain of consciousness, the notion of time is relatively clear. Not only do we distinguish without difficulty present sensation from the remembrance of past sensations or the anticipation of future sensations, but we know perfectly well what we mean when we say that of two conscious phenomena which we remember, one was anterior to the other; or that, of two foreseen conscious phenomena, one will be anterior to the other.

When we say that two conscious facts are simultaneous, we mean that they profoundly interpenetrate, so that analysis can not separate them without mutilating them.

The order in which we arrange conscious phenomena does not admit of any arbitrariness. It is imposed upon us and of it we can change nothing.

I have only a single observation to add. For an aggregate of sensations to have become a remembrance capable of classification in time, it must have ceased to be actual, we must have lost the sense of its infinite complexity, otherwise it would have remained present. It must, so to speak, have crystallized around a center of associations of ideas which will be a sort of label. It is only when they thus have lost all life that we can classify our memories in time as a botanist arranges dried flowers in his herbarium.

But these labels can only be finite in number. On that score, psychologic time should be discontinuous. Whence comes the feeling that between any two instants there are others? We arrange our recollections in time, but we know that there remain empty compartments. How could that be, if time were not a form pre-existent in our minds? How could we know there were empty compartments, if these compartments were revealed to us only by their content?

2. But that is not all; into this form we wish to put not only the phenomena of our own consciousness, but those of which other consciousnesses are the theater. But more, we wish to put there physical facts, these I know

not what with which we people space and which no consciousness sees directly. This is necessary because without it science could not exist. In a word, psychologic time is given to us and must needs create scientific and physical time. There the difficulty begins, or rather the difficulties, for there are two.

Think of two consciousnesses, which are like two worlds impenetrable one to the other. By what right do we strive to put them into the same mold, to measure them by the same standard? Is it not as if one strove to measure length with a gram or weight with a meter? And besides, why do we speak of measuring? We know perhaps that some fact is anterior to some other, but not *by how much* it is anterior.

Therefore two difficulties: (1) Can we transform psychologic time, which is qualitative, into a quantitative time? (2) Can we reduce to one and the same measure facts which transpire in different worlds?

3. The first difficulty has long been noticed; it has been the subject of long discussions and one may say the question is settled. *We have not a direct intuition of the equality of two intervals of time.* The persons who believe they possess this intuition are dupes of an illusion. When I say, from noon to one the same time passes as from two to three, what meaning has this affirmation?

The least reflection shows that by itself it has none at all. It will only have that which I choose to give it, by a definition which will certainly possess a certain degree of arbitrariness. Psychologists could have done without this definition; physicists and astronomers could not; let us see how they have managed.

To measure time they use the pendulum and they suppose by definition that all the beats of this pendulum are of equal duration. But this is only a first approximation; the temperature, the resistance of the air, the barometric pressure, make the pace of the pendulum vary. If we could escape these sources of error, we should obtain a much closer approximation, but it would still be only an approximation. New causes, hitherto neglected, electric, magnetic or others, would introduce minute perturbations.

In fact, the best chronometers must be corrected from time to time, and the corrections are made by the aid of astronomic observations; arrangements are made so that the sidereal clock marks the same hour when the same star passes the meridian. In other words, it is the sidereal day, that is, the duration of the rotation of the earth, which is the constant unit of time. It is supposed, by a new definition substituted for that based on the beats of the pendulum, that two complete rotations of the earth about its axis have the same duration.

However, the astromoners are still not content with this definition. Many of them think that the tides act as a check on our globe, and that the rotation of the earth is becoming slower and slower. Thus would be explained the apparent acceleration of the motion of the moon, which would seem to be going more rapidly than theory permits because our watch, which is the earth, is going slow.

4. All this is unimportant, one will say; doubtless our instruments of measurement are imperfect, but it suffices that we can conceive a perfect instrument. This ideal can not be reached, but it is enough to have conceived it and so to have put rigor into the definition of the unit of time.

The trouble is that there is no rigor in the definition. When we use the pendulum to measure time, what postulate do we implicitly admit? *It is that the duration of two identical phenomena is the same;* or, if you prefer, that the same causes take the same time to produce the same effects.

And at first blush, this is a good definition of the equality of two durations. But take care. Is it impossible that experiment may some day contradict our postulate?

Let me explain myself. I suppose that at a certain place in the world the phenomenon α happens, causing as consequence at the end of a certain time the effect α'. At another place in the world very far away from the first, happens the phenomenon β, which causes as consequence the effect β'. The phenomena α and β are simultaneous, as are also the effects α' and β'.

Later, the phenomenon α is reproduced under approximately the same conditions as before, and *simultaneously* the phenomenon β is also reproduced at a very distant place in the world and almost under the same circumstances. The effects α' and β' also take place. Let us suppose that the effect α' happens perceptibly before the effect β'.

If experience made us witness such a sight, our postulate would be contradicted. For experience would tell us that the first duration $\alpha\alpha'$ is equal to the first duration $\beta\beta'$ and that the second duration $\alpha\alpha'$ is less than the second duration $\beta\beta'$. On the other hand, our postulate would require that the two durations $\alpha\alpha'$ should be equal to each other, as likewise the two durations $\beta\beta'$. The equality and the inequality deduced from experience would be incompatible with the two equalites deduced from the postulate.

Now can we affirm that the hypotheses I have just made are absurd? They are in no wise contrary to the principle of contradiction. Doubtless they could not happen without the principle of sufficient reason seeming violated. But to justify a definition so fundamental I should prefer some other guarantee.

5. But that is not all. In physical reality one cause does not produce a given effect, but a multitude of distinct causes contribute to produce it, without our having any means of discriminating the part of each of them.

Physicists seek to make this distinction; but they make it only approximately, and, however they progress, they never will make it except approximately. It is approximately true that the motion of the pendulum is due solely to the earth's attraction; but in all rigor every attraction, even of Sirius, acts on the pendulum.

Under these conditions, it is clear that the causes which have produced a certain effect will never be reproduced except approximately. Then we should modify our postulate and our definition. Instead of saying: 'The same causes take the same time to produce the same effects,' we should say: 'Causes almost identical take almost the same time to produce almost the same effects.'

Our definition therefore is no longer anything but approximate. Besides, as M. Calinon very justly remarks in a recent memoir:[1]

One of the circumstances of any phenomenon is the velocity of the earth's rotation; if this velocity of rotation varies, it constitutes in the reproduction of this phenomenon a circumstance which no longer remains the same. But to suppose this velocity of rotation constant is to suppose that we know how to measure time.

Our definition is therefore not yet satisfactory; it is certainly not that which the astronomers of whom I spoke above implicitly adopt, when they affirm that the terrestrial rotation is slowing down.

What meaning according to them has this affirmation? We can only understand it by analyzing the proofs they give of their proposition. They say first that the friction of the tides producing heat must destroy *vis viva*.

They invoke therefore the principle of *vis viva*, or of the conservation of energy.

They say next that the secular acceleration of the moon, calculated according to Newton's law, would be less than that deduced from observations unless the correction relative to the slowing down of the terrestrial rotation were made. They invoke therefore Newton's law. In other words, they define duration in the following way: time should be so defined that Newton's law and that of *vis viva* may be verified. Newton's law is an experimental truth; as such it is only approximate, which shows that we still have only a definition by approximation.

If now it be supposed that another way of measuring time is adopted, the experiments on which Newton's law is founded would none the less have the same meaning. Only the enunciation of the law would be different, because it would be translated into another language; it would evidently be much less simple. So that the definition implicitly adopted by the astronomers may be summed up thus: Time should be so defined that the equations of mechanics may be as simple as possible. In other words, there is not one way of measuring time more true than another; that which is generally adopted is only more *convenient*. Of two watches, we have no right to say that the one goes true, the other wrong; we can only say that it is advantageous to conform to the indications of the first.

The difficulty which has just occupied us has been, as I have said, often pointed out; among the most recent works in which it is considered, I may mention, besides M. Calinon's little book, the treatise on mechanics of Andrade.

6. The second difficulty has up to the present attracted much less attention; yet it is altogether analogous to the preceding; and even, logically, I should have spoken of it first.

Two psychological phenomena happen in two different consciousnesses; when I say they are simultaneous, what do I mean? When I say that a physical phenomenon, which happens outside of every consciousness, is before or after a psychological phenomenon, what do I mean?

In 1572, Tycho Brahe noticed in the heavens a new star. An immense conflagration had happened in some far distant heavenly body; but it had happened long before; at least two hundred years were necessary for the light from that star to reach our earth. This conflagration therefore happened before the discovery of America. Well, when I say that; when, considering this gigantic phenomenon, which perhaps had no witness, since the satellites of that star were perhaps uninhabited, I say this phenomenon is anterior to the formation of the visual image of the isle of Española in the consciousness of Christopher Columbus, what do I mean?

A little reflection is sufficient to understand that all these affirmations have by themselves no meaning. They can have one only as the outcome of a convention.

7. We should first ask ourselves how one could have had the idea of putting into the same frame so many worlds impenetrable to one another. We should like to represent to ourselves the external universe, and only by so doing could we feel that we understood it. We know we never can attain this representation: our weakness is too great. But at least we desire the ability to conceive an infinite intelligence for which this representation could be possible, a sort of great consciousness which should see all, and which should classify all *in its time*, as we classify, *in our time*, the little we see.

This hypothesis is indeed crude and incomplete because this supreme intelligence would be only a demigod; infinite in one sense, it would be limited in another, since it would have only an imperfect recollection of the past; and it could have no other, since otherwise all recollections would be equally present to it and for it there would be no time. And yet when we speak of time, for all which happens outside of us, do we not unconsciously adopt this hypothesis; do we not put ourselves in the place of this imperfect god; and do not even the atheists put themselves in the place where god would be if he existed?

What I have just said shows us, perhaps, why we have tried to put all physical phenomena into the same frame. But that can not pass for a definition of simultaneity, since this hypothetical intelligence, even if it existed, would be for us impenetrable. It is therefore necessary to seek something else.

8. The ordinary definitions which are proper for psychologic time would suffice us no more. Two simultaneous psychologic facts are so closely bound together that analysis can not separate without mutilating them. Is it the same with two physical facts? Is not my present nearer my past of yesterday than the present of Sirius?

It has also been said that two facts should be regarded as simultaneous when the order of their succession may be inverted at will. It is evident that this definition would not suit two physical facts which happen far from one another, and that, in what concerns them, we no longer even understand what this reversibility would be; besides, succession itself must first be defined.

9. Let us then seek to give an account of what is understood by simultaneity or antecedence, and for this let us analyze some examples.

I write a letter; it is afterward read by the friend to whom I have addressed it. There are two facts which have had for their theater two different consciousnesses. In writing this letter I have had the visual image of it, and my friend has had in his turn this same visual image in reading the letter. Though these two facts happen in impenetrable worlds, I do not hesitate to regard the first as anterior to the second, because I believe it is its cause.

I hear thunder, and I conclude there has been an electric discharge; I do not hesitate to consider the physical phenomenon as anterior to the auditory image perceived in my consciousness, because I believe it is its cause.

Behold then the rule we follow, and the only one we can follow: when a phenomenon appears to us as the cause of another, we regard it as anterior. It is therefore by cause that we define time; but most often, when two facts appear to us bound by a constant relation, how do we recognize which is the cause and which the effect? We assume that the anterior fact, the antecedent, is the cause of the other, of the consequent. It is then by time that we define cause. How save ourselves from this *petitio principii?*

We say now *post hoc, ergo propter hoc;* now *propter hoc, ergo post hoc;* shall we escape from this vicious circle?

10. Let us see, not how we succeed in escaping, for we do not completely succeed, but how we try to escape.

I execute a voluntary act A and I feel afterward a sensation D, which I regard as a consequence of the act A; on the other hand, for whatever reason, I infer that this consequence is not immediate, but that outside my consciousness two facts B and C, which I have not witnessed, have happened, and in such a way that B is the effect of A, that C is the effect of B, and D of C.

But why? If I think I have reason to regard the four facts *A*, *B*, *C*, *D*, as bound to one another by a causal connection, why range them in the causal order *A B C D*, and at the same time in the chronologic order *A B C D*, rather than in any other order?

I clearly see that in the act *A* I have the feeling of having been active, while in undergoing the sensation *D* I have that of having been passive. This is why I regard *A* as the initial cause and *D* as the ultimate effect; this is why I put *A* at the beginning of the chain and *D* at the end; but why put *B* before *C* rather than *C* before *B*?

If this question is put, the reply ordinarily is: we know that it is *B* which is the cause of *C* because we always see *B* happen before *C*. These two phenomena, when witnessed, happen in a certain order; when analogous phenomena happen without witness, there is no reason to invert this order.

Doubtless, but take care; we never know directly the physical phenomena *B* and *C*. What we know are sensations *B'* and *C'* produced respectively by *B* and *C*. Our consciousness tells us immediately that *B'* precedes *C'* and we suppose that *B* and *C* succeed one another in the same order.

This rule appears in fact very natural, and yet we are often led to depart from it. We hear the sound of the thunder only some seconds after the electric discharge of the cloud. Of two flashes of lightning, the one distant, the other near, can not the first be anterior to the second, even though the sound of the second comes to us before that of the first?

11. Another difficulty; have we really the right to speak of the cause of a phenomenon? If all the parts of the universe are interchained in a certain measure, any one phenomenon will not be the effect of a single cause, but the resultant of causes infinitely numerous; it is, one often says, the consequence of the state of the universe a moment before. How enunciate rules applicable to circumstances so complex? And yet it is only thus that these rules can be general and rigorous.

Not to lose ourselves in this infinite complexity, let us make a simpler hypothesis. Consider three stars, for example, the sun, Jupiter and Saturn; but, for greater simplicity, regard them as reduced to material points and isolated from the rest of the world. The positions and the velocities of three bodies at a given instant suffice to determine their positions and velocities at the following instant, and consequently at any instant. Their positions at the instant t determine their positions at the instant $t+h$ as well as their positions at the instant $t-h$.

Even more; the position of Jupiter at the instant t, together with that of Saturn at the instant $t+a$, determines the position of Jupiter at any instant and that of Saturn at any instant.

The aggregate of positions occupied by Jupiter at the instant $t+e$ and Saturn at the instant $t+a+e$ is bound to the aggregate of positions occupied by Jupiter at the instant t and Saturn at the instant $t+a$, by laws as precise as that of Newton, though more complicated. Then why not regard one of these aggregates as the cause of the other, which would lead to considering as simultaneous the instant t of Jupiter and the instant $t+a$ of Saturn?

In answer there can only be reasons, very strong, it is true, of convenience and simplicity.

12. But let us pass to examples less artificial; to understand the definition implicitly supposed by the savants, let us watch them at work and look for the rules by which they investigate simultaneity.

I will take two simple examples, the measurement of the velocity of light and the determination of longitude.

When an astronomer tells me that some stellar phenomenon, which his telescope reveals to him at this moment, happened, nevertheless, fifty years ago, I seek his meaning, and to that end I shall ask him first how he knows it, that is, how he has measured the velocity of light.

He has begun by *supposing* that light has a constant velocity, and in particular that its velocity is the same in all directions. That is a postulate without which no measurement of this velocity could be attempted. This postulate could never be verified directly by experiment; it might be contradicted by it if the results of different measurements were not concordant. We should think ourselves fortunate that this contradiction has not happened and that the slight discordances which may happen can be readily explained.

The postulate, at all events, resembling the principle of sufficient reason, has been accepted by everybody; what I wish to emphasize is that it furnishes us with a new rule for the investigation of simultaneity, entirely different from that which we have enunciated above.

This postulate assumed, let us see how the velocity of light has been measured. You know that Roemer used eclipses of the satellites of Jupiter, and sought how much the event fell behind its prediction. But how is this prediction made? It is by the aid of astronomic laws: for instance Newton's law.

Could not the observed facts by just as well explained if we attributed to the velocity of light a little different value from that adopted, and supposed Newton's law only approximate? Only this would lead to replacing Newton's law by another more complicated. So for the velocity of light a value is adopted, such that the astronomic laws compatible with this value may be as simple as possible. When navigators or geographers determine a longitude, they have to solve just the problem we are discussing; they must, without being at Paris, calculate Paris time. How do they accomplish it? They carry a chronometer set for Paris. The qualitative problem of simultaneity is made to depend upon the quantitative problem of the measurement of time. I need not take up the difficulties relative to this latter problem, since above I have emphasized them at length.

Or else they observe an astronomic phenomenon, such as an eclipse of the moon, and they suppose that this phenomenon is perceived simultaneously from all points of the earth. That is not altogether true, since the propagation of light is not instantaneous; if absolute exactitude were desired, there would be a correction to make according to a complicated rule.

Or else finally they use the telegraph. It is clear first that the reception of the signal at Berlin, for instance, is after the sending of this signal from Paris. This is the rule of cause and effect analyzed above. But how much after? In general, the duration of the transmission is neglected and the two events are regarded as simultaneous. But, to be rigorous, a little correction would still have to be made by a complicated calculation; in practise it is not made, because it would be well within the errors of observation; its theoretic necessity is none the less from our point of view, which is that of a rigorous definition. From this discussion, I wish to emphasize two things: (1) The rules applied are exceedingly various. (2) It is difficult to separate the qualitative problem of simultaneity from the quantitative problem of the measurement of time; no matter whether a chronometer is used, or whether account must be taken of a velocity of transmission, as that of light, because such a velocity could not be measured without *measuring* a time.

13. To conclude: We have not a direct intuition of simultaneity, nor of the equality of two durations. If we think we have this intuition, this is an illusion. We replace it by the aid of certain rules which we apply almost always without taking count of them.

But what is the nature of these rules? No general rule, no rigorous rule; a multitude of little rules applicable to each particular case.

These rules are not imposed upon us and we might amuse ourselves in inventing others; but they could not be cast aside without greatly complicating the enunciation of the laws of physics, mechanics and astronomy.

We therefore choose these rules, not because they are true, but because they are the most convenient, and we may recapitulate them as follows: "The simultaneity of two events, or the order of their succession, the equality of two durations, are to be so defined that the enunciation of the natural laws may be as simple as possible. In other words, all these rules, all these definitions are only the fruit of an unconscious opportunism."

NOTES

* From 'The Value of Science' in *The Foundations of Physics*, The Science Press, New York, 1913, pp. 223–234.
[1] *Étude sur les diverses grandeurs*, Paris, Gauthier-Villars, 1897.

A. EINSTEIN

THE INADEQUACY OF CLASSICAL MODELS OF AETHER*

RELATIVITY AND THE ETHER

Why is that alongside of the notion derived by abstraction from everyday life, of ponderable matter the physicists set the notion of the existence of another sort of matter, the ether? The reason lies no doubt in those phenomena which gave rise to the theory of forces acting at a distance, and in those properties of light which led to the wave-theory. Let us shortly consider these two things.

Non-physical thought knows nothing of forces acting at a distance. When we try to subject our experiences of bodies by a complete causal scheme, there seems at first sight to be no reciprocal interaction except what is produced by means of immediate contact, e.g., the transmission of motion by impact, pressure or pull, heating or inducing combustion by means of a flame, etc. To be sure, gravity, that is to say, a force acting at a distance, does play an important part in every day experience. But since the gravity of bodies presents itself to us in common life as something constant, dependent on no variable temporal or spatial cause, we do not ordinarily think of any cause in connection with it and thus are not conscious of its character as a force acting at a distance. It was not till Newton's theory of gravitation that a cause was assigned to it; it was then explained as a force acting at a distance, due to mass. Newton's theory certainly marks the greatest step ever taken in linking up natural phenomena causally. And yet his contemporaries were by no means satisfied with it, because it seemed to contradict the principle derived from the rest of experience that reciprocal action only takes place through direct contact, not by direct action at a distance, without any means of transmission.

Man's thirst for knowledge only acquiesces in such a dualism reluctantly. How could unity in our conception of natural forces be saved? People could either attempt to treat the forces which appear to us to act by contact as acting at a distance, though only making themselves felt at very small distances; this was the way generally chosen by Newton's successors, who were completely under the spell of his teaching. Or they could take the line that Newton's forces acting at a distance only *appeared* to act thus directly; that they were really transmitted by a medium which permeated space, either by motions or by an elastic deformation of this medium. Thus the desire for unity in our view of the nature of these forces led to the hypothesis of the ether. It certainly led to no advance in the theory of gravitation or in physics generally to begin with, so that people got into the habit of treating Newton's law of force as an irreducible axiom. But the ether hypothesis was bound always to play a part, even if it was mostly a latent one at first in the thinking of physicists.

When the extensive similarity which exists between the properties of light and those of the elastic waves in ponderable bodies was revealed in the first half of the nineteenth century, the ether hypothesis acquired a new support. It seemed beyond a doubt that light was to be explained as the vibration of an elastic, inert medium filling the whole of space. It also seemed to follow necessarily from the polarizability of light that this medium, the ether, must be of the nature of a solid body, because transverse waves are only possible in such a body and not in a fluid. This inevitably led to the theory of the 'quasi-rigid' luminiferous ether, whose parts are incapable of any motion with respect to each other beyond the small deformations which correspond to the waves of light.

This theory, also called the theory of the stationary luminiferous ether, derived strong support from the experiments, of fundamental importance for the special theory of relativity too, of Fizeau, which proved conclusively that the luminiferous ether does not participate in the motions of bodies. The phenomenon of aberration also lent support to the theory of the quasi-rigid ether.

The evolution of electrical theory along the lines laid down by Clerk Maxwell and Lorentz gave a most peculiar and unexpected turn to the development of our ideas about the ether. For Clerk Maxwell himself the ether was still an entity with purely mechanical properties, though of a far more complicated kind than those of tangible solid bodies. But neither Maxwell nor his successors succeeded in thinking out a mechanical model for the ether capable of providing a satisfactory mechanical interpretation of Maxwell's laws of the electro-dynamic field. The laws were clear and simple, the mechanical interpretations clumsy and contradictory. Almost imperceptibly theoretical physicists adapted themselves to this state of affairs, which was a most depressing one from the point of view of their mechanistic program, especially under influence of the electro-dynamic researches of Heinrich Hertz. Whereas they had formerly demanded of an ultimate theory that it should be based upon fundamental concepts of a purely mechanical kind (e.g., mass-densities, velocities, deformations, forces of gravitation), they gradually became accustomed to admitting strengths of electrical and magnetic fields as fundamental concepts alongside of the mechanical ones, without insisting on a mechanical interpretation of them. The purely mechanistic view of nature was thus abandoned. This change led to a dualism in the sphere of fundamental concepts which was in the long run intolerable. To escape from it people took the reverse line of trying to reduce mechanical concepts to electrical ones. The experiments with β-rays and high-velocity cathode rays did much to shake confidence in the strict validity of Newton's mechanical equations.

Heinrich Hertz took no steps towards mitigating this dualism. Matter appears as the substratum not only of velocities, kinetic energy, and mechanical forces of gravity, but also of electro-magnetic fields. Since such fields are also found in a vacuum – i.e., in the unoccupied ether – the ether also appears as the substratum of electro-magnetic fields, entirely similar in nature to ponderable matter and ranking alongside it. In the presence of matter it shares in the motions of the latter and has a velocity everywhere in empty space; the etheric velocity nowhere changes discontinuously. There is no fundamental distinction between the Hertzian ether and ponderable matter (which partly consists of the ether).

Hertz's theory not only suffered from the defect that it attributed to matter and the ether mechanical and electrical properties, with no rational connection between them; it was also inconsistent with the result of Fizeau's famous experiment on the velocity of the propagation of light in a liquid in motion and other well authenticated empirical facts.

Such was the position when H. A. Lorentz entered the field. Lorentz brought theory into harmony with experiment, and did it by a marvelous simplification of basic concepts. He achieved this advance in the science of electricity, the most important since Clerk Maxwell, by divesting the ether of its mechanical, and matter of its electro-magnetic properties. Inside material bodies no less than in empty space the ether alone, not atomically conceived matter, was the seat of electro-magnetic fields. According to Lorentz the elementary particles of matter are capable *only* of executing movements; their electro-magnetic activity is entirely due to the fact that they carry electric charges. Lorentz thus succeeded in reducing all electro-magnetic phenomena to Maxwell's equations for a field in vacuo.

As regards the mechanical nature of Lorentz's ether, one might say of it, with a touch of humor, that immobility was the only mechanical property which Lorentz left it. It may be added that the whole difference which the special theory of relativity made in our conception of the ether lay in this, that it divested the ether of its last mechanical quality, namely immobility. How this is to be understood I will explain immediately.

The Maxwell-Lorentz theory of the electro-magnetic field served as the model for the space-time theory and the kinematics of the special theory of relativity. Hence it satisfies the conditions of the special theory of relativity; but looked at from the standpoint of the latter, it takes on a new aspect. If C is a co-ordinate system in respect to which the Lorentzian ether is at rest, the Maxwell-Lorentz equations hold good first of all in regard to C. According to the special theory of relativity these same equations hold good in exactly the same sense in regard to any new co-ordinate system C, which is in uniform translatory motion with respect to C. Now comes the anxious question, Why should I distinguish the system C, which is physically perfectly equivalent to the systems C', from the latter by assuming that the ether is at rest in respect to it? Such an asymmetry of the theoretical structure, to which there is no corresponding asymmetry in the system of empirical facts, is intolerable to the theorist. In my view the physical equivalence of C and C' with the assumption that the ether is at rest in respect to C but in motion with respect to C', though not absolutely wrong from a logical point of view, is nevertheless unsatisfactory.

The most obvious line to adopt in the face of this situation seemed to be the following: – There is no such thing as the ether. The electro-magnetic fields are not states of a medium but independent realities, which cannot be reduced to terms of anything else and are bound to no substratum, any more than are the atoms of ponderable matter. This view is rendered the more natural by the fact that, according to Lorentz's theory, electromagnetic radiation carries impulse and energy like ponderable matter, and that matter and radiation, according to the special theory of relativity, are both of them only particular forms of distributed energy, inasmuch as ponderable matter loses its exceptional position and merely appears as a particular form of energy.

In the meantime more exact reflection shows that this denial of the existence of the ether is not demanded by the restricted principle of relativity. We can assume the existence of an ether; but we must abstain from ascribing a definite state of motion to it, i.e., we must divest it by abstraction of the last mechanical characteristic which Lorentz left it. We shall see later on that this way of looking at it, the intellectual possibility of which I shall try to make clearer by a comparison that does not quite fit at all points, is justified by the results of the general theory of relativity.

Consider waves on the surface of water. There are two quite different things about this phenomenon which may be described. One can trace the progressive changes which take place in the undulating surface where the water and the air meet. One can also – with the aid of small floating bodies, say – trace the progressive changes in the position of the individual particles. If there were in the nature of the case no such floating bodies to aid us in tracing the movement of the particles of liquid, if nothing at all could be observed in the whole procedure except the fleeting changes in the position of the space occupied by the water, we should have no ground for supposing that the water consists of particles. But we could none the less call it a medium.

Something of the same sort confronts us in the electro-magnetic field. We may conceive the field as consisting of lines of force. If we try to think of these lines of force as something material in the ordinary sense of the word, there is a temptation to ascribe the dynamic phenomena involved to their motion, each single line being followed out through time. It is, however, well known that this way of looking at the matter leads to contradictions.

Generalizing, we must say that we can conceive of extended physical objects to which the concept of motion cannot be applied. They must not be thought of as consisting of particles, whose course can be followed out separately through time. In the language of Minkowski this is expressed as follows: – Not every extended entity in the four-dimensional world can be regarded as composed of world-lines. The special principle of relativity forbids us to regard the ether as composed of particles the movements of which can be followed out through time, but the theory is not incompatible with the ether hypothesis as such. Only we must take care not to ascribe a state of motion to the ether.

From the point of view of the special theory of relativity the ether hypothesis does certainly seem an empty one at first sight. In the equations of an electro-magnetic field, apart from the density of the electrical charge nothing appears except the strength of the field. The course of electromagnetic events in a vacuum seems to be completely determined by that inner law, independently of other physical quantities. The electro-magnetic field seems to be the final irreducible reality, and it seems superfluous at first sight to postulate a homogeneous, isotropic etheric medium, of which these fields are to be considered as states.

On the other hand, there is an important argument in favor of the hypothesis of the ether. To deny the existence of the ether means, in the last analysis, denying all physical properties to empty space. But such a view is inconsistent with the fundamental facts of mechanics. The mechanical behavior of a corporeal system floating freely in empty space depends not only on the relative positions (intervals) and velocities of its masses, but also on its state of rotation which cannot be regarded physically speaking as a property belonging to the system as such. In order to be able to regard the rotation of a system at least formally as something real, Newton regarded space as objective. Since he regards his absolute space as a real thing, rotation with respect to an absolute space is also something real to him. Newton could equally well have called his absolute space 'the ether'; the only thing that matters is that in addition to observable objects another imperceptible entity has to be regarded as real, in order for it to be possible to regard acceleration, or rotation, as something real.

Mach did indeed try to avoid the necessity of postulating an imperceptible real entity, by substituting in mechanics a mean velocity with respect to the totality of masses in the world for acceleration with respect to absolute space. But inertial resistance with respect to the relative acceleration of distant masses presupposes direct action at a distance. Since the modern physicist does not consider himself entitled to assume that, this view brings him back to the ether, which has to act as the medium of inertial action. This conception of the ether to which Mach's approach leads, differs in important respects from that of Newton, Fresnel and Lorentz. Mach's ether not only *conditions* the behavior of inert masses but is also conditioned, as regards its state, by them.

Mach's notion finds its full development in the ether of the general theory of relativity. According to this theory the metrical properties of the space-time continuum in the neighborhood of separate space-time points are different and conjointly conditioned by matter existing outside the region in question. This spatio-temporal variability of the relations of scales and clocks to each other, or the knowledge that 'empty space' is, physically speaking, neither homogeneous nor isotropic, which compels us to describe its state by means of ten functions, the gravitational potentials $g\mu v$, has no doubt finally disposed of the notion that space is physically empty. But this has also once more given the ether notion a definite content – though one very different from that of the ether of the mechanical wave-theory of light. The ether of the general theory of relativity is a medium which is itself free of *all* mechanical and kinematic properties, but helps to determine mechanical (and electro-magnetic) happenings.

The radical novelty in the ether of the general theory of relativity as against the ether of Lorentz lies in this, that the state of the former at every point is determined by the laws of its relationship with matter and with the state of the ether at neighboring points expressed in the form of differential equations, whereas the state of Lorentz's ether in the absence of electro-magnetic fields is determined by nothing outside it and is the same everywhere. The ether of the general theory of relativity can be transformed intellectually into Lorentz's through the substitution of constants for the spatial functions which describe its state, thus neglecting the causes conditioning the latter. One may therefore say that the ether of the general theory of relativity is derived through relativization from the ether of Lorentz.

The part which the new ether is destined to play in the physical scheme of the future is still a matter of uncertainty. We know that it determines both material relations in the space-time continuum, e.g., the possible configurations of solid bodies, and also gravitational fields; but we do not know whether it plays a material part in the structure of the electric particles of which matter is made up. Nor do we know whether its structure only differs materially from that of Lorentz's in the proximity of ponderable masses, whether, in fact, the geometry of spaces of cosmic extent is, taken as a whole, almost Euclidean. We can, however, maintain on the strength of the relativistic equations of gravitation that spaces of cosmic proportions must depart from Euclidean behavior if there is a positive mean density of matter, however small, in the Universe. In this case the Universe must necessarily form a closed space of finite size, this size being determined by the value of the mean density of matter.

If we consider the gravitational field and the electro-magnetic field from the standpoint of the ether hypothesis, we find a notable fundamental difference between the two. No space and no portion of space without gravitational potentials; for these give it its metrical properties without which it is not thinkable at all. The existence of the gravitational field is directly bound up with the existence of space. On the other hand a portion of space without an electro-magnetic field is perfectly conceivable, hence the electro-magnetic field, in contrast to the gravitational field, seems in a sense to be connected with the ether only in a secondary way, inasmuch as the formal nature of the electro-magnetic field is by no means determined by the gravitational ether. In the present state of theory it looks as if the electro-magnetic field, as compared with the gravitational field, were based on a completely new formal motive; as if nature, instead of endowing the gravitational ether with fields of the electro-magnetic type, might equally well have endowed it with fields of a quite different type, for example fields with a scalar potential.

Since according to our present-day notions the primary particles of matter are also, at bottom, nothing but condensations of the electro-magnetic field, our modern schema of the cosmos recognizes two realities which are conceptually quite independent of each other even though they may be causally connected, namely the gravitational ether and the electro-magnetic field, or – as one might call them – space and matter.

It would, of course, be a great step forward if we succeeded in combining the gravitational field and the electro-magnetic field into a single structure. Only so could the era in theoretical physics inaugurated by Faraday and Clerk Maxwell be brought to a satisfactory close.

The antithesis of ether and matter would then fade away, and the whole of physics would become a completely enclosed intellectual system, like geometry, kinematics and the theory of gravitation, through the general theory of relativity. An exceedingly brilliant attempt in this direction has been made by the mathematician H. Weyl; but I do not think that it will stand in the face of reality. Moreover, in thinking about the immediate future of theoretical physics we cannot unconditionally dismiss the possibility that the facts summarized in the quantum theory may set impassable limits to the field theory.

We may sum up as follows: According to the general theory of relativity space is endowed with physical qualities; in this sense, therefore, an ether exists. In accordance with the general theory of relativity space without an ether is inconceivable. For in such a space there would not only be no propagation of light, but no possibility of the existence of scales and clocks, and therefore no spatio-temporal distances in the physical sense. But this ether must not be thought of as endowed with the properties characteristic of ponderable media, as composed of particles the motion of which can be followed; nor may the concept of motion be applied to it.

NOTE

* From *The World as I see It* (Covici, Friede, New York, 1934) pp. 121-137 [tr. of *Mein Weltbild*, Amsterdam 1934; abridged reprint ed., Philosophical Library, N.Y. 1944 under the title *Essays in Science*, with this selection pp. 98-111].

H. MINKOWSKI

THE UNION OF SPACE AND TIME*

The views of space and time which I wish to lay before you have sprung from the soil of experimental physics, and therein lies their strength. They are radical. Henceforth space by itself, and time by itself, are doomed to fade away into mere shadows, and only a kind of union of the two will preserve an independent reality.

1. First of all I should like to show how it might be possible, setting out from the accepted mechanics of the present day, along a purely mathematical line of thought, to arrive at changed ideas of space and time. The equations of Newton's mechanics exhibit a two-fold invariance. Their form remains unaltered, firstly, if we subject the underlying system of spatial co-ordinates to any arbitrary *change of position;* secondly, if we change its state of motion, namely, by imparting to it any *uniform translatory motion*; furthermore, the zero point of time is given no part to play. We are accustomed to look upon the axioms of geometry as finished with, when we feel ripe for the axioms of mechanics, and for that reason the two invariances are probably rarely mentioned in the same breath. Each of them by itself signifies, for the differential equations of mechanics, a certain group of transformations. The existence of the first group is looked upon as a fundamental characteristic of space. The second group is preferably treated with disdain, so that we with untroubled minds may overcome the difficulty of never being able to decide, from physical phenomena, whether space, which is supposed to be stationary, may not be after all in a state of uniform translation. Thus the two groups, side by side, lead their lives entirely apart. Their utterly heterogeneous character may have discouraged any attempt to compound them. But it is precisely when they are compounded that the complete group, as a whole, gives us to think.

We will try to visualize the state of things by the graphic method. Let x, y, z be rectangular co-ordinates for space, and let t denote time. The objects of our perception invariably include places and times in combination. Nobody has ever noticed a place except at a time, or a time except at a place. But I still respect the dogma that both space and time have independent significance. A point of space at a point of time, that is, a system of values x, y, z, t, I will call a *world-point*. The multiplicity of all thinkable x, y, z, t systems of values we will christen the *world* With this most valiant piece of chalk I might project upon the blackboard four world-axes. Since merely one chalky axis, as it is, consists of molecules all a-thrill, and moreover is taking part in the earth's travels in the universe, it already affords us ample scope for abstraction; the somewhat greater abstraction associated with the number four is for the mathematician no infliction. Not to leave a yawning void anywhere, we will imagine that everywhere and everywhen there is something perceptible. To avoid saying 'matter' or 'electricity' I will use for this something the word 'substance'. We fix our attention on the substantial point which is at the world-point x, y, z, t, and imagine that we are able to recognize this substantial point at any other time. Let the vatiarions dx, dy, dz of the space co-ordinates of this substantial point correspond to a time element dt. Then we obtain, as an image, so to speak, of the everlasting career of the substantial point, a curve in the world, a *world-line*, the points of which can be referred unequivocally to the parameter t from $-\infty$ to $+\infty$. The whole universe is seen to resolve itself into similar world-lines, and I would fain anticipate myself by saying that in my opinion physical laws might find their most perfect expression as reciprocal relations between these world-lines.

The concepts, space and time, cause the x, y, z-manifold $t=0$ and its two sides $t>0$ and $t<0$ to fall asunder. If, for simplicity, we retain the same zero point of space and time, the first-mentioned group signifies in mechanics that we may subject the axes of x, y, z at $t=0$ to any rotation we choose about the origin, corresponding to the homogeneous linear transformations of the expression

$$x^2 + y^2 + z^2.$$

But the second group means that we may – also without changing the expression of the laws of mechanics – replace x, y, z, t by $x-\alpha t$, $y-\beta t$ $z-\gamma t$, t with any constant values of α, β, γ. Hence we may give to the time axis whatever direction we choose towards the upper half of the world, $t>0$. Now what has the requirement of orthogonality in space to do with this perfect freedom of the time axis in an upward direction?

To establish the connexion, let us take a positive parameter c, and consider the graphical representation of

$$c^2t^2 - x^2 - y^2 - z^2 = 1.$$

It consists of two surfaces separated by $t=0$, on the analogy of a hyperboloid of two sheets. We consider the sheet in the region $t>0$, and now take those homogeneous linear transformations of x, y, z, t into four new variables x', y', z', t', for which the expression for this sheet in the new

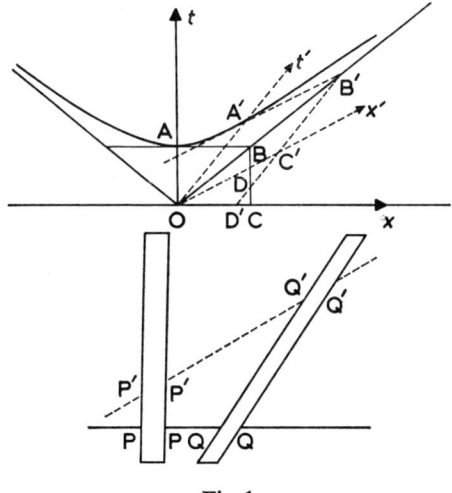

Fig. 1.

variables is of the same form. It is evident that the rotations of space about the origin pertain to these transformations. Thus we gain full comprehension of the rest of the transformations simply by taking into consideration one among them, such that y and z remain unchanged. We draw (Figure 1) the section of this sheet by the plane of the axes of x and t – the upper branch of the hyperbola $c^2t^2 - x^2 = 1$, with its asymptotes. From the origin O we draw any radius vector OA' of this branch of the hyperbola; draw the tangent to the hyperbola at A' to cut the asymptote on the right at B'; complete the parallelogram OA'B'C'; and finally, for subsequent use, produce B'C' to cut the axis of x at D'. Now if we take OC' and OA' as axes of oblique co-ordinates x', t', with the measures OC' = 1, OA' = 1/c

then that branch of the hyperbola again acquires the expression $c^2 t'^2 - x'^2 = 1$, $t' > 0$, and the transition from x, y, z, t to x', y', z', t' is one of the transformations in question. With these transformations we now associate the arbitrary displacements of the zero point of space and time, and thereby constitute a group of transformations, which is also, evidently, dependent on the parameter c. This group I denote by G_c.

If we now allow c to increase to infinity, and $1/c$ therefore to converge towards zero, we see from the figure that the branch of the hyperbola bends more and more towards the axis of x, the angle of the asymptotes becomes more and more obtuse, and that in the limit this special transformation changes into one in which the axis of t' may have any upward direction whatever, while x' approaches more and more exaclty to x. In view of this it is clear that group G_c in the limit when $c = \infty$, that is the group G_∞, becomes no other than that complete group which is appropriate to Newtonian mechanics. This being so, and since G_c is mathematically more intelligible than G_∞, it looks as though the thought might have struck some mathematician, fancy-free, that after all, as a matter of fact, natural phenomena do not possess an invariance with the group G_∞, but rather with a group G_c, c being finite and determinate, but in ordinary units of measure, *extremely great*. Such a premonition would have been an extraordinary triumph for pure mathematics. Well, mathematics, though it now can display only staircase-wit, has the satisfaction of being wise after the event, and is able, thanks to its happy antecedents, with its senses sharpened by an unhampered outlook to far horizons, to grasp forthwith the far-reaching consequences of such a metamorphosis of our concept of nature.

I will state at once what is the value of c with which we shall finally be dealing. It is the velocity of the propagation of light in empty space. To avoid speaking either of space or of emptiness, we may define this magnitude in another way, as the ratio of the electromagnetic to the electrostatic unit of electricity.

The existence of the invariance of natural laws for the relevant group G_c would have to be taken, then, in this way:

From the totality of natural phenomena it is possible, by successively enhanced approximations, to derive more and more exactly a system of reference x, y, z, t, space and time, by means of which these phenomena then presents themselves in agreement with definite laws. But when this is done, this system of reference is by no means unequivocally determined by the phenomena. *It is still possible to make any change in the system of reference that is in conformity with the transformations of the group G_c, and leave the expression of the laws of nature unaltered.*

For example, in correspondence with the figure described above, we may also designate time t', but then must of necessity, in connexion therewith, define space by the manifold of the three parameters x', y, z, in which case physical laws would be expressed in exactly the same way by means of x', y, z, t' as by means of x, y, z, t. We should then have in the world no longer *space*, but an infinite number of spaces, analogously as there are in three-dimensional space an infinite number of planes. Three-dimensional geometry becomes a chapter in four-dimensional physics. Now you know why I said at the outset that space and time are to fade away into shadows, and only a world in itself will subsist.

2. The question now is, what are the circumstances which force this changed conception of space and time upon us? Does it actually never contradict experience? And finally, is it advantageous for describing phenomena?

Before going into these questions, I must make an important remark. If we have in any way individualized space and time, we have, as a world-line corresponding to a stationary substantial point, a straight line parallel to the axis of t; corresponding to a substantial point in uniform motion, a straight line at an angle to the axis of t; to a substantial point in varying motion, a world-line in some form of a curve. If at any world-point x, y, z, t we take the world-line passing through that point, and find it parallel to any radius vector OA' of the above-mentioned hyperboloidal sheet, we can introduce OA' as a new axis of time, and with the new concepts of space and time thus given, the substance at the world-point concerned appears as at rest. We will now introduce this fundamental axiom:

The substance at any world-point may always, with the appropriate determination of space and time, be looked upon as at rest.

The axiom signifies that at any world-point the expression

$$c^2 \, dt^2 - dx^2 - dy^2 - dz^2$$

always has a positive value, or, what comes to the same thing, that any velocity v always proves less than c. Accordingly c would stand as the upper limit for all substantial velocities, and that is precisely what would reveal the deeper significance of the magnitude c. In this second form the first impression made by the axiom is not altogether pleasing. But we must bear in mind that a modified form of mechanics, in which the square root of this quadratic differential expression appears, will now make its way, so that cases with a velocity greater than that of light will henceforward play only some such part as that of figures with imaginary co-ordinates in geometry.

Now the impulse and true motive for assuming the group G_c came from the fact that the differential equation for the propagation of light in empty space possesses that group G_c.[1] On the other hand, the concept of rigid bodies has meaning only in mechanics satisfying the group G_∞. If we have a theory of optics with G_c, and if on the other hand there were rigid bodies, it is easy to see that one and the same direction of t would be distinguished by the two hyperboloidal sheets appropriate to G_c and G_∞, and this would have the further consequence, that we should be able, by employing suitable rigid optical instruments in the laboratory, to perceive some alteration in the phenomena when the orientation with respect to the direction of the earth's motion is changed. But all efforts directed towards this goal, in particular the famous interference experiment of Michelson, have had a negative result. To explain this failure, H. A. Lorentz set up an hypothesis, the success of which lies in this very invariance in optics for the group G_c. According to Lorentz any moving body must have undergone a contraction in the direction of its motion, and in fact with a velocity v, a contradiction in the ratio

$$1 : \sqrt{1 - v^2/c^2}.$$

This hypothesis sounds extremely fantastical, for the contraction is not to be looked upon as a consequence of resistances in the ether, or anything of that kind, but simply as a gift from above, – as an accompanying circumstance of the circumstance of motion.

I will now show by our figure that the Lorentzian hypothesis is completely equivalent to the new conception of space and time, which, indeed, makes the hypothesis much more intelligible. If for simplicity we disregard y and z, and imagine a world of one spatial dimension, then a parallel band, upright like the axis of t, and another inclining to the axis of t (see Figure 1) represent, respectively, the career of a body at rest or in uniform motion, preserving in each case a constant spatial extent. If OA' is parallel to the second band, we can introduce t' as the time, and x' as the space co-ordinate, and then the second body appears at rest, the first in uniform motion. We now assume that the first body, envisaged as at rest, has the length l, that is, the cross section PP of the first band on the axis of x is equal to l. OC, where OC denotes the unit of measure on the axis of x; and on the other hand, that the second body, envisaged as at rest, has the same length l, which then means that the cross section Q'Q' of the second band, measured parallel to the axis of x', is equal to l. OC'. We now have in these two bodies images of two equal Lorentzian electrons, one at rest and one in uniform motion. But if we retain the original co-ordinates x, t, we must give as the extent of the second electron the cross section of its appropriate band parallel to the axis of x. Now since Q'Q' $=l$.OC', it is evident that QQ$=l$.OD'. If dx/dt for the second band is equal to v, an easy calculation gives

$$\mathrm{OD'} = \mathrm{OC}\sqrt{1 - v^2/c^2},$$

therefore also PP:QQ$=1:\sqrt{1-v^2/c^2}$. But this is the meaning of Lorentz's hypothesis of the contraction of electrons in motion. If on the other hand we envisage the second electron as at rest, and therefore adopt the system of reference $x't'$, the length of the first must be denoted by the cross section P'P' of its band parallel to OC', and we should find the first electron in comparison with the second to be contracted in exactly the same proportion; for in the figure

$$\mathrm{P'P'}:\mathrm{Q'Q'} = \mathrm{OD}:\mathrm{OC'} = \mathrm{OD'}:\mathrm{OC} = \mathrm{QQ}:\mathrm{PP}.$$

Lorentz called the t' combination of x and t the local time of the electron in uniform motion, and applied a physical construction of this concept, for the better understanding of the hypothesis of contraction. But the credit of first recognizing clearly that the time of the one electron is just as good as that of the other, that is to say, that t and t' are to be treated identically, belongs to A. Einstein.[2] Thus time, as a concept unequivocally determined by phenomena, was first deposed from its high seat. Neither Einstein nor Lorentz made any attack on the concept of space, perhaps because in the above-mentioned special transformation, where the plane of x', t' coincides with the plane of x, t, an interpretation is possible by saying that the x-axis of space maintains its position. One may expect to find a corresponding violation of the concept of space appraised as another act of audacity on the part of the higher mathematics. Nevertheless, this further step is indispensable for the true understanding of the group G_c, and when it has been taken, the word *relativity-postulate* for the requirement of an invariance with the group G_c seems to me very feeble. Since the postulate comes to mean that only the four-dimensional world in space and time is given by phenomena, but that the projection in space and in time may still be undertaken with a certain degree of freedom, I prefer to call it the *postulate of the absolute world* (or briefly, the world-postulate).

3. The world-postulate permits identical treatment of the four co-ordinates x, y, z, t. By this means, as I shall now show, the forms in which the laws of physics are displayed gain in intelligibility. In particular the idea of acceleration acquires a clear-cut character.

I will use a geometrical manner of expression, which suggests itself at once if we tacitly disregard z in the triplex x, y, z. I take any world-point O as the zero-point of spacetime. The cone $c^2t^2 - x^2 - y^2 - z^2 = 0$ with apex 0 (Figure 2) consists of two parts, one with values $t<0$, the other with values $t>0$. The former, the front cone of O, consists, let us say, of all the world-points which 'send light to O,' the latter, the back cone of O, of all the world-points which 'receive light from O.' The territory bounded by the front cone alone, we may call 'before' O, that which is bounded by the back cone alone, 'after' O. The hyperboloidal sheet already discussed

$$F = c^2t^2 - x^2 - y^2 - z^2 = 1, t > 0$$

lies after O. The territory between the cones is filled by the one-sheeted hyperboloidal figures

$$-F = x^2 + y^2 + z^2 - c^2t^2 = k^2$$

for all constant positive values of k. We are specially interested in the hyperbolas with O as centre, lying on the latter figures. The single branches of these hyperbolas may be called briefly the internal hyperbolas with centre O. One of these branches, regarded as a world-line, would represent a motion which, for $t = -\infty$ and $t = +\infty$, rises asymptotically to the velocity of light, c.

If we now, on the analogy of vectors in space, call a directed length in the manifold of x, y, z, t a vector, we have to distinguish between the time-like vectors with directions from O to the sheet $+F=1, t>0$, and the space-like vectors

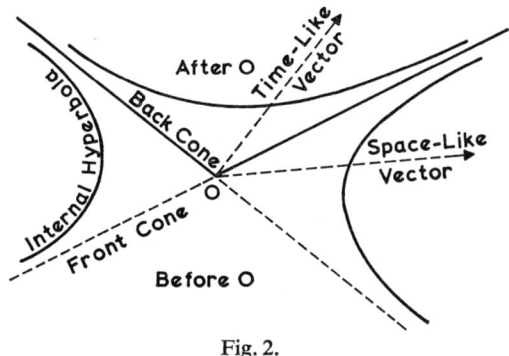

Fig. 2.

with directions from O to $-F=1$. The time axis may run parallel to any vector of the former kind. Any world-point between the front and back cones of O can be arranged by means of the system of reference so as to be simultaneous with O, but also just as well so as to be earlier than O or later than O. Any world-point within the front cone of O is necessarily always before O; any world-point within the back cone of O necessarily always after O. Corresponding to passing to the limit, $c=\infty$, there would be a complete flattening out of the wedge-shaped segment between the cones into the plane manifold $t=0$. In the figures this segment is intentionally drawn with different widths.

We divide up any vector we choose, e.g. that from O to x, y, z, t, into the four components x, y, z, t. If the directions of two vectors are, respectively, that of a radius vector OR from O to one of the surfaces $\mp F=1$, and that of a tangent RS at the point R of the same surface, the vectors are said to be normal to one another. Thus the condition that the vectors with components x, y, z, t, and x_1, y_1, z_1, t_1 may be normal to each other is
$$c^2 tt_1 - xx_1 - yy_1 - zz_1 = 0.$$

For the measurement of vectors in different directions the units of measure are to be fixed by assigning to a space-like vector from O to

$-F=1$ always the magnitude 1, and to a time-like vector from O to $+F=1$, $t>0$ always the magnitude $1/c$.

If we imagine at a world-point $P(x, y, z, t)$ the worldline of a substantial point running through that point, the magnitude corresponding to the time-like vector dx, dy, dz, dt laid off along the line is therefore

$$d\tau = \frac{1}{c}\sqrt{c^2 \, dt^2 - dx^2 - dy^2 - dz^2}.$$

The integral $\int d\tau = \tau$ of this amount, taken along the worldline from any fixed starting-point P_0 to the variable endpoint P, we call the proper time of the substantial point at P. On the world-line we regard x, y, z, t – the components of the vector OP – as functions of the proper time τ; denote their first differential coefficients with respect to τ by $\dot{x}, \dot{y}, \dot{z}, \dot{t}$; their second differential coefficients with respect to τ by $\ddot{x}, \ddot{y}, \ddot{z}, \ddot{t}$; and give names to the appropriate vectors, calling the derivative of the vector OP with respect to τ the velocity vector at P, and the derivative of this velocity vector with respect to τ the acceleration vector at P. Hence, since

$$c^2\dot{t}^2 - \dot{x}^2 - \dot{y}^2 - \dot{z}^2 = c^2,$$

we have
$$c^2\dot{t}\ddot{t} - \dot{x}\ddot{x} - \dot{y}\ddot{y} - \dot{z}\ddot{z} = 0,$$

i.e. the velocity vector is the time-like vector of unit magnitude in the direction of the world-line at P, and the acceleration vector at P is normal to the velocity vector at P, and is therefore in any case a space-like vector.

Now, as is readily seen, there is a definite hyperbola which has three infinitely proximate points in common with the world-line at P, and whose

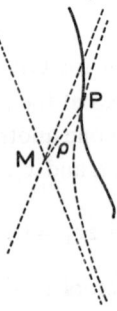

Fig. 3.

asymptotes are generators of a 'front cone' and a 'back cone' (Figure 3). Let this hyperbola be called the hyperbola of curvature at P. If M is the centre of this hyperbola, we here have to do with an internal hyperbola with centre M. Let ρ be the magnitude of the vector MP; then we recognize the acceleration vector at P as the vector in the direction MP of magnitude c^2/ρ.

If $\ddot{x}$, $\ddot{y}$, $\ddot{z}$, $\ddot{t}$ are all zero, the hyperbola of curvature reduces to the straight line touching the world-line in P, and we must put $\rho = \infty$.

4. To show that the assumption of group G_c for the laws of physics never leads to a contradiction, it is unavoidable to undertake a revision of the whole of physics on the basis of this assumption. This revision has to some extent already been successfully carried out for questions of thermodynamics and heat radiation,[3] for electromagnetic processes, and finally, with the retention of the concept of mass, for mechanics.[4]

For this last branch of physics it is of prime importance to raise the question – When a force with the components X, Y, Z parallel to the axes of space acts at a world-point P (x, y, z, t), where the velocity vector is $\dot{x}$, $\dot{y}$, $\dot{z}$, $\dot{t}$, what must we take this force to be when the system of reference is in any way changed? Now there exist certain approved statements as to the ponderomotive force in the electromagnetic field in the cases where the Group G_c is undoubtedly admissible. These statements lead up to the simple rule: – When the system of reference is changed, the force in question transforms into a force in the new space co-ordinates in such a way that the appropriate vector with the components $\dot{t}X$, $\dot{t}Y$, $\dot{t}Z$, $\dot{t}T$, where

$$T = \frac{1}{c^2}\left(\frac{\dot{x}}{\dot{t}}X + \frac{\dot{y}}{\dot{t}}Y + \frac{\dot{z}}{\dot{t}}Z\right)$$

is the rate at which work is done by the force at the world-point divided by c, remains unchanged. This vector is always normal to the velocity vector at P. A force vector of this kind, corresponding to a force at P, is to be called a 'motive force vector' at P.

I shall now describe the world-line of a substantial point with constant mechanical mass m, passing through P. Let the velocity vector at P, multiplied by m, be called the 'momentum vector' at P, and the acceleration vector at P, multiplied by m, be called the 'force vector, of the motion at P.

With these definitions, the law of motion of a point of mass with given motive force vector runs thus: [5] *The Force Vector of Motion is Equal to the Motive Force Vector*. This assertion comprises four equations for the components corresponding to the four axes, and since both vectors mentioned are *a priori* normal to the velocity vector, the fourth equation may be looked upon as a consequence of the other three. In accordance with the above signification of T, the fourth equation undoubtedly represents the law of energy. Therefore the component of the momentum vector along the axis of t, multiplied by c, is to be defined as the kinetic energy of the point mass. The expression for this is

$$mc^2 \frac{dt}{d\tau} = mc^2/\sqrt{1 - v^2/c^2},$$

i.e., after removal of the additive constant mc^2, the expression $\frac{1}{2}mv^2$ of Newtonian mechanics down to magnitudes of the order $1/c^2$. It comes out very clearly in this way, how the energy depends on the system of reference. But as the axis of t may be laid in the direction of any time-like vector, the law of energy, framed for all possible systems of reference, already contains, on the other hand, the whole system of the equations of motion. At the limiting transition which we have discussed, to $c = \infty$, this fact retains its importance for the axiomatic structure of Newtonian mechanics as well, and has already been appreciated in this sense by I. R. Schütz.[7]

We can determine the ratio of the units of length and time beforehand in such a way that the natural limit of velocity becomes $c = 1$. If we then introduce, further, $\sqrt{-1}\,t = s$ in place of t, the quadratic differential expression

$$d\tau^2 = -dx^2 - dy^2 - dz^2 - ds^2$$

thus becomes perfectly symmetrical in x, y, z, s; and this symmetry is communicated to any law which does not contradict the world-postulate. Thus the essence of this postulate may be clothed mathematically in a very pregnant manner in the mystic formula

$$3.10^5 \text{ km} = \sqrt{-1} \text{ s}.$$

NOTES

* From 'Space and Time', in *The Principle of Relativity* (ed. by A. Sommerfeld), Eng. tr. W. Perrett and G. B. Jeffery, London 1923 and Dover, N.Y. n.d., pp. 75–80.
[1] An application of this fact in its essentials has already been given by W. Voigt, *Göttinger Nachrichten*, 1887, p. 41.
[2] A. Einstein, *Ann. d. Phys.* **17** (1905), 891; *Jahrb. d. Radioaktivität und Elektronik* **4** (1907), 411.
[3] M. Planck, 'Zur Dynamik bewegter Systeme', *Berliner Berichte* (1907), 542; also in *Ann. d. Phys.* **26** (1908), 1.
[4] H. Minkowski, 'Die Grundgleichungen für die elektromagnetischen Vorgänge in bewegten Körpern', *Göttinger Nachrichten* (1908), 53.
[5] H. Minkowski, *loc. cit.*, p. 107. Cf. also M. Planck, *Verhandlungen der physikalischen Gesellschaft* **4** (1906), 136.
[6] I. R. Schütz, 'Das Prinzip der absoluten Erhaltung der Energie', *Göttinger Nachr.* (1897), 110.

E. MEYERSON

ON VARIOUS INTERPRETATIONS OF THE RELATIVISTIC TIME*

MINKOWSKI'S VIEW

Now let us turn back to the aspect of the new theory which we had temporarily set aside, namely the manner in which it tends to modify the usual concept of time. We will readily recognize that here again it is properly a question of spatialization. "Henceforth," Minkowski says, in setting forth the fundamentals of his conception, "space by itself, and time by itself, are doomed to fade away into mere shadows, and only a kind of union of the two will preserve an independent reality." A little later in the same fundamental exposition of his theory he repeats that "Space and time are to fade away into shadows, and only a world in itself will subsist."[1]

THE VIEWS OF LANGEVIN AND WIEN

It could doubtless be pointed out that this is only the personal opinion of a talented mathematician and that, despite the important role he played in creating the theory (one might say that the theory of general relativity would have been inconceivable without the formulas he introduced[2]), other advocates of the theory frequently profess less extreme opinions. Langevin, for example, actually writes: "We do not at all mean to say that time is a fourth dimension of space; that would make no sense."[3] Likewise, Wien declares that,

"although the relationship between space and time as revealed by relativity theory is very important, one must always bear in mind that we are dealing with a purely formal connection in this case, as follows from the fact that it is not time itself which plays this role, but imaginary time.[4]

SOMMERFELD, CASSIRER AND WEYL

Others, however, are less cautious, and relativists often express themselves in a way which suggests that there is a deep and inherent tendency toward

Minkowski's interpretation. Thus Sommerfeld, whose considerable works are well known, particularly the admirable theory by which he succeeded in connecting the theory of relativity with quantum theory, explicitly subscribes to Minkowski's view: "Minkowski's conception of space-time remains completely intact even today."[5] Among the philosophers, Cassirer expresses an altogether analogous opinion, stating that the assumption formulated by Minkowski in the above quotations "appears at the present time to be realized in all particulars."[6] Weyl's attitude is more ambiguous. He insists on the distinction between the three properly spatial dimensions and the temporal dimension and would have us speak rather of a universe with (3+1) dimensions. However, upon other occasions he himself uses the expression which he seems to have excluded, and, what is even more significant, he declares that

> subjectively there is an abyss between our modes of perception of time and space, but no trace of this difference remains in the objective universe which physics seeks to purge of immediate intuition. This universe is a four-dimensional continuum. There is neither 'space' nor 'time' but only consciousness which, moving in the objective universe, records the section as it comes to it and leaves it behind as *history*, like a process which unfolds itself in space and opens out into time.

Weyl is perfectly aware to what extent this conception upsets all our notions of reality. Nevertheless, our immediate sensation to the contrary, he believes this doctrine to be sound, for "it is only by the theory of relativity··· that we acquire for the first time a completely consistent knowledge of the nature of motion and change in the world." He is so convinced of this that he finds it "quite remarkable" that three-dimensional geometry seems so obvious, while four-dimensional geometry is so difficult.[7] Minkowski clearly did not overstate the case when he termed the tendencies of the relativistic concept "radical."[8]

EINSTEIN, EDDINGTON AND CUNNINGHAM

Moreover, Weyl is far from alone in his position. Quite apart from those who, like Sommerfeld, support Minkowski's statements without reservation, Einstein himself asserts that in relativistic physics "becoming in a three-dimensional space is somehow transformed into being in a four-dimensional world."[9] Eddington maintains that in relativity theory "the continuum formed of space and imaginary time is completely isotropic

VARIOUS INTERPRETATIONS OF RELATIVISTIC TIME 355

for all measurements; no direction can be picked out in it as fundamentally distinct from any other." Thus "events do not happen: they are just there and we come across them. 'The formality of taking place'··· of the event in question··· has no important significance."[10] According to Henri Marais,

> Let us say that the separation into time and space is evidently an artificial process which creates the 'illusion' of succession and expresses some sort of relationship – fundamental for us, accidental from the standpoint of reality – between our private perception and the world line of our organism.[11]

We must also quote Cunningham's statement which brings out this aspect of the theory in a very clear and plain-spoken way:

> The distinction (between space and time). as separate modes of correlating and ordering phenomena, is lost, and the motion of a point in time is represented as a stationary curve in four-dimensional space... The whole history of a physical system is laid out as a changeless whole.

He also writes:

> There is perhaps an analogy to be drawn between the analysis which lays out the whole history of phenomena as a single whole, and the things in themselves, the natural phenomena considered apart from the intelligence, for which consciousness of time and space does not exist...[and] in which, insofar as they are mechanically determined, the past and the future are interchangeable. Such a view of the universe is... the view of an intelligence which would comprehend at one glance the whole of time and space. But the limitations of the intellect resolve this changeless whole into its temporal and spatial aspects, and the past and the future of the physical world is the past and the future of the intelligence which perceives it.[12]

THE SPATIALIZATION OF TIME IN RELATIVITY

It must be noted that, if henceforth time and space are to be more or less merged into a single continuum, space will clearly be favored in the process. This is apparent from the evidence we have just cited, for if becoming is to be transformed into being (according to Einstein), so that the act of occurring becomes a simple unimportant formality for an event (according to Eddington); if succession is only an illusion (according to Marais), and if every physical system constitutes a changeless whole (according to Cunningham) – that can mean only one thing: the abolition or disappearance of time. Therefore Cunningham does not hesitate to speak of "Minkowski's timeless universe.[13] Let us observe, moreover, that this already follows from the very fact that the construction one arrives at is a *geometry*. And one need only open an exposition of this doctrine to note that, where time is concerned, it always speaks of one dimension, obviously conceived of as spatial, while no attempt is ever made to represent properly spatial dimensions in terms of time. Thus Marais, in the preface to the fine book we have already had occasion to quote in several contexts, does not hesitate to affirm that relativity aims at "incorporating time into space,"[14] and his testimony carries even more weight because Marais, whose essay is entirely mathematical, is nevertheless, as we know, a very competent philosopher.

THE IRREVERSIBILITY OF PHENOMENA

The tendency toward assimilation of time and space – which is really, as we have just noted, a transformation of time into space – sometimes seems to go rather far. Not only does it go farther than would seem to be authorized by our immediate sensation (a consideration for which the relativists, not without reason, care very little); it even exceeds the authority of the most clearly established facts and the most basic foundations of science. Can we really merge time with space as Minkowski assumed? Is it accurate to say with Eddington that in physical reality "the continuum formed of space and imaginary time is completely isotropic for all measurements" and that "no direction can be picked out in it as fundamentally distinct from any other"? On the contrary, it is clear that, taken literally, these are completely extravagant propositions having no connection with phenomena. The temporal dimension is by nature different from the spatial dimensions; we know this with sure and immediate knowledge, with a certainty against which any assault by intellect, no matter how seductive, would fail at the very onset. Indeed, as far as spatial dimensions are concerned, we can in large measure move about in them at will. This is the axiom of free mobility, to use Bertrand Russell's phrase,[15] and if we but ask ourselves, we will realize that it is an integral part of our idea of space. And we can in the same way assure ourselves that the notion of time contains no such element. To be sure, it would no longer be quite exact to state, as was done before Einstein and Minkowski, that we all progress continually along the temporal dimension in one and the same direction with a necessary and uniform movement. Langevin, in one of the admirable expositions he devoted to the theory, showed us that a traveler who left the earth with a speed equalling half the speed of light would find upon his return that two centuries had passed, while he himself would have aged only two years. But it is nonetheless certain that the faculty of moving in time is extremely limited, even according to the new concepts themselves. Of course the relativistic physicists are perfectly aware of this and make the necessary reservations when it comes to examining these questions closely. Thus Einstein himself uses the argument that we "cannot telegraph into the past" and Eddington states that the notion of entropy has survived the Einsteinian revolution and that (with the principle of least action) it constitutes one of the two generalizations toward which physics is converging.[16] Now the notion of entropy, which grew out of Carnot's principle, is nothing more than the expression of the irreversibility of phenomena, that is of continual progress in time. That is certainly the immediate conviction given to us by our innermost feelings in creating the notion of time – *certa scientia et clamante conscientia*, according to the scholastic expression Maine de Biran liked to quote. This conviction is continually confirmed by countless observations: we *know* that the chicken will not go back into the egg and that we are not any more likely in Einstein's world than in Newton's walk backwards or to digest before we have eaten. Wien continues the passage we quoted above: "Neither the theory of relativity nor any other theoretical concept can alter the fact that time is something totally different from a spatial dimension."

THE SOURCE OF THE RELATIVISTIC EXAGGERATIONS

How does it happen then that so many authorized explanations of the theory of relativity seem to want to imply the contrary? Sometimes they seem to wish to affirm that time is henceforth indistinguishable from space, since the four dimensions of 'Minkowski's world' are supposed to be perfectly isotropic. Other times, when they admit having reservations on this point (like Weyl, in stipulating that one must not speak of the four dimensions of the universe but of 3 + 1 dimensions, thereby granting that the temporal dimension is not of the same nature as the spatial dimensions), these reservations are clearly insufficient, referring only to the fact that in Minkowski's formula time seems to be qualified by an imaginary factor, and never referring to the irreversibility which we know to be so fundamental. Is this to be interpreted as an anomaly produced purely and simply by Einstein's theory, a sort of vicious tendency his doctrine inculcates in the minds of his followers? Not at all, since we need only examine a little more closely the evolution of our scientific knowledge to realize that, on the contrary, we are dealing here with a general tendency inherent in our reasoning, a tendency which relativity is apt to render more visible by the very fact that it pushes its explanations so far.

IDENTITY IN TIME

Common sense, although it recognizes that all things are subject to the conditions of space and time, does not treat time and space in quite the same way. Space appears, by its nature, totally indifferent to things: they undergo no modification as a result of having changed place.[17] It is true that if I took the puppy I hold in my arms to the top of Mont Blanc he would suffer, and that if I plunged him in water he would be asphyxiated, but this is the result of a change in the visible material conditions of his surroundings and not the result of mere spatial change. On the other hand, moving forward in time, he will undergo modification by this very fact. If twenty years from now one presented me with a dog resembling this one absolutely, and if one tried to make me believe it was the same one, I would not believe it in the least.

Our reason, however, does not remain fixed in such an attitude. On the contrary, it seeks to explain the modifications time brings to things, which means that, all things considered, our reason assumes that there should be no change simply as the result of the passing of time. Behind this search for explanations or *causes*, therefore, there is a conception which makes objects indifferent to their displacement in time, that is, which treats this displacement as if it were displacement in space. It is just as obvious that from the moment we bring time into our calculations, even if it is only for the purpose of simple prediction, we are forced to yield to the same tendency. For we represent time by a symbol, as a magnitude. What characterizes magnitudes is that they can increase or diminish – while time never goes backward and only in our imagination can we endow it with regressive movement.

THE SPATIALIZATION OF TIME IN THE PAST

Nevertheless, to modern physicists the assimilation of time to spatial magnitude has seemed to go almost without saying. Thus for Descartes time is one *dimension* and he defines the latter term as follows: "All that I understand by dimension is the mode and aspect according to which something is considered to be measurable."[18] D'Alembert considered that "not knowing time in itself and not even having a precise measure of it, we cannot represent the relationship between its parts any more clearly than by the relationship between the segments of an idefinite line." Elsewhere he writes: "A wise acquaintance of mine believes that one could regard duration as a fourth dimension."[19] Lagrange agrees that "mechanics can be regarded as a four-dimensional geometry, and mechanical analysis as an extension of geometrical analysis."[20] Moreover, scientists have not been the only ones to follow this natural turn of mind. As L. Brunschwicg rightly pointed out,

in general, modern philosophy has presupposed a correspondance between space and time. Spinoza's letter to Louis Meyer, the formulations in the Leibniz-Clarke correspondance and the key notions of the *Transcendental Esthetic* show the extent to which time has shared the destiny of space in classical rationalism.[21]

Thus the relativists are only extending – rather far, to be sure – a characteristic movement of science. It is quite simply the perennial attempt to explain becoming, change in time, by the negation of this change. Their theory about this is, as Bergson said, "the metaphysics inherent in the spatial representation of time. It is inevitable. Clear or confused, it was always the metaphysics of the mind speculating upon becoming."[22] Indeed, from the beginning of Greek philosophy this thought finds expression – in many aspects the definitive expression – in the image of the sphere of Parmenides. But it is frequently found in the most diverse forms to this very day. Hegel approaches it throughout his work and comes so close to it that one of his disciples and most faithful followers, rigorously developing the master's teaching, makes it almost completely explicit (cf. *De l'explication*, II, p. 63ff.).

THE DISSYMMETRY BETWEEN TIME AND SPACE

Furthermore, in the absence of any other proof, the very exaggerations of the relativists and the fact that they can sustain the illusion of having accomplished what no theory could possibly accomplish – namely the assimilation of time to space – would be sufficient to demonstrate this.

It is also demonstrated by the related but less controversial fact that this assimilation, when the new theory achieves it (more or less incompletely, it is to be understood), seems to them to be significant progress. Thus Langevin stresses the point that "Galileo's [equations], which characterize ordinary kinematics, introduce a dissymmetry between distance in space and the interval of time between two events, a dissymmetry which disappears in the new kinematics"; he seems to consider this a considerable advantage which the latter conception has to offer.[23] Jean Becquerel goes even further, for, upon observing that "space and time play different roles in the old conception" (which could not be more accurate), he adds: "We shall see this dissymmetry disappear in the Space-Time of the new theory."[24]

CARNOT'S PRINCIPLE AND PLAUSIBILITY

Finally, it is advisable to call attention again in this context to a third fact: the relativistic exaggerations in this sphere have been most often accepted without too much protest. Quite obviously this is the consequence of a state of mind closely related to the one by virtue of which, while the principles of conservation were readily accepted as *plausible* when they were only anticipated as well as after they had been proclaimed, everything relating to the irreversibility of phenomena has long remained obscure. Because of this, Sadi Carnot's teaching is still without an echo, and even in 1875, twenty years after the publication of Clausius' basic work, a thinker as penetrating and well-informed in scientific matters as Cournot could proclaim that there were only simple 'losses' in the transformation of heat into kinetic energy, constituting an "interfering cause more and more restricted in influence as the apparatus is perfected."[25]

Thus there is a real analogy between the processes by which relativity theory treats time and gravitation. In both cases, concepts which were in no way geometric, for common sense as well as for pre-Einsteinian physics, are formulated in terms of geometry, that is reduced to geometry. An attempt is made to spatialize them, to grasp them, understand them and explain them, by means of spatial concepts.

NOTES

* From *La déduction relativiste*, Payot, Paris, 1925, pp. 97–110; transl. by David A. and Mary-Alice Sipfle.

[1] H. A. Lorentz, A. Einstein, and H. Minkowski, *Das Relativitätsprincip*, Leipzig and Berlin, 1913, p. 56, 59; English transl. by W. Perrett and G. B. Jeffery in *The Principle of Relativity*, London, 1923, pp. 75, 80.
[2] Max Born, *La théorie de la relativité d'Einstein*, transl. by Finkelstein and Verdier, Paris, 1923, p. 283. For the importance of Minkowski's works, cf. also Laue, *Die Relativitätstheorie*, vol. I, 3rd ed., Brunswick, 1919, pp. 118, 169, 196; and Cunningham, *The Principle of Relativity*, Cambridge, 1914, p. 86.
[3] P. Langevin, 'L'aspect général de la théorie de la relativité', *Bulletin scientifique des étudiants de Paris* 2 (April-May 1922), p. 6.
[4] W. Wien, *Aus der Welt der Wissenschaft*, Leipzig, 1921, p. 271; cf. *ibid.*, p. 277.
[5] Lorentz, Einstein, Minkowski, *Das Relativitätsprincip*, p. 69. Cf. also what we have said on this subject in *De l'explication*, II, p. 377, note 3.
[6] Cassirer, *Die Einstein'sche Relativitätstheorie*, Berlin, 1921, p. 192; cf. *ibid.*, p. 119: "Spatial and temporal determinations become interchangeable. The past and the future are no longer distinguished except as + and − directions in space."
[7] H. Weyl, *Mathematische Analyse des Raumproblems*, Berlin, 1923, pp. 249, 189, 190 (we had cited part of these passages in § 47). It can be seen, *ibid.*, p. 241 and 249, that the author does not shrink before the very strange consequences of these conceptions.
[8] Lorentz, etc., *loc. cit.*, p. 56.
[9] Cf. Marcolongo, *Relatività*, Messina, 1921, p. 98. – In a more recent work Einstein declares that "neither the point in time at which a thing takes place nor the point in space at which a thing takes place have any physical reality, but only the event itself," so that "neither an absolute spatial relation nor an absolute temporal relation exists between two events, but only an absolute spatio-temporal relation." He adds that "it is impossible to divide the four-dimensional continuum into a three-dimensional spatial continuum and a one-dimensional temporal continuum in any way which makes sense from the objective point of view. For this reason the laws of nature appear in their most satisfying form from the logical point of view only if one expresses them as laws in the spatio-temporal continuum." At the same time he does recognize, however, that we "must keep in mind the fact from the physical point of view the temporal coordinate and the spatial coordinates are defined in quite different ways" (*Vier Vorlesungen ueber Relativitaetstheorie*, Brunswick, 1922, pp. 20–21).
[10] Eddington, *Space, Time and Gravitation*, New York, 1959, pp. 48, 51.
[11] Henri Marais, *Introduction géométrique à l'étude de la relativité*, Paris, 1923, p. 96. This statement, significantly enough, follows those cited above, § 20 and 47, where he affirms the reality of the entities defined by physical theory in general and relativity theory in particular.
[12] Cunningham, *The Principle of Relativity*, Cambridge, 1914, pp. 191, 213.
[13] *Ibid.*, p. 214.
[14] Henri Marais, *Introduction à l'étude géométrique de la relativité*, Paris, 1923, p. 6.
[15] B. Russell, *Essai sur les fondements de la géométrie*, transl. by Cadenat, Paris, 1901, p. 191.
[16] Eddington, *loc. cit.*, p. 149.
[17] Here and in the following we are only summing up the considerations we presented in *Identité et réalité*, p. 29ff., *De l'explication*, I, p. 150ff., and in the meeting of the Société française de philosophie, April 6, 1922 (*Bulletin*, p. 107ff.).
[18] Descartes, *Regulae ad directionem ingenii*, XIV.
[19] D'Alembert, *Dynamique*, p. 7, and *Encyclopédie*, under the word 'Dimension,' vol. IV, p. 1010. Here is the complete text of the second of these passages, which I find most

curious: "I have said above that it was not possible to conceive of more than three dimensions. A wise acquaintance of mine believes, however, however, that duration could be regarded as a fourth dimension and that the product of time and solidity would be in some way a product of four dimensions; that idea can be contested, but it seems to me that it has some merit, if only that of novelty."

[20] Lagrange, *Oeuvres*, Paris, 1867–1892, vol. IX, p. 337.
[21] L. Brunschwicg, *L'expérience humaine et la causalité physique*, Paris, 1922, p. 499.
[22] H. Bergson, *Durée et simultanéité*, Paris, 1922, p. 82.
[23] P. Langevin, *Le principe de relativité*, Paris, 1922, p. 10; cf. *ibid.*, p. 35.
[24] Jean Becquerel, *Le Principe de relativité et la gravitation*, Paris, 19922, pp. 8–9; cf. *ibid.*, p. 36. – E. Bauer likewise insists on the fact that "in classical theory" there remains "a complete dissymmetry" between time and space, which "somewhat compromises the rigor and elegance of classical kinematics" (*La théorie de la relativité*, Paris, 1922, pp. 23–24.)
[25] Cournot, *Matérialisme*, etc., Paris, 1875, p. 93.

A. EINSTEIN

COMMENT ON
MEYERSON'S 'LA DÉDUCTION RELATIVISTE'*

It is easy to show what gives this book its unique character. Its author is a man who not only comprehended the manner of thinking characteristic of modern physics but also had a profound understanding of the history of philosophy and the exact sciences – a man whose sure insight into psychology allowed him to uncover the internal connections and motives which make the mind work. Here we find happily combined the finesse of a logician, the instinct of a psychologist, vast knowledge, and simplicity of expression.

Meyerson's fundamental and guiding principle seems to be that it is not through an analysis of thought and speculation of a logical nature that a theory of knowledge can be reached, but only through the consideration and intuitive understanding of empirical observations. The 'empirical observations' to which he refers are constituted by the body of scientific results actually presented to us and by the historical account of their origin. The author seems to have felt that the principal problem lay in the relationship between scientific knowledge and experimental data: to what extent can one speak of an inductive method, to what extent of a deductive method, in the sciences?

He rejects both pragmatism and pure positivism; he even attacks them with some passion. Although the events and the facts of experience are at the base of all science, they do not constitute its content and its very essence; they are only the data which make up the subject matter for this science. Simple observation of empirical relations between experimental facts could not, for him, represent the sole purpose of science. Indeed, to begin with, the connections of a general nature which are expressed in our 'laws of nature' are not established simply by observations; they cannot be formulated and deduced unless one begins with rational constructions which cannot be the result of experience alone. Moreover, science is not content to formulate laws of experience; it seeks rather to construct a logical system which is based on a minimum number of premisses and which encompasses all the laws of nature in its conclusions. This system – or rather the body of concepts it represents – is coordinated with the objects of experience. On the other hand, this system, which reason tries to make conform to the totality of experimental data or to all that we experience, must correspond to the prescientific conception of the world of real things. All science is therefore founded on a system of philosophical realism. And the reduction of all experimental laws to logically deducible propositions is, according to Meyerson, the ultimate aim of all scientific research, a goal toward which we always move despite our conviction that only partial success is possible.

From this point of view Meyerson is a rationalist, not an empiricist. His position is to be distinguished from critical idealism in the Kantian sense, however. Indeed, no feature, no point in the system we are seeking can be known *a priori* to belong to it necessarily because of the very nature of our thought. And this is equally true with regard to the forms of logic and causality. We have no right to ask ourselves how the scientific system *must* be constructed but only how it actually *has been* constructed at the stages of its evolution already completed. Its logical foundations, as well as its internal structure, are therefore 'conventional' from the logical point of view. They find their justification only in the usefulness of the system when confronted by the facts, in its unity of thought and in the small number of premisses which it requires.

'*Relativism*' is the term Meyerson uses to designate the system deduced from the theory of relativity. One must be careful not to mistake this system for a new mode of thought distinct from that of classical physics (as might be suggested by certain passages of the book). The theory of relativity has never had such pretentions. Starting from the idea, rendered plausible by numerous observations concerning light, inertia and gravitation, that there is no physically privileged state of motion – the principle of relativity –, it posits this principle in the form of the following proposition: "The equations of physics must be covariant with respect to any transformation of points in the four-dimensional spatio-temporal continuum." The fundamental laws of physics as they were previously known are adapted to this principle with the least possible modification. By itself, the principle of relativity – or better, of covariance – would be much too general a base for the edifice of theoretical physics to be built on this foundation alone. It is not the physical theory as a whole which is new; it is only its adaptation to the principle of relativity. It seems to me that, everything considered, the author completely shares this point of view, for he often insists that relativistic thought is essentially in conformity with the laws and general tendencies science had already manifested (*La déduction relativiste*, pp. xi, 61 and 227ff, esp. 247 and 251).

On the other hand, the principle of relativity, considered in itself, seems much better established by experience than the present form of the theory, which has resulted from an adaptation of previous science. We are not certain, at this time, but we have reason to believe that the concepts of 'the metrical field' and 'the electromagnetic field' prove to be insufficient to interpret the facts in regard to quantum theory. But the idea that the principle of relativity itself could be refuted because of this scarcely deserves serious consideration.

The important thing for Meyerson, however, is that by its adaptation to the principle of relativity the intellectual structure of physics takes on, to a degree hitherto unknown, the character of a strictly logical and deductive system. Meyerson does not find this deductive and highly abstract character to be cause for censure; rather he sees here an application of the general tendency manifested by the history of the development of the exact sciences: the convenience – in the psychological sense of the word – of axioms and methods is seen to be sacrificed more and more for the sake of the logical unity of the entire system.

This deductive and constructive character allows Meyerson to make an extremely ingenious comparison of Relativity to the systems of Hegel and Descartes. He attributes the success of these three theories in their own times to the rigor of their logical connections and deductive derivation. The human mind is not content to posit relations; it wishes to *understand*. The superiority of Relativity over the previous two theories is due, according to Meyerson, to its quantitative precision and to its capacity to accommodate numerous experimental facts. Another point which he finds the theories of Descartes to have in common with those of the relativists is the assimilation of physical concepts with spatial, that is to say geometrical, concepts. It must be noted that this could be completely realized in Relativity only after (geometrical) derivation of the electromagnetic field, as in the theories of Weyl and Eddington.

Here again one must avoid a confusion in interpreting some of Meyerson's statements, particularly the following: "Relativity reduces physics to geometry." It is quite correct that, with this theory, (metrical) geometry, if regarded as distinct from the disciplines hitherto classified as 'physics,' has lost its independent existence. And Meyerson is able to quote (*La Déduction relativiste*, p. 137) a passage from Eddington where he speaks of "geometrical theory" of the universe. Even before the theory of relativity, there was no justification for considering geometry, as opposed to physics, an *a priori* science. Those who adopted this point of view were forgetting that geometry is the science of the possibilities of the displacement solids. According to the general theory of relativity, the metrical tensor defines the behavior of rods and clocks as well as the movement of freely displaceable solids, in the absence of electromagnetic effects.

This tensor is called 'geometrical' because the corresponding mathematical form appeared first in the science designated by the word 'geometry'. This is not sufficient reason to justify the application of the name 'geometry' to every science where this form plays a role, even when one illustrates it by means of comparisons employing symbols with which geometry has made us familiar. A similar argument would have allowed Maxwell and Hertz to qualify the equations of electromagnetism in the void as 'geometrical' since the geometrical concept of vector appears in these equations.

On the contrary, the essence of Weyl's and Eddington's theories for the representation of the electromagnetic field is not to be found in the annexation of this field to geometry but in the fact that they show a possible way to arrive at representing gravitation and electromagnetism from a single point of view, while the two corresponding fields had until that time been considered to be forms logically independent of one another. Consequently, I believe that the term 'geometrical' used in this order of ideas is entirely devoid of meaning. Furthermore, the analogy which Meyerson sets forth between relativistic physics and geometry is much more profound. Examining the revolution caused by the new theories from the philosophical point of view, he sees in it the manifestation of a tendency already indicated by previous scientific progress, but even more visible here – a tendency toward reduction of the 'diversity' to its simplest expression, that is, to its dissolution in *space*. What Meyerson shows in the theory of relativity itself is that this complete reduction, which was the dream of Descartes, is in reality impossible. Thus he rightly insists on the error of many expositions of Relativity which refer to the 'spatialization of time'. Time and space are fused in one and the same *continuum*, but this continuum is not isotropic. The element of spatial distance and the element of duration remain distinct in nature, distinct even in the formula giving the square of the world interval of two infinitely near events.

The tendency he denounces, although often latent, is nonetheless real and profound in the mind of the physicist, as is unequivocally shown by the extravagances of the vulgarizers and even of many scientists in their expositions of Relativity.

Meyerson's book is, I am convinced, one of the most remarkable ever written on the theory of relativity from the point of view of the theory of knowledge. I only regret that he did not seem to know the works of Schlick and Reichenbach, the value of which he would certainly have been able to appreciate.

NOTE

* From 'A propos de "La déduction relativiste" de M. Emile Meyerson', transl. by Mary-Alice and David A. Sipfle from the *Revue philosophique de la France et de l'étranger* **105** (1928), 161–166. The review was rendered in French by André Metz.

A. A. ROBB

THE CONICAL ORDER OF TIME-SPACE*

The study of Time and Space is one which in certain respects is extremely elusive and involves a number of difficulties which in ordinary daily life we are apt to overlook.

In scientific work, however, it is all-important to have clear ideas and to know exactly what our statements mean.

This is by no means always an easy task, for it frequently happens that our crude ideas on certain things may be sufficiently precise for certain purposes, but not precise enough for others.

Thus in the ordinary elementary teaching of plane geometry there are certain difficulties which are generally passed over, largely because they are real difficulties and a proper understanding of them could hardly be expected from a beginner.

For instance the use of ruler and compasses and the method of superposition.

The use of the ruler conveys a somewhat crude idea of what we mean in the physical world by points lying in a straight line, while the use of compasses conveys an equally crude idea of what we mean by points in a plane being equally distant from a given point in the plane.

The method of superposition involves ideas which are closely akin to those involved in the supposed use of compasses, but of a more elaborate character.

Both sets of ideas may be described as *ideas of congruence*.

Although there are other difficulties besides these to be overcome, still these will suffice for our present purpose, which is to show that certain points have been slurred over when we first began the study of geometry, which later on may require further elucidation.

Now let us approach this subject as a beginner of sufficient intelligence might be supposed to do.

There is one thing which we might observe, namely: that though we make use of figures drawn on paper to assist us in keeping the facts in mind, yet in proving a theorem, as distinguished from making use of the result, there is no necessity that the figure should be accurately drawn. A very rough figure will suffice and, if we are fairly expert, and the theorem not too complicated, we can dispense with a figure altogether.

Next let us suppose the figures to be accurately drawn on a plane sheet of paper (whatever the expressions 'accurately drawn' and 'plane' may mean) and then suppose the sheet of paper to be rolled up into a spiral, we could still make use of the figures on the curved sheet as mental images in proving our theorems, although our original straight lines would now (with certain exceptions) be no longer straight.

We could however substitute for our ruler a flexible string, drawn taut, so as to lie in contact with the curved surface of the paper and similarly we could make use of a flexible inextensible tape line or string instead of our original compasses and all our theorems would work out as before, except that lines would be curved which had originally been straight and lengths would be measured along such lines instead of 'directly' between points.

With such modifications, to every theorem concerning figures on the plane sheet there will be a corresponding theorem concerning figures on the curved sheet and *vice versa*, and similar methods of proof may be employed in the two cases.

Though the objects about which we are reasoning in the two cases are different, yet the logical processes are formally the same.

We can, however, go still further and consider the case where the figures are accurately drawn (whatever that may mean) on a plane sheet of india-rubber which is then stretched in any way.

In this case straight lines on the unstretched rubber would become lines, straight or curved, on the stretched rubber and a closed curve such as a circle would remain closed after the stretching.

Further, curves which intersected would still intersect and curves which did not intersect would not intersect after the stretching.

A point which lay inside a closed curve such as a circle, would become a point inside a closed curve on the stretched rubber.

Again, a point which lay between two other points in a line of some sort on the unstretched rubber would become a point between two corresponding points on the corresponding line on the stretched rubber.

The distances between the points would of course have altered according to our original standard and two lengths which were originally equal might no longer be equal, but nevertheless certain correspondences would still hold and could be traced between theorems involving equality of lengths on the unstretched rubber and theorems on the stretched rubber.

Perhaps the simplest way of seeing this is to introduce a system of coordinates (say Cartesian coordinates) on the unstretched rubber, by which any point of it would be represented by two numbers.

If then we imagine the rubber to be stretched, the same pairs of numbers could be taken to represent the same points of the rubber after stretching as before. The axes would now, generally speaking, become curved lines and the parallels to them would also in general become curved lines.

The points equidistant from a given point on the unstretched rubber would lie in a circle, and if the equation of this circle be taken as

$$(x - a)^2 + (y - b)^2 = r^2,$$

then this equation would represent also some curve on the stretched rubber. The radii of the circles would become some sort of lines all passing through one point and intersecting the distorted circle.

We should in this way get lines which had been straight, curves which had been circles, lengths which had been equal, etc., and we could deal with these algebraically in the same way as we did with the straight lines, circles and equal lengths on the unstretched rubber.

We notice that the things which actually do remain permanent are the particles of the rubber and certain features of their order.

If we consider the coordinate system we observe that, although the axes and the parallels to them are in general no longer straight after the stretching, yet as either set of parallel lines did not intersect before stretching, so the corresponding lines do not intersect after stretching and they preserve their original order.

We know however that, after a proper foundation has been laid, any geometrical theorem may be proved by coordinate methods and so it is evident that all reasoning which is done after coordinates have once been introduced will apply equally in dealing with certain other things than lines which are truly straight and lengths which are truly equal.

Thus though the sheet of rubber may have originally been plane, yet after stretching it may be curved in innumerable different ways and yet there are certain features which remain invariant throughout.

It is thus evident that although for purposes of mathematical reasoning the actual straightness of lines or actual equality of lengths in the ordinary sense of the terms is not essential, yet when we wish to make use of geometry to describe the physical world the meanings of 'straightness' and 'equality of length' are all important.

It is not sufficient that we should say that "there are such things as straight lines," or that "there are lengths which are equal." but it is necessary to have criteria by which we can say (at least approximately) "here are points which lie in a straight line" and "here is a length which is equal to yonder length."

The ruler and compasses give us rough standards of straightness and equality of length in the sense in which these terms are used in ordinary life, but, if we wish to go in for extreme accuracy, other standards must be employed and we must get more precise ideas as to what we really wish to convey when we make use of such expressions.

Consider first the question of what we mean when we say that two bodies are of equal length.

The ordinary method of comparing them is to make use of a measuring rod which we regard as *rigid*; or an *inextensible* tape line. But what do we mean by these words *'rigid'* and *'inextensible'*?

We find that it is by no means easy to say exactly what we do mean.

Approximate rigidity and inextensibility are common enough properties of solid bodies, but by the application of force all bodies are found to be more or less elastic, while change of temperature will also change the length of a rod compared with a parallel rod.

Again, if we wish to compare lengths which are not parallel, the usual mode of procedure would be to turn a measuring rod round from parallelism with the one length into parallelism with the other.

The possibility then arises that during the motion the standard may alter and give us results which indicate the lengths as equal when in reality (whatever that may mean) they are different.

Thus for example, if we wished to compare the lengths OA and OB where A and B are, say, the extremities of the major and minor axes of an ellipse whose centre is O, and suppose we had an elastic tape line which we place first with one end at O and the other at B.

If then keeping the one end fixed at O we move the other round the ellipse we should apparently get the same length for OA as for OB.

Now although this seems fantastic, yet the famous experiment of

Michelson and Morley seemed to show that just this sort of thing did happen when a body was turned round from a position such that its length was parallel to the direction of the earth's motion in its orbit into a position such that its length was perpendicular to that direction.

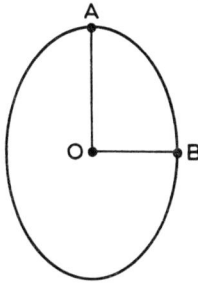

Fig. 1.

The experiment, which was an optical one, consisted in dividing a beam of light into two portions which travelled, the one in one direction, and the other in a transverse direction, and were reflected back again by mirrors.

If we adopt ordinary ideas for the moment and suppose the light to be propagated with a velocity v through a medium and the apparatus to move through that medium with a velocity u, it is easy to calculate the time of the double journey for the two portions of the beam.

For the case of a part of the beam which travels in the direction of motion of the apparatus and back again the time of the double journey is found to be

$$t_1 = \frac{2va_1}{v^2 - u^2},$$

where a_1 is the distance between the point of the apparatus where the beam divides and the corresponding reflector. For the case of the transverse portion of the beam the time of the double journey is found to be

$$t_2 = \frac{2a_2}{\sqrt{v^2 - u^2}},$$

where a_2 is the distance between the point of the apparatus where the beam divides and the other reflector.

Now it is possible to arrange things so that $t_1 = t_2$ and this can be done with extreme accuracy by means of the interference bands which are produced.

We should then have

$$\frac{2va_1}{v^2 - u^2} = \frac{2a_2}{\sqrt{v^2 - u^2}},$$

giving

$$a_1 = \sqrt{1 - (u/v)^2}\, a_2.$$

Thus a_1 would be slightly less than a_2.

It was found however that, when the apparatus was caused to rotate at a uniform slow rate, and the times of the double journey were equal for one position of the apparatus, then they were equal for all positions. This seemed to indicate that the dimensions of the apparatus in different directions changed as it rotated and the view was put forward by Fitzgerald and Lorentz that a material solid body contracted in the direction of its motion so that a sphere moving through space with a velocity u became a spheroid whose major and minor axes were in the ratio

$$1 : \sqrt{1 - (u/v)^2},$$

where v is the velocity of light.

It is clear that this once more raises the question as to the real meaning of 'equality of length' from which we started out.

Solid bodies apparently do not provide us with standards sufficiently permanent for dealing with such problems.

But the subject of motion raises a number of other difficulties.

There is in particular the question of 'absolute motion' and whether this expression has any precise meaning.

The underlying idea of those who believe in 'absolute motion' is that, if we consider a definite point of space at any instant, then that point preserves its identity at all other instants. The difficulty of identifying a point of space at two different instants is freely admitted, but for all that (so it is contended) the identity persists.

It was however noticed that, in the classical Newtonian Mechanics, the equations of motion preserved the same form for a system of bodies whose centre of inertia was in uniform motion in a straight line as for a similar system whose centre of inertia was 'at rest,' so that purely mechanical phenomena could not be expected to show up any difference between the two cases.

The question then naturally arose whether any difference could be detected by optical or electrical means, but experiment failed to show any.

Nextly it was pointed out by Larmor and Lorentz that the electromagnetic equations could also be transformed by a linear substitution so that they preserved the same form for a system moving with uniform velocity as they had for a system 'at rest.'

In order to do this, however, a *'local time'* had to be introduced.

We are all familiar with the use of 'local time' on the earth's surface, but the cases are different in one important respect.

The idea underlying the use of 'local time' on the earth's surface is simply that of having different names in different parts of the world for what is regarded as the same instant. Thus noon at Greenwich and noon at New York are both described as 12 o'clock local time, although the instants referred to are clearly different. On the other hand the use of chronometers in navigation is regarded as a method of approximately identifying the same instant at different parts of the earth.

But, as previously remarked, the 'local time' used in transforming the electromagnetic equations is of a different character and events which are regarded as simultaneous according to one 'local time' would not be simultaneous, in general, when compared by the 'local time' of a system which was in motion with respect to the first.

We might of course regard the one 'local time' as the true time and the other as a mathematical fiction, but there is no reason known why we should select the one rather than the other, just as there is no way of distinguishing a body 'at rest' from one moving uniformly in a straight line.

In fact it appears that, just as we have no method of distinguishing the same point of space at two distinct instants of time, so we cannot strictly identify the same instant of time at two distinct points of space.

It is to be observed that though we started out by trying to give a precise meaning to the idea of equality of length, in which we seemed to be concerned only with space, yet in our attempt to do so, we find difficulties with regard to time intruding themselves.

We can see, however, that even in our original use of compasses the time element intrudes, since in comparing lengths by the use of compasses, the compasses are moved and the idea of motion involves that of time.

Also in the Michelson and Morley experiment, since light takes a finite interval of time in getting from an object P to an object Q and back again to P, we are introducing time relations in comparing lengths.

The question now arises: suppose we imagine a flash of light sent out at an instant A from a particle P to a distant particle Q and arriving there at an instant B and suppose it reflected back to P where it arrives at an instant C; how are we to identify the instant B with any instant at P between A and C?

If we regard P as being 'at rest' we might reasonably think to identify B with the instant at P which is midway between A and C, but this would imply that we had some means of measuring intervals of time, and that brings us up against all the same sort of difficulties which we encountered in trying to find a satisfactory method of measuring space intervals.

On the other hand, if P be in uniform motion in the direction PQ it would seem that B is not identical with the instant at P which is midway between A and C.

In any case we do not know of any means of telling whether P is 'at rest' or not.

Having thus been led on from the consideration of spatial relations to those of time we seem at first sight to have increased our difficulties instead of solving them, but if we persevere in our task we shall find that we have made an appreciable advance towards solving our problem.

From the consideration of figures drawn on a sheet of rubber which was afterwards stretched in any way, we were led to recognise the importance of *order* in the study of the logic of geometry, and since order also plays a part in time relations, it seems worth while to consider order in time.

Now here we find an interesting and very important thing.

If I consider two distinct instants of which I am *directly conscious*[1], I notice that the one is *after* the other.

Noon to-day is *after* noon yesterday and I cannot invert the order.

There is in fact what is called an asymmetrical relation between the two instants, such that if B be *after* A, then A is not *after B*.

If we consider two points or two particles in space, say P and Q, there is nothing analogous to this and we have no reason to say that Q is *after* P rather than that P is *after* Q.

We might, of couse, give them an order by means of some convention, but such convention would be quite arbitrary, whereas in the case of the instants, it is a matter of fact and not of convention, quite independently of what words we may employ to express that fact.

The simplest relation of order among points is a relation of *between* which involves three terms instead of two.

This relation of *between* has been employed by various mathematicians in investigating the foundations of geometry, but the relation of equality of lengths then appears as something extraneous, grafted on to the system.

The use of an asymmetrical relation such as *after* appears to have great advantages over a relation such as *between* in constructing a theory of order and I have found it possible, by means of such a relation, to construct a system of geometry of space and time. It might perhaps more correctly be described as a geometry of time, of which spacial geometry forms a part.

In constructing this system it is necessary to modify certain currently accepted notions, but the modifications required all appear to be capable of justification and the structure, when completed, will be found closely to resemble our ordinary conceptions.

We shall regard an instant as a fundamental concept which, for present purposes, it is unnecessary further to analyse, and shall consider the relations of order among the instants of which I am directly conscious.

Thus for such instants we find the following properties:

(1) If an instant B be *after* an instant A, then the instant A is not *after* the instant B, and is said to be *before* it.

(2) If A be any instant, I can conceive of an instant which is *after* A and also of one which is *before* A.

(3) If an instant B be *after* an instant A, I can conceive of an instant which is both *after* A and before B.

(4) If an instant B be *after* an instant A and an instant C be *after* the instant B, the instant C is *after* the instant A.

(5) If an instant A be neither *before* nor *after* an instant B, the instants A and B are identical.

The set of instants of which I am directly conscious have thus got a linear order.

But now let us consider the fifth of these properties.

It might at first sight be supposed that it was a necessary consequence of our ideas of *before* and *after*. That it is really logically independent of the other properties may be shown by the help of a geometrical illustration. This illustration is very suggestive and we purpose to make further use of it, but the logic of our theory is independent of the illustration.

Suppose we have a set of cones having their axes parallel and having equal vertical angles, and further, suppose each cone to terminate at the vertex, which is however to be regarded as a point of the cone.

We shall call such a cone having its opening pointed upwards an α cone, and one with the opening pointed downwards a β cone.

Thus corresponding to any point of space there is an α cone of the set having the point as vertex, and similarly there is a β cone of the set having the point as vertex.

Now it is possible by using such cones and making a convention with respect to the use of the words *before* and *after* to set up a type of order of the points of space.

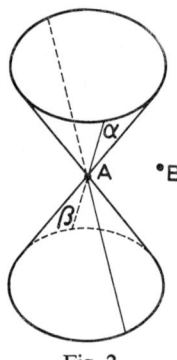

Fig. 2.

For the purposes of this illustration we shall make the convention that, if A_1 be any point and α_1 and β_1 be the corresponding α and β cones, then any point A_2 will be said to be *after* A_1 provided it is distinct from A_1 and lies either on or inside the cone α_1 and will be said to be *before* A_1 provided it is distinct from A_1 and lies either on or inside the cone β_1.

It is easy to see that with this convention we have the following:

(1) If a point B be *after* a point A, then the point A is not *after* the point B and is said to be *before* it.

(2) If A be any point, there is a point which is *after* A and also a point which is *before* A.

(3) If a point B be *after* a point A there is a point which is both *after* A and *before* B.

(4) If a point B be *after* a point A and a point C be *after* the point B, the point C is *after* the point A.

We cannot however assert that if a point A be neither *before* nor *after* a point B, that the points A and B need be identical.

This is easily seen since the point B might lie in the region outside both the α and β cones of A. (Figure 2.)

This illustration shows that the fifth condition is logically independent of the other four.

The type of order which we have illustrated by means of the cones, we shall speak of as *conical order*, but the logical development of the subject is independent of this illustration.

We may note however in passing that, if A and B be distinct points one of which is neither *before* nor *after* the other, then there are points which are *after* both A and B and also points which are *before* both A and B.

This follows since in this case the α cones of A and B intersect, as do also the β cones of A and B.

It should further be noted that if we have any line straight or curved in space, but whose tangent nowhere makes a greater angle with the axes of the cones than their semi-vertical angle, then if we confine our attention to the points of any one such line, we can assert that: if a point A be neither *before* nor *after* a point B, the points A and B are identical.

Thus provided we confine our attention to the points of such a line, the whole five conditions are satisfied.

Returning now to the consideration of instants, we observed that there was a difficulty in identifying the same instant at different places.

The relations of *before* and *after*, however, enable us to say in certain cases that instants at a distance are distinct. Thus if I can send out any influence or material particle from a particle P at the instant A so as to reach a distant particle Q at the instant B then this is sufficient to show that B is *after* and therefore distinct from A.

If now the influence or material particle be reflected back to P and arrives there at the instant C, then C is *after* and therefore distinct from B, while moreover, C is *after* A.

Now suppose the influence be a flash of light or other instantaneous electromagnetic disturbance and we appear to have reached a limit.

We do not seem to be able to send out any influence or material particle from P at any instant *after A* so as to arrive at Q at the instant B, and we do not seem to be able to send out any influence or material particle from Q at the instant B so as to arrive at P *before* the instant C.

In fact the range of instants at P which are *after A* and *before C* appear to be quite separated from the instant B so far as any influence is concerned.

Now let us suppose that light has this property.

It may or may not be strictly true of light but, provided there be some influence which has this property (and others which we shall specify later), such influence will serve for the purpose in hand, and we shall, provisionally at any rate, ascribe it to light.

Now B could at most be identical with only one of the instants at P, and such instant would require to be *after A* and *before C*, but we have no means of saying which instant it is.

The other instants in this range would then all be either *before* or *after B*.

But what now do we really mean when we say that one instant is *after* another or one event *after* another?

If I at the instant A can produce any effect however slight at the instant B, this is *sufficient* to imply that B is *after A*.

A present action of mine may produce some effect to-morrow, but nothing which I may do now can have any effect on what occurred yesterday.

It appears to me that we have here the essential features of what we really mean when we use the word *after*, and that the abstract power of a person at the instant A to produce an effect at a distinct instant B is not merely a *sufficient*, but also a *necessary* condition that B is *after A*.

If we accept this as the meaning of *after* it would then appear that no instant at P which is *after A* and *before C* is either *before* or *after B*.

We have already seen that the idea of an element being neither *before* nor *after* another element, and yet distinct from it, involves no logical absurdity, and so if we give up the attempt to identify the instant B with any instant at P we get a logically consistent view of things.

Thus according to the view here adopted *there is no identity of instants at different places at all.*

We may express the idea in this form: *the present instant, properly speaking, does not extend beyond here.*

THE CONICAL ORDER OF TIME-SPACE

Thus there are instants at a distance *before* the present instant and *after* it, and also instants neither *before* nor *after* it, but such instants are to be regarded as being all quite distinct from the present instant here.

Thus, according to the view here adopted, *the only really simultaneous events are events which occur at the same place.*

The theory which we desire to expound with regard to time and space may now briefly be described as follows:

Taking the above view of instants and the relations of *before* and *after*, we express in terms of these relations the conditions that the set of instants should have a *conical order* of a certain type. We then find that we have got a description not only of time but also of space such as that with which we are already familiar.

In fact we may be said to analyze spatial relations in terms of the time relations of before and after.

In first approaching this subject it is a great assistance to have some concrete way of representing the facts to our minds even though such representation may make use of some of the conceptions which we are trying to analyze.

In doing so one must remember however that the justification of our theory lies in the logical procedure and not in the representation.

Thus in trying to convey a general idea of what we are doing we shall find it both convenient and suggestive to make use of our mental images of cones, in the way already described, in order to picture what we mean by *conical order*.

The idea of *conical order* is not at all dependent on this representation, but is built up by a rather lengthy piece of reasoning from the relations of *before* and *after*.

The representation by means of cones may be compared to the rough scaffolding used in the erection of a building, which is removed when the building is complete and its component parts in position.

We must, however, be certain that the building is not supported by the scaffolding, or it will not be able to stand alone.

In order to make sure of this in our theory, great care has to be taken, and, for details on this matter, I must refer readers to my larger work.

Moreover, the representation by means of cones gives only a three-dimensional conical order, whereas the conical order with which we are really concerned is a four-dimensional one.

The representation also introduces a sort of distortion, but this need not trouble us when we deal only with descriptive features.

Now in ordinary mathematical physics we are accustomed to localize an instantaneous event by means of four numbers x, y, z, t. Of these numbers x, y and z are called spacial coordinates while t is referred to as the 'time'.

But now having come to regard all instants at different places as distinct, we regard these four numbers as really representing four coordinates of an instant.

The coordinate t, however, has different *before* and *after* relations from those associated with the other three coordinates x, y, z, which are made clear by the conception of *conical order*.

In order to avoid confusion therefore, we shall speak of the former not as 'time' but as a t coordinate.

We are not yet, however, in a position to introduce coordinates except for 'scaffolding' purposes.

Neither again are we at liberty to speak of 'velocity' except for scaffolding purposes until we have defined the meaning of the word.

Moreover, in the actual proof of theorems, we must not employ the ideas of equality of lengths or angles until these ideas are seen to be definable in terms of *before* and *after* relations.

We may, however, make use of such terms in the 'scaffolding' which is mere poetry and rhetoric.

Let us therefore first consider this pictorial representation in which we have to confine ourselves to three coordinates instead of four, which we shall take to be x, y and t, and shall regard as rectangular.

Now by taking suitable units we may regard the 'velocity of light' as unity and under these circumstances if we imagine a flash of light starting from the position $x=a$, $y=b$, $t=c$, the rays of light would be represented by the generators of the upper half of the cone whose equation is

$$(x-a)^2 + (y-b)^2 - (t-c)^2 = 0,$$

which we take as the α cone corresponding to (a, b, c).

The lower portion of this same locus constitutes the β cone of (a, b, c).

The point (a, b, c) itself is regarded as belonging to both the α and β cones.

The successive positions of a material particle would be represented by some line straight or curved, but since it appears that a particle of matter never quite attains to the 'velocity of light' the tangent to this curve would make an angle with the axis of t which is always less than 45°: the semi-vertical angle of the cones.

The successive positions of a particle which remains at rest with respect to the system of axes would be represented by a straight line parallel to, or coincident with, the axis of t.

The successive positions of a particle which remains in uniform motion with respect to the system of axes would also be represented by a straight line, but one inclined to the axis of t.

The successive positions of an accelerated particle would be represented by a curved line.

The set of instants of which any one individual is *directly conscious* would also be represented by some line straight or curved, whose tangent always makes an angle with the axis of t less than the semi-vertical angle of the cones.

We thus see that for the set of instants of which any one individual is directly conscious, or the set of instants which any one particle occupies, we can assert that: if an instant A be neither *before* nor *after* an instant B, the instants A and B are identical.

We cannot, however, assert this of the instants of which two individuals are directly conscious, or which two distinct and separate particles occupy.

It may be that an instant of which I am directly conscious is neither *before* nor *after* some instant of which you are directly conscious, but they are not identical, and our illustration shows that this involves no logical contradiction.

It is to be noted, however, that if A and B be two distinct instants, one of which is neither *before* nor *after* the other, then since the α cones intersect and also the β cones, there are instants which are *after* both A and B and also instants which are *before* both A and B, so that we may both speak of to-morrow or of yesterday, though strictly speaking we have no common present.

Thus instead of regarding ourselves as, so to speak, swimming along in an ocean of space (as we usually do), we are to think of ourselves rather as swimming along in an ocean of time, while *spatial relations are to be regarded as the manifestation of the fact that the elements of time form a system in conical order: a conception which may be analyzed in terms of the relations of after and before.*

The view that time relations are fundamental appears to have an important bearing on what Professor William James called the theory of a 'block universe': by which name he referred to the theory that the universe is something like a cinematograph film in which the photographs have already been taken and which is merely in process of being exhibited to us.

Most writers on this subject treat time as if it were merely a fourth dimension of space: an attitude which encourages one to favour the 'block universe' idea.

When instead, we regard *before* and *after* relations as fundamental, and analyse spatial relations up in terms of these, the whole subject appears in a very different light and the 'block universe' theory does not commend itself so strongly.

If the universe were in this way like a cinematograph film which is merely being displayed before us, then its innumerable details must have been fixed through all eternity and there would be complete determinism as to the future.

But have we really any grounds for thinking that the universe is of this nature: or, reverting to the cinematograph analogy, is it any simpler to suppose that the film has already been taken than to suppose that the play is in process of being acted?

If the *after* relation has the significance which I suggested and if what we call time and space may be analysed in terms of *before* and *after* then it would seem that instead of having grounds for belief in a 'block universe' we have actually got grounds for an opposite view.

It seems therefore that the question turns on the significance of the *after* relation and its asymmetric character.

It is interesting to note that recently, on quite different grounds, some physicists are coming round to the view that the universe is not strictly deterministic.

Scientific predictions as to future events are made on the assumption that certain uniformities will continue.

If they do continue the prediction may be a logical consequence of their doing so, but, if the uniformities do not continue, the conclusion may be unwarranted.

The continuance of the uniformities is only an assumption for which we have no absolute guarantee, and, should they cease, no promise is broken, since none was ever made. A departure from uniformity initiated at an instant A may extend to an instant B which is *after* A; and this would be an *effect* at B of the departure from uniformity initiated at A.

All applied mathematics becomes pure mathematics when we get away from our fundamental assumptions and begin to draw logical conclusions from them.

Now I have ascribed certain characteristics to instants and to *before* and *after* relations which may or may not be strictly correct, but which serve as the basis by means of which one may apply a certain type of pure geometry to map out time and spacial relations.

The geometry, as I have already pointed out, is a logical structure built up from certain postulates which I shall formulate.

As a logical structure a geometry may have more than one application, as for instance, ordinary Euclidean plane geometry might be taken primarily as applying to figures on what we call a plane and again to geodesic lines drawn on a developable surface.

For the purposes of physical science, however, it is not sufficient merely that we should say, for instance, that *there are* such things as 'straight lines' or that *there are* lengths which are equal, but it is necessary to have criteria by which we can say (at least approximately) '*here are* points which lie in a straight line' and '*here is* a length which is equal to yonder length'.

In other words we must have more or less clear ideas of the physical things to which we apply our abstract theory.

The abstract theory itself does not require this, but the physical application does; and for this reason, I have tried to make clear the sort of physical meaning which I ascribe to the notions of an instant, the *before* and *after* relations and the criteria given by light flashes.

If we should discover, for instance, that the formal properties which we provisionally ascribe to light actually hold for some other influence; then the geometry which I propose to develop would apply with this new interpretation of its postulates.

Now I have made use of ordinary geometric cones in order to enable us to form a concrete picture of what I mean by '*conical order*', but the idea of conical order is not at all dependent upon this graphic representation, but is built up by a rather lengthy piece of reasoning from the asymmetrical relations which I denote by the words *before* and *after*.

The representation by means of cones may be compared to the rough scaffolding used in the erection of a building which is removed when the building is complete and its component parts in position.

We must, however, be certain that the building is not supported by the scaffolding, or it will not be able to stand alone.

In order to make sure of this in our theory, great care must be taken not to take things for granted because they hold in our models.

In the first place we are not at liberty to introduce coordinates except for scaffolding purposes until we have defined them. Neither are we at liberty to speak of 'velocity' except for scaffolding purposes till its meaning is defined.

Moreover in the actual proof of theorems we must not employ the ideas of equality of lengths or angles until these ideas are seen to be definable in terms of *before* and *after* relations.

We may however, and actually do, make use of such non-permissible ideas in our graphic representation.

Thus in the models we supposed the cones to have their axes parallel (or identical) and to have equal vertical angles, and neither the idea of *cone*, of *parallel*, of *axis*, of *angle*, nor of *equal* has been analysed in terms of *before* and *after* and therefore must be excluded in defining the α and β sub-sets, which are the names which I shall hereafter apply to the entities corresponding to the α and β cones.

The *before* and *after* relations are converse asymmetrical relations and either may be defined in terms of the other; so that it is a matter of indifference which of them we take as fundamental.

I actually take the relation of *after* as fundamental and define *before* in terms of it.

NOTES

* From *The Absolute Relations of Time and Space*, Cambridge Univ. Press, 1921, pp. 1–16; 22–24.
[1] The fact that I am *directly conscious* of the two instants is very important, in view of later developments.

P. FRANK

IS THE FUTURE ALREADY HERE?*

The lack of logical clarity in the formulation of the theory of relativity does not show its dangerous character for clear thinking very distinctly until we consider its consequences outside the special field of physics. And, as everybody knows, we have occasion only too often to observe such consequences. It can be said without exaggeration that there is no philosophical congress, no philosophical textbook, not even an issue of a philosophical journal, where we do not encounter examples of attempts to draw arguments in favor of metaphysical opinions from the statements of the theory of relativity. And in all these cases it can be shown that the source of all such arguments lies in the fact that the 'physical' sentences from which they started were themselves formulated not in a scientifically physical manner, but in a metaphysical manner resulting from a distortion of the correct logical mode of expression.

One of the most wide-spread metaphysical interpretations of the theory of relativity, to the consideration of which we will confine ourselves here, is that which undertakes to prove the fatalistic conception of the world, that is to say, the view that everything that happens is determined from all eternity, and that there is no development and nothing really new in the world.

In order to find a formulation of this opinion by a philosopher, we may turn at random at any collection of utterances of philosophers about questions of the day, for instance, to the proceedings of the Eight International Congress of Philosophy (Prague, 1934). There we find the following passage from a lecture by F. Lipsius (Leipzig):

The question arises whether, thereby, space has lost its independence of time altogether. The theory of relativity and Minkowski's sentence, "Henceforth, space by itself and time by itself are doomed to fade away into mere shadows and only a kind of union of the two will preserve an independent reality" suggest this view. The question remains whether the mathematico-formalistic theory which desires to make time the fourth dimension of space, can be maintained, although this consideration reintroduces into philosophy the doctrine of the unreality of change and declares all development to be illusion.

I do not intend to argue about this quotation. I have cited it only as an example of the views which likewise occur in hundreds of other philosophical writers. What I should like to show here is rather the fact that these metaphysical interpretations of the theory of relativity have their origin in the insufficiently clear formulations which can be found in treatises of physics themselves. Thus, analogously to the statements in which it is said that an iron rod may have different lengths 'for two observers', we find similar modes of expression concerning the simultaneity of events. For instance, it is frequently said that two events E and F happen at the same time for the observer B_1, while for the observer B_2, F occurs later than E. Everybody will have read sentences of this kind in textbooks on physics. No great misfortune will follow from this, for the physicists, among themselves, know very well that although one speaks of two observers B_1 and B_2, one does not employ two observers in the actual application of this mode of expression but two kinds of measurements, and that speaking of two observers is merely a form of expression which is somehow believed to be particularly clear and interesting and attractive to the non-physicist. The irony of the matter lies in the fact that it is just the non-physicist and the philosopher, for whose sake this mode of expression is introduced, who are misled, while for the physicist himself it is comparatively irrelevant whether this or that expression is chosen.

This subjective formulation of the theory of relativity, which speaks of two observers B_1 and B_2, has given rise to consequences which belong to the field of moral philosophy and contain advice on human conduct. From this subjective formulation arose that utilization of the theory of relativity as an argument in favor of fatalism of which I spoke at the beginning of this section. I fear that some readers will accuse me of exaggeration in this connection. I will therefore confess immediately that it was not only the philosophers who misunderstood this subjective mode of expression, but that some of the physicists have done so themselves. Often the philosophers have not misinterpreted the formulations of the physicists, but have simply taken over the formulations together with their misinterpretations ready for use.

As an example I will therefore not quote a philosopher but the well-known physicist and astronomer Sir James Jeans who, in his Sir Halley Stewart Lecture of 1935, 'Man and the Universe', also refers to the theory of relativity as an argument in favor of fatalism. Jeans starts out from the

IS THE FUTURE ALREADY HERE? 389

fact that, according to this theory, two events E and F may happen simultaneously for an observer B_1, and yet one after the other for a different observer B_2. But then obviously the following may occur. For the observer B_1, events E and F both happen in the present, but for the other observer B_2, F happens in the present, while E happened earlier, that is, in the past. Thus one and the same event E, which B_1 only expects to occur, may already have happened and be over for the observer B_2. This, to cite Jeans literally, is formulated as follows: "It is meaningless to speak of the facts which are apt to come... and it is futile to speak of trying to alter them, because, although they may be yet to come for us, they may already have come for others." And he at once draws conclusions from this for human behavior. He says in fact: "Such a view reduces living beings to automata." And he describes in poetical style how man has changed from a participant into a mere spectator in the world theatre.

To see quite clearly through this kind of argument, let us illustrate it in terms of a concrete example. Suppose the event F occurs for both observers B_1 and B_2 at the same time, say the present; let it consist in a watch showing the time of ten o'clock on its face. Let the event E be the collision of two motor-cars. For B_1 this happens simultaneously with the event F, that is to say, the watch shows ten o'clock. But for B_2, according to what has been said before, the event E occurs before the event F, i.e., the collision before ten o'clock, say at one minute to ten. Immediately before ten o'clock (say, just one second before) the impact has not yet occurred as far as B_1 is concerned, while, for B_2 it has already taken place, namely at 9:59. Therefore, so Jeans argues, the observer B_1 cannot prevent the motor-cars from colliding, although, for him, they have as yet not collided. There is no use in making any effort, since for his fellow observer B_2 the accident has already happened at one minute to ten. In this way, the ancient fatalistic belief of the Mussulmans is justified with the help of the newest physics of the 20th century.

But however evident this argument may appear to many, and however often and in however many variations it may be used by philosophers of our time, it is in reality void of any logical justification. It can even be safely said that with a correct logical formulation of the theory of relativity it dissolves into thin air.

A sentence of the form 'the collision of these motor-cars will take place at 10 o'clock for the observer B_1, but has already taken place at one minute to 10 for the observer B_2' is certainly a sentence conforming to the customary formation rules of relativistic syntax, and thus significant. But we have yet to examine which statement about empirically verifiable facts is meant by this sentence according to the language of the theory of relativity. If this sentence is formulated in this way, then the two observers, as men of flesh and blood, no longer occur, and the sentence referred to has the meaning: 'at the time of the collision of the motor-cars a certain watch shows exactly ten o'clock, while another watch, which moves relatively to the first with a certain velocity v, only shows one minute to ten.' Here the three events, collision of the motor-cars, position of the hands on the first watch to ten, and position of the hands on the second watch at one minute to ten, coincide spatially and temporally.

If this mode of expression is chosen, it becomes at once perfectly obvious that in relativity as well as in classical mechanics only one collision of motor-cars takes place, and that two different watches moving with different velocities merely show different times at the moment of impact. These different time data result in the following way from the theory of relativity: in every rigid system of reference, the watches are regulated at any of its points synchronously 'with respect to this system'. Here, in every system, the position of the hands of one single watch at a certain time point is still arbitrarily choosable. Let us choose the watch at the origin of the coordinates. If, say, S_1 and S_2 are the two systems of reference in respect to which the observers B_1 and B_2 mentioned above and the watches by which we replaced them are at rest, then we will assume of the watches at the origins of S_1 and S_2 that they show the same time while passing one another, in which case S_2 may have the velocity v with respect to S_1. If then x denotes the projection of the distance of a certain watch from the origin of the system S_1 in which it rests, e.g., of the watch showing ten o'clock at the moment of the motor-car accident, then – as shown first by Einstein – at the coincidence of both watches, the position of the hands of the first watch differs from the watch resting in S_2 by xv/c^2, on account of the synchronous regulation with respect to two different systems (namely, S_1 and S_2). And that is the one minute which the watch in S_2 (in inexact terms: for the observer B_2) lacks of ten o'clock.

But in this formulation it is quite impossible to express the following assertion: 'the observer B_1 cannot prevent the automobile accident, although for him it has not yet happened, because it has already happened for B_2'. This sentence can only be made expressible by introducing the observing subject in an incorrect manner into the language of the theory of relativity. The real role of the observer is here, just as in classical physics (as we have shown in the first section), only that of reading off a scale the number with which the pointer coincides. But as already mentioned, there is no objection against one and the same observer reading both watches. Then, in the above case, he will simply find that, at the moment of the automobile collision, one of the two watches coinciding with the impact, namely, the one resting in S_1, shows ten o'clock, while the one resting in S_2 shows one minute to ten. But in that case it will be difficult to formulate the assertion that the impact which is only going to happen, has yet, in a different sense, already taken place. There exists no such moment when the watch in S_1 does not yet show ten o'clock (but, say, just one second before ten) while the watch resting in S_2 shows only one minute to ten, because, at the place of the impact, there is no time interval between those two points where the one watch shows ten o'clock and the other one minute to ten, since both time points coincide at that place.

This argument leading to fatalism is usually formulated in a manner more general and abstract and even more congenial to metaphysical misinterpretation. It contemplates all events which ever happened or ever will happen as being already contained in Minkowski's spatio-temporal world. Every present is merely a three-dimensional cross-section through this four-dimensional world. Moreover, this section is arbitrary, inasmuch as it is laid through all time points which are simultaneous with respect to an arbitrary system S of reference. This system of reference may be chosen at liberty. That which we call development is therefore nothing but a wandering through the eternally existing four-dimensional continuum. And this wandering may be carried out in many ways. We may start out from an arbitrary three-dimensional plane section, as we have seen. Then the development of the world merely consists in a parallel translation of this section vertically to itself. But nothing new can arise. Everything has existed forever.

This argument is found in variant forms in innumerable philosophical writings. But even physicists apply it occasionally. Again, I will quote only a few passages from the lecture by Jeans already mentioned:

Then the theory of relativity came and taught that there is no clear-cut distinction between space and time; time is so interwoven with space that it is impossible to divide it up into past, present, and future in any absolute manner. This being so, the tapestry cannot consistently be divided into those parts which are already woven and those which are still to be woven. Such a distinction can have no objective reality behind it...

And now we get to the point. Jeans next proceeds to make a hypothesis, an assumption, which is intended to provide a suitable explanation of this state of affairs, namely, of the relativity of the subdivision into past and future. He says literally:

The shortest cut to logical consistency was to suppose that the tapestry is already woven throughout its full extent, both in space and time, so that the whole picture exists, although we only become conscious of it bit by bit – like separate flies crawling over a tapestry.

This assumption is already of such a kind as to be untranslatable into a scientific language; it is irreducible to verifiable sentences. Words are here combined according to formation rules contradicting the syntax of every scientific language. The jump into metaphysics has been made. This can easily be shown. For what, in Jeans' sentence, do the words 'already woven' mean? In ordinary syntax, the word 'already' means in such a context the same as 'at an earlier time'. But here it is applied to the four-dimensional continuum, and with reference to this continuum a time point means merely a three-dimensional cross-section. 'The four-dimensional continuum subsists at a certain time point' is a meaningless word combination, if considered from the standpoint of customary syntax. And 'already woven in space and time' is only another formulation of this meaningless word combination. But this meaningless sentence only serves to pave the way for another meaningless sentence, namely, that 'the whole picture exists'. Thereby the transition has been made to a kind of metaphysical sentences particularly dear to philosophers, namely, to sentences containing the word 'exists' in a manner that precludes any reduction to verifiable sentences.

People talk as if the four-dimensional world continuum might 'exist' in the same way as a real empirical body exists. They go so far as to say: "the really existing is not the three-dimensional world of bodies but the four-dimensional space-time world." If a real body, say, the table at which I am sitting, 'does not exist', but a four-dimensional continuum does, then the word 'exists' is deprived of its reducibility to verifiable sentences. For the testing of the sentence 'P exists' always happens in such a way that I can convince myself of the existence of P in just the same manner as I do of the existence of the table at which I am sitting.

If one wants to attach a sense to the sentence 'the four-dimensional space-time continuum exists', then the word 'exists' must here be taken in an entirely different sense, namely, in the sense in which it is said of a mathematical formula: 'a formula exists by means of which the area of a plane figure is calculable', or 'there exists a solution for a certain type of differential equations'. This second meaning can be thought of as being reduced to the first, if the formula is imagined as something corporeal, say an accumulation of ink marks on paper. It may then be said that this formula, namely, this accumulation of ink, exists and supplies us with an instrument for solving a problem, just as an axe exists and supplies us with an instrument for splitting wood. Then the 'existence of the four-dimensional continuum' is tantamount to the existence of a formula for the calculation of the future state of the world from the present one. And the predetermination of the future, which is supposed to be supported by this view, is the same as was already implied in classical mechanics and symbolized in the Laplacean mind: the future is calculable in advance with the help of mechanics.

If we wish to employ this mode of expression, that everything which is calculable beforehand is really already there and does not yield anything 'new', then one ought to say with respect to classical mechanics that there is no such thing as development. For in relativity mechanics we also know the four-dimensional continuum only by virtue of a calculation from an initial state by means of the equations of motion. 'Existence of this continuum' means just this calculability.

The only difference consists in the fact that relativity mechanics permits us to establish an additional connection which classical physics does not mention. There are formulas by which we can calculate the various three-dimensional cross-sections from one another. In the simplest case this is afforded by the Lorentz transformation. But with this nothing is gained as far as predetermination is concerned, for not until we know the whole four-dimensional continuum are we in a position to lay cross-sections through it. Just as in classical mechanics, we derive our knowledge of the four-dimensional world only from an integration of the equations of motion. Relativity theory merely teaches us that these equations must have certain invariant properties. The illusion of the reality of the four-dimensional continuum arises only from the metaphysical employment of the word 'exists', which, in turn, as we have shown, has been prepared for by the metaphysical employment of the word 'already.'

How strong the suggestive power of the metaphysical language is may best be seen from the way in which Jeans, in his already mentioned lecture, subordinates his interpretation of the theory of relativity to a history, especially constructed for this purpose, of the human acquisition of knowledge of nature. After explaining how deeply human self-consciousness has suffered through Copernicus' discovery of the earth being but one body like any other among millions of bodies, and how, through Darwin's theory, man lost his exceptional position among the animals and became an animal like any other, he continues:

It is difficult to imagine human importance being rated lower than this; yet many thought that the physical theory of relativity, which Einstein advanced in 1905, exhibited human life in a still more ignoble light. Hitherto, the scientist and the plain man had been at one in thinking that events came to maturity with the passage of time, somewhat as the pattern of a tapestry is woven out of a loom... The pattern of the yet unwoven part of the picture may be inevitably determined by the way in which the loom is set, or it may not; at any rate, this part of the picture is not yet in existence. And so long as the weaving is not yet an accomplished fact, it is at least conceivable that something may still happen to modify it. The operator who works the loom can still alter the setting of the loom, and so, within limits, modify those parts of the tapestry which are still to come, according to his choice. In the same way, it seemed possible that humanity, and life in general, might be able to exercise some influence, however slight, on events that had not yet emerged from the womb of time.

As opposed to this view, in the opinion of Jeans, through his and many other philosophers' interpretation of the theory of relativity, a fundamental change has taken place. The future is not merely predetermined by the present initial states and the mechanical laws, i.e. by the 'setting of the loom'; the whole four-dimensional continuum, and thus the future too, is already there. From this juxtaposition it can be seen quite clearly that here the physicist Jeans has in mind exactly that metaphysical formulation of the theory of relativity which, as we have shown, is in no way reducible to verifiable sentences, and thus has no scientifically formulable content. That Jeans considers this metaphysical interpretation a real explanation can be gathered from the fact that he has retained a certain doubt concerning it. He says:

I do not think that such a view is absolutely forced upon us in any compelling manner by the facts of physics. At one time it seemed plausible because it gave a simple explanation of these facts, but no one would maintain that it is the unique explanation.

Thus the confusion of metaphysics and science comes to such a pass that Jeans treats the metaphysical interpretation as if it were a physical hypothesis, which is capable of being confirmed or disconfirmed by scientific progress. In reality, the metaphysical statements are not verifiable through any observational sentences, and hence they can neither be confirmed by physical or any other research, nor, likewise, can they be refuted.

NOTE

* From *Interpretations and Misinterpretations of Modern Physics*, Paris, Hermann et Cie, 1938, pp. 46-55.

THE PRINCIPLE OF EQUIVALENCE*

We now turn to the consequences of dynamic relativity, which go beyond the epistemological relativity. For this purpose we must analyze Einstein's theory of gravitation, since Einstein adopted Mach's idea of dynamic relativity and developed it further. Whereas Mach restricted his investigations to rotation, Einstein applied the principle to all kinds of motions; consequently his formulation is superior. He was able to give this general formulation by transforming the ideas of Mach into a differential principle.

Einstein expressed his *principle of equivalence* in the form of a thought experiment. Let a mass m be suspended by a spring in a closed compartment such as an elevator (Figure 1). A physicist in this compartment observes suddenly that the spring expands. He can easily verify this expansion by using a measuring rod. The increase in the tension of the spring indicates a stronger pull of the mass m. How can the physicist find the cause of this pull? He could give two explanations.

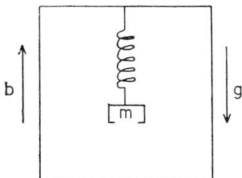

Fig. 1. Equivalence of acceleration and gravitation.

Explanation I. The compartment has received an upward acceleration (in the direction of arrow b) from some external force. The effect of the inertia of the mass m is therefore a downward pull opposite to the direction of the acceleration.

Explanation II. The compartment has remained at rest, but a downward directed gravitational field g (arrow g) has arisen and therefore exerts a stronger pull on the mass m.

It is impossible to decide experimentally between these two explanations inside the compartment. This is still true if we permit the physicist to look out of a window, since he will observe only kinematic phenomena, and these do not enable him to decide between the two explanations. It might be objected that explanation II requires the appearance of large observable masses below the compartment, but this is true only if static gravitational fields are assumed. As soon as we admit dynamic fields in Mach's sense, the gravitational field g can be attributed to a motion of the surrounding masses.

What is the basis of this indistinguishability? According to Einstein, its empirical basis is the equality of gravitational and inertial mass. This new distinction must be added to the usual distinction between mass and weight. There are therefore three concepts: inertial mass, gravitational mass, and weight.

The first distinction originated with Newton's discovery that the weight of a body depends not only on the body itself but also on the distance at which the body is located relative to the attracting mass. A mass m (Figure 2) resting on a spring balance will exert a different force (measurable by the tension of the spring, which is indicated by its length) on the support, according to the distance of the apparatus from the center of the earth. This fact is expressed by the formula

(1) $\quad \mathbf{F} = m.\mathbf{g}$

which resolves the force $\mathbf{F}$ exerted by a body on its support into the intensity $\mathbf{g}$ of the earth's gravitational field (the vectorial nature of $\mathbf{g}$ is usually ignored and it is written 'g') and a proportionality factor m due to the body itself. The structure of formula (1) is analogous to that of the formula

(2) $\quad \mathbf{F} = e.\mathbf{E}$

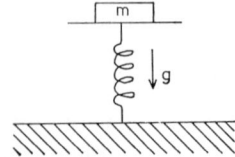

Fig. 2. Measurement of gravitational mass.

of electrostatics, where the mechanical force **F** results on the one hand from an intensity **E**, which is independent of the attracted body and characterizes the field, and on the other hand a proportionality factor *e* which is interpreted as the electric charge of the body. Correspondingly we might call *m* the *gravitational charge*.[1] This factor *m* is the *gravitational mass* of the body, i.e., the constant that expresses the effect of gravity upon it.

The mass of the body has also a quite different effect. If a carriage supporting the mass *m* is put in motion on a horizontal plane by the release of a compressed spring (Figure 3), then the force **F** of the spring will produce a certain acceleration **b** which determines the velocity with which the carriage continues to roll horizontally after the push. The following equation applies to this relation

(3) $\mathbf{F} = m.\mathbf{b}$

It turns out that in this equation *m* has the same value as that in equation (1). This is an empirical statement which we can imagine to be tested as follows. Assume objects of different materials, which, according to Figure 2, show the same compression of the spring and are then pushed, according to Figure 3, by a spring under the same tension. It can be shown that the push will give them equal velocities. This result is not self-evident. It is conceivable, for example, that volume would have an influence on inertia and that among the masses of equal weights those having a greater volume would receive a smaller velocity in the experiment of Figure 3. This question can be decided only by experience.

The principle of the equality of inertial and gravitational mass, which incidentally is also the reason for the equality of the velocities of falling bodies (body which is more strongly attracted by gravity has to overcome a correspondingly greater inertia) has been confirmed to a high degree by experiments. It is mentioned explicitly by Einstein as an empirical principle constituting the basis of his principle of equivalence.

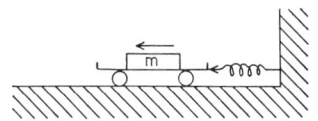

Fig. 3. Measurement of inertial mass.

The equivalence of inertia and gravity is the strict formulation of Mach's principle in the narrower sense. It implies that every phenomenon of inertia observable in an accelerated system can also be explained as a gravitational phenomenon; therefore it cannot be interpreted to indicate uniquely a state of motion. Conversely, we can use the principle of equivalence to *transform away* that gravitational field which was considered an absolute datum in classical mechanics. A freely falling elevator is a system in which the gravitation of the earth is transformed away. Any object in it when pushed would assume a rectilinear, force-free motion in the sense of the law of inertia.

The possibility of 'transforming away' is subject to certain essential restrictions. Generally speaking, we can transform away gravitational fields only in infinitesimal regions. Let us consider for example the radial field of the earth (Figure 4). If we let a rigid system of cells (the dotted lines of the figure) move in the direction of arrow b with an acceleration $g = 981$ cm/sec^2, the earth field will be transformed away in cell a but not in any of the others. We can now make the following statement: for any given small region b we can always specify for the system of cells an accelerated motion which will transform away the gravitational field at b. We may therefore say that any gravitational field can always be transformed away in any given region, but not in all regions at the same time by the same transformation.

This principle takes the place of the Newtonian concept of inertial system. By inertial system [2] Newton understands those astronomically determined systems in which the law of inertia applies, i.e., those systems

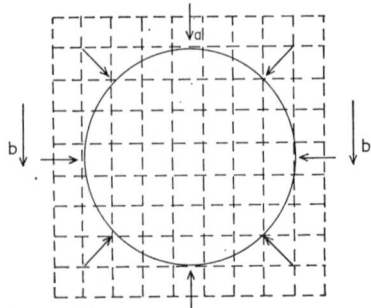

Fig. 4. Local 'transforming away' of the gravitational field.

that move uniformly relative to absolute space. It can be shown within the framework of Newton's theory that one can obtain local inertial systems by transforming away the gravitational field, although these systems are in a different state of motion provided that the equivalence of inertial and gravitational mass is presupposed. The gravitational field, which as such is still present, is compensated in these local systems by their acceleration relative to absolute space and the resulting inertial forces. According to Einstein, however, only these local systems are the actual inertial systems. In them the field, which generally consists of a gravitational and an inertial component, is transformed in such manner that the gravitational component disappears and only the inertial component remains. There are, strictly speaking, only local inertial systems. The astronomical inertial systems of Newton can at best be approximations which gradually change in the neighborhood of stars. Only because distances in space are large compared to the masses of the stars, and because the stars have very low speeds, are astronomical inertial systems possible as approximations.

We must now formulate this idea more precisely. Above all we have to state exactly what is meant by an 'actual' inertial system, which for the time being has only a more or less intuitive meaning. Let us investigate first how the local inertial systems result according to Newton. Newton's equation for the motion of a mass point in a gravitational field is given by

(1) $\quad \ddot{x} = g$

If we now relate the x-coordinate to a freely falling system, i.e., introduce the transformation

(2) $\quad x = x' + \dfrac{g}{2} t^2$

$\qquad y = y'$

then

$\qquad \ddot{x} = \ddot{x}' + g$

and (1) becomes

(3) $\quad \ddot{x}' = 0$

which is the equation of motion in an inertial system. Within mechanics there exists no difference between the two kinds of inertial systems, and it would be a play on words to argue that one or the other of the two is an 'actual' inertial system. If we take into account, however, *extra-mechanical phenomena, there will be a difference: whereas according to Newton the astronomical inertial systems form the normal systems for all phenomena, Einstein maintains that it is the local inertial systems which form the normal systems*. We shall study the resulting difference in the example of the motion of light.

According to the Newtonian theory only the astronomical inertial systems are the normal systems for the propagation of light. Only in them does light travel in straight lines, while its path is curved in a local inertial system. The motion of a light ray which moves parallel to the y-axis is given in the Newtonian inertial system by the differential equation

(4) $\quad \dot{x} = 0$
$\quad\quad\quad \dot{y} = c$

These equations are valid according to Newton even if there is a gravitational field, as for example on the surface of the earth. The earth is embedded (for short intervals of time) in an astronomical inertial system upon which the gravitational field of the earth is only locally superimposed. With respect to light, this gravitational field does not exist at all. If we now apply transformation (2) to these equations, they become

(5) $\quad \dot{x}' = -gt$
$\quad\quad\quad \dot{y}' = c$

Relative to K', light no longer travels along straight lines, because its x'-coordinate is no longer a linear function of time.

According to Einstein, however, the local inertial systems are the actual inertial systems for all other phenomena. In the case of the light ray, for instance, the equation of motion must be linear in the local inertial system K', and the differential equations must therefore be:

(6) $\quad \dot{x}' = 0$
$\quad\quad\quad \dot{y}' = c$

If we now go in turn with transformation (2) to the system K which is stationary on the earth's surface and consequently at rest in the astronomical inertial system, the equations will become

(7) $\quad \dot{x} = gt$
$\quad\quad\quad \dot{y} = c$

It is relative to *this* system that light is now curved.

We shall illustrate the train of thought that leads from (6) to (7) by the path of a light ray; this will bring out the purely kinematic basis of the inference. Let us imagine a compartment (Figure 5) at rest on the earth. Relative to the local inertial system it will perform an upward accelerated motion. Let us also assume that a light ray enters the compartment

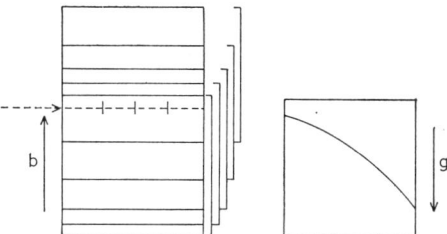

Fig. 5. Bending of a light ray as a consequence of the principle of equivalence.

through a slit on the left-hand side. We can now determine its path within the compartment if we assume that the local inertial system is at rest, and if we construct the motion of the light ray relative to the compartment by superimposing the straight line path of the light ray upon the accelerated motion of the compartment. The different consecutive positions assumed by the compartment are indicated by the square brackets of Figure 5. The end of the light ray is a little farther to the right for each successive position of the compartment, corresponding to the marks on the dotted line. It can now easily be seen that these marks have different positions relative to the compartment in its various locations. On the right-hand side we have drawn the same process relative to the compartment as a rest system and indicated the marks this time in their relative positions in the compartment. The path of the light ray is therefore a *curved line* relative to the compartment. This is a purely kinematic effect. It derives from the fact that the horizontal motion of the light is uniform, while the vertical motion of the compartment is accelerated. Since we have started from the assumption, however, that light travels in straight lines relative to the local inertial system which falls freely relative to the earth, we have now arrived at the far-reaching physical consequence that light assumes a curved path relative to a system which rests on the earth: there is a curvature of light in the gravitational field of a mass center.

It is irrelevant in this case whether the mass center itself is resting in an astronomical inertial system, since this inertial system no longer constitutes a normal system in the neighborhood of the mass center. Indeed, it is no longer reasonable to speak here of an inertial system with a superimposed gravitational field. The astronomical inertial system is destroyed in the neighborhood of the mass center and cannot be extended from the surrounding space to the region of the mass field without losing its inertial character. Its functions have been taken over by the local inertial system to which it cannot be rigidly attached.

In these assumptions we find the core of the general theory of relativity. It is a genuine physical principle which, with the inclusion of all non-mechanical phenomena in the characterization of the local inertial system, states a *physical hypothesis* that goes far beyond the experience stated in the equivalence of inertial and gravitational mass. Einstein's hypothesis corresponds to a methodological procedure frequently used in physics. Although the hypothesis does not follow logically from the empirical evidence but claims much more, it is assumed in the hope that the observation of further derivable consequences will confirm it. After the special theory or relativity had formulated the laws of clocks, measuring rods, the motion of light, etc., for inertial systems, the new hypothesis could now be formulated by the statement that it is not the astronomical inertial systems, but the local inertial systems, for which the special theory of relativity holds. The gravitation-free ideal case required for the special theory of relativity is therefore not realized in the astronomical inertial systems, but in the local inertial systems. We may thus speak of the *principle of local inertial systems*, which states that the *local inertial systems are those systems in which the light- and matter-axioms are satisfied.*[3] With this hypothesis Einstein introduces the general theory of relativity, and the special theory of relativity thus becomes the limiting case of the general theory.

For the sake of completeness, we shall now show how the same inferences that lead to physical consequences regarding light also lead to similar consequences regarding clocks. We shall again consider a kinematic effect that results from the accelerated motion of a clock relative to an inertial system, and infer from it an effect in the gravitational field. The kinematic effect with which we are concerned is the *Doppler effect*.

Let us first consider the Doppler effect that results from uniform motion

THE PRINCIPLE OF EQUIVALENCE

(Figure 6). Let us assume that an observer is moving in a straight line with uniform velocity away from U_1. Whenever the clock U_1 completes a period, it sends out a signal which will reach the observer at increasingly distant points. The intervals between the various light signals are therefore longer for the observer than the unit intervals of the clock U_2 which he

Fig. 6. Doppler effect as a result of uniform motion.

carries with him. For him clock U_1 runs slower than U_2. Let us now consider a similar process in the case of accelerated motion (Figure 7). The two clocks U_1 and U_2 are connected by a rigid rod, and the system which they form has an accelerated motion. U_1 again sends signals after each unit period. The first signal leaves A_1 and reaches U_2 when U_2 has reached A_2. The second signal leaves U_1 at B_1 and reaches U_2 at B_2, etc. The distances A_1A_2, B_1B_2, C_1C_2... will become longer and longer, and an observer who moves with U_2 will thus experience a Doppler effect in the sense of a retardation of U_1. In either case there is therefore a retardation of one clock relative to the signals which arrive from the other clock. Whereas in the case of uniform motion, only one of the clocks is in motion while the other is at rest, the effect will appear in the case of accelerated motion even when the two clocks are at rest relative to each other, provided the rigid system which they form moves as a whole. The latter case permits reinterpretation in terms of the principle of equivalence. Two clocks which are at rest in the gravitational field of a mass center are in an accelerated motion relative to the corresponding local inertial system. Our consideration will therefore lead directly to the assertion that a gravita-

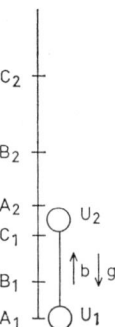

Fig. 7. Doppler effect as a result of accelerated motion.

tional field produces a retardation of those clocks which are located in regions that have a higher absolute value of the gravitational potential. In the case of atom clocks, there would be a red shift of the spectral lines, because a retardation of the frequency manifests itself as a shift of the wave-length in the direction of the red end of the spectrum.

It should be noted that this effect is independent of the retardation of clocks. We have used for its derivation nothing but the Doppler effect. The Doppler effect was also known in the classical theory of time, which does not include however the retardation of clocks. The retardation of clocks in a gravitational field must therefore occur if the principle of equivalence alone is correct, regardless whether there is an Einsteinian retardation of clocks for uniform motion. This latter effect shows only in the *quantitative* calculations of the retardation of clocks in a gravitational field, where it appears as a small correction factor.

This last result is due to the fact that the Doppler effect can be calculated as the superposition of two effects, namely, the classical Doppler effect and the Einsteinian retardation of clocks. Conversely, we can recognize from this result that the Einsteinian retardation of clocks in uniform motion has nothing to do with the Doppler effect.

The bending of light and the retardation of clocks are direct consequences of the principle of equivalence, and they demonstrate very clearly the hypothetical character of the principle since they are empirically confirmable phenomena. The third of the so-called Einstein effects, namely the advance of the perihelion of planetary orbits, does not follows immediately from the principle of equivalence, but from Einstein's theory of gravitation based upon it.

NOTES

* From *Philosophy of Space and Time*, Dover N.Y., 1956, pp. 222–232.
[1] H. Weyl, *op. cit.*, p. 225.
[2] This term was introduced by L. Lange, 'Über die wissenschaftliche Fassung des Galileischen Beharrungsgesetzes', *Wundts Philos. Studien,* 1885, Vol. II.
[3] Strictly speaking this should read: "in which these axioms are satisfied to a higher degree of approximation." Cf. A., § 34.

H. P. ROBERTSON

GEOMETRY AS A BRANCH OF PHYSICS*

Is space really curved? That is a question which, in one form or another, is raised again and again by philosophers, scientists, T. C. Mits and readers of the weekly comic supplements. A question which has been brought into the limelight above all by the genial work of Albert Einstein, and kept there by the unceasing efforts of astronomers to wrest the answer from a curiously reluctant Nature.

But what is the meaning of the question? What, indeed, is the meaning of each word in it? Properly to formulate and adequately to answer the question would require a critical excursus through philosophy and mathematics into physics and astronomy, which is beyond the scope of the present modest attempt. Here we shall be content to examine the roles of deduction and observation in the problem of physical space, to exhibit certain high points in the history of the problem, and in the end to illustrate the viewpoint adopted by presenting a relatively simple caricature of Einstein's general theory of relativity. It is hoped that this, certainly incomplete and possibly naïve, description will present the essentials of the problem from a neutral mathematico-physical viewpoint in a form suitable for incorporation into any otherwise tenable philosophical position. Here, for example, we shall not touch directly upon the important problem of form versus substance – but if one wishes to interpret the geometrical substratum here considered as a formal backdrop against which the contingent relations of nature are exhibited, one should be able to do so without distorting the scientific content.

First, then, we consider geometry as a deductive science, a brach of mathematics in which a body of theories is built up by logical processes from a postulated set of axioms (not 'self-evident truths'). In logical position geometry differs not in kind from any other mathematical discipline – say the theory of numbers or the calculus of variations. As mathematics, it is not the science of measurement, despite the implications of its name – even though it did, in keeping with the name, originate in the codification of rules for land surveying. The principal criterion of its validity as a mathematical discipline is whether the axioms as written down are self-consistent, and the sole criterion of the truth of a theorem involving its concepts is whether the theorem can be deduced from the axioms. This truth is clearly relative to the axioms; the theorem that the sum of the three interior angles of a triangle is equal to two right angles, true in Euclidean geometry, is false in any of the geometries obtained on replacing the parallel postulate by one of its contraries. In the present sense it suffices for us that geometry is a body of theorems, involving among others the concepts of point, angle and a unique numerical relation called distance between pairs of points, deduced from a set of self-consistent axioms.

What, then, distinguishes Euclidean geometry as a mathematical system from those logically consistent systems, involving the same category of concepts, which result from the denial of one or more of its traditional axioms? This distinction cannot consist in its 'truth' in the sense of observed fact in physical science; its truth, or applicability, or still better appropriateness, in this latter sense is dependent upon observation, and not upon deduction alone. The characteristics of Euclidean geometry, as mathematics, are therefore to be sought in its internal properties, and not in its relation to the empirical.

First, Euclidean geometry is a *congruence geometry*, or equivalently the space comprising its elements is *homogeneous and isotropic*; the intrinsic relations between points and other elements of a configuration are unaffected by the position or orientation of the configuration. As an example, in Euclidean geometry all intrinsic properties of a triangle – its angles, area, etc. – are uniquely determined by the lengths of its three sides two triangles whose three sides are respectively equal are 'congruent'; either can by a 'motion' of the space into itself be brought into complete coincidence with the other, whatever its original position and orientation may be. These motions of Euclidean space are the familiar translations and rotations, use of which is made in proving many of the theorems of Euclid. That the existence of these motions (the axiom of 'free mobility') is a desideratum, if not indeed a necessity, for a geometry applicable to physical space, has been forcibly argued on *a priori* grounds by von Helmholtz, Whitehead, Russell and others; for only in a homogeneous and isotropic space can the traditional concept of a rigid body be maintained.[1]

But the Euclidean geometry is only one of several congruence geo-

GEOMETRY AS A BRANCH OF PHYSICS 411

metries; there are in addition the 'hyperbolic' geometry of Bolyai and Lobachewsky, and the 'spherical' and 'elliptic' geometries of Riemann and Klein. Each of these geometries is characterized by a real number K, which for the Euclidean geometry is zero, for the hyperbolic negative, and for the spherical and elliptic geometries positive. In the case of 2-dimensional congruence spaces, which *may* (but need not) be conceived as surfaces embedded in a 3-dimensional Euclidean space, the constant K may be interpreted as the *curvature* of the surface into the third dimension – whence it derives its name. This name and this representation are for our purposes at least psychologically unfortunate, for we propose ultimately to deal exclusively with properties intrinsic to the space under consideration – properties which in the later physical applications can be measured within the space itself – and are not dependent upon some extrinsic construction, such as its relation to an hypothesized higher dimensional embedding space. We must accordingly seek some determination of K – which we nevertheless continue to call curvature – in terms of such inner properties.

In order to break into such an intrinsic characterization of curvature, we first relapse into a rather naïve consideration of measurements which may be made on the surface of the earth, conceived as a sphere of radius R. This surface is an example of a 2-dimensional congruence space of positive curvature $K = 1/R^2$ on agreeing that the abstract geometrical concept 'distance' r between any two of its points (not the extremities of a diameter) shall correspond to the lesser of the two distances *measured on the surface* between them along the unique great circle which joins the two points.[2] Consider now a 'small circle' of radius r (measured on the surface!) about a point P of the surface; its perimeter L and area A (again measured on the surface!) are clearly less than the corresponding measures $2\pi r$ and πr^2 of the perimeter and area of a circle of radius r in the Euclidean plane. An elementary calculation shows that for sufficiently small r (i.e., small compared with R) these quantities on the sphere are given approximately by:

(1) $L = 2\pi r \left(1 - Kr^2/6 + ...\right),$

  $A = \pi r^2 \left(1 - Kr^2/12 + ...\right).$

Thus, the ratio of the area of a small circle of radius 400 miles on the surface of the earth to that of a circle of radius 40 miles is found to be only 99.92, instead of 100.000 as in the plane.

Another consequence of possible interest for astronomical applications is that in spherical geometry the sum σ of the three angles of a triangle (whose sides are arcs of great circles) is *greater* than 2 right angles; it can in fact be shown that this 'spherical excess' is given by

(2) $\quad \sigma - \pi = K\delta,$

where δ is the area of the spherical triangle and the angles are measured in radians (in which $180° = \pi$). Further, each full line (great circle) is of finite length $2\pi R$, and any two full lines meet in two points – there are no parallels!

In the above paragraph we have, with forewarning, slipped into a non-intrinsic quasi-physical standpoint in order to present the formulae (1) and (2) in a more or less intuitive way. But the essential point is that these formulae are in fact independent of this mode of presentation; they are relations between the mathematical concepts distance, angle, perimeter and area which follow as logical consequences from the axioms of this particular kind of non-Euclidean geometry. And since they involve the space-constant K, this 'curvature' may in principle at least be determined *by measurements made on the surface*, without recourse to its embedment in a higher dimensional space.

Further, these formulae may be shown to be valid for a circle or triangle in the hyperbolic plane, a 2-dimensional congruence space for which $K<0$. Accordingly here the perimeter and area of a circle are *greater*, and the sum of the three angles of a triangle *less*, than the corresponding quantities in the Euclidean plane. It may also be shown that each full line is of infinite length, that through a given point outside a given line an infinity of full lines may be drawn which do not meet the given line (the two lines bounding the family are said to be 'parallel' to the given line), and that two full lines which meet do so in but one point.

The value of the intrinsic approach is especially apparent in considering 3-dimensional congruence spaces, where our physical intuition is of little use in conceiving them as 'curved' in some higher dimensional space. The intrinsic geometry of such a space of curvature K provides formulae for the surface area S and the volume V of a 'small sphere' of radius r, whose leading terms are

(3)
$$S = 4\pi r^2 \left(1 - Kr^2/3 + \ldots\right),$$
$$V = 4/3\pi r^3 \left(1 - Kr^2/5 + \ldots\right).$$

It is to be noted that in all these congruence geometries, except the Euclidean, there is at hand a natural unit of length $R = 1/|K|^{\frac{1}{2}}$; this length we shall, without prejudice, call the 'radius of curvature' of the space.

So much for the congruence geometries. If we give up the axiom of free mobility we may still deal with the geometry of spaces which have only limited or no motions into themselves.[3] Every smooth surface in 3-dimensional Euclidean space has such a 2-dimensional geometry; a surface of revolution has a 1-parameter family of motions into itself (rotations about its axis of symmetry), but not enough to satisfy the axiom of free mobility. Each such surface has at a point $P(x, y)$ of it an intrinsic 'total curvature' $K(x, y)$, which will in general vary from point to point; knowledge of the curvature at all points essentially determines all intrinsic properties of the surface.[4] The determination of $K(x, y)$ by measurements on the surface is again made possible by the fact that the perimeter L and area A of a closed curve, every point of which is at a given (sufficiently small) distance r from $P(x, y)$, are given by the formulae (1), where K is no longer necessarily constant from point to point. Any such variety for which $K = 0$ throughout is a ('developable') surface which may, on ignoring its macroscopic properties, be rolled out without tearing or stretching onto the Euclidean plane.

From this we may go on to the contemplanation of 3-or higher dimensional ('Riemannian') spaces, whose intrinsic properties vary from point to point. But these properties are no longer describable in terms of a single quantity, for the 'curvature' now acquires at each point a directional character which requires in 3-space 6 components (and in 4-space 20) for its specification. We content ourselves here to call attention to a single combination of the 6, which we call the 'mean curvature' of the space at the point $P(x, y, z)$, and which we again denote by K – or more fully by $K(x, y, z)$; it is in a sense the mean of the curvatures of various surfaces passing through P, and reduces to the previously contemplated space-constant K when the space in question is a congruence space.[5] This concept is useful in physical applications, for the surface area S and the volume V of a sphere of radius r about the point $P(x, y, z)$ as center are again given by formulae (3), where now K is to be interpreted as the mean curvature $K(x, y, z)$ of the space at the point P. In four and higher dimensions similar concepts may be introduced and similar formulae developed, but for them we have no need here.

We have now to turn our attention to the world of physical objects about us, and to indicate how an ordered description of it is to be obtained in accordance with accepted, preferably philosophically neutral, scientific method. These objects, which exist for us in virtue of some pre-scientific concretion of our sense-data, are positioned in an extended manifold which we call physical space. The mind of the individual, retracing at an immensely accelerated pace the path taken by the race, bestirs itself to an analysis of the interplay between object and extension. There develops a notion of the permanence of the object and of the ordering and the change in time – another form of extension, through which object and subject appear to be racing together – of its extensive relationships. The study of the ordering of actual and potential relationships, the physical problem of space and time, leads to the consideration of geometry and kinematics as a branch of physical science. To certain aspects of this problem we now turn our attention.

We consider first that proposed solution of the problem of space which is based upon the postulate that space is an *a priori* form of the understanding. Its geometry must then be a congruence geometry, independent of the physical content of space; and since for Kant, the propounder of this view, there existed but one geometry, space must be Euclidean – and the problem of physical space is solved on the epistemological, pre-physical, level.

But the discovery of other congruence geometries, characterized by a numerical parameter K, perforce modifies this view, and restores at least in some measure the objective aspect of physical space; the *a posteriori* ground for this space-constant K is then to be sought in the contingent. The means for its intrinsic determination is implicit in the formulae presented above; we have merely (!) to measure the volume V of a sphere of radius r or the sum σ of the angles of a triangle of measured area δ, and from the results to compute the value of K. On this modified Kantian view, which has been expounded at length by Russell,[6] it is inconceivable that K might vary from point to point – for according to this view the very possibility of measurement depends on the constancy of space-structure, as guaranteed by the axiom of free mobility. It is of interest to mention in passing, in view of recent cosmological findings, the possibility raised by Calinon (in 1889!) that the space-constant K might vary with time.[7] But this possibility is rightly ignored by Russel, for the same arguments which would on this *a priori* theory require the constancy of K in space would equally require its constancy in time.

In the foregoing sketch we have dodged the real hook in the problem of measurement. As physicists we should state clearly those aspects of the physical world which are to correspond to elements of the mathematical system which we propose to employ in the description ('realisation' of the abstract system). Ideally this program should prescribe fully the operations by which numerical values are to be assigned to the physical counterparts of the abstract elements. How is one to achieve this in the case in hand of determining the numerical value of the space-constant K?

Although Gauss, one of the spiritual fathers of non-Euclidean geometry, at one time proposed a possible test of the flatness of space by measuring the interior angles of a terrestrial triangle, it remained for his Göttingen successor Schwarzschild to formulate the procedure and to attempt to evaluate K on the basis of astronomical data available at the turn of the century.[8] Schwarzschild's pioneer attempt is so inspiring in its conception and so beautiful in its expression that I cannot refrain from giving here a few short extracts from his work. After presenting the possibility that physical space may, in accordance with the neo-Kantian position outlined above, be non-Euclidean, Schwarzschild states (in free translation):

> One finds oneself here, if one but will, in a geometrical fairyland, but the beauty of this fairy tale is that one does not know but what it may be true. We accordingly bespeak the question here of how far we must push back the frontiers of this fairyland; of how small we must choose the curvature of space, how great its radius of curvature.

In furtherance of this program Schwarzschild proposes:

> A triangle determined by three points will be defined as the paths of light-rays from one point to another, the lengths of its sides a, b, c, by the times it takes light to traverse these paths, and the angles α, β, γ will be measured with the usual astronomical instruments.

Applying Schwarzschild's prescription to observations on a given star, we consider the triangle ABC defined by the position A of the star and by two positions B, C of the earth – say six months apart – at which the angular positions of the star are measured. The base $BC=a$ is known, by measurements within the solar system consistent with the prescription, and the interior angles β, γ which the light-rays from the star make with the base-line are also known by measurement. From these the *parallax* $p=\pi-(\beta+\gamma)$ may be computed; in Euclidean space this parallax is simply the inferred angle α subtended at the star by the diameter of the earth's orbit. In the other congruence geometries the parallax is seen, with the aid of formula (2) above, to be equal to

$$(2') \qquad p = \pi - (\beta + \gamma) = \alpha - K\delta,$$

where α is the (unknown) angle at the star A, and δ is the (unknown) area of the triangle ABC. Now in spite of our incomplete knowledge of the elements on the far right, certain valid conclusions may be drawn from this result. First, if space is hyperbolic ($K<0$), for distant stars (for which $\alpha\sim0$), the parallax p will remain positive; hence if stars are observed whose parallax is zero to within the errors of observation, this estimated error will give an upper limit to the absolute value $-K$ of the curvature. Second, if space is spherical ($K>0$), for a sufficiently distant star (more distant than one-quarter the circumference of a Euclidean sphere of radius $R=1/K^{\frac{1}{2}}$, as may immediately be seen by examining a globe) the sum $\beta+\gamma$ will exceed two right angles; hence the parallax p of such a star should be negative, and if no stars are in fact observed with negative parallax, the estimated error of observation will give an upper limit to the curvature K. Also, in this latter case the light sent out by the star must return to it after traversing the full line of length $2\pi R$, (πR in elliptic space) and hence we should, but for absorption and scattering, be able to observe the returning light as an anti-star in a direction opposite to that of the star itself!

On the basis of the evidence then available, Schwarzschild concluded that if space is hyperbolic its radius of curvature $R=1/(-K)^{\frac{1}{2}}$ cannot be less than 64 light-years (i.e., the distance light travels in 64 years), and that if the space is elliptic its radius of curvature $R=1/K^{\frac{1}{2}}$ is at least 1600 light-years. Hardly imposing figures for us today, who believe on other astronomical grounds that objects as distant as 500 million light-years have been sighted in the Mt Wilson telescope, and who are expecting to find objects at twice that distance with the new Mt Palomar mirror!

GEOMETRY AS A BRANCH OF PHYSICS

But the value for us of the work of Schwarzschild lies in its sound operational approach to the problem of physical geometry – in refreshing contrast to the pontifical pronouncement of Poincaré who, after reviewing the subject, stated:[9]

> If therefore negative parallaxes were found, or if it were demonstrated that all parallaxes are superior to a certain limit, two courses would be open to us; we might either renounce Euclidean geometry, or else modify laws of optics and suppose that light does not travel rigorously in a straight line.
> It is needless to add that all the world would regard the latter solution as the more advantageous.
> The Euclidean geometry has, therefore, nothing to fear from fresh experiments. [!]

So far we have tied ourselves into the neo-Kantian doctrine that space must be homogeneous and isotropic, in which case our proposed operational approach is limited in application to the determination of the numerical value of the space-constant K. But the possible scope of the operational method is surely broader than this; what if we do apply it to triangles and circles and spheres in various positions and at various times and find that the K so determined is in fact dependent on position is space and time? Are we, following Poincaré, to attribute these findings to the influence of an external force postulated for the purpose? Or are we to take our findings at face value, and accept the geometry to which we are led as a natural geometry for physical science?

The answer to this methodological question will depend largely on the *universality* of the geometry thus found – whether the geometry found in one situation or field of physical discourse may consistently be extended to others – and in the end partly on the predilection of the individual or of his colleagues or of his times. Thus Einstein's special theory of relativity, which offers a physical kinematics embracing measurements in space and time, has gone through several stages of acceptance and use, until at present it is a universal and indispensable tool of modern physics. Thus Einstein's general theory of relativity, which offers an extended kinematics which includes in its geometrical structure the universal force of gravitation, was long considered by some comtemporaries to be a *tour de force*, at best amusing but in practice useless. And now, in extending this theory to the outer bounds of the observed universe, the kind of geometry suggested by the present marginal data seems to many so repugnant that they would follow Poincaré in postulating some *ad hoc* force, be it a double standard of time or a secular change in the velocity of light or Planck's constant, rather than accept it.

But enough of this general and historical approach to the problem of physical geometry! While we should like to complete this discussion with a detailed operational analysis of the solution given by the general theory of relativity, such an undertaking would require far more than the modest mathematical background which we have here presupposed. Further, the field of operations of the general theory is so unearthly and its *experimenta crusis* so delicate that an adequate discussion would take us far out from the familiar objects and concepts of the workaday world, and obscure the salient points we wish to make in a welter of unfamiliar and esoteric astronomical and mathematical concepts. What is needed is a homely experiment which could be carried out in the basement with parts from an old sewing machine and an Ingersoll watch, with an old file of *Popular Mechanics* standing by for reference! This I am, alas, afraid we have not achieved, but I do believe that the following example of a simple theory of measurement in a heat-conducting medium is adequate to expose the principles involved with a modicum of mathematical background. The very fact that it will lead to a rather bad and unacceptable physical theory will in itself be instructive, for its very failure will emphasize the requirement of universality of application – a requirement most satisfactorily met by the general theory of relativity.

The background of our illustration is an ordinary laboratory, equipped with Bunsen burners, clamps, rulers, micrometers and the usual miscellaneous impedimenta there met – at the turn of the century, no electronics required! In it the practical Euclidean geometry reigns (hitherto!) unquestioned, for even though measurements are there to be carried out with quite reasonable standards of accuracy, there is no need for sophisticated qualms concerning the effect of gravitational or magnetic or other general extended force-fields on its metrical structure. Now that we feel at home in these familiar, and disarming, surroundings, consider the following experiment:

Let a thin, flat metal plate be heated in any way – just so that the temperature T is not uniform over the plate. During the process clamp or otherwise constrain the plate to keep it from buckling, so that it can reasonably be said to remain flat by ordinary standards. Now proceed to make simple geometrical measurements on the plate with a short metal rule, which has a certain coefficient of expansion c, taking care that the rule is allowed to come into thermal equilibrium with the plate at each setting before making the measurement. The question now is, what is the geometry of the plate *as revealed by the results of these measurements*?

It is evident that, unless the coefficient of expansion c of the rule is zero, the geometry will not turn out to be Euclidean, for the rule will expand more in the hotter regions of the plate than in the cooler, distorting the (Euclidean) measurements which would be obtained by a rule whose length did not change according to the usual laboratory standards. Thus the perimeter L of a circle centered at a point at which a burner is applied will surely turn out to be greater than π times its measured diameter $2r$, for the rule will expand in measuring through the hotter interior of the circle and hence give a smaller reading than if the temperature were uniform. On referring to the first of formulae (1) above it is seen that the plate would seem to have a negative curvature K at the center of the circle – the kind of structure exhibited by an ordinary twisted surface in the neighborhood of a 'saddle-point'. In general the curvature will vary from point to point in a systematic way; a more detailed mathematical analysis of the situation shows that, on removing heat sources and neglecting radiation losses from the faces of the plate, K is everywhere negative and that the 'radius of curvature' $R=1/(-K)^{\frac{1}{2}}$ at any point P is inversely proportional to the rate s at which heat flows past P. (R is in fact equal to k/cs, where k is the coefficient of heat conduction *of the plate* and c is as before the coefficient of expansion *of the rule*.) The hyperbolic geometry is accordingly realized when the heat flow is constant throughout the plate, as when the long sides of an elongated rectangle are kept at different fixed temperatures.[10]

And now comes the question, what is the true geometry of the plate? The flat Euclidean geometry we had uncritically agreed upon at the beginning of the experiment, or the non-Euclidean geometry reavealed by measurement? It is obvious that the question is improperly worded; the geometry is determinate only when we prescribe the method of measurement, i.e., when we set up a correspondence between the physical aspects (here readings on a definite rule obtained in a prescribed way) and the elements (here distances, in the abstract sense) of the mathematical system. Thus our original common-sense requirement that the plate not buckle, or that it be measured with an invar rule (for which $c \sim 0$), leads to Euclidean geometry, while the use of a rule with a sensible coefficient of expansion leads to a locally hyperbolic type of Riemannian geometry, which is in general not a congruence geometry.

There is no doubt that anyone examining this situation will prefer Poincaré's common-sense solution of the problem of the physical geometry of the plate – i.e., to attribute to it Euclidean geometry, and to consider the measured deviations from this geometry as due to the action of a force (thermal stresses in the rule). Most compulsive to this solution is the fact that this disturbing force lacks the requirement of universality; on employing a brass rule in place of one of steel we would find that the local curvature is trebled – and an ideal rule ($c=0$) would, as we have noted, lead to the Euclidean geometry.

In what respect, then, does the general theory of relativity differ in principle from this geometrical theory of the hot plate? The answer is: *in its universality*; the force of gravitation which it comprehends in the geometrical structure acts equally on all matter. There is here a close analogy between the gravitational mass M of the field-producing body (Sun) and the inertial mass m of the test-particle (Earth) on the one hand, and the heat conduction k of the field (plate) and the coefficient of expansion c of the test-body (rule) on the other. *The success of the general relativity theory of gravitation as a physical geometry of space-time is attributable to the fact that the gravitational and inertial masses of any body are observed to be rigorously proportional for all matter.* Whereas in our geometrical theory of the thermal field the ratio of heat conductivity to coefficient of expansion varies from substance to substance, resulting in a change of the geometry of the field on changing the test-body.

From our present point of view the great triumph of the theory of relativity lies in its absorbing the universal force of gravitation into the geometrical structure; its success in accounting for minute discrepancies in the Newtonian description of the motions of test-bodies in the solar field, although gratifying, is nevertheless of far less moment to the philosophy of physical science.[11] Einstein's achievements would be substantially as great even though it were not for these minute observational tests.

Our final illustration of physical geometry consists in a brief reference to the cosmological problem of the geometry of the observed universe as a whole – a problem considered in greater detail elsewhere in this volume. *If* matter in the universe can, taken on a sufficiently large scale (spatial gobs millions of light-years across), be considered as uniformly distributed, and if (as implied by the general theory of relativity) its geometrical structure is conditioned by matter, then to this approximation our 3-dimensional astronomical space must be homogeneous and isotropic, with a spatially-constant K which may however depend upon time. Granting this hypothesis, how do we go about measuring K, using of course only procedures which can be operationally specified, and to which congruence geometry are we thereby led? The way to the answer is suggested by the second of the formulae (3), for if the nebulae are by-and-large uniformly distributed, then the number N within a sphere of radius r must be proportional to the volume V of this sphere. We have then only to examine the dependence of this number N, as observed in a sufficiently powerful telecope, on the distance r to determine the deviation from the Euclidean value. But how is r operationally to be defined?

If all the nebulae were of the same intrinsic brightness, then their apparent brightness as observed from the Earth should be an indication of their distance from us; we must therefore examine the exact relation to be expected between apparent brightness and the abstract distance r. Now it is the practice of astronomers to assume that brightness falls off inversely with the square of the 'distance' of the object – as it would do in Euclidean space, if there were no absorption, scattering, and the like. We must therefore examine the relation between this astronomer's 'distance' d, as inferred from apparent brightness, and the distance r which appears as an element of the geometry. It is clear that *all* the light which is radiated at a given moment from the nebula will, after it has traveled a distance r, lie on the surface of a sphere whose area S is given by the first of the formulae (3). And since the practical procedure involved in determining d is equivalent to assuming that all this light lies on the surface of a Euclidean sphere of radius d, it follows immediately that the relationship between the 'distance' d used in practice and the distance r dealt with in the geometry is given by the equation

$$4\pi d^2 = S = 4\pi r^2 (1 - Kr^2/3 + ...);$$

whence, to our approximation

(4)
$$d = r(1 - Kr^2/6 + ...), \text{ or}$$
$$r = d(1 + Kd^2/6 + ...).$$

But the astronomical data give the number N of nebulae counted out to a given inferred 'distance' d, and in order to determine the curvature from them we must express N, or equivalently V, to which it is assumed proportional, in terms of d. One easily finds from the second of the formulae (3) and the formula (4) just derived that, again to the approximation here adopted,

(5) $V = 4/3\pi d^3 (1 + 3/10 Kd^2 + ...)$.

And now on plotting N against inferred 'distance' d and comparing this empirical plot with the formula (5), it should be possible operationally to determine the 'curvature' K.[12]

The search for the curvature K indicates that, after making all known corrections, the number N seems to increase faster with d than the third power, which would be expected in Euclidean space, hence K is *positive*. The space implied thereby is therefore bounded, of finite total volume, and of a present 'radius of curvature' $R = 1/K^{\frac{1}{2}}$ which is found to be of the order of 500 million light-years. Other observations, on the 'red-shift' of light from these distant objects, enable us to conclude with perhaps more assurance that this radius is increasing in time at a rate which, if kept up, would double the present radius in something less than 2000 million years.

With this we have finished our brief account of geometry as a branch of physics, a subject to which no one has contributed more than Albert Einstein, who by his theories of relativity has brought into being physical geometries which have supplanted the tradition-steeped *a priori* geometry and kinematics of Euclid and Newton.

NOTES

* In *Albert Einstein: Philosopher-Scientist*, Evanston 1949, pp. 315–332.
[1] Technically this requirement, as expressed by the axiom of free mobility, is that there exist a motion of the 3-dimensional space into itself which takes an arbitrary configuration, consisting of a point, a direction through the point, and a plane of directions containing the given direction, into a standard such configuration. For an excellent presentation of this standpoint see B. A. W. Russell's *The Foundations of Geometry*, Cambridge 1897, or Russell and A. N. Whitehead's 'Geometry VI: Non-Euclidean Geometry', 11th ed., *Encyclopaedia Britannica*.
[2] The motions of the surface of the earth into itself, which enable us to transform a point and a direction through it into any other point and direction, as demanded by the axiom

of free mobility, are here those generated by the 3-parameter family of rotations of the earth about its center (not merely the 1-parameter family of diurnal rotations about its 'axis'!).

[3] We are here confining ourselves to metric (Riemannian) geometries, in which there exists a differential element ds of distance, whose square is a homogeneous quadratic form in the coordinate differentials.

[4] That is, the 'differential', as opposed to the 'macroscopic', properties. Thus the Euclidean plane and a cylinder have the same differential, but not the same macroscopic, structure.

[5] The quantities here referred to are the six independent components of the Riemann-Christoffel tensor in 3 dimensions, and the 'mean curvature' here introduced (not to be confused with the mean curvature of a surface, which is an extrinsic property depending on the embedment) is $K = -R'=6$, where R' is the contracted Ricci tensor. I am indebted to Professor Herbert Busemann, of the University of Southern California, for a remark which suggested the usefulness for my later purposes of this approach. A complete exposition of the fundamental concepts involved is to be found in L. P. Eisenhart's *Riemannian Geometry*, Princeton 1926.

[6] In the works already referred to in footnote 1 above.

[7] 'Les espaces géometriques', *Revue Philosophique* **27** (1889) 588–595. The possibilities at which Calinon arrives are, to quote in free translation:

"1. Our space is and remains rigorously Euclidean.

2. Our space realizes a geometrical space which differs very little from the Euclidean, but which always remains the same.

3. Our space realizes successively in time different geometrical spaces; otherwise said, our spatial parameter varies with the time, whether it departs more or less away from the Euclidean parameter or whether it oscillates a definite parameter very near to the Euclidean value."

[8] 'Über das zulässige Krümmungsmaass des Raumes', *Vierteljahrsschrift der astronomischen Gesellschaft* **35** (1900) 337–347. The *annual parallax*, as used in practice, is one-half that defined below.

[9] *Science and Hypothesis*, p. 81; transl. by G. B. Halsted, Science Press, 1929.

[10] This case, in which the geometry is that of the Poincaré half-plane, has been discussed in detail by E. W. Barankin, 'Heat Flow and Non-Euclidean Geometry', *American Mathematical Monthly* **49** (1942) 4–14. For those who are numerically-minded it may be noted that for a steel plate ($k = 0.1$ cal/cm deg) 1 cm thick, with a heat flow of 1 cal/cm^2 sec, the natural unit of length R of the geometry, as measured by a steel rule ($c = 10^{-5}$/deg), is 10^4 cm $\sim$ 328 ft!

[11] Even here an amusing and instructive analogy exists between our caricature and the relativity theory. On extending our notions to a 3-dimensional heat-conducting medium (without worrying too much about how our measurements are actually to be carried out!), and on adopting the standard field equation for heat conduction, the 'mean curvature' introduced above is found at any point to be $-(cs/k)^2$, which is of second order in the characteristic parameter c/k. (The case in which the temperature is proportional to $a^2 - r^2$, which requires a continuous distribution of heat sources, has been discussed in some detail by Poincaré, *Loc. cit.* pp. 76–78, in his discussion of non-Euclidean geometry.) The field equation may now itself be given a geometrical formulation, at least to first approximation, by replacing it by the requirement that the mean curvature of the space *vanish* at any point at which no heat is being supplied to the medium – in complete analogy with the procedure in the general theory of relativity by

which the classical field equations are replaced by the requirement that the Ricci contracted curvature tensor vanish. Here, as there, will now appear certain deviations, whose magnitude here depends upon the ratio c/k, between the standard and the modified theories. One curious consequence of this treatment is that on solving the modified field equation for a spherically-symmetric source (or better, sink) of heat, one finds precisely the same spatial structure as in the Schwarzschild solution for the gravitational field of a spherically-symmetric gravitational mass – the correspondence being such that the geometrical effect of a sink which removes 1 cal per sec from the medium is equivalent to the gravitational effect of a mass of 10^{23} gm, e.g., of a chunk of rock 200 miles in diameter!

[12] This is, of course, an outrageously over-simplified account of the assumptions and procedures involved. All nebulae are *not* of the same intrinsic brightness, and the modifications required by this and other assumptions tacitly made lead one on merry astronomical chase through the telescope, the Earth's atmosphere, the Milky Way and the Magellanic Clouds to Andromeda and our other near extra-galactic neighbors, and beyond. The story of this search has been delightfully told by E. P. Hubble in his *The Realm of the Nebulae*, Yale 1936, and in his *Observational Approach to Cosmology*, Oxford 1937, the source of the data mentioned below.

E. MEYERSON

THE RELATIVISTIC EXPLANATION
OF GRAVITATION*

1. THE ENIGMA OF NEWTONIAN GRAVITATION

For our present purpose, we need only note that in the general theory of relativity, gravitation, like inertia, depends on the essential nature of space, and that the particular property of space which accounts for gravitation – namely its *curvature* – is clearly of a mathematical kind.

Perhaps the implications of this observation will become somewhat clearer if we review very briefly the status of gravitation before the introduction of the new concepts. At the height of the Middle Ages, in the twelfth century, when gravitation was of course considered to be a purely terrestrial phenomenon, John of Salisbury declared that the downward motion of bodies is a *true* phenomenon but not one which could be understood as *necessary*. He used this example to show (contrary to the prevailing doctrine) the possibility of a science of that which simply exists – a science which we would call positivistic.[1] With Newton's formulation of his law the problem acquired new urgency, and from that time on we find scientists treating the force of gravity as a mystery which has to be explained. It is in these terms that Leibniz objected to the Newtonian theory, claiming that it was a *do-nothing* hypothesis and accusing it of "destroying both our philosophy, which seeks reasons, and the divine wisdom which furnishes them."[2] Although there were many philosophers after Leibniz who accepted the new notion and even made it the basis of a general theory of reality (Cf. *Identité et Réalité*, pp. 80–81), physicists continued to resist it. Their attitude is shown by the continuous outpouring of theories, often by renowned scientists, in an attempt to understand the mystery. There were reputedly more than two hundred of them. The most widely accepted theory was that of Le Sage. It is of utmost significance in the present context to note that quite reputable physicists have devoted time and effort to developing this theory, in spite of the extravagant assumptions it entails. One must conclude, indeed, that they were obsessed by the need for science to solve the enigma at any price. Among those who held this opinion in the nineteenth century was the illustrious Maxwell, who declared that if a scientific theory could with some probability lead to an explanation of gravitation, men of science would devote the rest of their lives to it.[3] Among our contemporaries, Borel proclaimed Newton's law "quite absurd in its classical form", and added that a person who could explain his law (by statistics, for example) would be admired as much as Newton.[4] Similarly, Brillouin observed that it was "every physicist's dream, too often unrealized", to finally see "universal gravitation come out of its magnificent isolation.[5]

2. The relativistic solution

Einstein's solution is, essentially, to make gravitation and inertia interdependent, in some sense to blend them together. It had been known for a long time that the same mass was involved in both kinds of phenomena, but that was a bare statement of fact. Newton had believed that it was necessary to verify this experimentally by comparing pendulums made of different materials. Quite recently, after the discovery of radioactive substances, which, as we know, behave differently in many respects from all other known substances, it seemed necessary to demonstrate experimentally that the rule was equally valid in their case. For Einstein, on the contrary, this question no longer arises, and this is something new.[6]

Thus, for a follower of Einstein the problem which tormented earlier physicists so much, as we have just seen, no longer exists. Weyl is quite right, from the point of view of relativity, to speak of "the solution of the enigma of gravity",[7] not only because *instantaneous* action at a distance is henceforth replaced by action operating at the speed of light – it had long been noted that there were inadmissible presuppositions in Laplace's famous calculation attributing to gravitation a velocity 50 or even 100 million times greater than that of light[8] – but also because the *nature* of this action turns out to be determined, and there is no longer any need to search for an intermediate cause such as Le Sage's 'ultramundane corpuscles' were supposed to be. This undoubtedly is the most remarkable achievement of the theory, as Einstein himself states in one of his most recent publications:[9]

The possibility of reducing the numerical equality of inertia and gravitation to one and the same fundamental character gives the general theory of relativity such a great advantage over classical mechanics that all the difficulties must be considered inconsequential beside such progress.

3. It is spatial

The Einsteinian physicist obviously understands in a spatial way. As a matter of fact, the identification between inertial and gravitational mass which we have just discussed is not the only one. There is also in relativity a fusion between the strictly mathematical properties of space and its physical properties, which were previously thought to belong to clearly distinct domains. This inertia was obviously considered a property of physical space, space containing bodies. *A priori* demonstrations such as that of Kant could cast doubt on this to a certain extent, since they apparently concern only space in general. But upon more careful examination, one is convinced that such demonstrations presuppose the presence of material bodies in order to indicate direction as well as velocity. Kant's deduction, moreover, expressly stipulates that all motion must be related to material bodies placed in space. Newton had made this point explicit in giving the principle of inertia its definitive form. There cannot be the slightest doubt that *absolute* space as he understands it is physical space. The *confusion* between physical and mathematical space – it goes without saying that we do not use the term *confusion* in any pejorative sense – is a peculiarity of recent concepts and clearly distinguishes them from their predecessors. "Gravitation," Langevin tells us, "is understood to be an aspect of geometry." Henceforth, "geometry and physics form a single whole," but this whole is "a geometry of a higher order," that is, it has the form of a science of space, "gravitation being but one of the aspects of this geometry." Such a geometry consequently "seems destined to absorb the whole of physics."[10]

4. Analogy with previous theories

Thus it is essential to note that the relativistic explanation is above all a geometrical explanation. Borel even seems to think that this fact constitutes an essential distinction between his theory and the ones which preceded it; indeed he predicts that, whereas the nineteenth century was the century of mechanistic explanations, the "twentieth century will perhaps be the century of geometrical explanations."[11] Strictly speaking, this formula is not entirely accurate. Actually, any true scientific explanation is, and at bottom can only be, a spatial explanation; therefore, it must necessarily involve geometry. Considered, then, from this point of view, relativity is not quite as revolutionary as it perhaps seems at first glance. It merely continues the work of providing spatial explanations while giving physics a new method it had not hitherto used.

However, Borel's statement is not without merit. For the present, let us merely note that relativists are able to understand this because for them space is no longer exactly what is spontaneously presented by common sense perception; it exhibits a property – *curvature* – which this perception does not reveal in any way.

5. Projectiles and Gravitation

A striking analogy is revealed when the evolution of the concept of gravitation brought about by the Einsteinian explanation is compared to that brought about by the introduction of the principle of inertia. This analogy helps us to grasp in some way, if not how an Einsteinian physicist truly understands gravitation, at least how he can arrive at an understanding of it. As a matter of fact, the reason that we at first found it surprising and unbelievable that an Einsteinian physicist should understand gravitation in this way is certainly because we imagine gravitation to be a kind of force. For the same reason, it seemed completely plausible to accept Le Sage's explanation, which replaced Newtonian action at a distance with the impact of invisible particles. But let us reconsider these phenomena which we attribute to the mutual attraction of bodies, this time in terms of pure movement or displacement. What we really know about a body supposedly attracted by another – what I see, for example, when I remove the support which holds a body above the ground – is that it is displaced, even though nothing seems to be pushing it.[12] Now when Aristotle observed that an object thrown by the hand continued to move even after the hand had ceased to touch it, he found this to be a mysterious phenomenon in exactly the same sense that gravitation was a mystery before general relativity. This is why the theory which he imagined bears so strong a resemblance to Le Sage's theory of gravity; like Le Sage, he evokes the impact of invisible material or quasi-material particles to explain a visible movement. But Galileo and Descartes have made all explanation of this sort superfluous, simply substituting the relationship of the body to space: the body changes its place because space possesses the peculiar property of being indifferent to movement so long as it is uniform and rectilinear. Therefore it is not so paradoxical that Einstein, by modifying the properties of space and the relationship of the body to space, could in this way explain the movement which Le Sage attributed to the intervention of a material agent.

NOTES

* From *La déduction relativiste*, Paris, Payot, 1925, Ch. 6, transl. by David and Mary Alice Sipfle.

[1] Joannis Sarisberiensis... *Opera omnia*. Patrologie Migne, Second Series, v. CXIX, Paris, 1855, *Polycraticus*, Bk. II, Ch. XXI: Scio equidem lapidem vel sagittam, quam in nubes jaculatus sum, exigente natura recasurum in terram, in quam feruntur nutu suo pondera, nec tamen simpliciter recidere in terram, aut quia novi necesse est. Potest enim recidere et non recidere. Alterum tamen, etsi non necessario, verum tamen est... Ceterum etsi non esse possit, nihil impedit esse scientiam, quae non necessariorum tamen, sed quorumlibet existentium est, nisi forte et tu cum stoicis existentia censeas necessariis comparanda.

[2] Leibniz, *Nouveaux essais*, *Opera* (ed. by Erdmann), Berlin 1840, p. 203.

[3] Clerk Maxwell, *Scientific Papers*, Cambridge 1890, Vol. II, p. 155 ff., 311.

[4] E. Borel, *Le hasard*, 2nd ed., Paris 1914, p. 3, 300. Borel goes on to explain, in his more recent book, that "it is because it was able to find a place for universal gravitation in a more general conception of the universe... that Einstein's general theory of relativity was received with admiration and curiosity in scientific circles all over the world". (*L'espace et le temps*, Paris 1922, p. 39).

[5] M. Brillouin, 'Propos sceptiques au sujet du principe de relativité', *Scientia* 13 (1913) 23. Cf. analogous statements made somewhat earlier by M. O. Lodge, 'The Aether of Space', *Nature* 79 (1909) 323, and M. W. Ritz, 'Die Gravitation', *Scientia* 9 (1909) 255.

[6] At a time when the general theory of relativity existed only in some kind of nascent state, Einstein cited the importance of Eötvös' experiments for the theory of relativity. In these experiments Eötvös demonstrated with a high degree of exactitude the identity of inertial and gravitational mass, "the definitions of which are, from the logical point of view, independent of one another". ('Zum Relativitaetsproblem', *Scientia* 15 (May 1914) 342.) Weyl was equally insistent on this point (*Raum, Zeit, Materie*, 4th ed., Berlin 1921, p. 197).

[7] H. Weyl, *Loc. cit.*, p. 199. E. Bauer, in his excellent little book *La théorie de la relativité* (Paris 1922, 63), makes this aspect of the theory quite clear: "Thus the notion of the force of gravitation, of universal attraction – a property of matter which seemed to make itself felt at a distance, instantaneously, without any conceivable mechanism – disappears from science".

[8] Cf. *Identité et réalité*, p. 80. Cf., however, the reservations formulated below, § 192.

[9] A. Einstein, *Vier Vorlesungen ueber Relativitaetstheorie*, etc., Brunswick 1922, p. 38.

[10] P. Langevin, 'L'aspect général de la théorie de la relativité', *Bulletin scientifique* etc., No. 2, p. 19, 22; and Preface to Eddington, *Espace, temps et gravitation* (transl. by

Rossignol), Paris 1920, p. 11. In order to further justify this complete fusion [*confusion*] of physics and geometry, which is one of the most essential characteristics of the new doctrine, the relativists take pains to emphasize that "only when taken together are physics and geometry susceptible to an empirical verification", given that assumptions, such as the one which equates the path followed by light rays to a straight line, are constantly involved in the observations. (H. Weyl *loc. cit.*, p. 80.) Thus, "the existence of a geometry independent of physics is compromised once and for all" (*ibid.*, p. 292) and "the metric field depends on the material realities which fill the universe" (p. 193), while, on the other hand, "gravitation appears to be an emanation from the metric field" (p. 198), so that "geometry, mechanics and physics form... an indissoluble theoretical whole, which must be conceived of *en bloc.*" (p. 57). In a more recent work this physicist declares, moreover, that the metrical structure is not given *a priori* in any rigid manner, "but constitutes a *field of the state* of physical reality causally dependent upon the state of matter." He adds this picturesque image: "Like the snail, matter itself constructs and forms its own home." (H. Weyl, *Mathematische Analyse des Raumproblems*, Berlin 1923, p. 44.)

[11] E. Borel, *L'espace et le temps*, p. 208.

[12] This is the meaning of the famous thought experiment based on the image of a *suspended chest*. Since it was first set forth by Einstein himself (cf. *La théorie de la relativité restreinte et généralisée* (transl. by J. Rouvière), Paris 1921, p. 57), it has also been presented in a great number of expositions of the theory of general relativity, particularly in those which are not purely mathematical. This line of reasoning is indeed especially appropriate to demonstrate that it is impossible to distinguish a gravitational field from a field of accelerated motion. By understanding this, by assimilating the action of gravity to a pure and simple motion, it becomes possible to explain gravitation in terms of the properties of space.

V. LENZEN

GEOMETRICAL PHYSICS*

The geometrical theory of gravitation characterizes gravitational force in terms of a curvature of space-time. We mean by the curvature of space that unit measuring rods will not serve to build a Cartesian lattice system. The curvature of space-time means that the separation of two world points can not be expressed as

$$ds^2 = dx^2 + dy^2 + dz^2 - c^2 dt^2,$$

where dx, dy, dz, dt are determined by unit rods, and clocks which are regulated by the principle of the constancy of the velocity of light. In curved space-time one introduces a Gaussian coordinate system which is characterized by the components of the fundamental metrical tensor, g_{11}, g_{12}, etc.

In the Newtonian theory the motion of a body is determined by the principle of inertia and a force of gravitation; the force explains the departure of the motion from that in a straight line with constant velocity. In the geometrical theory the motion is such that the world line is a geodesic in space-time. The geodesic is determined by the g's, and thus the g's play the role of potentials of a force. The geometry of a space-time region is determinable from space-time coincidences of all sorts of bodies. Thus, the geometrical theory of gravitation defines it in terms which are independent of the special properties of bodies.

Now, assuming that we have a concept of matter, the geometry of a region is found to be determined by the distribution of matter. The geometry is determinable from the observation of coincidences; hence the law of gravitation may be transformed into a definition of matter. It is then an experimental fact that what we have previously called matter satisfies this definition. The transformation of the experimentally verified law of gravitation into a definition has been made by Eddington. Einstein has expressed this result in a popular fashion by the statement that space is eating up matter.

The g's which define the geometry of a space-time region are the potentials of the gravitational field. The thought occurs that the coefficients of a still more general geometry might serve to characterize the electromagnetic field.

The first proposal in this direction was made by Weyl. In Riemannian geometry the length of a line is invariant; Weyl invented a geometry in which length depends upon position. He then correlated the potentials of the electromagnetic field with the coefficients which determine change in length. This was the beginning of the program for a unitary geometrical theory of physics.

Inasmuch as Weyl's theory leads to consequences which do not agree with experience, a distinction has therefore been drawn between natural geometry and world geometry. Natural geometry is descriptive of the behavior of measuring rods and clocks; world geometry is a graphical representation of physical theory. Eddington has also developed a system of world geometry. However, since world geometry is merely of mathematical interest I shall not expound it in further detail.

Einstein has continued the search for a unitary theory which would enable one to express electromagnetic phenomena in terms of the behavior of real rods and clocks, but the development has not yet been brought to a conclusion.

The conception of a geometrical system of physics was envisaged by Descartes. In his analysis of the concept of matter he concluded that the essence of matter is extension, for only of extension did he have a clear and distinct idea. Descartes thus discovered the epistemological basis for a geometrical theory of physics. But the accomplishment of this goal required the construction of a space-time geometry and the invention of several branches of mathematics, and so physics was developed in terms of the physical quantities mass and force by Newton and his followers. In recent years, however, the mathematical technique has been developed which offers hope to the prospect of constructing a geometrical system.

The geometrical conception of nature is based upon the notion of space-time coincidence – a concept which is subject to criticism in quantum mechanics.

NOTE

* From *The Nature of Physical Theory*, New York, J. Wiley, New York 1931, pp. 230–231.

H. BERGSON

DISCUSSION WITH BECQUEREL OF THE PARADOX OF THE TWINS*

We have stated but cannot repeat often enough: in the theory of relativity, the slowing of clocks is only as real as the shrinking of objects by distance. The shrinking of receding objects is the way the eye takes note of their recession. The slowing of the clock in motion is the way the theory of relativity takes note of its motion: this slowing measures the difference, or 'distance', in speed between the speed of the moving system to which the clock is attached and the speed, assumed to be zero, of the system of reference, which is motionless by definition; it is a perspective effect. Just as upon reaching a distant object we see it in its true size and then see shrink the object we have just left, so the physicist, going from system to system, will always find the same real time in the systems in which he installs himself and which, by that very fact, he immobilizes, but will always, in keeping with the perspective of relativity, have to attribute more or less slowed times to the systems which he vacates, and which, by that very fact, he sets in motion at greater or lesser speeds. Now, if I reasoned about someone far away, whom distance has reduced to the size of a midget, as about a genuine midget, that is, as about someone who is and acts like a midget, I would end in paradoxes or contradictions; as a midget, he is 'phantasmal', the shortening of his figure being only an indication of his distance from me. No less paradoxical will be the results if I give to the wholly ideal, phantasmal clock that tells time in the moving system in the perspective of relativity, the status of a real clock telling this time to a real observer. My distantly-removed individuals are real enough and, as real, retain their size; it is as midgets that they are phantasmal. In the same way, the clocks that shift with respect to motionless me are indeed real clocks; but insofar as they are real, they run like mine and tell the same time as mine; it is insofar as they run more slowly and tell a different time that they become phantasmal, like people who have degenerated into midgets.

Let us imagine a normal-sized Peter and Paul conversing. Peter stays where he is, next to me; I see him and he sees himself in his true size. But

Paul moves off and becomes midget-sized in Peter's eyes and mine. If I now go around thinking of Peter as normal-sized and of Paul as a midget, picturing him that way back with Peter and resuming his conversation, I shall necessarily end in absurdities of paradoxes; I have no right to bring Peter, who has remained normal, in contact with Paul turned midget, to imagine that the latter can speak with the former, see him, listen to him, perform any action at all, because Paul, as midget, is only a mental view, an image, a phantom. Nevertheless, this is exactly what both partisan and adversary of the theory of relativity did in the debate, begun at the Collège de France in April 1922, on the implications of special relativity.[1] The former merely kept pointing to the perfect mathematical coherence of the theory, but then retained the paradox of multiple and real times – as if one were to say that Paul, having returned to the vicinity of Peter, had been changed into a midget. The latter probably wanted no paradox, but he could have avoided it only by showing that Peter is a real being and that Paul turned midget is a mere phantom, that is, by making a distinction that belongs no longer to mathematical physics but to philosophy. Remaining, on the contrary, on his opponents' ground, he only succeeded in furnishing them with an occasion for reinforcing their position and confirming the paradox. The truth is that the paradox vanishes when we make the distinction that is indispensable. The theory of relativity remains intact, with its infinite multiplicity of imaginary times and a single, real time.

This is exactly our argument. That there has been some difficulty in grasping it, and that it is not always easy, even for the relativist physicist, to philosophize in terms of relativity, is to be gathered from a very interesting letter addressed to us by a most distinguished physicist.[2] Inasmuch as other readers may have encountered the same difficulty and as none, surely, will have formulated it more clearly, we are going to quote the main points in this letter. We shall then reproduce our reply.

Let AB be the trajectory of the projectile plotted in the system earth. Starting from point A on the earth, where Peter will remain, the projectile carrying Paul heads toward B at speed v; having arrived at B, the projectile turns around and heads back to point A at speed $-v$. Peter and Paul meet again, compare measurements, and exchange impressions. I say that they are not in agreement about the duration of the journey: if Peter asserts that Paul has stayed away a given length of time, which he has estimated at A, Paul will reply that he is quite sure he has not spent that much time on the trip, because he has himself calculated its duration with a unit of time defined in the same way and has found it shorter. Both will be right.

DISCUSSION OF THE PARADOX OF THE TWINS 435

I am assuming that the trajectory has been staked out with identical clocks, borne along with the earth, hence belonging to the system earth, and that they have been synchronized by light signals. In the course of his journey, Paul can read the time shown by the particular clock near which he is passing, and can compare this time with that indicated by an identical clock in his projectile.

You can already see how I am orienting the question: the point is to compare adjacent events, to observe a simultaneity of clock readings *at the same place*. We are not straying from the psychological conception of simultaneity, for, in accord with your own expression, an event E occurring beside clock C is given in simultaneity with a reading on clock C in the psychologist's sense of the word simultaneity.

At the event 'departure of the projectile', Peter's and Paul's clocks both point to 0^e. I am assuming, of course, that the projectile attains its speed instantaneously. There, then, is the projectile that constitutes a system S' travelling in rectilinear and uniform motion with respect to the system earth, at speed v. For the sake of clarity, I shall assume that $v = 259{,}807$ km sec^{-1} so that the factor $\sqrt{(1-(v^2/c^2))} = \frac{1}{2}$.

I shall assume that at the end of an hour, recorded on the clock of the projectile, the latter passes the middle M of the distance AB. Paul reads the time both on his clock (1^e) and, simultaneously, on the system earth's clock located at M. What time will he read on the latter? One of the Lorentz equations supplies the answer.

We know that the Lorentz formulae give the relations linking the space and time co-ordinates of an event *measured by Peter* with the space and time co-ordinates of the same event *measured by Paul*. In the present case, the event is the meeting of the projectile with the system earth's clock at M; its co-ordinates in the projectile system S' are $x' = 0$, $t' = 1^e$; the formula

$$t = \frac{1}{\sqrt{(1-(v^2/c^2))}} \left(t' + \frac{vx'}{c^2}\right)$$

gives $t = 2t'$ (since

$$\frac{1}{\sqrt{(1-(v^2/c^2))}} = 2).$$

The clock at point M therefore records 2^e.

Paul therefore notes that the system earth's clock before which he is passing is one hour ahead of his; of course, he does not have to push his clock ahead; he records the disagreement. Continuing on his journey, he notes that the time differences between his clock and those he successively encounters increase in such proportion to his own clock-time that, on arriving at B, his clock points to 2^e; but the system earths clock at B points to 4^e.

Having arrived at B, the projectile turns back along BA at speed $-v$. Now there is a *change in system of reference*. Paul abruptly leaves the system moving with speed $+v$ with respect to the earth and passes into the system of speed $-v$. Everything starts over again on the return trip. Let us imagine that the clock in the projectile and the one at B are automatically moved back to zero, and that the other earth-linked clocks are synchronized with the one at B. We can begin the preceding argument all over again: at the end of one hour's journey, recorded on Paul's clock, he will again find as he passes M that his clock reads 1^e, whereas the earth clock reads 2^e, etc.

But why imagine the clocks set back to zero? It was useless to interfere with them. We know there is an initial-shifting from zero to take into account; this shifting amounts to 2^e for the projectile's clock and 4^e for the system earth's clock; they are constants to be added to the times that would be shown had all the clocks been pushed back to zero. Thus, if we have not interfered with the clocks, when the projectile recrosses M, Paul's clock will show $1+2=3^c$, the one at point M, $2+4=6^c$, and Peter's $4+4=8^c$.

Behold the result! For Peter, who has remained at A on the earth, it is indeed eight hours that have elapsed between Paul's departure and return. But, if we ask 'living conscious' Paul, he will say that his clock read 0^c at departure and reads 4^c upon return, that it has recorded a duration of 4^c, and that he has really been travelling 4^c and not 8^c.

So goes the objection. As we stated, it is impossible to present it in clearer terms. That is why we have reproduced it just as it was addressed to us, without reformulation. Here, then is our reply:

Two important remarks must be made at the outset.

(1) If we take a stand outside the theory of relativity, we conceive of absolute motion and, therewith, absolute immobility; there will be really motionless systems in the universe. But, if we assume that all motion is relative, what becomes of immobility? It will be the state of the system of reference, the system in which the physicist imagines himself located, inside which he is seen taking measurements and to which he relates every point in the universe. One cannot move with respect to oneself; and, consequently, the physicist-builder of Science, is motionless by definition, once the theory of relativity is accepted. It unquestionably occurs to the relativist physicist, as to any other physicist, to set in motion the system of reference in which he had at first installed himself; but then, willy-nilly, consciously or unconsciously, he adopts another, if only for an instant; he locates his real personality within this new system, which thus becomes motionless by definition; and it is then no more than an image of himself that he mentally perceives in what was just now, in what will in a moment again become, his system of reference.

(2) If we stand outside the theory of relativity, we can quite readily conceive of an absolutely motionless individual, Peter, at point A, next to an absolutely motionless cannon; we can also conceive of an individual, Paul, inside a projectile launched far out from Peter, moving in a straight line with absolutely uniform motion toward point B and then returning, still in a straight line with absolutely uniform motion, to point A. But, from the standpoint of the theory of relativity, there is no longer any absolute motion or absolute immobility. The first of the two phases just mentioned then becomes simply an increasing distance apart between Peter and Paul; and the second, a decreasing one. We can therefore say, at will, that Paul is moving away from and then drawing closer to Peter, or that Peter is moving away from and then drawing closer to Paul. If I am with Peter, who then chooses himself as system of reference, it is Peter who is motionless; and I explain the gradual widening of the gap by saying that the projectile is leaving the cannon, and the gradual narrowing, by saying that the projectile is returning to it. If I am with Paul, now adopting himself as system of reference, I explain the widening and narrowing by saying that it is Peter, together with the cannon and the earth, who is leaving and then returning to Paul. The symmetry is perfect.[3] We are dealing, in short, with two systems, S and S', which nothing prevents us from assuming to be identical; and one sees that since Peter and Paul regard themselves, each respectively, as a system of reference and are thereby immobilized, their situations are interchangeable.

I come now to the essential point.

If we stand outside the theory of relativity, there is no objection to expressing ourselves like anyone else, to saying that both Peter and Paul, the one absolutely motionless and the other absolutely in motion, exist at the same time as conscious beings, even physicists. But, from the standpoint of the theory of relativity, immobility is of our decreeing: that system becomes immobile which we enter mentally. A 'living, conscious' physicist then exists in it by hypothesis. In short, Peter is a physicist, a living, conscious being. But what of Paul? If I leave him living and conscious, all the more if I make him a physicist like Peter, I thereupon imagine him taking himself as system of reference, I immobilize him. But Peter and Paul cannot both be motionless at one and the same time, since, by hypothesis, there is first a steadily increasing and then a steadily decreasing distance between them. I must therefore choose between them; and, in point of fact, I did choose, since I said that it was Paul who was shot into space and thereby immobilized Peter's system into a system of reference.[4] But then, Paul is clearly a living, conscious being at the moment of leaving Peter; he is still clearly a living, conscious being at the moment of returning to Peter (he would even remain a living, conscious being in the interval if, during this interval, we agreed to lay aside all questions of measurement and, more especially, all relativist physics); but, for Peter the physicist, making measurements and reasoning about them, accepting the laws of physico-mathematical perspective, Paul, once launched into space, is no more than a mental view, an image – what I have called a 'phantom' or, again, an 'empty puppet'. It is this Paul en route (neither conscious nor living, reduced to the state of an image) who exists in a slower time than Peter's. It would therefore be useless for Peter, attached to the motionless system that we call earth, to try to question this particular Paul at the moment of his re-entering the system, about his travel impressions: this Paul has noted nothing and had no impressions, since he exists only in Peter's mind. What is more, he vanishes the moment he touches Peter's system. The Paul who has impressions is a Paul who has lived in the interval, and the Paul who has lived in the interval is a Paul who was interchangeable with Peter at every moment, who occupied a time identical with Peter's and aged just as much as Peter. Everything the physicist will tell us about Paul's findings on his journey will have to be understood as being about findings that *the physicist Peter attributes to Paul* when he makes himself a referrer and considers Paul no more than a referent – findings that Peter is obliged to attribute to Paul as soon as he seeks a picture of the world that is independent of any system of reference. The Paul who gets out of the projectile on returning from his journey and then again becomes part of Peter's system, is something like a flesh-and-blood person stepping out of the canvas upon which he had been painted: it was to the portrait, not the person, to Paul referent, not referrer, that Peter's arguments and calculations applied while Paul was on his journey. The person replaces the portrait, Paul referent again becomes Paul referrer or capable of referring, the moment he passes from motion to immobility.

But I must go into more detail, as you yourself have done. You imagine the projectile impelled by speed v such that we have $\sqrt{1 - v^2/c^2} = \frac{1}{2}$. Let AB then be the trajectory of the projectile plotted in the system earth, and M the middle of the straight line AB. "I shall assume," you say, "that at the end of an hour recorded on the clock in the projectile, the latter passes the middle M of the distance AB. Paul reads the time both on his clock (1^e) and, simultaneously, on the system earth's clock located at M. What time will he read on the latter, if both clocks pointed to 0^e at departure? One of the Lorentz equations gives the answer: the clock at M points to 2^e"

I reply: Paul is incapable of reading anything at all; for, insofar as, according to you, he is in motion with respect to motionless Peter, whom you have made referrer, he is nothing more than a blank image, a mental view. Peter alone will henceforth have to be treated as a real, conscious being (unless you renounce the physicist's standpoint, which here is one of measurement, to return to the standpoint of common sense or ordinary perception). Hence we must not say, "Paul reads the time..." We must say, "Peter, that is, the physicist, pictures Paul reading the time..." And, since Peter applies, and must apply, the Lorentz equations, he naturally pictures Paul reading 1^e on his moving clock at the moment when, in Peter's view, this clock passes in front of the clock of the motionless system, which, in Peter's eyes, points to 2^e. But, you will tell me: "Nonetheless, does there not exist in the moving system, a moving clock that records its own particular time independently of anything Peter can imagine of it?" Without any doubt. The time of this real clock is exactly what Paul would read on it if he became real again, I mean, alive and conscious. But, at this precise moment, Paul would become the physicist; he would take his system as the system of reference and immobilize it. His clock would then point to 2^e – exactly the time to which Peter's clock pointed. I use the past tense because already Peter's clock no longer points to 2^e but to 1^e, being now the clock of Peter referent and no longer referrer.

I need not pursue the argument. Everything you said about the times read by Paul on his clock when he arrives at B, then when he comes back to M, and, finally, when he is about to touch A and re-enter the system earth, all this applies not to living, conscious Paul, actually looking at his moving clock, but to a Paul whom physicist Peter *pictures as watching* this clock (and whom the physicist *must* picture in this way and need not distinguish from a living, conscious Paul: this distinction is the philosopher's concern). It is for this merely imagined and referred-to Paul that four – imagined – hours will have elapsed while eight – lived – hours will have elapsed for Peter. But Paul, conscious and therefore referrer, will have lived eight hours, since we shall have to apply to him everything we just said about Peter.

To sum up, in this reply we once more gave the meaning of the Lorentz equations. We have described this meaning in many ways; we have sought by many means to present a concrete vision of it. One could just as easily have established it *in abstracto* in the standard step-by-step deduction of these equations.[5] One would recognize that the Lorentz equations quite clearly express what the measurements *attributed* to S must be in order that the physicist in S may see the physicist *imagined by him* in S', finding the same speed for light as he does.

NOTES

* From *Duration and Simultaneity* (transl. by L. Jacobson), Bobbs-Merrill, 1965, pp. 163–172.
[1] We are alluding to an objection to the theory of relativity voiced by M. Painlevé.
[2] [Bergson tactfully refrains from naming this physicist, but he is identified as Jean Becquerel (1878–1953) by André Metz in 'Le temps d'Einstein et la nouvelle édition de l'ouvrage de M. Bergson, *Durée et simultanéité*', *Revue de philosophie* 31 (1924) 241–260.
[3] It is perfect, we repeat, between Peter and Paul as the referrers, as it is between Peter and Paul as the referents. Paul's turning back has nothing to do with the matter, since

Peter turns back as well if Paul is the referrer. We shall, moreover, directly demonstrate the reciprocity of acceleration in the next two appendixes.

[4] It is clearly by extension that use has been made of the expression 'system of reference' in the passage from the above-quoted letter, in which it was stated that Paul, in turning back, "changes his system of reference." Paul is really, by turns, in systems that *can become* systems of reference; but neither of these two systems, while it is considered in motion, is a system of reference.

[5] Albert Einstein, *La théorie de la relativité restreinte et généralisée*, pp. 101–107; Jean Becquerel, *Le principe de relativité et la théorie de la gravitation*, pp. 29–33.

A. N. WHITEHEAD

COMMENT ON THE PARADOX OF THE TWINS

A critical examination of the trains of thought contained in the preceding papers of this symposium would far exceed the limits of space at my disposal. Accordingly in concentrating on one or two points I must not be presumed to undervalue the importance or interest of the remainder.

In the first place, I feel some doubt as to the adequacy of Prof. Wildon Carr's explanation of the traveller's experience when he returned from his journey to a star.[1] According to the traveller's count of time he had lived for two years, but when he returned home he found that two hundred years of earthly history had passed. As I understand Professor Carr – and indeed the mathematicians whom he follows – the reason is that he had been travelling so very fast. But I thought that, according to the relativity theory, there is no absolute space for him to travel through, and that either Jack or Jill or the Hill have equal reason to claim the prerogative of being at rest. For example, Professor Wildon Carr writes: "First, I am always at the centre of the universe, co-ordinating it from an unchanging position." Accordingly the traveller in the meteorite, from the earth to the star and back, is "always at the centre of the universe, co-ordinating it from an unchanging position"; also the chronologer on the earth is "always at the centre of the universe, co-ordinating it from an unchanging position." Why should the chronologer reckon two hundred years to the traveller's two years? Why should it not be the other way round? Apparently it is not a matter of chance; for we are all quite certain that it is the man who travels to the star who will go slow in his reckoning. Yet the traveller has an equal right to say that the earth suddenly started with great velocity on a voyage in space, and suddenly stopped and came back again, and that, curiously enough, just as the earth stopped at its furthest distance, a star came up to him and then retreated.

My own explanation is that there is a universe, of which both the traveller and the chronologer have diverse experiences dominated by the diverse histories of their bodies as elements in that universe. The real diversity of relations of their bodies to the universe is the cause of their discordance in time-reckoning. I do not complain of Professor Wildon Carr for speaking of a "universe", when according to his theory there is no such thing, in the sense of an effective physical totality; for we are put into difficulties by the inadequacy of language. But I do consider him to have been arbitrary – so far as his own theory is concerned – in assigning all the motion to the traveller on the meteorite. We are both agreed as to the mathematical theory; accordingly I doubt whether he, and many mathematicians, have been completely successful in interpreting that theory in terms of the realities of our experience.

Let us scrutinize more narrowly the problem of the Earth and the traveller in space. When the traveller reckons time by days, what does he count? The rotations of the Earth? Certainly not. At least, certainly not, if the traveller is to count twice 365 revolutions to the chronologer's count of two hundred times 365 revolutions. For if the traveller counts the Earth's revolutions, he presumably uses his own definition of simultaneity, and will count 3·65 revolutions of the Earth on his way out; and on his way back he will adopt another definition of simultaneity and will count another 3·65 revolutions of the Earth; in all, 7·3 revolutions of the Earth. What has happened to the remaining 72992·7 revolutions which have occurred between his departure and return? He dropped those out of account in his sudden change of space-time systems at the star, when he ceased his outward journey and commenced his return.

The annexed figure will elucidate the argument: E_1E_2 is the time-axis for the Earth, E_1S for the traveller on his outward journey, SE_2 for the traveller on his return journey. The dotted lines are moments of simultaneity according to the various space-time systems, and are therefore diagrammatically symbolical of instantaneous three-dimensional spaces;

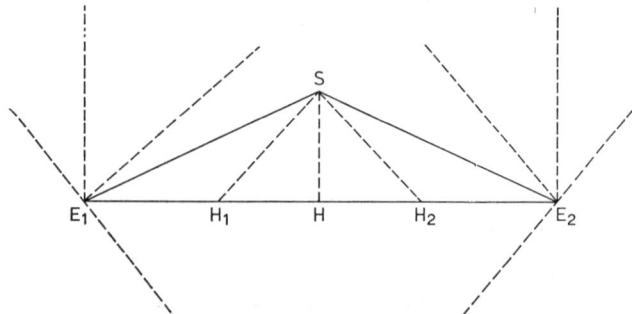

SH is the moment of simultaneity (according to Earth-reckoning) of the arrival at the star and the corresponding Earth-instant (H). Thus E_1H comprises 36 500 revolutions of the Earth. Again SH_1 is the moment of simultaneity (according to the outward traveller's reckoning) of the arrival at the star and the corresponding Earth-instant (H_1). Thus E_1H_1 comprises 3·65 revolutions of the Earth, which the traveller has counted. When the traveller at once starts to return, he changes his moment of simultaneity to SH_2, to correspond to this new meaning of himself as at rest; and he counts the 3·65 revolutions comprised in the portion H_2E_2 of the Earth's time axis. In the flurry of an instantaneous change of motion at S, the traveller dropped out of account the 72992·7 revolutions between H_1 and H_2. If he had noticed them, he would have counted them; and would then have agreed with the Earth-chronologer on his return. It will be noted that I have simplified my arithmetic by assuming exactly 365 days to the year.

There is an obvious criticism to be made against the basis of this calculation. The traveller cannot know in any direct way what is simultaneously (according to his definition) happening on Earth. He can only receive signals transmitted to him with some finite velocity. For example, suppose he counts the days on Earth by means of a signal transmitted from Greenwich each day at noon. If the velocity of transmission of the signals, reckoned in the Earth space-time, be greater than that of the traveller, he will receive less than half on his outward voyage and more than half on his return. If the velocity of transmission be less than that of the traveller, he will receive them all on the return voyage. In either case he receives all the signals sent by the Astronomer Royal, and there is no disagreement between the traveller and the chronologer on Earth.

But we are all agreed that the traveller will have counted 730 days (two years). He accomplishes this by not attending to the Earth at all. He takes his clock with him. Suppose that the hour hand of his clock makes one revolution per day, and that it is rated so as to run truly before he starts, and that the works are not disarranged by the sudden jolt involved in the immense accession of initial velocity. We also assume that its works are in no way dependent on gravity. There is no inherent difficulty about these assumptions, which also hold at the star for the transition from the outward to the inward journey. The clock then runs truly during the outward and inward journey, its casing being at rest relatively to the traveller.

In these circumstances the traveller will count 730 revolutions for the double journey. The fact of the clock running truly means that the time of one revolution of the hour hand is congruent to the time of one revolution of the Earth. But the lapse of clock time is a lapse of time according to the traveller's meaning, and this meaning differs from that for Earth time. But, though the meanings for time are different in the two cases, the lapses of time according to the different meanings are comparable as to congruence. Of course if this be denied, there can be no sense in comparing two such lapses so as to say that one lapse is 730 days and the other is 73 000 days. For in that case a day in one sense is not comparable to magnitude with a day in the other sense. But we are agreed that the days are of equal length; though there are more of them, between the departure and return of the traveller, according to the Earth chronologer's meaning for time than for the traveller's two meanings as he journeys outwards and inwards.

So far, I presume, we are all agreed. But I want to ask Professor Wildon Carr and his friends, the orthodox relativists, what they mean by the congruence of the Earth's day and the traveller's day? What do they mean by the congruence of two of the Earth's days with each other, or of two of the traveller's days with each other? We are told to lay one measure rule – clock, for instance – alongside another, and that their coincidence is the *meaning* of congruence. But how am I to lay two successive revolutions of a clock hand along side each other? There seems therefore to be no *meaning* in the assertion of the congruence of two successive clock days. Still less can there be any meaning to this mode of comparison of the days registered by two clocks which are whizzing past each other with a velocity not far short of that of light. I echo the saying of Berkeley's Philonous, which I have already quoted in a previous book: "I am not for imposing any sense on your words; you are at liberty to explain them as you please. Only, I beseech you, make me understand something by them."

All the difficulty arises from the denial of simultaneity as a fundamental fact of awareness. If this be admitted, for each mode of time, and if the equality of relative motions be admitted (*i.e.* of B with respect to A in comparison with that of A with respect to B), then all difficulty as to congruence vanishes. I have considered this question at length – though with an unnecessarily clumsy analysis – in my "Principles of Natural Knowledge," and the details are too lengthy for reproduction here. I expect to be told by some that the comparison of equal times is a convention, founded upon an arbitrary selection of a certain type of recurring phenomena as periodic in equal lapses of time. For example, molecules will be brought into evidence, and we shall then found ourselves on Einstein's dictum that a molecule is a natural clock. Let it be noted that on this theory the assumption is purely arbitrary: there can be no sense in saying that it is nearly true and approximately verified, for the very *meaning* of equality of time-lapses is involved. But if this be the case, the explanation of the identity of colours of light emitted by molecules of the same type, as being due to vibrations in equal periods, becomes mere nonsense. For it cannot be due to the fact that we *call* the times equal. Furthermore half-periods and uniform velocity have no meaning: for they are all bound up with the intrinsic comparability of time-lapses. Accordingly we cannot explain the sub-division of periods by the uniform velocity of light signals, nor can we explain the uniform velocity of light signals by reference to the sub-division of molecular periods. Perhaps it will be claimed that we are driven to the *convention* that light, as our quickest system of signals, is moving with uniform velocity. But our only experience of light is as moving through media, the air, glass, water, etc. We believe that these media affect the velocity of light. Accordingly we should be reduced to founding the very *meaning* of equal lapses of time upon the behaviour of light under circumstances of which we have no direct experience. Please note that I am speaking of the *meaning* of equality of time-lapses, on the assumption (which I do not share) that such meaning is a purely arbitrary convention. My own belief is that the congruence of time-lapses expresses an important relation of time-lapses founded upon the intrinsic character of time exhibited in the fundamental fact of simultaneity.

In conclusion, let us come back to the traveller and the Earth-chronologist, with their discordance of two years for the traveller and two hundred years for the Earth-chronologist. At the beginning of this paper I explained it in general terms by the diverse histories of their bodies as elements in the universe. We are now prepared to consider the exact diversity of history which produces the discordance in chronology. Both the Earth-chronologer and the traveller have been at rest from their own point of view, in Professor Wildon Carr's excellent phrase, "always at the centre of the Universe, co-ordinating it from an unchanging position." This is true of each of them at each instant. But there is this difference between the traveller and the Earth-chronologer: The Earth-chronologer is at rest during every instant in the same sense of the term. But the traveller changes the sense of the term in which he is at rest. Neglecting the start and the final return, there is an essential change at the star. In other words, there is no difficulty about the explanation if you admit that acceleration and deceleration (as distinct from uniform velocity) express an essential fact of the life history of any body, and is not merely an accidental outcome of the arbitrary choice of co-ordinates. But this admission as to acceleration is just what orthodox relativists deny, if I rightly apprehend their statements.

NOTES

* From 'The Problem of Simultaneity', *Aristotelian Society, Suppl. Vol.* 3 (1923), 34–41.
[1] The author refers here to Wildon Carr's paper read in the same symposium of Aristotelian Society (*loc. cit.*, 15–25).

COMMENT ON THE CLOCK PARADOX*

An objection to the relativistic theory of clocks that plays an important role in the literature, and may therefore be discussed, is given by the so-called *clock paradox*. The clock U' (Figure 1) is slow compared to the time in K, and when it reaches U_2 will show an earlier time than U_2. Let us imagine that U' is stopped in its path at this instant and turned around so that it will travel back to U_1. The time taken in turning the clock around may be ignored, since it is negligible compared to the time of the round trip. The same retardation occurs during the return trip, and

Fig. 1. Retardation of clocks.

U' must therefore be slow compared to U_1 when it reaches U_1. The last statement is independent of both the simultaneity definition for K and the behavior of U_2. We can therefore say: if a clock U' is first moved away from U_1 and then returned, it will be slow relative to U_1.

It seems that according to the theory of relativity the process can be interpreted in the opposite manner. We consider U' to be at rest, while U_1 is moved (to the left) and then returned. On the basis of this description we should conclude that U_1 is slow relative to U', since U_1 is the moved clock. This result constitutes a contradiction, because a neighborhood comparison, independently of the definition of simultaneity, can tell us which of the clocks is slow when they meet. Only one of the two assertions can be correct.

The contradiction is quite striking and may under no circumstances be solved by considering the following two statements compatible: 'when brought together U' is slow relative to U_1' and 'when brought together U_1 is slow relative to U''.[1] The comparison of the two clocks is independent of the definition of simultaneity. If the two above statements were both considered to be true, such a conception would contradict the basic rule of the theory of relativity that a point-event (a coincidence) has an objective significance. A solution can be given only if we can show that one of the apparently equally correct inferences is incorrect. In fact, it is the second one.

The error lies in a misconception of relativity, which can be explained as follows. The theory of gravitation shows that the special theory of relativity is applicable only because the distant masses of the fixed stars (drawn as a circle in Figure 2) determine a particular metrical field. If we take account of the masses of the fixed stars F, the apparent equivalence of the two interpretations vanishes. According to the first interpretation U' is moving, while U and the fixed stars F remain at rest. According to the second interpretation U' is at rest, and U and the fixed stars F are in motion. This analysis eliminates the symmetry of the two processes; the second is an entirely different process from the first because of the effect of the moving fixed stars, which produce a gravitational field at the instant of the reversal of the motion and thus cause a retardation of U'. Due to the gravitational field U' is the retarded clock even according to the second interpretation. Calculations prove this conclusion quantitatively correct.[2] The mistake that led to the paradox therefore resulted from the fact that the considerable effects of gravitation were ignored.

A remark may be added concerning the extension of Einstein's theory of clocks to living organisms. The retardation of clocks has often been illustrated by the example of the twins: of two newborn twins, one makes a cosmic trip with a velocity slightly below the velocity of light and returns as a boy, while the other twin has in the meantime become an old man. This consequence, which many people have regarded as

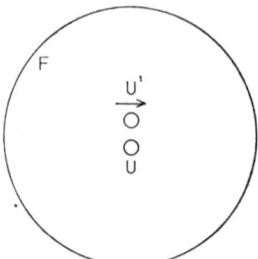

Fig. 2. The asymmetry in the clock paradox.

absurd, actually contains nothing impossible or inconceivable and agrees with the theory of relativity in every respect. In fact, a similar case has been described in W. Müller's poem *Der Mönch von Heisterbach*, which describes a monk going on a walk and returning after three hundred years to his monastery, where nobody recognizes him. The poet's imagination thus created ideas which modern physics no longer regards as impossible.

If the objection is raised that the theory of relativity as a physical theory applies to physical processes alone and not to living organisms, one forgets that there are many basic principles of physics which also apply directly to living beings. Galileo's law of falling bodies governs a falling stone as well as a falling egg or a falling human being. The laws of gravitation apply in general to animate and inanimate objects in equal fashion. After the discovery of the spherical shape of the earth, it was immediately inferred from this physical theory that human beings who live on opposite sides of the spherical surface nonetheless have the subjective feeling of upright posture, because living organisms adjust themselves to the physical gravitational field. It is a similar claim which is made by the theory of relativity in the example of the twins, namely, that living organisms, like clocks, adjust themselves to the metrical field. It would be an unjustifiable hypothesis to assume that they would behave differently, since the principle that the time scale of natural clocks is identical with the time scale of living organisms (insofar as it can be defined) is one of the oldest principle of natural science. The example of the twins is explained by the fact that the ultimate constituents of living organisms are atoms. If every atomic period, i.e., the period of the electron within the atom, is retarded to the same degree under the influence of motion or of a metrical field, physiological phenomena would show the same retardation, since they result from the integration of many atomic periods.

NOTES

* From *Philosophy of Space and Time*, Dover 1956, pp. 192–194.
[1] This is the opinion of J. Petzold, *Die Stellung der Relativitätstheorie in der geistigen Entwicklung der Menschheit*, Dresden 1921, p. 104.
[2] Cf. e.g., A. Kopff, *Grundzüge der Einsteinschen Relativitätstheorie*, Leipzig 1921, p. 117 and 189.

COMMENT ON THE PARADOX
OF THE TWINS*

Consider a pair of 'identical' twins. Let one of them take a voyage in a rocket ship, which we suppose to be capable of attaining a speed close to that of light, while the other remains on the Earth. When the twin who has made the voyage returns to the Earth, the clocks in his rocket ship will show the elapse of a time interval,

$$\Delta t_0 = \int_{t_1}^{t_2} \sqrt{1 - \frac{v^2(t)}{c^2}}\, dt$$

while similarly constructed clocks of the twin who remained on the Earth will show the elapse of a time interval $t_2 - t_1 > \Delta t_0$. But as we have seen earlier, all physical, chemical, nervous, psychological, etc., processes will be subject to the same Lorentz transformation that applies to clocks. Therefore, the twin who took the journey will in every way have experienced less time than did the one who remained on the Earth. And if the speed of the rocket ship was close to that of light, this time difference could be quite appreciable. For example, if 20 years passed for a man who remained on the Earth, only one or two years might have passed for the man who was in the rocket ship.

Before proceeding to discuss the significance of this conclusion, let us first note that it does not violate the principle of relativity, which asserts that the laws of physics must constitute the same relationships, independent of how the frame of reference moves. For as we pointed out in the previous chapter we have thus far restricted ourselves to the special theory of relativity, in which the laws of physics are invariant only for observers moving at a constant speed. The conclusions of this theory evidently cannot be applied symmetrically in the frames of both observers, since one of them is accelerated and the other is not. For this reason it is not legitimate to interchange observers, and to say, for example, that the observer in the rocket ship should equally well see his twin in the laboratory as having aged less than he has. Rather, as long as we remain within the special theory of relativity, we must give the unaccelerated reference frame a unique role in the expression of the laws of physics; and in this way we explain how observers who have suffered different kinds of movements can, on meeting again, find that they have experienced different amounts of time.

To obtain laws that are the same for accelerated as for unaccelerated observers, we must go on to the general theory of relativity. But to do this we must bring in the *gravitational field*. As Einstein has shown, in an accelerated frame of reference, new effects must occur, which are equivalent to those that would be produced by a gravitational field. Indeed, from the point of view of the accelerated observer, one could say that there is an additional effective gravitational field, which acts on the general environment (stars, planets, Earth, etc.) and explains its acceleration relative to the rocket ship.

According to the general theory of relativity, two clocks running at places of different gravitational potential will have different rates. If the observer on the rocket ship uses the same laws of general relativity that are used by the observer on the Earth, but considers the different gravitational potentials that are appropriate in his frame of reference, he will then predict a difference of the rates of the two kinds of clocks. And, as a further calculation shows, he will come to the same conclusions about this time difference as are obtained by the observer on the Earth (for whom the laws of general relativity reduce to those of special relativity because he is not accelerated). So the different degree of 'agings' of the two twins is fully compatible with the principle of relativity, when the theory is generalized sufficiently to apply to accelerated frames of reference.

Why does the different aging of the two twins seem paradoxical to most people, when they first hear of it? The answer is basically in the habitual mode of thought, whereby we automatically regard all that is co-present in our sense perceptions as happening at the same time, which we call 'now'. Thus, on looking out at the stars in the night sky we cannot avoid seeing the whole firmament as existing 'now', simultaneous with our act of perception. As a result we are led, almost without further conscious thought, to the supposition that if a rocket ship went out in space, we could keep on watching it, or otherwise remain in immediate contact with it, comparing each event that happened to it (e.g., the ticking of a clock) with corresponding events that are happening to us at the same time. When it returned, it would then be seen to have experienced the same amount of time, as indeed does happen with all systems with which we are familiar (which latter of course move at speeds that are very low in relation to that of light).

It is of course by now very well known to us that what we see in the night sky is not actually happening at the same moment at which we perceive it, but rather that all that we see is past and gone (the distant nebulae, for example, are seen as they were a hundred million years ago or more). Moreover, our judgement as to *when* what we see actually did exist is based on the correction, $\Delta t = r/c$, for the time light takes to reach us. And, as we have brought out in earlier chapters, this correction is not the same for all observers, but depends on their speeds. As a result, our habitual procedure of assigning a unique time to each event no longer has much meaning. And if distant events do not have a unique time of occurrence, the same for all valid methods of measuring it, then there is no longer any good reason to suppose that two observers who separate and then meet will necessarily have experienced the same amount of time.

NOTE

* From *The Special Theory of Relativity*, W. A. Benjamin, New York, pp. 165–167.

K. GÖDEL

STATIC INTERPRETATION OF SPACE-TIME WITH EINSTEIN'S COMMENT ON IT*

One of the most interesting aspects of relativity theory for the philosophical-minded consists in the fact that it gave new and surprising insights into the nature of time, of that mysterious and seemingly self-contradictory[1] being which, on the other hand, seems to form the basis of the world's and our own existence. The very starting point of special relativity theory consists in the discovery of a new and very astonishing property of time, namely the relativity of simultaneity, which to a large extent implies[2] that of succession. The assertion that the events A and B are simultaneous (and, for a large class of pairs of events, also the assertion that A happened before B) loses its objective meaning, in so far as another observer, with the same claim to correctness, can assert that A and B are not simultaneous (or that B happened before A).

Following up the consequences of this strange state of affairs one is led to conclusions about the nature of time which are very far-reaching indeed. In short, it seems that one obtains an unequivocal proof for the view of those philosophers who, like Parmenides, Kant, and the modern idealists, deny the objectivity of change and consider change as an illusion or an appearance due to our special mode of perception.[3] The argument runs as follows: Change becomes possible only through the lapse of time. The existence of an objective lapse of time,[4] however, means (or, at least, is equivalent to the fact) that reality consists of an infinity of layers of 'now' which come into existence successively. But, if simultaneity is something relative in the sense just explained, reality cannot be split up into such layers in an objectively determined way. Each observer has his own set of 'nows', and none of these various systems of layers can claim the prerogative of representing the objective lapse of time.[5]

This inference has been pointed out by some, although by surprisingly few, philosophical writers, but it has not remained unchallenged. And actually to the argument in the form just presented it can be objected that the complete equivalence of all observers moving with different (but uniform) velocities, which is the essential point in it, subsists only in the abstract space-time scheme of special relativity theory and in certain empty worlds of general relativity theory. The existence of matter, however, as well as the particular kind of curvature of space-time produced by it, largely destroy the equivalence of different observers [6] and distinguish some of them conspicuously from the rest, namely those which follow in their notion the mean motion of matter.[7] Now in all cosmological solutions of the gravitational equations (i.e., in all possible universes) known at present the local times of all *these* observers fit together into one world time, so that apparently it becomes possible to consider this time as the 'true' one, which lapses objectively, whereas the discrepancies of the measuring results of other observers from this time may be conceived as due to the influence which a motion relative to the mean state of motion of matter has on the measuring processes and physical processes in general.

From this state of affairs, in view of the fact that some of the known cosmological solutions seem to represent our world correctly, James Jeans has concluded[8] that there is no reason to abandon the intuitive idea of an absolute time lapsing objectively. I do not think that the situation justifies this conclusion and am basing my opinion chiefly[9] on the following facts and considerations:

There exist cosmological solutions of another kind[10] than those known at present, to which the aforementioned procedure of defining an absolute time is not applicable, because the local times of the special observers used above cannot be fitted together into one world time. Nor can any other procedure which would accomplish this purpose exist for them; i.e., these worlds possess such properties of symmetry, that for each possible concept of simultaneity and succession there exist others which cannot be distinguished from it by any intrinsic properties, but only by reference to individual objects, such as, e.g., a particular galactic system.

Consequently, the inference drawn above as to the non-objectivity of change doubtless applies at least in these worlds. Moreover it turns out that temporal conditions in these universes (at least in those referred to in the end of note 10) show other surprising features, strengthening futher the idealistic viewpoint. Namely, by making a round trip on a rocket ship in a sufficiently wide curve, it is possible in these worlds to travel into any region of the past, present, and future, and back again, exactly as it is possible in other worlds to travel to distant parts of space.

This state of affairs seems to imply an absurdity. For it enables one e.g., to travel into the near past of those places where he has himself lived. There he would find a person who would be himself at some earlier period of his life. Now he could do something to this person which, by his memory, he knows has not happened to him. This and similar contradictions, however, in order to prove the impossibility of the worlds under consideration, presuppose the actual feasibility of the journey into one's own past. But the velocities which would be necessary in order to complete the voyage in a reasonable length of time[11] are far beyond everything that can be expected ever to become a practical possibility. Therefore it cannot be excluded *a priori*, on the ground of the argument given, that the space-time structure of the real world is of the type described.

As to the conclusions which could be drawn from the state of affairs explained for the question being considered in this paper, the decisive point is this: that for *every* possible definition of a world time one could travel into regions of the universe which are passed according to that definition.[12] This again shows that to assume an objective lapse of time would lose every justification in these worlds. For, in whatever way one may assume time to be lapsing, there will always exist possible observers to whose experienced lapse of time no objective lapse corresponds (in particular also possible observers whose whole existence objectively would be simultaneous). But, if the experience of the lapse of time can exist without an objective lapse of time, no reason can be given why an objective lapse of time should be assumed at all.

It might, however, be asked: Of what use is it if such conditions prevail in certain *possible* worlds? Does that mean anything for the question interesting us whether in *our* world there exists a objective lapse of time? I think it does. For, (1) Our world, it is true, can hardly be represented by the particular kind of rotating solutions referred to above (because these solutions are static and, therefore, yield no red-shift for distant objects); there exist however also *expanding* rotating solutions. In such universes an absolute time also might fail to exist,[13] and it is not impossible that our world is a universe of this kind. (2) The mere compatibility with the laws of nature[14] of worlds in which there is no distinguished absolute time, and, therefore, no objective lapse of time can exist, throws some light on the meaning of time also in those worlds in which an absolute time *can* be defined. For, if someone asserts that this absolute time is lapsing, he accepts as a consequence that, whether or not an objective lapse of time exists (i.e., whether or not a time in the ordinary sense of the word exists), depends on the particular way in which matter and its motion are arranged in the world. This is not a straightforward contradiction; nevertheless, a philosophical view leading to such consequences can hardly be considered as satisfactory.

EINSTEIN'S REPLY TO GÖDEL

Kurt Gödel's essay constitutes, in my opinion, an important contribution to the general theory of relativity, especially to the analysis of the concept of time. The problem here involved disturbed me already at the time of the building up of the general theory of relativity, without my having succeeded in clarifying it. Entirely aside from the relation of the theory of relativity to idealistic philosophy or to any philosophical formulation of questions, the problem presents itself as follows (Figure 1):

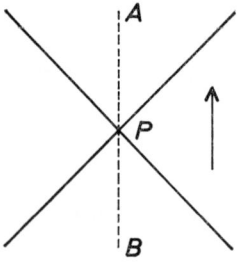

Fig. 1.

If P is a world point, a 'light cone' ($ds^2 = 0$) belongs to it. We draw a 'time-like' world line through P and on this line observe the close world-points B and A, separated by P. Does it make any sense to provide the world line with an arrow, and to assert that B is *before P, A after P*? Is what remains of temporal connection between world points in the theory of relativity an asymmetrical relation, or would one be just as much justified, from the physical point of view, to indicate the arrow in the opposite direction and to assert that A is *before P, B after P*?

In the first instance the alternative is decided in the negative, if we are justified in saying: If it is possible to send (to telegraph) a signal (also passing by in the close proximity of P) from B to A, but not from A to B, then the one-sided (asymmetrical) character of time is secured, i.e., there exists no free choice for the direction of the arrow. What is essential in this is the fact that the sending of a signal is, in the sense of thermodynamics, an irreversible process, a process which is connected with the growth of entropy (whereas, *according to our present knowledge*, all elementary processes are reversible).

If, therefore, B and A are two, sufficiently neighboring, world points, which can be connected by a time-like line, then the assertion: 'B is before A', makes physical sense. But does this assertion still make sense, if the points, which are connectable by the time-like line, are arbitrarily far separated from each other? Certainly not, if there exist point-series connectable by time-like lines in such a way that each point precedes temporally the preceding one, *and if the series is closed in itself*. In that case the distinction 'earlier – later' is abandoned for world points which lie far apart in a cosmological sense, and those paradoxes, regarding the *direction* of the causal connection, arise, of which Mr Gödel has spoken.

Such cosmological solutions of the gravitation equations (with not vanishing Λ-constant) have been found by Mr Gödel. It will be interesting to weigh whether these are not to be excluded on physical grounds.

NOTES

* From *Albert Einstein: Philosopher-Scientist*, Evanston 1949, pp. 557–562; 687–688.
[1] Cf., e.g., J. M. E. McTaggart, 'The Unreality of Time', *Mind* **17** (1908).
[2] At least if it is required that any two point events are either simultaneous or one succeeds the other, i.e., that temporal succession defines a complete linear ordering of all point events. There exists an absolute partial ordering.
[3] Kant (in the *Critique of Pure Reason*, 2. ed. (1787) p. 54) expresses this view in the following words: "those affections which we represent to ourselves as changes, in beings with other forms of cognition, would give rise to a perception in which the idea of time, and therefore also of change, would not occur at all." This formulation agrees so well with the situation subsisting in relativity theory, that one is almost tempted to add: such as, e.g., a perception of the inclination relative to each other of the world lines of matter in Minkowski space.
[4] One may take the standpoint that the idea of an objective lapse of time (whose essence is that only the present really exists) is meaningless. But this is no way out of the dilemma; for by this very opinion one would take the idealistic viewpoint as to the idea of change, exactly as those philosophers who consider it as self-contradictory. For in both views one denies that an objective lapse of time is a possible state of affairs, *a fortiori* that it exists in reality, and it makes very little difference in this context, whether our idea of it is regarded as meaningless or as self-contradictory. Of course for those who take either one of these two viewpoints the argument from relativity theory given below

is unnecessary, but even for them it should be of interest that perhaps there exists a second proof for the unreality of change based on entirely different grounds, especially in view of the fact that the assertion to be proved runs so completely counter to common sense. A particularly clear discussion of the subject independent of relativity theory is to be found in: Paul Mongré, *Das Chaos in kosmischer Auslese*, 1898.

[5] It may be objected that this argument only shows that the lapse of time is something relative, which does not exclude that it is something objective; whereas idealists maintain that it is something merely imagined. A relative lapse of time, however, if any meaning at all can be given to this phrase, would certainly be something entirely different from the lapse of time in the ordinary sense, which means a change in the existing. The concept of existence, however, cannot be relativized without destroying its meaning completely. It may furthermore be objected that the argument under consideration only shows that time lapses in different ways for different observers, whereas the lapse of time itself may nevertheless be an intrinsic (absolute) property of time or of reality. A lapse of time, however, which is not a lapse in some definite way seems to me as absurd as a coloured object which has no definite colours. But even if such a thing were conceivable, it would again be something totally different from the intuitive idea of the lapse of time, to which the idealistic assertion refers.

[6] Of course, according to relativity theory all observers are equivalent in so far as the laws of motion and interaction for matter and field are the same for all of them. But this does not exclude that the structure of the world (i.e., the actual arrangement of matter, motion, and field) may offer quite different aspects to different observers, and that it may offer a more 'natural' aspect to some of them and a distorted one to others. The observer, incidentally, plays no essential role in these considerations. The main point, of course, is that the world itself has certain distinguished directions, which directly define certain distinguished local times.

[7] The value of the mean motion of matter may depend essentially on the size of the regions over which the mean is taken. What may be called the 'true mean motion' is obtained by taking regions so large, that a further increase in their size does not any longer change essentially the value obtained. In our world this is the case for regions including many galactic systems. Of course a true mean motion in this sense need not necessarily exist.

[8] Cf. *Man and the Universe*, Sir Halley Stewart Lecture (1935) 22–23.

[9] Another circumstance invalidating Jeans' argument is that the procedure described above gives only an approximate definition of an absolute time. No doubt it is possible to refine the procedure so as to obtain a precise definition, but perhaps only by introducing more or less arbitrary elements (such as, e.g., the size of the regions or the weight function to be used in the computation of the mean motion of matter). It is doubtful whether there exists a precise definition which has so great merits, that there would be sufficient reason to consider exactly the time thus obtained as the true one.

[10] The most conspicuous physical property distinguishing these solutions from those known at present is that the compass of inertia in them everywhere rotates relative to matter, which in our world would mean that it rotates relative to the totality of galactic systems. These worlds, therefore, can fittingly be called 'rotating universes'. In the subsequent considerations I have in mind a particular kind of rotating universes which have the additional properties of being static and spatially homogeneous, and a cosmological constant < 0. For the mathematical representation of these solutions, cf. my paper forthcoming in *Rev. Mod. Phys.*

[11] Basing the calculation on a mean density of matter equal to that observed in our

world, and assuming one were able to transform matter completely into energy, the weight of the 'fuel' of the rocket ship, in order to complete the voyage in t years (as measured by the traveller), would have to be of the order of magnitude of $10^{22}/t^2$ times the weight of the ship (if stopping, too, is effected by recoil). This estimate applies to $t \ll 10''$. Irrespective of the value of t, the velocity of the ship must be at least $1/\sqrt{2}$ of the velocity of light.

[12] For this purpose incomparably smaller velocities would be sufficient. Under the assumptions made in note 11 the weight of the fuel would have to be at most of the same order of magnitude as the weight of the ship.

[13] At least if it required that successive experiences of one observer should never be simultaneous in the absolute time, or (which is equivalent) that the absolute time should agree in direction with the times of all possible observers. Without this requirement an absolute time always exists in an expanding (and homogeneous) world. Whenever I speak of an 'absolute' time, this of course is to be understood with the restriction explained in note 9, which also applies to other possible definitions of an absolute time.

[14] The solution considered above only proves the compatibility with the general form of the field equations in which the value of the cosmological constant is left open; this value, however, which at present is not known with certainty, evidently forms part of the laws of nature. But other rotating solutions might make the result independent of the value of the cosmological constant (or rather of its vanishing or non-vanishing and of its sign, since its numerical value is of no consequence for this problem). At any rate these questions would first have to be answered in an unfavourable sense, before one could think of drawing a conclusion like that of Jeans mentioned above. *Note added Sept. 2, 1949:* I have found in the meantime that for *every* value of the cosmological constant there do exist solutions, in which there is no world-time satisfying the requirement of note 13.

A. S. EDDINGTON

THE ARROW OF TIME, ENTROPY AND THE EXPANSION OF THE UNIVERSE*

Setting aside the guidance of consciousness, we discover a signpost for time in the physical world itself. The signpost is a rather peculiar one, and I would not venture to say that the discovery of the signpost amounts to the same thing as the discovery of an objective 'going on of time' in the universe. But at any rate it provides a unique criterion for discriminating between past and future, whereas there is no corresponding absolute distinction between right and left. The signpost depends on a certain measurable physical quantity called entropy. Take an isolated system and measure its entropy at two instants t_1 and t_2: the rule is that the instant which corresponds to the greater entropy is the later. We can thus find out by purely physical measurements whether t_1 is before or after t_2 without trusting to the intuitive perception of the direction of progress of time in our consciousness. In mathematical form the rule is that the entropy S fulfils the law: dS/dt is always positive. This is the famous Second Law of Thermodynamics.

Entropy may most conveniently be described as a measure of the disorganisation of a system. I do not intend that to be taken as a definition, because disorganisation is a flexible term depending to some extent on our point of view; but in all those processes which increase the entropy of a system we can see chance creeping in where formerly it was excluded, so that conditions which were specialised or systematised become chaotic. Many examples can be given of natural processes which break up an organised system into a random distribution. Plane waves of sunlight all travelling in one direction fall on a white sheet of paper and are scattered in all directions. The direction of the waves becomes disorganised; accordingly there is an increase of entropy. When a solid body moves as a whole, its molecules travel forward together; when it is stopped by hitting something, the molecules begin to move in all directions indiscriminately. It is as though the disciplined march of a regiment suddenly stopped, and it became a jostling throng of individuals all trying to go in different directions. This random motion of the molecules is identified with the heat-energy of the body. Quantitatively the heat produced by impact is the exact equivalent of the lost energy of motion of the body as a whole, but it has a less organised from. Nature keeps strict account of all these little wastages of organisation which are continually occurring; each is debited against the total stock of organisation contained in the universe. The balance is always growing less. One day it will all be used up.

Heat, when concentrated, is not *fully* disorganised energy. A further decrease of organisation occurs when the heat diffuses evenly so as to bring the body and its surroundings to a uniform temperature. In other words heat-energy suffers loss of organisation when it flows from a hotter body to a colder body. This is one of the most common occasions of increase of entropy (disorganisation), for unless the temperature is everywhere uniform heat is always leaking from hotter to colder regions. The fact that a certain amount of organisation is retained in a concentrated store of heat enables us partially to convert heat into visible motion – the reverse of what happens at impact. But only partially. To drive a train we must put into the engine more heat-energy than will appear as energy of motion of the train, the extra quantity being needed to make up for its inferior organisation. In that way without any creation of organisation we furnish enough organised energy to the train; the excess energy, which has been drained of organisation as far as practicable, is turned out as waste into the condenser of the engine.

In using entropy as a signpost for time we must be careful to treat a properly *isolated* system. Isolation is necessary because a system can gain organisation by draining it from other contiguous systems. Evolution shows us that more highly organised systems develop as time goes on. This may be partly a question of definition, for it does not follow that organisation from an evolutionary point of view is to be reckoned according to the same measure as organisation from the entropy point of view. But in any case these highly developed systems may obtain their organisation by a process of collection, not by creation. A human being as he grows from past to future becomes more and more highly organised – or so he fondly imagines. At first sight this appears to contradict the signpost law that the later instant corresponds to the greater disorganisation. But to apply the law we must make an isolated system of him. If we prevent him from acquiring organisation from external sources, if we cut off his consumption of food and drink and air, he will ere long come to a state which everyone would recognise as a state of extreme 'disorganisation'.

It is possible for the disorganisation of a system to become complete. The state then reached is called thermodynamic equilibrium. Entropy can increase no further and, since the second law of thermodynamics forbids a decrease, it remains constant. Our signpost for time then disappears; and, so far as that system is concerned, time ceases to go on. That does not mean that time ceases to exist; it exists and extends just as space exists and extends, but there is no longer any dynamic quality in it. A state of thermodynamic equilibrium is necessarily a state of death, so that no consciousness will be present to provide an alternative indicator of 'time's arrow'.

There is no other independent signpost for time in the physical world – at least no other local signpost; so that if we discredit or explain away this property of entropy the distinction of past and future disappears altogether. I base this statement on a law which has become universally accepted in atomic physics, which is called 'the Principle of Detailed Balancing'[1]

Having found our signpost, let us look around. Ahead there is ever-increasing disorganisation in the universe. Although the sum total of organisation is diminishing, certain local structures exhibit a more and more highly specialised organisation at the expense of the rest; that is the phenomenon of evolution. But ultimately these must be swallowed up by the advancing tide of chance and chaos, and the whole universe will reach the final state in which there is no more organisation to lose. A few years ago we should have said that it would end as a uniform featureless mass in thermodynamic equilibrium; but that does not take into account what we have recently learnt as to the expansion of the universe. The theory of the expanding universe introduces some differences of description but, I think, no essential difference of principle, and it will be convenient to consider it later, adhering for the present to the older ideas. When the final heat-death overtakes the universe time will *extend* on and on, presumably to infinity, but there will be no definable sense in which it can be said to *go on*. Consciousness must have disappeared from the physical world before this stage is reached and, dS/dt having vanished, there will remain nothing to point the way of progress of time. This is the end of the world.

Now let us look in the opposite direction towards the past. Following time backwards we find more and more organisation in the world. If we are not stopped earlier, we go back to a time when the matter and energy of the world had the maximum possible organisation. To go back further is impossible. We have come to another end of space-time – an abrupt end – only according to our orientation we call it 'the beginning'.

I have no philosophical axe to grind in this discussion. I am simply stating the results to which our present fundamental conceptions of physical law lead. I am much more concerned with the question whether the existing scheme of science is built on a foundation firm enough to stand the strain of extrapolation throughout all time and all space, than with prophecies of the ultimate destiny of material things or with arguments for admitting an act of Creation. I find no difficulty in accepting the consequences of the present scientific theory as regards the future – the heat-death of the universe. It may be billions of years hence, but slowly and inexorably the sands are running out. I feel no instinctive shrinking from this conclusion. From a moral standpoint the conception of a cyclic universe, continually running down and continually rejuvenating itself, seems to me wholly retrograde. Must Sisyphus for ever toll his stone up the hill only for it to roll down again every time it approaches the top? That was a description of Hell. If we have any conception of progress as a whole reaching deeper than the physical symbols of the external world, the way must, it would seem, lie in excape from the Wheel of things. It is curious that the doctrine of the running-down of the physical universe is so often looked upon as pessimistic and contrary to the aspirations of religion. Since when has the teaching that "heaven and earth shall pass away" become ecclesiastically unorthodox?

The extrapolation towards the past raises much graver difficulty. Philosophically the notion of an abrupt beginning of the present order of Nature is repugnant to me, as I think it must be to most; and even those who would welcome a proof of the intervention of a Creator will probably consider that a single winding-up at some remote epoch is not really the kind of relation between God and his world that brings satisfaction to the mind. But I see no excape from our dilemma. One cannot say definitely that future developments of science will not provide an escape; but it would seem that the difficulty arises not so much from a fault in the present system of physical law as in the whole relation of the method of analysis of experience employed in physical science to the actualities with which it deals. The dilemma is this: Surveying our surroundings we find them to be far from a "fortuitous concourse of atoms". The picture of the world as drawn in existing physical theories shows an arrangement of the individual atoms and photons which if it originated by a chance coincidence would be excessively improbable. The odds against it are multillions to 1. (I use 'multillions' as a general term for a number which, if written out in full in the usual decimal notation, would fill all the books in a large library). This non-random feature of the world might possibly be identified with purpose or design; let us, however, non-committally call it anti-chance. We are unwilling to admit in physics that there is any anti-chance in the reactions between the billions of atoms and quanta in the inorganic systems that we study; and indeed all our experimental evidence goes to show that these are governed by the laws of chance. Accordingly we do not recognise anti-chance in the laws of physics, but only in the data to which those laws are applied. In the corresponding mathematical treatment we exclude anti-chance from the differential equations of physics and relegate it to the boundary conditions – for it has to be brought in somewhere. One cannot help feeling that this segregation of the chance from the anti-chance is a characteristic rather of our method of attacking the problem than of the objective universe itself. It is as though we ironed out a region large enough to include our more immediate experience at the cost of puckering in the regions outside. We have swept away the anti-chance from the field of our current physical problems, but we have not got rid of it. When some of us are so misguided as to try to get back milliards of years into the past we find the sweeping piled up like a high wall, forming a boundary – a beginning of time – which we cannot climb over.

Without insisting dogmatically on the finality of the second law of thermodynamics, we must emphasise that it is very deeply rooted in physics. The engineer dealing with the practical problems of the heat engine, the quantum physicist discussing the laws of radiation, the astronomer investigating the interior of a star, the student of cosmic rays tracing perhaps the disintegrations of atoms in space beyond the galaxy, have all pinned their faith to the rule that the disorganisation or random element can increase but never diminish. This faith is not unreasonable when we recall that to abandon the second law of thermodynamics means that we uproot the signpost of time.

I have sometimes been taken to task for not sufficiently emphasising in my discussions of these problems that the laws concerning entropy are a matter of probability, not of certainty. I said above that if we observe a system at two instants, the instant corresponding to the greater entropy is the later. Strictly speaking, I ought to have said that (for a smallish system) the chances are, say, 10^{20} to 1 that it is the later. For by a highly improbable coincidence the multitudinous particles might at the later instant accidentally arrange themselves in a distribution with as much organisation as at the earlier instant; just as in shuffling a pack of cards there is a possibility that we may accidentally arrange the cards in suits or sequences. Some critics seem to have been shocked at my lax morality in making the former statement when I was well aware of the 1 in 10^{20} chance of its being wrong. Let me make a confession. I have in the past twenty-five years written a number of scientific papers and books, broadcasting a good many statements about the physical world. I am afraid there are not many of these statements for which I can claim that the chance of being wrong is no more than 1 in 10^{20}. My average risk is more like 1 to 10 – or is that too boastful an estimate? Certainly if it turns out that nine-tenths of what I tell you in this book is correct, I am either very fortunate or else very platitudinous. I think that if we were not allowed to make statements which had a 1 in 10^{20} chance of being untrue, conversation would languish somewhat. Presumably the only persons entitled to open their lips would be the pure mathematicians.

The irreversible dissipation of energy in the universe has been a recognised doctrine of science since 1852 when it was formulated explicitly by Kelvin. Kelvin drew the same conclusions about the beginning and end of things as those given here – except that, since less attention was paid to the universe in those days, he considered the earth and the solar system. The general ideas have not changed much in eighty years; but the recognition of the finitude of space and the recent theory of the expanding universe now involve some supplementary considerations.

The conclusion that the total entropy of the universe at any instant is greater than at a previous instant dates from a time when an 'instant' was conceived to be an absolute time-partition extending throughout the universe. We have to reconsider the matter now that Einstein has abolished these absolute instants; but it appears that no change is required. I think I am right in saying that it is not necessary that the instants should be absolute, or that the time t referred to in dS/dt should be a form of absolute time. For the first instant we can choose any arbitrary space-like section of space-time (smooth or crinkled), and for the second instant any other space-like section which does not intersect the first. One of these instants will be later throughout than the other;[2] and the total entropy of the universe integrated over the later instant will be greater than over the earlier instant. This generalisation is made possible by the fact that the energy or matter which carries the disorganisation cannot travel from place to place faster than light.

The consequences of introducing the expansion of the universe are more difficult to foresee. Fundamental questions are raised as to the appropriate way of defining entropy when the background conditions are no longer invariable. I believe that the progress of the theory in other directions in the next few years will place us in a better position to treat the thermodynamical problem which it raises, and I prefer not to try to anticipate its conclusions.

Meanwhile it is important to notice that the expansion of the universe is another irreversible process. It is a one-way characteristic like the increase of disorganisation. Just as the entropy of the universe will never return to its present value, so the volume of the universe will never return to its present value. From the expansion of the universe we reach independently the same outlook as to the beginning and end of things that we have here reached by considering the increase of entropy. In particular the conclusion seems almost inescapable that there must have been a definite beginning of the present order of Nature. The theory of the expanding universe adds something new, namely an estimate of the date of this beginning. From the scientific point of view it is uncomfortably recent– scarcely more than 10,000 million years ago.

In the expanding universe we can decide which of two instants is the later by the criterion that the later instant corresponds to the larger volume of the universe. (The instants are defined as before to be two non-intersecting space-like sections of space-time.) This provides an alternative signpost for time. But it is only applicable to time taken throughout the universe as a whole. The position of entropy as the unique *local signpost* remains unaffected. The fact that the direction of time for the universe, regarded as a single system, is indicated both by increasing volume and by increasing entropy suggests that there is some undiscovered relation between the two criteria. That is one of the points on which we may expect more light in the next few years.

By accepting the theory of the expanding universe we are relieved of one conclusion which we had felt to be intrinsically absurd. It was argued that every possible configuration of atoms must repeat itself at some distant date. But that was on the assumption that the atoms will have only the same choice of configurations in the future that they have now. In an expanding space any particular congruence becomes more and more improbable. The expansion of the universe creates new possibilities of distribution faster than the atoms can work through them, and there is no longer any likelihood of a particular distribution being repeated. If we continue shuffling a pack of cards we are bound sometime to bring them into their standard order – but not if the conditions are that every morning one more card is added to the pack.

So I think after all there will not be a second (accidental) delivery of these Messenger Lectures this side of eternity.

NOTES

* From *New Pathways of Science*, Macmillan, New York, 1935, pp. 54–61; 66–68.
[1] *The Nature of the Physical World*, p. 79.
[2] That is to say, all observes, whatever their position and motion, will encounter them in the same order.

A. GRÜNBAUM

THE EXCLUSION OF BECOMING FROM THE PHYSICAL WORLD*

I. THE ISSUE OF THE MIND-DEPENDENCE OF BECOMING

In the common-sense view of the world, it is of the very essence of time that events occur now, or are past or future. Furthermore, events are held to change with respect to belonging to the future or the present. Our commonplace use of tenses codifies our experience that any particular present is superseded by another whose event-content thereby 'comes into being'. It is this occurring *now* or coming into being of previously future events and their subsequent belonging to the past which is called 'becoming'. But the past and the future can be characterized, respectively, as before and after the present. Hence I shall center my account of becoming on the status of the present or now as an attribute of events which is encountered in *perceptual* awareness.

Granted that becoming is a prominent feature of our temporal awareness, I ask: *Must* becoming therefore also be a feature of the temporal order of physical events *independently* of our awareness of them, as the common-sense view supposes it to be? And if not, is there anything within physical theory *per se* to warrant this common-sense conclusion?

It is apparent that the becoming of physical events in our temporal awareness does not itself guarantee that becoming has a mind-independent physical status. Common-sense color attributes, for example, surely *appear* to be properties of physical objects independently of our awareness of them and are held to be such by common sense. And yet scientific theory tells us that they are mind-dependent qualities, like sweet and sour are. Of course, if physical theory claims that, contrary to common sense, becoming is not a feature of the temporal order of physical events with respect to earlier and later, then a more comprehensive scientific and philosophical theory must take suitable cognizance of becoming as a conspicuous characteristic of our temporal awareness of both physical and mental events.

In this paper I aim to clarify the status of temporal becoming by dealing with each of the questions I posed. Clearly, an account of becoming which provides answers to these questions is *not* an *analysis* of what the common-sense man actually *means* when he says that a physical event belongs to the present, past, or future; instead, such an account sets forth how these ascriptions ought to be construed within the framework of a theory which would supplant the scientifically untutored view of common sense. That the common-sense view is indeed scientifically untutored is evident from the fact that *at a time t*, both of the following physical events qualify as occurring 'now' or 'belonging to the present' according to that view: (1) A stellar explosion that occurred several million years before time t but which is first seen on earth at time t, and (2) a lightning flash originating only a fraction of a second before t and observed at time t. If it be objected that present-day common-sense beliefs have begun to allow for the finitude of the speed of light, then I reply that they err at least to the extent of associating absolute simultaneity with the now.

The temporal relations of earlier (before) and later (after) can obtain between two physical events independently of the transient now, and of any minds. On the other hand, the classification of events into past, present, and future, which is inherent to becoming, requires reference to the transient now as well as to the relations of earlier and later. Hence the issue of the mind-dependence of becoming turns on the status of the transient now. And to assert in this context that becoming is mind-dependent is *not* to assert that the obtaining of the relation of temporal precedence among physical events is mind-dependent.

With these explicit understandings, I can state my thesis as follows: Becoming is mind-dependent because it is not an attribute of physical events per se but requires the occurrence of certain *conceptualized conscious experiences* of the occurrence of physical events. The doctrine that becoming is mind-dependent has been misnamed 'the theory of the block universe'. I shall therefore wish to dissociate the tenets of this doctrine both from serious misunderstandings by its critics and from the very misleading suggestions of the metaphors used by some of its exponents. Besides stating my positive reasons for asserting the mind-dependence of becoming, I shall defend this claim against the major objections which have been raised against it.

II. THE DISTINCTION BETWEEN TEMPORAL BECOMING AND THE ANISOTROPY OF TIME

To treat these various issues without risking serious confusions, we must sharply distinguish between the following two questions: (1) Do physical events *become* independently of any conceptualized awareness of their occurrence, and (2) are there any kinds of physical or biological processes which are *irreversible* on the strength of the laws of nature and/or of *de facto* prevailing boundary conditions? I shall first state how these two questions have come to be identified and will then explain why it is indeed an error of consequence to identify them. The second of these questions, which pertains to irreversibility, is often formulated by asking whether the time of physics and biology has an 'arrow'. But this formulation of question (2) can mislead by inviting misidentification of (2) with (1). For the existence of an arrow is then misleadingly spoken of as constituting a 'one-way forward flow of time', but so also is becoming on the strength of being conceived as the forward 'movement' of the present. And this misidentification is then used to buttress the false belief that an affirmative answer to the question about irreversibility entails an affirmative answer to the question about becoming. To see why I claim that there is indeed a weighty misidentification here, let us first specify what is involved logically when we inquire into the existence of kinds of processes in nature which are irreversible.

If the system of world lines, each of which represents the career of a physical object, is to exhibit a one-dimensional temporal order, relations of simultaneity between spatially separated events are required to define world states. For our purposes it will suffice to use the simultaneity criterion of some one local inertial frame of the special theory of relativity instead of resorting to the cosmic time of some cosmological model.

Assume now that the events belonging to *each* world line are invariantly ordered by a *betweenness* relation having the following *formal* property of the spatial betweenness of the points on a Euclidean straight line: Of any three elements, only one can be between the other two. This betweenness of the events is clearly temporal rather than spatial, since it *invariantly* relates the events belonging to each *individual world line* with respect to all inertial systems while no such *spatial* betweenness obtains invariantly.[1] So long as the temporal betweenness of the world lines is formally Euclidean in the specified sense, any two events on one of them or any two world states can serve to define two time senses which are *ordinally* opposite to each other with respect to the assumed temporal betweenness relations.[2] And the members of the simultaneity classes of events constituting one of these two opposite senses can then bear lower real number coordinates, while those of the other sense can bear higher coordinates. It is immaterial at this stage which of the two opposite senses is assigned the higher real numbers. All we require is that the real number coordinatization reflect the temporal betweenness relations among the events as follows: Events which are temporally between two given events E and E' must bear real number coordinates which are numerically between the time coordinates of E and E'. Employing some one time coordinatization meeting this minimal requirement, we can use the locutions 'initial state', 'final state', 'before', and 'after' on the basis of the magnitudes of the real number coordinates, entirely without prejudice to whether there are irreversible kinds of processes.[3] By an 'irreversible process' (à la Planck) we understand a process such that no counter-process is capable of restoring the original *kind* of state of the system at another time. Note that the temporal vocabulary used in this definition of what is *meant* by an irreversible kind of process does *not* assume tacitly that there *are* irreversible processes: As used here, the terms 'original state', 'restore', and 'counter-process' presuppose only the coordinatization based on the assumed betweenness.

It has been charged that one is guilty of an illicit spatialization of time if one speaks of temporal betweenness while leaving it *open* whether there are irreversible kinds of processes. But this charge overlooks that the *formal* property of the betweenness on the Euclidean line which I invoked is abstract and, as such, neither spatial nor temporal. And the meaningful attribution of this formal property to the betweenness relation among the events belonging to each world line without any assumption of irreversibility is therefore *not* any kind of illicit spatialization of time. We might as well say that since temporal betweenness does have this abstract property, the ascription of the latter to the betweenness among the points on a line of space is a temporalization of space![4]

Thus, the assumption that the events belonging to each world line are invariantly ordered by an abstractly Euclidean relation of temporal betweenness does not entail the existence of irreversible kinds of processes, but allows every kind of process to be reversible.[5] If there are irreversible processes, then the two ordinally opposite time senses are indeed *further* distinguished structurally as follows: There are certain kinds of sequences of states of systems specified in the order of increasing time coordinates such that these same kinds of sequences do *not* likewise exist in the order of decreasing time coordinates. Or, equivalently, the existence of irreversible processes *structurally* distinguishes the two opposite time senses as follows: There are certain kinds of sequences of states of systems specified in the order of *decreasing* time coordinates such that these same kinds of sequences do *not* likewise obtain in the order of increasing time coordinates. Accordingly, if there are irreversible kinds of processes, then time is *anisotropic*.[6] When physicists say with Eddington that time has an 'arrow', it is this anisotropy to which they are referring metaphorically. Specifically, the spatial opposition between the head and the tail of the arrow represents the structural anisotropy of time.

Note that we were able to characterize a process as irreversible and time as anisotropic without any explicit or tacit reliance on the transient now or on tenses of past, present, and future.[7] By the same token, we are able to assert metaphorically that time has an 'arrow' without any covert or outright reference to events as occurring *now*, happening at present, or coming into being. Nonetheless, the anisotropy of time symbolized by the arrow has been falsely equated in the literature with the transiency of the now or becoming of events via the following steps of reasoning: (1) The becoming of events is described by the kinematic metaphor 'the flow of time' and is conceived as a *shifting* of the now which *singles out the future direction of time* as the sense of its 'advance', and (2) although the physicist's arrow does not involve the transient now, his assertion that there is an arrow of time is taken to be equivalent to the claim that there is a *flow* of time in the direction of the future; this is done by attending to the head of the arrow *to the neglect of its tail* and identifying the former with the direction of 'advance' of the now. The physicist's assertion that time has an 'arrow' discerningly codifies the empirical fact that the two ordinally opposite time senses are *structurally different* in specified respects. But in thus codifying this empirical fact, the physicist does *not* invoke the transient now to single out one of the two time senses as preferred over the other. By contrast, the claim that the present or now shifts in the direction of the future does invoke the transient now to single out one of the two time senses and – as we are about to see – is a mere truism like 'All bachelors are males.' For the terms 'shift' or 'flow' are used in their literal kinematic senses in such a way that the *spatial* direction of a shift or flow is specified by where the shifting object is at *later* times. Hence when we speak metaphorically of the now as 'shifting' temporally in a particular *temporal* direction, it is then simply a matter of definition that the now shifts or advances in the direction of the future. For this declaration tells us no more than that the nows corresponding to later times are later than those corresponding to earlier ones, which is just as uninformative as the truism that the earlier nows precede the later ones.[8]

It is now apparent that to assert the existence of irreversible processes in the sense of physical theory by means of the metaphor of the arrow does not entail at all that there is a mind-independent becoming of physical events as such. Hence those wishing to assert that becoming is independent of mind cannot rest this claim on the anisotropy of physical time.

Being only a tautology, the kinematic metaphor of time flowing in the direction of the future does not itself render any empirical fact about the time of our experience. But the role played by the present in becoming is a feature of the experienced world codified by common-sense time in the following informative sense: To each of a great diversity of events which are ordered with respect to earlier and later by physical clocks, there corresponds one or more particular experiences of the event as occurring *now*. Hence we shall say that our experience exhibits a *diversity of 'now-contents'* of awareness which are temporally ordered with respect to each of the relations earlier and later. Thus, it is a significant feature of the experienced world codified by common-sense time that there is a sheer diversity of nows, and in that sheer diversity the role of the future is no greater than that of the past. In this *directionally neutral* sense, therefore, it is informative to say that there is a *transiency* of the now or a coming into being of different events. And of course, in the context of the respective relations of earlier and later, this flux of the present makes for events being past and future.

In order to deal with the issue of the mind-dependence of becoming, I wish to forestall misunderstandings that can arise from using the terms 'become', and 'come into being' in senses which are *tenseless*. These senses do *not* involve belonging to the present or occurring now as understood in tensed discourse, and I must emphasize strongly that my thesis of the mind-dependence of becoming pertains only to the *tensed* variety of becoming. Examples of tenseless uses of the terms 'come into being', 'become', and 'now' are the following: (1) A child *comes into being* as a legal entity the moment it is conceived biologically. What is meant by this possibly false assertion is that for legal purposes, the career of a child *begins* (tenselessly) at the moment at which the ovum is (tenselessly) fertilized. (2) If gunpowder is suitably ignited at any particular time t, an explosion comes into being at that time t. The species of coming into being meant here involves a common-sense event which is here asserted to *occur tenselessly at time t*. (3) When heated to a suitable temperature, a piece of iron *becomes* red. Clearly, this sentence asserts that after a piece of iron is (tenselessly) suitably heated, it is (tenselessly) red for an unspecified time interval. (4) In Minkowski's two-dimensional spatial representation of the space-time of special relativity, the event shown by the origin point is called the 'Here-NOW', and correlatively certain event classes in the diagram are respectively called 'Absolute PAST' and 'Absolute FUTURE'; but Minkowski's 'Here-NOW' denotes an arbitrarily chosen event of reference which can be chosen *once and for all* and continues to qualify as 'now' at various times independently of when the diagram is used. Hence there is no transiency of the now in the relativistic scheme depicted by Minkowski, and his absolute past and absolute future are simply absolutely earlier and absolutely later than the arbitrarily chosen fixed reference event called 'Here-NOW'.[9] Accordingly, we must be mindful that there are tenseless senses of the words 'becoming' and 'now'.

But conversely we must realize that some important *seemingly* tenseless uses of the terms 'to exist', 'to occur', 'to be actual', and 'to have being or reality' are in fact laden with the present tense. Specifically, all these terms are often used in the sense of to occur NOW. And by tacitly making the *nowness* of an event a necessary condition for its occurrence, existence, or reality, philosophers have argued fallaciously as follows. They first assert that the universe can be held to exist only to the extent that there are present events. Note that this either asserts that only present events exist now (which is trivial) or it is false. Then they invoke the correct premise that the existence of the physical universe is not mind-dependent and conclude (from the first assertion) that being present, occurring now, or becoming is *independent* of mind or awareness. Thus, Thomas Hobbes wrote: "The present only has a being in nature; things past have a being in the memory only, but things to come have no being at all, the future being but a fiction of the mind..."[10] When declaring here that only present events or present memories of past events 'have being', Hobbes *appears* to be appealing to a sense of 'to have being' or of 'to exist' which is *logically independent* of the concept of existing-NOW. But his claim depends for its plausibility on the tacit invocation of *present* occurrence as a logically necessary condition for having being or existing. Once this fact is recognized, his claim that "The present only has a being in nature" is seen to be the mere tautology that "Only what exists now does indeed exist now." And by his covert appeal to the irresistible conviction carried by this triviality, he makes plausible the utterly unfounded conclusion that nature can be held to exist only to the extent that there are *present* events and *present* memories of past events. Clearly the fact that an event does not occur now does not justify the conclusion that it does not occur at some time or other.

III. THE MIND-DEPENDENCE OF BECOMING

Being cognizant of these logical pitfalls, we can turn to the following important question: If a physical event occurs *now* (at present, in the present), what attribute or relation of its occurrence can warrantedly be held to qualify it as such?

In asking this question, I am being mindful of the following fact: if at a given clock time t_0 it is true to say of a particular event E that it is occurring now or happening at present, then this claim could not also be truly made at all other clock times $t \neq t_0$. And hence we must distinguish the tensed assertion of *present* occurrence from the tenseless assertion that the event E occurs at the time t_0: namely the latter tenseless assertion, if true at all, can truly be made at all times t other than t_0 no less than at the time t_0. By the same token we must guard against identifying the tensed assertion, made at some particular time t_0, that the event E happens *at present* with the tenseless assertion made at *any* time t, that the event E occurs or 'is present' at time t_0. And similarly for the distinction between the tensed senses of being past or being future, on the one hand, and the tenseless senses *of being past at time t_0*, or being future at time t_0, on the other. To be future at time t_0 just means to be later than t_0, which is a tenseless relation. Thus our question is: what *over* and *above its otherwise tenseless occurrence at a certain clock time t*, in fact at a time t characterizes a physical event as *now* or as belonging to the present? It will be remembered why my construal of this question does *not* call for an analysis of the common-sense meaning of 'now' or of 'belonging to the present' but for a critical assessment of the status which common sense attributes to the present.[11] Given this construal of the question, my reply to it is: What qualifies a physical event at a time t as belonging to the present or as now is not some physical attribute of the event or some relation it sustains to other *purely physical* events; instead what is necessary so to qualify the event is that at the time t at least one human or other *mind-possessing* organism M experiences the event at the time t such that at t, M is *conceptually aware* of experiencing at that time either the event itself or another event simultaneous with it in M's reference frame.[13] And that awareness does not, in general, comprise information concerning the date and numerical clock time of the occurrence of the event. What then is the content of M's conceptual awareness at time t that he *is experiencing* a certain event *at that time*? M's experience of the event at time t is coupled with an awareness of the temporal coincidence of his experience of the event with a state of *knowing* that he has that experience at all. In other words, M experiences the event at t *and* knows that he is experiencing it. Thus, presentness or nowness of an event requires conceptual awareness of the presentational immediacy of either the experience of the event or, if the event is itself *un*perceived, of the *experience* of another event simultaneous with it. For example, if I just hear a noise at a time t, then the noise does not qualify at t as *now* unless at t I am judgmentally aware of the fact of my hearing it at all and of the temporal coincidence of the hearing with that awareness.[14] If the event at time t is itself a mental event (e.g., a pain), then there is no distinction between the event and our experience of it. With this understanding, I claim that the nowness at a time t of either a physical or a mental event requires that there be an *experience* of the event or of another event simultaneous with it which satisfies the specified requirements. And by satisfying these requirements, the *experience* of a physical event qualifies at the time t as occurring *now*. Thus, the fulfillment of the stated requirements by the *experience* of an event at time t is also *sufficient* for the nowness of that *experience* at the time t. But the mere fact that the experience of a physical event qualifies as now at a clock time t allows that in point of physical fact the physical event itself occurred millions of years before t, as in the case of now seeing an explosion of a star millions of light years away. Hence, the mere presentness of the experience of a physical event at a time t does *not* warrant the conclusion that the clock time of the event is t or some *particular* time before t. Indeed, the occurrence of an external physical event E can never be simultaneous in any inertial system with the direct perceptual registration of E by a conceptualizing organism. Hence if E is presently experienced as happening at some particular clock time t, then there is no inertial system in which E occurs at that *same* clock time t. Of course, for *some* practical purposes of daily life, a nearby terrestrial flash in the sky can be held to be simultaneous with someone's experience of it with impunity, whereas the remote stellar explosion of a supernova or an eclipse of the sun, for example, may not. But this kind of practical impunity of common-sense perceptual judgments of the presentness of physical events cannot detract from their scientific falsity. And hence I do not regard it as incumbent upon myself to furnish a philosophical account of the status of nowness which is compatible with the now-verdicts of common sense. In particular, I would scarcely countenance making the nowness of the *experience* of a physical event *sufficient* for the nowness of the event, and even informed common sense might balk at this in cases such as a stellar explosion. But all that is essential to my thesis of mind-dependence is that the nowness of the *experience* of at least one member of the simultaneity class to which an event E belongs is *necessary* for the nowness of the event E itself. And hence my thesis would allow a compromise with common sense to the following extent: allowing ascriptions of nowness to those physical events which have the very vague relational property of occurring only 'slightly earlier' than someone's appropriate experience of them.

Note several crucial commentaries on my characterization of the now:

(1) My characterization of *present* happening or occurring *now* is intended to *deny* that belonging to the present is a physical attribute of a physical event E which is *independent* of any *judgmental awareness* of the occurrence of either E itself or of another event simultaneous with it. But I am *not* offering any kind of *definition* of the adverbial attribute now, which belongs to the conceptual framework of tensed discourse, solely in terms of attributes and relations drawn from the tenseless (Minkowskian) framework of temporal discourse familiar from physics. In particular, I avowedly invoked the present tense when I made the nowness of an event

E at time *t* dependent on someone's knowing at *t* that he *is experiencing E*. And this is tantamount to someone's judging at *t*: I am experiencing *E now*. But this formulation is *non* viciously circular. For it serves to articulate the mind-dependence of nowness, *not* to claim erroneously that nowness has been eliminated by explicit definition in favor of tenseless temporal attributes or relations. In fact, I am very much less concerned with the adequacy of the specifics of my characterization that with its thesis of mind-dependence.

(2) It makes the nowness of an event at time *t* depend on the existence of conceptualized awareness that an experience of the event or of an event simultaneous with it is being had at *t*, and points out the insufficiency of the mere having of the experience. Suppose that at time *t* I express such conceptualized awareness in a linguistic utterance, the utterance being quasi-simultaneous with the experience of the event. Then the utterance satisfies the condition necessary for the *present* occurrence of the experienced event.[15]

(3) *In the first instance*, it is only an experience (i.e., a mental event) which can ever qualify as occurring now, and moreover a mental event (e.g., a pain) must meet the specified awareness requirements in order to qualify. A *physical* event like an explosion can qualify as now at some time *t* only *derivatively* in one of the following two ways: (a) it is necessary that someone's *experience* of the physical event does so qualify, or (b) if unperceived, the physical event must be simultaneous with another physical event that does so qualify in the derivative sense indicated under (a). For the sake of brevity, I shall refer to this complex state of affairs by saying that physical events belonging to regions of space-time wholly devoid of conceptualizing percipients at no time qualify as occurring now and hence as such do not become.

(4) My characterization of the now is narrow enough to exclude past and future events. It is to be understood here that the *reliving* or anticipation of an event, however vivid it may be, is *not* to be misleadingly called 'having an experience' of the event when my characterization of the now is applied to an experience.

My claim that nowness is mind-dependent does not assert at all that the nowness of an event is arbitrary. On the contrary, it follows from my account that it is not at all arbitrary what event or events qualify as being *now* at any given time *t*. To this extent, my account accords with common sense. But I repudiate much of what common sense conceives to be the status of the now. Thus, when I wonder in thought (which I *may* convey by means of an interrogative verbal utterance) whether it is 3 p.m. Eastern Standard Time now, I am asking myself the following: Is the particular percept of which I am now aware when asking this question a member of the simultaneity class of events which qualify as occurring at 3 p.m., E.S.T., on this particular day? And when I wonder in thought about what is happening now, I am asking the question: What events of which I am not aware are simultaneous with the particular now-percept of which I *am* aware upon asking this question?

That the nowness attribute of an occurrence, when ascribed non-arbitrarily to an event, is inherently mind-dependent seems to me to emerge from a consideration of the kind of information which the judgment 'It is 3 p.m., E.S.T., now' can be warrantedly held to convey. Clearly such a judgment is informative, unlike the judgment 'All bachelors are males.' But if the word 'now' in the informative temporal judgment does not involve reference to a particular content of conceptualized awareness or to the linguistic utterance which renders it at the time, then there would seem to be nothing left for it to designate other than either the time of the events already identified as occurring at 3 p.m., E.S.T., or the time of those identified as occurring at some other time. In the former case, the initially informative temporal judgment 'It is 3 p.m., E.S.T., now' turns into the utter triviality that the events of 3 p.m., E.S.T. occur at 3 p.m., E.S.T.! And in the latter case, the initially informative judgment, if false in point of fact, becomes self-contradictory like 'No bachelors are male'.

What of the retort to this objection that independently of being perceived, physical events themselves possess an unanalyzable property of nowness (i.e., presentness) over and above merely occurring at these clock times? I find this retort wholly unavailing for several reasons:

(1) It must construe the assertion 'It is 3 p.m., E.S.T., now' as claiming *non-trivially* that when the clock strikes 3 p.m. on the day in question, this clock event and all of the events simultaneous with it intrinsically have the unanalyzable property of nowness or presentness. But I am totally at a loss to see that anything non-trivial can possibly be asserted by the claim that at 3 p.m. nowness (presentness) inheres in the events of 3 p.m. For all I am able to discern here is that the events of 3 p.m. are indeed those of 3 p.m. on the day in question!

(2) It seems to me of decisive significance that nowness (in the sense associated with becoming) plays no role as a property of physical events themselves in any of the extant theories of physics. There have been allegations in the literature (most recently in H. A. C. Dobbs, 'The "Present" in Physics', *BJPS* **19**, 317–24 (1968–69)) that such branches of statistical physics as meteorology and indeterministic quantum mechanics implicitly assert the existence of a physical counterpart to the human sense of the present. But both below and elsewhere (Reply to Dobbs, *BJPS* **20**, 145–53 (1969)), I argue that these allegations are mistaken. Hence I maintain that if nowness were a mind-*in*dependent property of physical events themselves, then it would be very strange indeed that it could go unrecognized in all extant physical theories *without detriment to their explanatory success*. And I hold with Reichenbach[15] that "If there is Becoming [independently of awareness] the physicist must know it."

(3) As we shall have occasion to note near the end of §4, the thesis that nowness is *not* mind-dependent poses a serious perplexity pointed out by J. J. C. Smart, and the defenders of the thesis have not even been able to hint how they might resolve that perplexity without utterly trivializing their thesis.

The claim that an event can be now (present) only upon either being experienced or being simultaneous with a suitably experienced event accords fully, of course, with the common-sense view that there is no more than one time at which a particular event is present and that this time cannot be chosen arbitrarily. But if an event is ever experienced at all such that there is simultaneous awareness of the fact of that experience, then there exists a time at which the event does qualify as being now provided that the event occurs only 'slightly earlier' than the experience of it.

The relation of the conception of becoming espoused here to that of common sense may be likened to the relation of relativity physics to Newtonian physics. My account of nowness as mind-dependent disavows rather than vindicates the common-sense view of its status. Similarly, relativity physics entails the falsity of the results of its predecessor. Though Newtonian physics thus cannot be reduced to relativity physics (in the technical sense of reducing one theory to another), the latter enables us to see why the former works as well as it does in the domain of low velocities: Relativity theory shows (via a comparison of the Lorentz and Galilean transformations) that the observational results of the Newtonian theory in that domain are sufficiently correct numerically for some practical purposes. In an analogous manner, my account of nowness enables us to see why the common-sense concept of becoming can function as it does in serving the pragmatic needs of daily life.

A *now-content* of awareness can comprise awareness that one event is later than or succeeds another, as in the following examples: (1) When I perceive the 'tick-tock' of a clock, the 'tick' is not yet part of my past when I hear the 'tock'.[16] As William James and Hans Driesch have noted, melody awareness is another such case of quasi-instantaneous awareness of succession.[17] (2) Memory states are contained in now-contents when we have awareness of other events as being earlier than the event of our awareness of them. (3) A now-content can comprise an envisionment of an event as being later than its ideational anticipation.

IV. CRITIQUE OF OBJECTIONS TO THE MIND-DEPENDENCE OF BECOMING

Before dealing with some interesting objections to the thesis of the mind-dependence of becoming, I wish to dispose of some of the caricatures of that thesis with which the literature has been rife under the misnomer 'the theory of the block universe'. The worst of these is the allegation that the thesis asserts the timelessness of the universe and espouses in M. Čapek's words, the "preposterous view ... that ... time is merely a huge and chronic [sic!] hallucination of the human mind."[18] But even the most misleading of the spatial metaphors that have been used by the defenders of the mind-dependence thesis do not warrant the inference that the thesis denies the objectivity of the so-called 'time-like separations' of events known from the theory of relativity. To assert that nowness and thereby pastness and futurity are mind-dependent is surely *not* to assert that the earlier-later relations between the events of a world line are mind-dependent, let alone hallucinatory.

The mind-dependence thesis does deny that physical events themselves happen in the tensed sense of coming into being apart from anyone's awareness of them. But this thesis clearly avows that physical events do happen independently of any mind in the tenseless sense of merely occurring at certain clock times in the context of objective relations of earlier and later. Thus, it is a travesty to equate the objective *becominglessness* of physical events asserted by the thesis with a claim of *timelessness*. In this way the thesis of mind-dependence is misrepresented as entailing that all events happen simultaneously or form a *'totum simul'*.[19] But it is an egregious blunder to think that if the time of physics lacks *passage* in the sense of there not being a transient now, then physical events cannot be temporally separated but must all be simultaneous.

A typical example of such a misconstrual of Weyl's and Einstein's denial of physical passage is given by supposing them to have claimed "that the world is like a film strip: The photographs *are already there* (my italics) and are merely being exhibited to us."[20] But when photographs of a film strip "are already there," they exist *simultaneously*. Hence it is wrong to identify Weyl's denial of physical becoming with the pseudo-image of the 'block universe' and then to charge his denial with entailing the absurdity that all events are simultaneous. Thus, Whitrow says erroneously: "The theory of 'the block universe' ...implies that past (and future) events coexist with those that are present."[21] We shall see in §v that a corresponding error vitiates the allegation that determinism entails the absurd contemporaneity of all events. And it simply begs the question to declare in this context that "*the passage of time...* is the very essence of the concept."[22] For the undeniable fact that passage in the sense of transiency of the now is integral to the common-sense concept of time may show only that, in this respect, this concept is anthropocentric.

The becomingless physical world of the Minkowski representation is viewed *sub specie aeternitatis* in that representation in the sense that the relativistic account of time represented by it makes no reference to the particular times of anyone's *now*-perspectives. And, as J. J. C. Smart observed, "The tenseless way of talking does not therefore imply that physical things or events are eternal in the way in which the number 7 is."[23] We must therefore reject Whitrow's odd claim that according to the relativistic conception of Minkowski, "external events *permanently* (my italics) exist and we merely come across them."[24] According to Minkowski's conception, an event qualifies as a *becomingless* occurrence by occurring in a network of relations of earlier and later and thus can be said to occur "at a certain time t." Hence to assert tenselessly that an event exists (occurs) is to claim that there is a time or clock reading t with which it coincides. But surely this assertion does not entail the absurdity that the event exists (occurs) at *all* clock times or 'permanently'. To occur tenselessly at some time t or other is not at all the same as to exist 'permanently'.

Whitrow himself acknowledges Minkowski's earlier–later relations when he says correctly that "the relativistic picture of the world recognizes only a difference between earlier and later and not between past, present, and future." [25] But he goes on to query: "If no events *happen*, except our observations, we might well ask – why are the latter exceptional?" [26] I reply first of all: But Minkowski asserts that events happen tenselessly in the sense of occurring at certain clock times. And as for the exceptional status of the events which we register in observational awareness, I make the following obvious but only partial retort: Being registered in awareness, these events are *eo ipso* exceptional.

I say that this retort is only partial because behind Whitrow's question there lurks a more fundamental query. This query must be answered by those of us who claim with Russell that "Past, present and future arise from time-relations of subject and object, while earlier and later arise from time-relations of object and object." [27] That query is: Whence the becoming in the case of mental events that become and are causally dependent on physical events, given that physical events themselves do not become independently of being perceived but occur tenselessly? More specifically, the question is: If our *experiences* of (extra and/or intradermal) physical events are causally dependent upon these events, how is it that the former *mental* events can properly qualify as being 'now', whereas the eliciting physical events *themselves* do not so qualify, and yet both kinds of events are (severally and collectively) alike related by quasi-serial relations of earlier and later? [28]

But, as I see it, this question does not point to refuting evidence against the mind-dependence of becoming. Instead, its force is to demand (a) the recognition that the complex mental states of judgmental awareness as such have distinctive features of their own, and (b) that the articulation of these features as part of a theoretical account of 'the place of mind in nature' acknowledges *what may be peculiar to the time of awareness*. That the existence of features peculiar to the time of awareness does not pose perplexities militating against the mind-dependence of becoming seems to me to emerge from the following three counter-questions, which I now address to the critics:

(1) Why is the mind-dependence of becoming more perplexing than the mind-dependence of common-sense color attributes? That is, why is the former more puzzling than that physical events such as the reflection of certain kinds of photons from a surface causally induce mental events such as seeing blue, which are qualitatively fundamentally different in some respects? In asking this question, I am *not* assuming that nowness is a *sensory quality* like red or sweet but only that nowness and sensory qualities alike depend on awareness.

(2) Likewise assuming the causal dependence of mental on physical events, why is the mind-dependence of becoming more puzzling than the fact that the raw feel components of mental events, such as a particular event of seeing green, are not members of the *spatial* order of physical events?[29] Yet mental events and the raw feels ingredient in them are part of a time system that comprises physical events as well.[30]

(3) Mental events must differ from physical ones in some respect qua being mental, as illustrated by their not being members of the same system of spatial order. Why then should it be puzzling that on the strength of the *distinctive* nature of conceptualized awareness and self-awareness, mental events differ further from physical ones with respect to becoming, while both kinds of events sustain temporal relations of simultaneity and precedence?

What is the reasoning underlying the critics' belief that their question has the capability of pointing to the refutation of the mind-dependence of becoming? Their reasoning seems to me reminiscent of Descartes' misinvocation of the principle that there must be nothing more in the effect than is in the cause *à propos* of one of his arguments for the existence of God: The more perfect, he argued, cannot proceed from the less perfect as its efficient and total cause. The more perfect, i.e., temporal relations involving becoming, critics argue, cannot proceed from the less perfect, i.e., becomingless physical time, as its efficient cause. By contrast, I reason that nowness (and thereby pastness and futurity) are features of events *as experienced* conceptually, *not* because becoming is likewise a feature of the physical events which causally elicit our awareness of them, but because these elicited states are indeed specified states of *awareness*. Once we recognize the role of awareness here, then the diversity and order of the events of which we have awareness in the form of now-contents give rise to the transiency of the now as explained in

Section III, due cautions being exercised, as I emphasized there, that this transiency not be construed tautologically.

In asserting the mind-dependence of becoming, I allow fully that the kind of neurophysiological brain state which underlies our mere awareness of an event as simply occurring now differs in specifiable ways from the ones underlying tick-tock or melody awareness, memory awareness, anticipation awareness, and dream-free sleep. But I cannot see why the states of awareness which make for becoming must have physical event-counterparts which isomorphically become in their own right. Hence I believe I have coped with Whitrow's question as to why only perceived events become. Indeed, it seems to me that the thesis of mind-dependence is altogether free from an important perplexity which besets the opposing claim that physical events are inherently past, present, and future. This perplexity was stated by Smart as follows:

> 'If past, present, and future were real properties of events (i.e., properties possessed by physical events independently of being perceived), then it would require (non-trivial) explanation that an event which becomes present (i.e., qualifies as occurring *now*) in 1965 becomes present (now) at that date and not at some other (and this would have to be an explanation over and above the explanation of why an event of this sort occurred in 1965).[31]

It would, of course, be a complete trivialization of the thesis of the mind-*in*dependence of becoming to reply that *by definition* an event occurring at a certain clock time *t* has the unanalyzable attribute of nowness at time *t*.

Thus, to the question "Whence the becoming in the case of mental events that become and are causally dependent on physical events which do not themselves become?", I reply: "Becoming can characterize mental events qua their being both bits of *awareness* and sustaining relations of temporal order.

The awareness which each of several human percipients has of a given physical event can be such that all of them are alike prompted to give the same tensed description of the external event. Thus, suppose that the effects of a given physical event are simultaneously registered in the awareness of several percipients such that they each perceive the event as occurring at essentially the time of their first awareness of it. Then they may each think at that time that the event belongs to the present. The parity of access to events issuing in this sort of intersubjectivity of tense has prompted the common-sense belief that the nowness of a physical event is an intrinsic, albeit transient attribute of the event. But this kind of intersubjectivity does not discredit the mind-dependence of becoming; instead, it serves to show that the becoming present of an event, though mind-dependent no less than a pain, need not be *private* as a pain is. Some specific person's particular pain is private in the sense that this person has privileged access to its raw feel component.[32] The mind-dependence of becoming is no more refuted by such intersubjectivity as obtains in regard to tense than the mind-dependence of common-sense color attributes is in the least disproved by agreement among several percipients as to the color of a chair.

V. BECOMING AND THE CONFLICT BETWEEN DETERMINISM AND INDETERMINISM

If the doctrine of mind-dependence of becoming is correct, a very important consequence follows, which seems to have been previously overlooked: Let us recall that the nowness of events is generated by (our) conceptualized *awareness* of them. Therefore, *nowness is made possible by processes sufficiently macro-deterministic (causal) to assure the requisitely high correlation between the occurrence of an event and someone's being made suitably aware of it.* Indeed, the very concept of experiencing an external event rests on such macro-determinism, and so does the possibility of empirical knowledge. In short, insofar as there is a transient present, it is made possible by the existence of the requisite degree of macro-determinism in the physical world. And clearly, therefore, the transiency of the present can obtain in a completely deterministic physical universe, be it relativistic or Newtonian.

The theory of relativity has repudiated the uniqueness of the simultaneity slices within the class of physical events which the Newtonian theory had affirmed. Hence Einstein's theory certainly precludes the conception of 'the present' which some defenders of the objectivity of becoming have linked to the Newtonian theory. But it must be pointed out that the doctrine of the mind-dependence of becoming, being entirely compatible with the Newtonian theory as well, does not depend for its validity on the espousal of Einstein's theory as against Newton's.

Our conclusion that there can be a transient now in a completely deterministic physical universe is altogether at variance with the contention of a number of distinguished thinkers that the indeterminacy of the laws of physics is both a necessary and sufficient condition for becoming. And therefore I now turn to the examination of their contention.

According to such noted writers as A. S. Eddington, Henri Bergson, Hans Reichenbach, H. Bondi, and G. J. Whitrow, it is a distinctive feature of an *indeterministic* universe, as contrasted with a deterministic one, that physical events belong to the present, occur *now*, or come into being over and above merely becoming present in *awareness*. I shall examine the argument given by Bondi, although he no longer defends it, as well as Reichenbach's argument. And I shall wish to show the following: insofar as events do become, the indeterminacy of physical laws is neither sufficient nor necessary for conferring nowness or presentness on the occurrences of events, an attribute whereby the events come into being. And thus my analysis of their arguments will uphold my previous conclusion that far from depending on the indeterminacy of the laws of physics, becoming requires a considerable degree of macro-determinism *and* can obtain in a completely deterministic world. Indeed, I shall go on to point out that not only the becoming of any kind of event but the temporal order of earlier and later among physical events depends on the at least quasi-deterministic character of the macrocosm. And it will then become apparent in what way the charge that a deterministic universe must be completely *timeless* rests on a serious misconstrual of determinism.

Reichenbach contends: "When we speak about the progress of time [from earlier to later]..., we intend to make a synthetic [i.e., factual] assertion which refers both to an immediate experience and to physical reality".[33] And he thinks that this assertion about events coming *into* being independently of mind – as distinct from merely occurring tenselessly at a certain clock time – can be justified in regard to physical reality on the basis of indeterministic quantum mechanics by the following argument:[34] In classical deterministic physics, both the past and the future were determined in relation to the present by one-to-one functions even though they differed in that there could be direct observational records of the past and only predictive inferences concerning the future. On the other hand, while the results of past measurements on a quantum mechanical system are *determined* in relation to the present *records* of these measurements, a present measurement of one of two conjugate quantities does

EXCLUSION OF BECOMING FROM THE PHYSICAL WORLD 491

not uniquely determine in any way the result of a *future* measurement of the other conjugate quantity. Hence, Reichenbach concludes:

> The concept of "becoming" acquires significance in physics: the present, which separates the future from the past, is the moment at which that which was undetermined becomes determined, and "becoming" has the same meaning as "becoming determined."... it is with respect to "now" that the past is determined and that the future is not.[35]

I join Hugo Bergmann[36] in rejecting this argument for the following reasons. In the indeterministic quantum world, the relations between the sets of measurable values of the state variables characterizing a physical system at different times are, in principle, *not* the one-to-one relations linking the states of classically behaving closed systems. But I can assert correctly in 1966 that this holds for a given state of a physical system and its absolute future quite independently of whether that state occurs at midnight on December 31, 1800 or at noon on March 1, 1984. Indeed, if we consider *any one* of the temporally successive regions of space-time, we can veridically assert the following at *any* time: the events belonging to that particular region's absolute past could be (more or less) uniquely specified in records which are a part of that region, whereas its particular absolute future is thence quantum mechanically unpredictable. Accordingly, *every* event, be it that of Plato's birth or the birth of a person born in the year 2000 A.D., *at all times* constitutes a divide in Reichenbach's sense between its own recordable past and its unpredictable future, *thereby satisfying Reichenbach's definition of the 'present' or 'now' at any and all times!* And if Reichenbach were to reply that the indeterminacies of the events of the year of Plato's birth have already been transformed into a determinacy, whereas those of 2000 A.D. have not, then the rejoinder would be: this tensed conjunction holds for any state between sometime in 428 B.C. and 2000 A.D. that qualifies as now during that interval on grounds other than Reichenbach's asymmetry of determinedness; but the second conjunct of this conjunction does not hold for any state after 2000 A.D. which qualifies as now after that date. Accordingly, contrary to Reichenbach, the now of conceptualized awareness must be invoked tacitly at time *t*, if the instant *t* is to be nontrivially and nonarbitrarily singled out as present or now by Reichenbach's criterion, i.e., if the instant *t* is to be uniquely singled out at time *t* as being 'now' in virtue of being the threshold of the transition from indeterminacy to determinacy.

Turning to Bondi, we find him writing:

...the flow of time has no significance in the logically fixed pattern demanded by deterministic theory, time being a mere coordinate. In a theory with indeterminacy, however, the passage of time transforms statistical expectations into real events.[87]

If Bondi intended this statement to assert that the indeterminacy makes for our human inability to know in advance of their actual occurrence what particular kinds of events will in fact materialize, then, of course, there can be no objection. For in an indeterministic world, the attributes of specified kinds of events are indeed not uniquely fixed by the properties of earlier events and are therefore correspondingly unpredictable. But I take him to affirm beyond this the following traditional philosophical doctrine; in an indeterministic world, events come *into* being by becoming present with time, whereas in a deterministic world the status of events is one of merely occurring tenselessly at certain times. And my objections to his appeal to the transformation of statistical expectations into real events by the passage of time fall into several groups as follows.

(1) Let us ask: what is the character and import of the difference between a (micro-physically) indeterministic and a deterministic physical world in regard to the attributes of future events? The difference concerns only the type of functional connection linking the attributes of future events to those of present or past events. Thus, *in relation to the states existing at other times*, an indeterministic universe allows alternatives as to the attributes of an event that occurs at some given time, whereas a deterministic universe provides no corresponding latitude. But this difference does *not* enable (micro-physical) indeterminism – as contrasted with determinism to make for a difference in the *occurrence-status* of future events by enabling them to come *into* being. Hence in an indeterministic world, physical events no more *become* real (i.e., present) and are no more precipitated into existence, as it were, than in a deterministic one. In either a deterministic or indeterministic universe, events can be held to come into being or to become 'actual' by becoming *present in (our) awareness*; but becoming actual in virtue of occurring *now* in that way no more makes for a mind-independent coming into existence in an indeterministic world than it does in a deterministic one.

(2) Nor does indeterminacy as contrasted with determinacy make for any difference whatever at any time in regard to the *intrinsic attribute-specificity* of the future events themselves, i.e., to their being (tenselessly)

what they are. For in either kind of universe, it is a fact of logic that what will be, will be, no less than what is present or past is indeed present or past![38] The result of a future quantum mechanical measurement may not be definite prior to its occurrence in relation to earlier states, and thus our prior knowledge of it correspondingly cannot be definite. But a quantum mechanical event has a tenseless occurrence status at a certain time which is fully compatible with its intrinsic attribute-definiteness just as a measurement made in a deterministic world does. Contrary to a widespread view, this statement holds also for those events which are constituted by energy states of quantum mechanical systems, since energy *can* be measured in an arbitrarily short time in that theory.[39]

Let me remark parenthetically that the quantum theory of measurement has been claimed to show that the *consciousness* of the human observer is essential to the definiteness of a quantum mechanical event. I am not able to enter into the technical details of the argument for this conclusion, but I hope that I shall be pardoned for nonetheless raising the following question in regard to it. Can the quantum theory account for the relevant physical events which presumably occurred on the surface of the earth *before* man and his consciousness had evolved? If so, then these physical events cannot depend on human consciousness for their specificity. On the other hand, if the quantum theory cannot in principle deal with *pre*-evolutionary physical events, then one wonders whether this fact does not impugn its adequacy in a fundamental way.

In an indeterministic world, there is a lack of attribute-specificity of events *in relation to events at other times*. But this *relational* lack of attribute-specificity cannot alter the fact of logic that an event is intrinsically attribute-specific in the sense of tenselessly being what it is at a certain clock time t.[40]

It is therefore a far-reaching mistake to suppose that unless and until an event of an indeterministic world belongs to the present or past, the event must be *intrinsically* attribute-*in*definite. This error is illustrated by Čapek's statement that in the case of an event 'it is only its presentness [i.e., nowness] which creates its specificity... by eliminating all other possible features incompatible with it'.[41] Like Bondi, Čapek overlooks that it is only with respect to some now or other that an event can be future at all to begin with and that the lack of attribute-specificity or 'ambiguity' of a future event is not intrinsic but relative to the events of the prior now-perspectives.[42] In an indeterministic world, an event is intrinsically attribute-determinate by being (tenselessly) what it is (tenselessly), regardless of whether the time of its occurrence be now (the present) or not. What makes for the coming into being of a future event at a later time t is *not* that its attributes are indeterministic with respect to prior times but only that it is registered in the now-content of awareness at the subsequent time t.

(3) Two quite different things also seem to be confused when it is inferred that in an indeterministic quantum world, future physical events themselves distinctively come into being with the passage of time over and above merely occurring and becoming present to awareness, whereas in a deterministic universe they do not come into being: (i) the epistemic precipitation of the *de facto* event-properties of future events out of the wider matrix of the possible properties allowed in advance by the quantum-mechanical probabilities, a precipitation or becoming definite which is constituted by our getting to *know* these *de facto* properties at the later times, and (ii) a mind-independent coming into being over and above merely occurring and becoming present to awareness at the later time. The *epistemic* precipitation is indeed effected by the passage of time through the transformation of a merely statistical expectation into a definite piece of available information. But this does *not* show that in an indeterministic world there obtains any kind of becoming present ('real') with the passage of time that does not also obtain in a deterministic one. And in either kind of world, becoming as distinct from mere occurrence at a clock time requires conceptualized awareness.

We see then that the physical events of the indeterministic quantum world as such do not come into being anymore than those of the classical deterministic world but alike occur tenselessly. And my earlier contention that the transient now is mind-dependent and irrelevant to physical events as such therefore stands.

Proponents of indeterminism as a physical basis of objective becoming have charged that a deterministic world is timeless. Thus, Čapek writes:

...the future in the deterministic framework... becomes something *actually* existing, a sort of disguised and hidden present which remains hidden only from our limited knowledge, just as distant regions of space are hidden from our sight. "Future" is merely a label given by us to the unknown part of the *present* reality, which exists in the same degree as scenery hidden from our eyes. As this hidden portion of the present is contemporary with the portion accessible to us, the temporal relation between the present and the future is eliminated; the future loses its status of "futurity" because instead of succeeding the present it *coexists* with it.[43]

In the same vein, G. J. Whitrow declares:

> There is indeed a profound connection between the reality of time and the existence of an incalculable element in the universe. Strict causality would mean that the consequences pre-exist in the premisses. But, if the future history of the universe pre-exists logically in the present, why is it not already present? If, for the strict determinist, the future is merely "the hidden present," whence comes the illusion of temporal succession?[44]

But I submit that there is a clear and vast difference between the relation of one-to-one functional connection between two temporally-separated states, on the one hand, and the relation of temporal coexistence or simultaneity on the other. How, one must ask, does the fact that a future state is uniquely specified by a present state detract in the least from its being later and entail that it paradoxically exists at present? Is it not plain that Čapek trades on an ambiguous use of the terms 'actually existing' and 'coexists' to confuse the time sequential relation of being *determined* by the present with the simultaneity relation of contemporaneity with the present? In this way, he fallaciously saddles determinism with entailing that future events exist now just because they are determined by the state which exists now. When he tells us that according to determinism's view of the future, 'we are already dead without realizing it now',[45] he makes fallacious use of the correct premiss that according to determinism, the present state uniquely specifies at what later time any one of us shall be dead. For he refers to the determinedness of our subsequent deaths misleadingly as our 'already' being dead and hence concludes that determinism entails the absurdity that we are dead *now*! Without this ambiguous construal of the term 'already', no absurdity is deducible.

When Whitrow asks us why, given determinism, the future is not already present even though it 'pre-exists logically in the present', the reply is: precisely because existing at the present time is radically different in the relevant temporal respect from what he calls 'logical pre-existence in the present'. Whitrow ignores the fact that states hardly need to be simultaneous just because they are related by one-to-one functions. And he is able to claim that determinism entails the illusoriness of temporal succession (i.e., of the earlier-later relations) only because he uses the term 'hidden present' just as ambiguously as Čapek uses the term 'coexists'.

But, more fundamentally, we have learned from the theory of relativity that events sustain time-like separations to one another *because* of their *causal* connectibility or deterministic relatedness, *not* despite that deterministic relatedness. And nothing in the relativistic account of the temporal order depends on the existence of an indeterministic microphysical substratum! Indeed, in the absence of the causality assumed in the theory in the form of causal (signal) connectibility, it is altogether unclear how the system of relations between events would possess the kind of *structure* that we call the 'time' of physics.[46]

NOTES

* From 'The Meaning of Time', in E. Freeman and W. Sellars (eds.), *Basic Issues in the Philosophy of Time*, Open Court, La Salle, Ill., 1971, pp. 196–227.
[1] For example, consider the events in the careers of human beings or of animals who *return* to a spatially fixed terrestrial habitat every so often. These events occur at space points on the earth which certainly do *not* exhibit the betweenness of the points on a Euclidean straight line.
[2] For details cf. A. Grünbaum, *Philosophical Problems of Space and Time*, 2nd ed., *Boston Studies* XII, 1974, pp. 214–216. Hereafter this work will be cited as *PPST*.
[3] This non-committal character of the term 'initial state' seems to have been recognized by O. Costa de Beauregard in one part of his paper entitled 'Irreversibility Problems', in Proceedings of the 1964 International Congress for Logic, Methodology and Philosophy of Science, North-Holland Publishing Co., Amsterdam, 1965, p. 327. But when discussing my criticism of Reichenbach's account of irreversibility (*PPST*, pp. 261–263), Costa de Beauregard (*ibid.*, p. 331) overlooks that my criticism invokes initial states in only the non-committal sense set forth above.
[4] Thus, it is erroneous to maintain, as M. Čapek does (*The Philosophical Impact of Contemporary Physics*, D. Van Nostrand Co., Inc., Princeton, N.J., 1961, pp. 349, also 347 and 355) that the distinction between temporal betweenness and irreversibility is 'fallacious' by virtue of being "based on the superficial and deceptive analogy of 'the course of time' with a geometrical line" (*ibid.*, p. 349). If Čapek's condemnation of this distinction were correct, the following fundamental question of theoretical physics could not even be intelligibly and legitimately asked: Are the *prima facie* irreversible processes known to us indeed irreversible, and, if so, on the strength of what laws and/or boundary conditions are they so? For this question is predicated on the very distinction which Čapek rejects as 'fallacious'.

By the same token, Čapek errs (*ibid.*, p. 355) in saying that when Reichenbach characterizes entropically counter-directed epochs as 'succeeding each other', then irreversibility 'creeps in' along with the asymmetrical relations of before and after.
[5] On the basis of a highly equivocal of the term 'irreversible', M. Čapek (*The Philosophical Impact of Contemporary Physics*, pp. 166–167 and 344–345) has claimed incorrectly that the account of the space-time properties of world lines given by the special theory of relativity entails the irreversibility of physical processes represented by world lines. He writes: 'The world lines, which by definition are constituted by a succession of isotopic events, are *irreversible* in all systems of reference" (*ibid.*, p. 167) and "the rela-

tivistic universe is dynamically constituted by the network of causal lines *each of which is irreversible;...* this irreversibility is a topological invariant" (*ibid.*, pp. 344–345). But Čapek fails to distinguish between (1) the *non-inversion or invariance of time-order as between different Galilean frames*, which the Lorentz-transformation equations assert in the case of causally connectible events, and (2) the irreversibility of processes represented by world lines in the standard sense of the *non-restorability* of the same kind of state in any frame. Having applied the term 'irreversibility' to (1) no less than to (2) after failing to distinguish them, Čapek feels entitled to infer that the Lorentz transformations attribute irreversibility within any one frame to processes depicted by world lines, just because these transformations assert the invariance of time-order on the world lines as between different frames. That the Lorentz equations do not disallow the reversibility of physical processes becomes clear upon making each of the *two* replacements $t \to -t$ and $t' \to -t'$ in them: These replacements issue in the same set of equations except for the sign of the velocity term in each of the numerators; i.e., they merely reverse the direction of the motion. Therefore, these two replacements do *not* involve any violation of the theory's time-order invariance as between different frames S and S'. By contrast, different equations exhibiting a violation of time-order invariance on the world lines would be obtained by replacing *only* one of the two variables t and t' by its negative counterpart in the Lorentz equations.

[6] For a discussion of the various kinds of irreversible processes which make for the anisotropy of time and furnish specified criteria for the relations of temporal precedence and succession, see Costa de Beauregard, 'Irreversibility Problems', p. 327, and A. Grünbaum, *PPST*, Chapter 8.

[7] Some have questioned the possibility of stating what specific physical events do occur in point of fact at particular clock times without covert appeal to the transient now. In their view, any physical description will employ a time coordination, and any such coordination must ostensively invoke the now to designate at least one state as, say, the origin of the time coordinates. But I do not see a genuine difficulty here, for two reasons. First, it is not clear that the designation of the birth of Jesus, for example, as the origin of time coordinates, tacitly makes logically indispensable use of the now or of tenses. And second, in some cosmological models of the universe, an origin of time coordinates can clearly be designated non-ostensively: In the 'big bang' model, the big bang itself can be designated uniquely and *non*-ostensively as the state one having no temporal predecessor. For a defense of the view that the specification of dates involves essential logical use of indexical signs such as 'now', cf. R. Gale: 'Indexical Signs, Egocentric Particulars, and Token-Reflexive Words', *The Encyclopedia of Philosophy*, New York 1967. Gale's article also contains further references to some of the literature on this issue.

[8] The claim that the now advances in the direction of the future is a truism as regards both the correspondence between nows and *physically* later clock times and their correspondence with psychologically (introspectively) later contents of awareness. What is *not* a truism, however, is that the *introspectively* later nows are *temporally correlated* with states of our physical environment that are later as per criteria furnished by irreversible physical processes. This latter correlation depends for its obtaining on the laws governing the physical and neural processes necessary for the *mental* accumulation of memories and for the registry of information *in awareness*. (For an account of some of the relevant laws, see A. Grünbaum, *PPST*, Chapter 9, Sections A and B.) Having exhibited the aforementioned truisms as such and having noted the role of the empirical laws just mentioned, I believe I have answered Costa de Beauregard's complaint

(in 'Irreversibility Problem', p. 337) that "stressing that the arrows of entropy and information increase are parallel to each other is *not* proving that the flow of subjectivistic time has to follow the arrows!"

[9] A very illuminating account of the logical relations of Minkowski's language to tensed discourse is given by W. Sellars in 'Time and the World Order', *Minnesota Studies in the Philosophy of Science*, Vol. III (ed. by H. Feigl and G. Maxwell), University of Minnesota Press, Minneapolis, 1962, p. 571.

[10] Quoted from G. J. Whitrow, *The Natural Philosophy of Time*, Thomas Nelson and Sons Ltd., London, pp. 129–130.

[11] For a searching treatment of the ramifications of the contrast pertinent here, see W. Sellars, 'Philosophy and the Scientific Image of Man', in *Frontiers of Science and Philosophy*, (ed. by Robert G. Colodny), University of Pittsburgh Press, Pittsburgh, 1962, pp. 35–78.

[12] It will be noted that I speak here of the dependence of nowness on an organism M which is mind-possessing in the sense of having conceptualized or judgmental awareness, as contrasted with mere sentiency. Since biological organisms other than man (e.g., extra-terrestrial one) may be mind-possessing in this sense, it would be unwarrantedly restrictive to speak of the mind-dependence of nowness as its 'anthropocentricity'. Indeed, it might be that conceptualized awareness turns out not to require a *biochemical* substratum but can also inhere in a suitably complex 'hardware' computer. That a physical substratum of some kind is required would seem to be abundantly supported by the known dependence of the content and very existence of consciousness on the adequate functioning of the human body.

[13] The distinction pertinent here between *mere* hearing and judgmental awareness that it is being heard is well stated by R. Chisholm as follows:

"We may say of a man simply that he observes a cat on the roof. Or we may say of him that he observes *that* a cat is on the roof. In the second case, the verb 'observe' takes a 'that'-clause, a propositional clause as its grammatical object. We may distinguish, therefore, between a 'propositional' and a 'nonpropositional' use of the term 'observe', and we may make an analogous distinction for 'perceive', 'see', 'hear', and 'feel'.

If we take the verb 'observe' propositionally, saying of the man that he observes that a cat is on the roof, or that he observes a cat to be on the roof, then we may also say of him that he *knows* that a cat is on the roof; for in the propositional sense of 'observe', observation may be said to imply knowledge. But if we take the verb nonpropositionally, saying of the man only that he observes a cat which is on the roof, then what we say will not imply that he knows that there is a cat on the roof. For a man may be said to observe a cat, to see a cat, or to hear a cat, in the nonpropositional sense of these terms, without his knowing that a cat is what it is that he is observing, or seeing, or hearing. 'It wasn't until the following day that I found out that what I saw was only a cat'." (R. M. Chisholm, *Theory of Knowledge*, Prentice-Hall, Inc., (Englewood Cliffs, N. J.; 1966, p. 10) (I am indebted to Richard Gale for this reference.)

[14] The judgmental awareness which I claim to be essential to an event's qualifying as now may, of course, be expressed by a linguistic utterance, but it clearly need not be so expressed. I therefore consider an account of nowness which is *confined* to utterances as inadequate. Such an overly restrictive account is given in J. J. C. Smart's otherwise illuminating defense of the anthropocentricity of tense (*Philosophy and Scientific Realism*, Routledge and Kegan Paul Ltd, London, 1963, Chapter VII). But this undue restrictiveness is quite inessential to his thesis of the anthropocentricity of nowness. And the non-restrictive treatment which I am advocating in its stead would obviate his

having to rest his case on (1) *denying* that 'this utterance' can be analyzed as 'the utterance which is *now*', and (2) insisting that 'now' must be elucidated in terms of 'this utterance' (*ibid.*, pp. 139–140).

[15] H. Reichenbach, *The Direction of Time*, University of California Press, Berkeley, Calif., 1956, p. 16.

[16] P. Fraisse, *The Psychology of Time*, Eyre and Spottiswoode Ltd, London, 1964, p. 73.

[17] A. Grünbaum, *PPST*, p. 325.

[18] M. Čapek, *The Philosophical Impact of Contemporary Physics*, p. 337.

[19] On the basis of such a misunderstanding, M. Čapek incorrectly charges the thesis with a "spatialization of time" in which "successive moments already *coexist*" (The *Philosophical Impact of Contemporary Physics*, pp. 160–163) and in which "the universe with its whole history is conceived as a single huge and timeless bloc, given at once" (*ibid.*, p. 163). See also p. 355.

[20] G. J. Whitrow, *The Natural Philosophy of Time*, p. 228. For a criticism of another such misconstrual, see A. Grünbaum, *PPST*, pp. 327–328.

[21] G. J. Whitrow, *The Natural Philosophy of Time*, p. 88.

[22] *Ibid.*, pp. 227–228.

[23] J. J. C. Smart, *Philosophy and Scientific Realism*, p. 139.

[24] G. J. Whitrow, *The Natural Philosophy of Time*, p. 88, n. 2.

[25] G. J. Whitrow, *The Natural Philospphy of Time*, p. 293.

[26] *Ibid.*, p. 88, n. 2.

[27] B. Russell, 'On the Experience of Time', *The Monist* **25** (1915) 212.

[28] The need to deal with this question has been pointed out independently by Donald C. Williams and Richard Gale.

[29] Mental events, as distinct from the neurophysiological counterpart states which they require for their occurrence, are *not* heads in the way in which, say, a biochemical event in the cortex or medulla oblongata is.

[30] Thus a conscious state of elation induced in me by the receipt of good news from a telephone call C_1 could be *temporally between* the physical chain C_1 and another such chain C_2, consisting of my telephonic transmission of the good news to someone else.

[31] J. J. C. Smart, *Philosophy and Scientific Realism*, p. 135.

[32] I am indebted to Richard Gale for pointing out to me that since the term 'psychological' is usefully reserved for mind-dependent attributes which are private, as specified, it would be quite misleading to assert the mind-dependence of tense by saying that tense is 'psychological'. In order to allow for the required kind of intersubjectivity, I have therefore simply used the term 'mind-dependent'.

[33] Hans Reichenbach, *The Philosophy of Space and Time*, Dover Publications, Inc., New York, 1958, pp. 138–39.

[34] Hans Reichenbach, 'Les Fondements Logiques de la Mécanique des Quanta', *Annales de l'Institut Poincaré* **13** (1953), 154–57.

[35] *Ibid.*

[36] Cf. H. Bergmann, *Der Kampf um das Kausalgesetz in der jüngsten Physik*, Vieweg & Sohn, Braunschweig, 1929, 27–28 [English tr. in *Boston Studies* XIII].

[37] H. Bondi, 'Relativity and Indeterminacy', *Nature* **169** (1952), 660.

[38] I am indebted to Professor Wilfrid Sellars for having made clarifying remarks to me in 1956 which relate to this point. And Costa de Beauregard has reminded me of the pertinent French dictum *Ce qui sera, sera*. There is also the well-known (Italian) song *Che Sera, Sera*.

[39] Yakir Aharonov and David Bohm have noted that time does not appear in Schrö-

dinger's equation as an operator but only as a parameter and have pointed out the following: (1) The time of an energy state is a dynamical variable belonging to the measuring apparatus and therefore *commutes* with the energy of the observed system. (2) Hence the energy state and the time at which it exists do *not* reciprocally limit each other's well-defined status in the manner of the non-commuting conjugate quantities of the Heisenberg Uncertainty Relations. (3) Analysis of illustrations of energy measurement (e.g., by collision) which seemed to indicate the contrary shows that the experimental arrangements involved in these examples did not exhaust the measuring possibilities countenanced by the theory. Cf. their two papers on 'Time in the Quantum Theory' and 'The Uncertainty Relation for Time and Energy', *Physical Review* **122** (1961), 1649, and *Physical Review* **134** (1964), B1417. I am indebted to Professor A. Janis for this reference.

[40] A helpful account of the difference relevant here between being *determinate* (i.e., intrinsically attribute-specific) and being *determined* (in the relational sense of causally necessitated or informationally ascertained), is given by Donald C. Williams in *Principles of Empirical Realism*, Charles C. Thomas, 1966, Springfield, Ill, pp. 274ff.

[41] Čapek, *op. cit.*, p. 340.

[42] Čapek writes further: "As long as the ambiguity of the future is a mere appearance due to the limitation of our knowledge, the temporal character of the world remains necessarily illusory', and 'the principle of indeterminacy... means the *reinstatement of becoming in the physical world*" [*ibid.*, p. 334]. But granted that the indeterminacy of quantum theory is ontological rather than merely epistemological, this indeterminacy is nonetheless relational and hence unavailing as a basis for Čapek's conclusions.

[43] *Ibid.*, pp. 334–35, cf. also p. 164.

[44] G. J. Whitrow, *op. cit.*, p. 295.

[45] M. Čapek, *op. cit.*, p. 165.

[46] Accordingly, we must qualify the following statement by J. J. C. Smart, *op. cit.*, pp. 141–42: "We can now see also that the view of the world as a space-time manifold no more implies determinism than it does the fatalistic view that the future "is already laid up". It is compatible both with determinism and with indeterminism, i.e., both with the view that earlier time slices of the universe are determinately related by laws of nature to later time slices and with the view that they are not so related'. This statement needs to be qualified importantly, since it would not hold if 'indeterminism' here meant a macro-indeterminism such that macroscopic causal chains would not exist.

For a discussion of other facets of the issues here treated by Smart, see A. Grünbaum, 'Free Will and Laws of Human Behavior', *The American Philosophical Quarterly*, October 1971.

M. ČAPEK

THE INCLUSION OF BECOMING IN THE PHYSICAL WORLD*

On April 6, 1922, at the meeting of the French Philosophical Society, Meyerson, one of the outstanding philosophers of science at that time, asked Einstein a point-blank question: Is spatialization of time, i.e., the tendency to regard time, 'the fourth dimension', as not being essentially different from the spatial dimension, a legitimate interpretation of Minkowski's fusion of space and time? Meyerson was obviously prompted to ask this question by the fact that the above-mentioned static interpretation of space-time was – and, as we shall see, still is – present not only in numerous popular or semi-popular presentations of the relativity theory, but in serious scientific and philosophical treatises as well. For our purposes it will suffice to give only two illustrations from the period prior to 1922; some more recent examples will be given later.

With Minkowski space and time became particular aspects of a single four-dimensional concept; the distinction between them as separate modes of correlating and ordering phenomena is lost, and the motion of a point in time is represented as a *stationary curve in four-dimensional space*. Now if all motional phenomena are looked at from this point of view they become *timeless phenomena* in *four-dimensional space*. The whole history of a physical system is laid out as a *changeless whole*.[1]

The second illustration is perhaps even more characteristic:

There is thus far an intrinsic similarity, a kind of coordinateness, between space and time, or as the Time Traveller, in a wonderful anticipation of Mr Wells, puts it: "There is no difference between Time and Space except that our consciousness moves along it."[2]

And in the footnote on the same page the author added:

It is interesting that even the terms used by Minkowski to express these ideas as "Three-dimensional geometry becoming a chapter of the four-dimensional physics," are anticipated in Mr Wells's fantastic novel. Here is another example (*Time-Machine*, Tauchnitz, ed., p. 14) illustrative of what is now called a world-tube: "For instance, here is a portrait (or, say a statue) of a man at eight years old, another at fifteen, another at seventeen, another at twenty-one and so on. All these are evidently sections, as it were, three-dimensional representations of his Four-Dimensional Being which is a fixed and unalterable thing." Thus Mr Wells seems to perceive clearly the absoluteness, as it were, of the world tube and the relativity of its various sections.

This is explicit enough. All the ingredients of the static interpretation of space-time, as it still lingers in the minds of some physicists and philosophers, are contained in these two quotations.

In asking Einstein, Meyerson did not hide his own negative attitude toward the spatializing interpretation of relativity. His argument consisted of five logically related parts. The first three are merely three different aspects of one argument. He pointed out first that the privileged character of the temporal dimension is preserved in Einstein's cylindrical model of the universe in which, unlike the spatial dimensions, time is uncurved and unidirectional. He then recalled Einstein's own words that "we cannot send wire messages into the past." Thirdly, he pointed out that the law of entropy which 'guarantees' the irreversibility of time remained intact within the relativistic framework. Meyerson's fourth argument was very probably inspired by Bergson: the spatialization of time in the relativity theory is, according to him, merely the last manifestation of the perennial tendency to treat time in a space-like fashion. The fifth argument should probably have been put in the first place since, as we shall see, it underlies the rejection of the backward-moving causal actions. Meyerson recalled Weyl's proposal to speak of three-plus-one dimensions of space-time rather than of four dimensions, since in Minkowski's formula for the spatiotemporal interval the time variable is preceded by an algebraic sign different from that of the three spatial variables. Thus the heterogeneity of space and time is reflected even in the mathematical symbolism of the relativity theory.[3]

What is surprising in Einstein's answer to Meyerson is not so much his apparently complete agreement with him – Meyerson, after all, quoted Einstein himself to support his criticism of the spatialization of time – but the very briefness of his reply: "It is certain that in the four-dimensional continuum all dimensions are not equivalent."[4]

Today, in the light of Einstein's later utterances, it is clear that the very briefness of his reply was due to the fact that he was not especially interested in this question. Only thus can we explain the vacillations and ambiguities in his attitude toward this particular problem. These ambiguities were documented by Meyerson himself when he discussed the same problem in a more systematic way in his book *La Déduction relativiste* in 1928.[5] While he recalled again Einstein's statement that "we cannot send wire messages into the past," he also quoted another of Einstein's utter-

INCLUSION OF BECOMING IN THE PHYSICAL WORLD   503

ances according to which "the becoming in the three-dimensional space is somehow converted into a being in the world of four dimensions." One can hardly have a more radical formulation of static interpretation! Yet, Einstein's response to Meyerson's book was enthusiastically positive. He not only praised it as "one of the most remarkable books written about the relativity theory from the standpoint of epistemology," but he also explicitly agreed with its central thesis, that is, with his rejection of the spatializing interpretation of the world of Minkowski.[6]

This, however, was not Einstein's last word on this subject. In 1949 Gödel wrote a vigorous defense of the static interpretation of space-time.[7] According to him, the relativization of simultaneity destroys the objectivity of the time lapse and thus substantiates "the view of those philosophers who, like Parmenides, Kant and modern idealists consider change as an illusion or an appearance due to our special mode of perception." As an additional argument against the objectivity of time lapse Gödel adduced the mathematical possibility of certain cosmological models in which it would be possible "to travel into any region of the past, present, and future, and back again, exactly as it is possible in other worlds to travel to distant parts of space." Such a trip could be described by a world line similar to the F-$(H$-$N)$-G-F line of the diagram in this essay (Figure 1). Gödel even makes an estimate of the quantity of fuel and the velocity of a rocket ship needed to make such a fantastic trip. Such a rocket would be a realization of the time-machine of H.G. Wells and Gödel would subscribe to Silberstein's view that the famous British fiction writer anticipated relativistic physics.

We would expect that Einstein, who two decades before endorsed Meyerson's criticism of the static interpretation without reservations, would have rejected unequivocally such an extreme form of spatialization of time. But the very opposite happened: Einstein's comment on Gödel's essay was distinctly, though cautiously, sympathetic.[8] Did Einstein then forget his previous view that "we cannot send wire messages into he past"? Is not Gödel's hypothetical rocket merely an oversize form of signal traveling into the past? Such doubts are hardly fair if we read Einstein's comment carefully. Einstein indeed modified his view in the following way: it is impossible to send wire messages to the past on the macroscopic scale; but this is not necessarily true for microscopic phenomena which seemed to be reversible. Not only this; if we concede with Gödel the possibility of the closed world lines on a huge megacosmic scale, says Einstein, then the relation of succession itself becomes relativized; for on a circular world line it is a matter of convention to say that A precedes B rather than vice versa. In other words, Einstein as late as 1949 considered the possibility

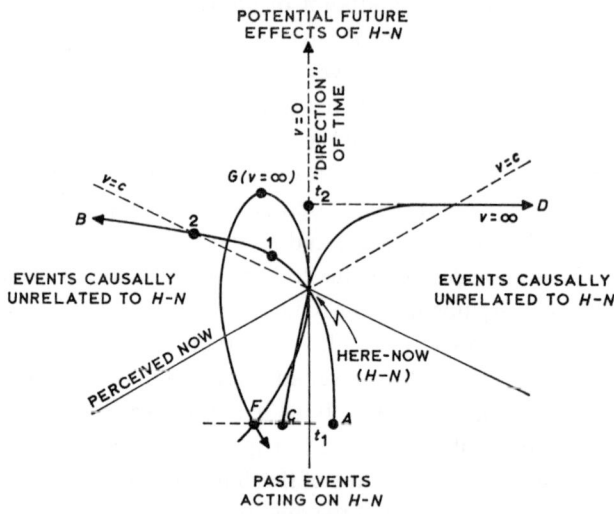

Fig. 1. *Three impossible world lines*. The diagram above, which is an elaboration of Figure 1 in Costa de Beauregard's essay, depicts three world lines, that is, four-dimensional orbits, whose existence is excluded by the limiting character of the velocity of light. They represent bodies moving in H-N ('here-now') with the admissible velocity $v < c$; but all of them would acquire eventually a velocity $v > c$. A body moving along the world line A-$(H$-$N)$-B would acquire it beyond the point event 1; at the point 2 it would overtake the photon emitted from H-N and would enter the 'elsewhere' region (see Costa de Beauregard, Figure 2). This would mean that an observer in the elsewhere region, contemporary in the relativistic sense with H-N, would perceive the signal from an event future with respect to H-N. The world line C-$(H$-$N)$-D is the world line of a body, or of a signal, moving eventually with infinite velocity. It would be equivalent to the realization of the Newtonian instantaneous space at time t_2. The third trajectory could be called a 'Gödel line', after the distinguished mathematician Gödel, who adduced the possibility of certain cosmological models in which such travel may be formally permissible. Moving along this line would require that a body reach infinite velocity at point G, would turn backwards and, after crossing the region of events causally unrelated to H-N, that is the 'elsewhere' region, it would enter the region of 'absolute past'. Absolute, that is, with respect to H-N and all events causally subsequent to it. Leaving point event G, the body would eventually cross itself at time t_1. The point of intersection, F, would represent an event both successive to and simultaneous with itself. In causal terms this would be an event affected by its own future effects.

that the irreversible time is confined to what Reichenbach called "the world of the middle dimensions" while it may be absent both on the cosmic scale and on the microphysical level. It is true that he added cautiously: "It will be interesting to weigh whether these (i.e., cosmological solutions) are not to be excluded on physical grounds."[9] Despite this reservation it is clear that Einstein was closer to the spatializing interpretation in 1949 than in 1928.

Einstein was not alone in his vacillations on this point. Weyl, Jeans, Reichenbach and others shifted their views on this subject, sometimes even within one and the same book. A more consistently negative attitude toward the static interpretation was shown by Langevin and, contrary to what Meyerson claimed, by Eddington;[10] and among philosophers by Bergson and Whitehead. It is true that the attitude of the latter two was mostly inspired by their general philosophical outlook, even though the effort to grasp the concrete physical meaning of the relativistic formulae was not lacking in either of them. Despite all criticisms the spatializing interpretation still lingers, though more in the minds of philosophers than in those of physicists. Besides the relatively recent essay by Gödel (1949), there was Williams' article with the challenging title "The Myth of Passage" (1951). Even more recently, Quine claimed that the discovery of the principle of relativity "leaves no reasonable alternative to treating time as space-like."[11] Among contemporary philosophers of science two most vigorous defenders of the becomingless view of space-time are Costa de Beauregard and Grünbaum. The former speaks of matter as "displayed statically in space-time," (*"statiquement deployée dans l'espace-temps"*), while the latter says explicitly that "coming into being is only coming into awareness."[12] Thus the opinion is still divided – sometimes divided within one and the same mind. This shows clearly how complex and difficult the problem of correct interpretation of the relativistic fusion of space and time still is.

1. THE ALLEGED ARGUMENT FOR THE STATIC
INTERPRETATION OF SPACE-TIME

The crucial issue which we face is as follows: are there any cogent reasons for the static interpretation of space-time or is the very opposite true? In other words, does an attentive analysis of the conceptual structure of the relativity theory support the becomingless view or does it suggest the very opposite?

The most frequent and superficially most plausible argument in favor of the becomingless view is based on the claim that the relativization of simultaneity definitely destroys the objectivity of temporal order. A pair of events appearing simultaneous in one frame of reference is no longer simultaneous in other inertial systems. Even worse, some events succeeding each other in one system may appear in a reversed order in another appropriately chosen system. Since there is no privileged frame of reference which would impart a mark of objectivity on any of these systems, what objective status can succession and becoming still retain? This is a standard argument and thus it is hardly surprising that we can find it in Gödel's essay to which we referred above:

> The argument runs as follows: Change becomes possible only through the lapse of time. The existence of an objective lapse of time, however, means (or at least, is equivalent to the fact) that reality consists of an infinity of layers of 'now' which come into existence successively. But, if simultancity is something relative in the sense just explained, reality cannot be split into such layers in an objectively determined way. Each observer has his own set of 'nows', and none of these various systems of layers can claim the prerogative of representing the objective lapse of time.[13]

Similarly Costa de Beauregard:

> In Newtonian kinematics the separation between past and future was objective, in the sense that it was determined by a single instant of universal time, the present. This is no longer true in relativistic kinematics: the separation of space-time at each point of space and instant of time is not a dichotomy but a trichotomy (past, future, elsewhere). Therefore there can no longer be any objective and essential (that is, not arbitrary) division of space-time between 'events which have already occurred' and 'events which have not yet occurred...'. This is why first Minkowski, then Einstein, Weyl, Fantappiè, Feynman, and many others have imagined space-time and its material contents as spread out in four dimensions. For those authors, of whom I am one, who take seriously the requirement of covariance, relativity is a theory in which everything is 'written' and where change is only relative to the perceptual mode of living beings.[14]

We have to consider carefully what is correct and what is questionable in these passages. Gödel and Costa de Beauregard correctly pointed out that while the classical Newtonian space-time possessed a stratified structure in the sense that it was regarded as a continuous succession of three-dimensional strata, each of which represented a particular cosmic 'now' or 'present', the relativistic space-time does not yield to such stratification. No common series of such cosmic 'nows' exist for different observers; the observers in different inertial frames split the four-dimensional continuum along different instantaneous 'cleavage planes.' Each such cleavage plane is a substratum of the events simultaneous for the corresponding observer, but – unlike in the Newtonian space-time – none of them possesses a privileged, objective character. This is the meaning of the relativity of simultaneity. But, contrary to what Gödel and Costa de Beauregard believe, from the relativization of simultaneity it does not follow that the lapse of time and change lose their objective status. Gödel's conclusion would have been correct if lapse of time or duration were completely synonymous with the classical even-flowing Newtonian time consisting of the succession of the world-wide instants. This had been accepted tacitly through the whole classical period in the same way that space and Euclidean space were regarded as synonymous. The fact that some critics of relativity in defending the objective status of universal time really defended the time of Newton merely added to this confusion.

What Gödel and modern neo-Eleatics do not consider at all is the possibility that the Newtonian time may be only a special case of the far broader concept of time or temporality in general in the same sense that the Euclidean space is a specific instance of space or spatiality in general. If we admit this possibility, then the negation of the Newtonian time entails an elimination of temporality and change in general as little as the giving up of the Euclidean geometry destroys the possibility of any geometry. Similarly, the present revision of classical determinism means merely a widening, not an abandoning of causation in general; despite the fears of some conservative philosophers, the probabilistic universe is not an irrational chaos, even though its rationality is of far broader kind than the restricted form of rationality characterizing the Newton-Laplacean determinism.[15]

2. Consequences of the Constancy of the World Interval

Let us now consider in detail the argument that the relativization of simultaneity implies without qualification a relativization of succession and thus destroys forever the objective status of 'lapse of time'. Its plausibility is undeniable: for if there is no objective 'now' unambiguously separating the past from the future, what objective status can succession still claim? In other words, if succession itself is relative, depending on the choice of our frame of reference, it cannot constitute an objective feature of reality. The last conclusion follows unquestionably from its premise; unfortunately (or rather, fortunately!) the premise itself is not correct. For it is simply not true that simultaneity and, in particular, succession of events are purely and without qualification relative. In making such claim we would be guilty of completely disregarding certain mathematical implications of Minkowski's formula for the constancy of the world interval. This formula follows from the Lorentz transformation and it shows in a condensed way the differences between classical and relativistic mechanics. In the former the spatial distance and the temporal interval separating two events E_1 and E_2 are *separately* invariant for each inertial frame $s = $ const, $t_2 - t_1 = $ const (where $s = \sqrt{(x_2 - x_1)^2 + (y_2 - y_1)^2 + (z_2 - z_1)^2}$, with x_1, y_1, z_1, t_1, x_2, y_2, z_2, t_2 being the spatial and temporal coordinates of E_1 and E_2 respectively). In Minkowski's space-time the constancy does not belong to the spatial distance and the temporal interval separately, but only to the quantity called 'world interval', which is defined in the following way:

$$I = s^2 - c^2(t_2 - t_1)^2 = \text{const} \ (c = 3 \times 10^{10} \text{ cm/sec}).$$

We can then distinguish three distinct groups of relations between two events according to whether the world interval is positive, zero or negative: $I > 0$, $I = 0$, $I < 0$. Each group should be considered separately.

(a) When $I > 0$, then $s^2 > c^2(t_2 - t_1)^2$; in other words, the spatial separation between the events E_1 and E_2 is greater than their separation in time multiplied by the velocity of the fastest causal action, i.e., the velocity of electromagnetic radiation. This means that no causal interaction can take place between such events; they are not only causally unconnected, but even unconnectible, that is, intrinsically mutually independent.[16] Since the interval should retain its positive sign in *all* inertial frames of reference, and since this sign remains unaffected when $t_1 = t_2$ or when the temporal interval $t_2 - t_1$ changes its sign, we can see the possibility that the events E_1, E_2, succeeding each other within one group of systems, will appear simultaneous in another group, and will appear in *a reversed order* in still other systems. In other words, the simultaneity and succession of causally unrelated events is fully and without qualification relative. But this statement is restricted to the specific case just considered.

(b) $I = 0$, or $s^2 = c^2(t_2 - t_1)^2$. Since the spatial distance is equal to the separation in time multiplied by the velocity c, it is clear that this is the case of a photon or more generally of any quantum of radiation, in two successive 'positions'. It is obvious that in this case the interval $t_2 - t_1$ can never become zero unless the spatial distance itself would vanish at the same time; but in that case the events E_1, E_2 would merge. In other words, each photon at every instant is simultaneous with itself. This statement can be generalized: every world point – or rather world event – is simultaneous with itself in every frame of reference. (As we shall see, this is not as trivial as it sounds.) But as long as the spatial distance does not vanish, the corresponding time interval does not vanish either. In other words, two events of this kind, successive in one frame of reference, must never appear simultaneous in any other system; a fortiori, they can never appear in a reversed order. This is only natural; for two successive positions of a photon or, to use the undulatory language, two successive states of the vibratory electromagnetic disturbance, are simple instances of causally related events; the reversion of their temporal order would be equivalent to the reversion of their causal order. This would mean that what appears as a cause of a certain event would appear as an effect of the same event in another system!

Such a case was possible in classical physics; when Flammarion imagined an observer moving away from the earth with a velocity greater than that of light and seeing the earthly history reversed so that "Waterloo would precede Austerlitz,"[17] it was 'science fiction' which, nevertheless, was compatible with the principles of Newtonian physics. Moreover, it did not contradict the unidirectional character of causal relations because the reversion mentioned above was only apparent. For physicists of the last century believed with Newton that it was possible, at least in principle, to distinguish the real temporal (and causal) order from the merely spurious or apparent one. The distortions of temporal and causal perspective produced by some relative motions, e.g., by the motion of Flammarion's observer with respect to light, disappear in the only true perspective of the privileged frame of reference – absolute motionless space. But in the relativity theory the situation would be far more serious: because of the absence of any privileged frame of reference there is no way – in the special relativity at least – to differentiate between 'apparent' and 'real' order, and thus a reversion of causal order due to an appropriate change of the system would result in most serious discrepancies and causal anomalies. Fortunately, Flammarion's fantasy is excluded by the very principle of constant velocity of light according to which no material body can attain the velocity equal to that of electromagnetic radiation. For this reason, the succession of two states of the electromagnetic disturbance in the void can never degenerate into an apparent simultaneity in any other system; nor can it ever appear in reversed order. As the consideration of the third case will show, this is true generally of any couple of causally related events.

(c) When $I<0$, then $s^2<c^2(t_2-t_1)^2$; the spatial distance is then smaller than the product of the separation in time and the velocity of light (and gravitation). This is the case of two events whose causal links are propagated with the velocity smaller than c; in other words, the connecting causal links are not the world lines of photons, but those of 'material particles'. Since the interval I must retain its negative sign in all frames of reference, the temporal interval of the events E_1, E_2 cannot vanish in any system: otherwise for $t_2=t_1$, $s^2<0$, i.e., their spatial distance would become imaginary. In other words, the succession of the causally related events can never degenerate into simultaneity in any other system; a fortiori, it can never be reversed. This is clearly a generalization of the result obtained in (b) and it can be summarized as follows: the succession of causally related events, whether they are joined by the world lines of photons or by those of material particles, is a topological invariant independent of our choice of system of reference. Or more concretely: *the world lines of any kind are irreversible.* Although this important conclusion was pointed out explicitly by Langevin as early as in 1911,[18] it was frequently overlooked not only by the authors of popular or semi-popular expositions of relativity, but sometimes also by serious thinkers.[19] This was undoubtedly due to the fact that the absence of the metrical invariance of temporal intervals obscured the aforesaid topological invariants and that the case (a) was not distinguished from the cases (b) and (c).

Before formulating our general conclusion, let us consider the conditions under which the spatial distance can vanish in the three aforementioned instances $I>0$, $I=0$, $I<0$. We already saw that in the case (b) the spatial separation can disappear only if the corresponding time interval vanishes: $s=0$, when and only when $t_2-t_1=0$. This is the case of two events merging into one; every photon coincides spatiotemporally with itself, and, more generally, every event coincides spatiotemporally with itself and no change of the frame of reference could produce its dislocation into two different events. (We shall see the full significance of this apparent truism later.) In the case (a) the possibility of the spatial separation becoming zero in any system is excluded since $s^2=0$ would imply $c^2(t_2-t_1)^2<0$, i.e., the time interval separating the events E_1 and E_2 would become imaginary. On the other hand, the spatial distance can vanish by an appropriate choice of the system in the case (c); the condition $s=0$ determines the frame of reference in which a particle appears motionless, and E_1 and E_2 are two successive events 'at the same place'. But 'the sameness of place' is completely relativized and has lost its original Newtonian connotation of a motionless part of the motionless space. Every moving particle can by an appropriate change of the frame of reference be converted into a motionless one, and no frame of reference has a privileged character: this is the meaning of the relativistic denial of the motionless Newtonian space. The only exceptions are quanta of radiation; they cannot be made motionless by any change of the standpoint; they are essentially and under all conditions in motion in all the systems. This is implied by the principle of constant velocity of light, and, more specifically, by the relativistic formula for the addition of velocities.

3. THE DYNAMIC CHARACTER OF TIME-SPACE

Our conclusions are then as follows:

(a) The succession of causally related events is preserved in *all* frames of reference. In other words, the irreversibility of the world lines, which are constituted by causal successions of events, is a *topological invariant*.

(b) The succession of causally unrelated events is completely relativized.

(c) Equally fully relativized is the simultaneity of all events with an apparently trivial exception of the simultaneity of each event with itself. The last part of this statement can be expressed in the following way: *absolute coincidences*, that is coincidences both in space and time, are as much topologically invariant as the temporal order of causally related events.

The propositions (b) and (c) are not logically independent. The relativization of the succession of causally unrelated events and the relativity of simultaneity of distant events are two related consequences of the fact that *the temporal order of all causally unrelated events remains undetermined*. A pair of such events can appear in certain temporal order in some systems, in a reversed order in other systems, and finally simultaneous in the third category of systems which constitute, so to speak, a boundary case separating the first two groups of frames of reference. No events can be judged simultaneous unless they are causally unrelated. Only if there were instantaneous causal connections in nature would the simultaneity of causally (in this case instantaneously) related events be possible.

This, indeed, was the case of Newtonian mechanics, and it is certainly not accidental that Galileo's transformation is obtained when we substitute an infinite value for the velocity c in the Lorentz transformation. Infinite velocity means instantaneous interaction. It is true that classical physics knew since Roemer's discovery in 1675 the finite velocity of light which in the nineteenth century was found equal to the velocity of electromagnetic waves; but it remained completely unaware of the limiting character of this velocity. No upper limit was imposed on the range of possible velocities, that is, on the speed of causal interactions. Thus for a considerable time the velocity of gravitation was believed to be infinite. Laplace still believed that it was at least 50 000 000 times larger than that of light.[20]

In truth, the assumed existence of the Newtonian space, spread instantaneously and orthogonally with respect to the 'axis of time', was an embodiment of instantaneous connections; every geometrical distance in such space can in virtue of its instantaneous character be regarded as a world line of a point moving with instantaneous velocity. This network of instantaneous geometrical relations, constituting 'space at an instant', was at the same time an objective substrate of absolutely simultaneous events. When we say, for instance: "Sirius is eight light years from the earth," it has in classical physics the following meaning: (1) that there is an instantaneous space at this particular moment in which the events both on the earth and on Sirius are located; (2) that because of its finite velocity the luminous message which I perceive now left Sirius eight years ago. It is clear that the difference between 'now' and 'seen now' was fully recognized by classical physics; but although the objective 'now' was by definition unperceivable, it was in principle inferable and calculable on the basis of the classical theorem for the addition of velocities which was applied to the relative motion of the luminous source and the observer.

The belief in the distinction between 'now' and 'seen now' was due to the fact that classical physics – unlike the general theory of relativity today – accepted the distinction between static geometrical space and its changing physical content. 'now', that is, absolute simultaneity, belonged to the former; 'seen now', that is, the perceived, spurious simultaneity, belonged to the latter. It was this distinction which inspired the search for an absolute frame of reference which would be the substrate of the objective simultaneity. It is sufficiently known how this search, carried on by the experiments of Michelson, Morley, Trouton, Noble, Tomaschek, and Chase, ended in the failure which inspired the most comprehensive and revolutionary revision of the traditional concepts of space and time. The profound and far-reaching meaning of this revision is still not always fully understood now, more than a half century after the formulation of the special theory of relativity.

Thus we read frequently, and not only in semi-popular treatises, that the simultaneity of distant events, absolute for Newton, "was made relative by Einstein." To use such a language is highly misleading. It suggests almost inevitably that behind the inherent relativity of the human frames of reference there lies hidden the true absolute simultaneity, the absolute 'now', even if it may remain forever inaccessible to our knowledge. It is far more accurate to say that the simultaneity of distant events was *eliminated* instead of being merely relativized. What objective status could possibly exist for an entity which is unobservable by definition, and which is an inferential construct different in different frames of reference, none of which possesses a privileged character? It is thus not sufficient to join the adjective 'relative' to the noun 'simultaneity'; the noun itself should be dropped because of its lurking ontological connotation. Einstein himself did not hesitate to do it:

There is no such thing as simultaneity of distant events; consequently there is also no such thing as immediate action at a distance in the sense of Newtonian mechanics.[21]

This correlation between simultaneity of distant events and the network of instantaneous connections can be expressed in a far more explicit way. The class of objectively simultaneous events constitutes the space of classical physics at a certain instant. Conversely, any instantaneous three-dimensional cut across the four-dimensional world process contains the events objectively simultaneous at that instant. Thus the simultaneity of distant events implies their juxtaposition and vice versa. This is what Newton had in mind when he claimed that "every indivisible moment of duration is everywhere."[22] The cosmic 'now', in virtue of its universality, is instantaneously spread everywhere; this is the meaning of the classical correlation of absolute simultaneity and absolute space.

But such instantaneous three-dimensional cuts, admissible in the physics of Newton and Laplace, are excluded by the physics of relativity. Contrary to Newton's belief, there is no moment of time which is present everywhere. This lack of correlation between 'now' and 'everywhere' was expressed by various thinkers in different ways. In Eddington's words, there are no "world-wide instants";[23] according to Whitehead, there is not such a thing as "nature at an instant";[24] or, as Robb said, "there is no identity of instants at different places at all"; in other words, "the present instant, properly speaking, does not extend beyond here."[25] *Since there is no absolute space correlated with each instant of time, there is no absolute juxtaposition which would serve as a substratum of absolutely simultaneous events.* But while there is no juxtaposition of events which would be a juxtaposition for all frames of reference, *there are certain types of succession which remain such in all systems.* As we have seen, these types of succession are represented by causal chains, that is, by the world lines of material and luminous 'particles'. Unlike spatial juxtaposition, the irreversibility of the world lines has an *absolute* significance, independent of the conventional choice of the system of reference. We can hardly have a more convincing illustration of the dynamic character of space-time.

We may anticipate the following objection: what about the relativization of the succession of the causally unrelated events? Is it not as fatal to the ontological status of time as the relativization of juxtaposition is to the ontological status of space? Not speaking of the fact that the succession of causally related events still remains invariant, we must not forget that the relativization of the simultaneity of distant events and the relativization of the succession of causally unrelated events entail each other (see above). If we substitute with Einstein the term 'elimination' for that of 'relativization', it becomes clear that the succession of causally unrelated events is as much devoid of concrete physical meaning as the simultaneity of remote events. Nothing in nature corresponds to either of them. To continue to refer to them as something 'real, though relative' betrays the pre-relativistic modes of thought. Such expressions result from an incongruous overlapping of two incompatible languages, the Newtonian and relativistic; it is the resistance of our Newtonian subconscious which prevents us from saying boldly and consistently that simultaneity of distant events as well as the succession of causally independent events simply does not exist.

For there are only two types of relations in the relativistic universe: that of successive causal connections and that of contemporary causal independence. Since the universe consists of the dynamical network of the irreversible causal lines, their irreversibility which remains absolute in the relativity theory is conferred to the universe as a whole. Needless to say, it is not the irreversibility of the Newtonian time. The world process according to Newton consisted of the irreversible series of the world-wide instants, that is, of the 'now-everywhere' planes; and we have seen that no such cleavage planes are admissible in the relativistic universe. We have seen that a three-dimensional space, at any moment, is an arbitrary instantaneous cut in the four-dimensional process and that such artificial cuts were superseded by the four-dimensional regions of causal independence ('elsehere' of Eddington, 'co-presence' of Whitehead) which separates the front cone of causal future from the rear cone of the causal past. But this does or at least should make clear two important points. First, the impossibility of three-dimensional instantaneous cuts radically transforms the classical concept of space; space now is incorporated into the four-dimensional world process in which the classical space of Newton is a mere artificial instantaneous cross-section. Second, the fact that the past and the future are now more effectively separated than in classical physics certainly does not weaken the objective status of succession. Thus, all these evidences point to one important conclusion: the relativistic union of space with time is far more appropriately characterized as a *dynamization of space* rather than a spatialization of time.

4. IMPOSSIBILITY OF THE BACKWARD FLOWING TIME AND OF THE SELF-INTERSECTING CAUSAL LINES

The impossibility of the backward-flowing time follows directly from the irreversibility of the causal lines and is embodied graphically in the relativistic time-space diagram. The world lines emanating from any 'here-now' event must be contained in the causal front cone of 'absolute future'. They can never radiate into the region of 'elsewhere', which is forbidden to them by virtue of the limiting character of the velocity of light. It means that their angle with the local time axis can never be 90°; this would imply the existence of infinite velocities and the resulting flattening of the frontward causal cone. With the concomitant flattening of the rearward causal cone the region of 'elsewhere' would be squeezed out of existence. This would be a return to classical physics which indeed admitted the existence of the world lines orthogonal to the time axis; in truth, every distance in the Newtonian space belonged to this category. The whole of instantaneous classical space may be regarded as an infinitely dense network of such orthogonal world lines. The backward-running local time would require that the corresponding world line would be bent by an angle greater than 90°; in other words, the corresponding causal front cone would be turned backwards like an upturned umbrella. Such a case was impossible even in classical physics; in the physics of relativity, for which the past and the future are even more effectively separated by the region of 'elsewhere', this is, so to speak, doubly impossible. No world line starting from 'here-now' can ever reach the region of 'elsewhere'; a fortiori none can be bent backwards to reach the rearward causal cone of the past. In the three-dimensional time-space diagram by which we symbolize the relations in the four-dimensional world process, the space angle of the frontward cone can never attain the value of 2π; a fortiori, it can never surpass it. This definitely excludes all Wellsian fantasies about the travelers visiting their own past or the past of their own ancestors.

The impossibility of the backward-bent world lines clearly entails the impossibility of any line recrossing its past course. Such a case would be in conflict with another essential idea of relativity: the absolute character of spatiotemporal coincidences. We have seen that every world event coincides spatiotemporally with itself; consequently, it is simultaneous with itself. This is not as silly a truism as it sounds, especially when we formulate it in the following way: each event is simultaneous with itself and only with itself; or, in Robb's formulation, "an instant cannot be in two places at once."[26] This evidently was not accepted by the physics of Newton, according to which every instant of time was present through the whole of space; each 'now' was everywhere. Nor is it accepted by those who, like Gödel, accept the possibility of self-intersecting world lines. In the latter case there would be some events which, besides being simultaneous with themselves, would also be simultaneous with other instants in time! In other words, a certain event, corresponding to a single

INCLUSION OF BECOMING IN THE PHYSICAL WORLD 517

point in which the corresponding world line recrosses itself, would be simultaneous with a remote future instant. In such a case we would be clearly on the brink of magic; indeed, some serious thinkers are tempted to interpret the alleged fact of precognition by a similar retroactive action of the future on the past! There is no place here to dwell on all the causal anomalies which would result and on the intrinsic discrepancies of the language in which similar situations are described; suffice it to say that the retroactive action of future events remains impossible as long as we adhere to the requirement of the relativity theory that no causal action can escape from the frontward causal cone of 'absolute future'.

One recent argument for the idea of backward-flowing time is its alleged usefulness in removing certain kinds of causal anomalies in the so-called creation and annihilation of microphysical 'particles'. Instead of saying that a pair of electrons was created, one positive and one negative – the positive one a moment later being converted into radiation – we can interpret this process as the world line of a single electron moving in a zigzag way through space-time. Reichenbach, who mentions this 'equivalent description' not without sympathy, is fully aware of its limited usefulness:

The anomalies of creating from nothing and vanishing into nothing are thus eliminated; however, in exchange for them another causal anomaly enters the description; the electron travels part of its path backward in time.[27]

To this we must add the following footnote. First, the alleged anomaly, which is supposed to be removed, is much less 'anomalous' than Reichenback believed. It is not true that the process described above involves 'creation from nothing' and 'vanishing into nothing'; the pair of particles arises from electromagnetic radiation, into which it can be reconverted, instead of coming miraculously from or vanishing into 'pure nothing'. Thus the whole process appears causally anomalous only to our Democritian logic of solid and permanent bodies which the whole trend of contemporary physics tends to discredit. Second, the whole process is ruled by Einstein's famous equation about the equivalence of mass and energy; the energy of the vanished quantum was converted into the rest mass of both particles and into their kinetic energy. This is only one example of many impressive confirmations of the relativistic dynamics on the microphysical scale. But then in the light of the very close connection between the dynamics and the kinematics of relativity, should we not expect that the relativistic space-time diagram is applicable to the microphysical scale as well; in other words, that the electrons cannot travel backwards in time any more than the macroscopic signals can?

The fallacious belief in the reversibility of time is another example of the unfortunate influence of false spatial analogies: if we can travel over the same road in an opposite direction, why could not we travel backward over the same 'path of time'? But, in truth, as Whitehead and even Russell stressed,[28] we never travel over the same road again. When I travel back to South Station in Boston in the afternoon, it is not the 'same road' on which I traveled in an opposite direction in the morning; the road itself is a number of hours older! An attentive analysis would show that the idea of unidirectional time-re-emerges in the very formulation which purports to deny it; for 'reversal of time' is supposed to take place *after* time flowed in its normal forward direction! Now the word 'after' means, if it means anything at all, 'continuing in the original forward direction of time'. In other words, to say that time at a certain one of its moments changes its direction is equivalent to the self-contradictory assertion that 'time flows backward while continuing to flow forward.'

5. The Status of 'Now' and the Potentiality of the Future

Insisting correctly on the relativization of simultaneity, our modern neo-Eleatics jump to the conclusion that 'now' does not have any objective status at all, being nothing but "a temporal mode of experiencing ego".[29] In such a view, becoming still exists on the subjective, psychological level; but the physical world outside of the stream of consciousness simply is, it does not become.[30] In other words, "coming into being is only coming into present awareness."[31] The apparent queerness of this view should not prevent us from analyzing it. Not speaking of its enormous epistemological difficulties (about which later), its main defect is that it not only does not follow from the relativity theory, but is even incompatible with it.

My physical 'here-now', which corresponds roughly to my psychological awareness of the present, precedes all events of my causal future and follows all events contained in the backward cone of my causal past. Since this 'before-after' relation is invariant in *all* systems, it follows that *in no frame of reference can my particular 'here-now' appear simultaneous with any event of my causal future or with any event in my causal past.*[32] This follows from the fact that the succession of the events constituting the world lines can never degenerate into simultaneity in any system: this obviously applies to the world line of my own body. In this sense my 'now' still remains absolute. It is not absolute in the classical Newtonian sense since it is confined to 'here' and does not spread instantaneously over the whole universe. Yet, it remains absolute in the sense that it is *anterior* to its own causal future in *any* frame of reference.

On the psychological level this anteriority of the present with respect to its subsequent future moments is embodied in the characteristic feeling which constitutes the central part of our awareness of time. It is the feeling of the present pointing beyond itself toward a not yet realized future. Future situations can be inferentially preconstructed and even imaginatively anticipated with a great degree of vividness, but by their own nature, they can never be perceived; they cannot become present as long as they are absent; after becoming present, they are no longer future. It is this *absence in time* which is symbolically represented by the characteristic 'not-yet' feeling. According to the static interpretation of space-time this feeling is illusory; what we call the future are merely those portions of the world lines which have not yet entered our awareness. We must resolutely get rid of the persistent illusion that they are coming into being; for this is merely an illicit objectification of the temporal order of our perception. For the correct construction of the objective view of the world our feeling of 'now' is as irrelevant as the feeling of 'here'; they both are merely accidental and shifting perspectives of the timeless four-dimensional whole.

Despite its superficial plausibility and its widespread popularity, hardly any other view is more incompatible both with the spirit and the letter of the relativity theory. It ignores completely another feature of the relativistic time-space: no event of my causal future can ever be contained in the causal past of any conceivably real observer. By 'conceivably real observer' we mean any frame of reference in any part of my causal past or anywhere in my present region of 'elsewhere'. In a more ordinary language, *no event which has not yet happened in my present 'here-now' system could possibly have happened in any other system*. To believe otherwise would mean to accept the existence either of the actions moving backward in time or at least of those moving with the velocity greater than light:

quod non. Since the inclusion into the causal past of the observer is the necessary condition for the perceivability of events, it means that the postulated existence of future events is unobservable in principle. If we continue to postulate it, then we face the following dilemma: either to believe in their observability, which would contradict relativity; or to admit their intrinsic unobservability, but still insist on their existence; this would contradict the most elementary rules of scientific methodology. It is far simpler and sounder to place the unobservable future events into the same category as phlogiston, caloricum, mechanical ether, and other discredited and useless fictions.

We can anticipate the following objection: Your denial of the reality of future events is admittedly based on the exclusion of future observers from the category of 'possibly existing observers'. Thus the whole argument is nothing but *petitio principii.* The division between the past, present and future frames of reference is relative and arbitrary, since it is continually shifting; what is still in the future for me now will be included (or, as it is fashionable to say, is tenselessly included) in the causal past of our posterity. No particular 'now' has a privileged character, being merely an accidental and passing perspective of the changeless spatio-temporal whole.[33]

It should be noted that this counter-argument has nothing to do with relativity. For *the transiency of 'now' is not a discovery of the relativistic physics; it is as old as human awareness of time.* Furthermore, the following points should be noted:

(1) The terms 'transient' and 'arbitrary' are not synonymous. On every individual world line, the 'here-now' moment separates unambiguously the past events from the unrealized potentialities of the future events, and *this separation holds in all other possibly existing frames of reference.* It certainly cannot be called arbitrary. In this precise sense each 'here-now' is absolute. Its transiency makes it neither ambiguous nor arbitrary. On the contrary, the specific character of each 'now' requires its transiency since without it each 'now' would lose its temporal characteristics.

(2) The reality of the psychological present and of its transiency is recognized, however reluctantly, even by the protagonists of the static view; otherwise their claim that "now is a temporal mode of experiencing ego"[34] would be devoid of any meaning. Since they claim at the same time that our transient now does not have any objective status in the

physical world, they face embarrassedly the following question: How is the transient psychological present intelligibly related to the allegedly becomingless whole? MacTaggart, who was one of the most vigorous opponents of the reality of time, was uneasily aware of this difficulty when he asked: Why do we not live in the reign of George III?[35] No answer can be given as long as we assume the timelessness of the world which puts all 'nows' on an equal, 'accidental' footing.

(3) If we say that my future events are 'located' in the causal past of my remote future descendants, then we merely say in the atrociously misleading tenseless language a mere trivial tautology; nothing is thus gained since my future descendants are as unreal and as intrinsically unobservable as any of my future events.

(4) This difficulty automatically disappears when we frankly accept becoming in the physical world which runs, so to speak, parallel to the stream of our experience. In other words, our present psychological present has an objective counterpart not, it is true, in the unreal cosmic 'now', but in the present moment of the history of our planet. There is an objective succession of moments in the objective world corresponding to the transient character of our mental present. Then MacTaggart's question is answered; in truth, it is otiose even to raise it. It is important to stress that although 'here-now' is not equivalent to the Newtonian 'everywhere-now', its significance is far from local. Thus the virtualities of our future history which our earthly now" separates from our causal past *remain potentialities for all contemporary observers in the universe.* Something which did not yet happen for us could not have happened 'elsewhere' in the universe. Similarly, transiency is not confined to any particular region; all world lines are irreversible and they are irreversible in the same sense. The idea of two world lines, no matter how remote from each other, running in the opposite direction always implies the absurdities of the backward-running or self-intersecting causal series.[36]

The discussion of epistemological difficulties inherent in the becomingless view is beyond the scope of this paper. Besides, the root difficulty of this view was indicated in previous paragraphs; it is the same difficulty which was concisely characterized by Lotze long before Bergson, James, Lovejoy, Broad, Meyerson, Eddington, Whitehead and Whitrow: "We must either admit Becoming or else explain the becoming of and unreal appearance of Becoming."[37] This untolerable dichotomy of two completely heterogeneous and unrelated realms, of changing 'appearance' and static 'Reality' (with capital R) is avoided if we acccept the genuine reality of becoming. The structure of the relativistic time-space certainly does not discourage us from doing it. The present broadening of the concept of causality in microphysics points in the same direction: the Laplacian static determinism yields its place to the view of the open world in which genuine novelties come into being.

NOTES

* From 'Time in Relativity Theory: Arguments for a Philosophy of Becoming', in *Voices of Time* (ed. by J. T. Fraser), Braziller, New York, 1966, pp. 434–454 (slightly enlarged).
[1] E. Cunningham, *The Principle of Relativity*, Cambridge University Press, Cambridge, 1914, p. 191.
[2] L. Silberstein, *The Theory of Relativity*, Macmillan, London, 1914, p. 134.
[3] *Bulletin de la société française de philosophie*, April, 1922, p. 108.
[4] *Ibid.*, p. 111: "Dans le continuum à quatre dimensions il est certain que toutes les directions ne sont pas équivalents."
[5] H. Meyerson, *La Déduction relativiste, cf.*, in particular, ch. VII, 'Le Temps'; on Einstein's different views *cf.* pp. 100 and 104.
[6] A. Einstein, 'A propos de *La Déduction relativiste* de M. E. Meyerson', *Rev. Philosophique* **105** (1928) 161.
[7] K. Gödel, 'A Remark About the Relationship Between Relativity Theory and Idealistic Philosophy', *Albert Einstein, Philosopher-Scientist*, Vol. 7, The Library of Living Philosophers (ed. by Paul Schilpp), Evanston 1949, p. 557.
[8] A. Einstein, 'Remarks Concerning the Essays Brought Together in This Cooperative Volume', *op cit.*, p. 687.
[9] *Op. cit.*, p. 688.
[10] *La Déduction relativiste, op. cit.*, p. 97. I extended Meyerson's survey of the divided opinion on this matter in the article 'Relativity and the Status of Space', *Rev. of Metaphysics* **9** (1955) 160, where I also pointed out that the quotation from A. S. Eddington was taken by Meyerson out of the context. The hesitancies on this point in H. Reichenbach's thought can be seen most clearly in his posthumous book, *The Direction of Time*, University of California Press, Berkeley, 1956; after rejecting the static interpretation of space-time which he traces correctly to the influence of the Eleatic tradition (p. 11) he virtually eliminated time later when he claims that there is no unique direction of time either on the cosmic scale (p. 128f.) or on the microphysical level (p. 262).
[11] D. Williams, 'The Myth of Passage', *J. of Philos.* **48** (1951) 457; W. V. O. Quine, *Word and Object*, M.I.T. Press, Cambridge, 1960, p. 172.
[12] O. Costa de Beauregard, *Le Second Principe de la science du temps*, Editions du Seuil, Paris, 1963, p. 132; A. Grünbaum, *Philosophical Problems of Space and Time*, Knopf, New York, 1963, p. 329.
[13] K. Gödel, *op cit.*, p. 557.
[14] See essay by O. Costa de Beauregard in this volume ['*Voices of Time*', pp. 417–433, esp. p. 430. *Ed.*]

[15] The distinction between 'determinism' and 'causality' made by Louis de Broglie prior to his reconversion to determinism: *Continu et Discontinu en Physique Moderne*, Editions A. Michel, Paris, 1959, p. 61. *Cf.* also M. Čapek, 'The Doctrine of Necessity Reexamined', *Rev. of Metaphysics* 5 No. 1 (1951) 11, and 'Toward a Widening of the Notion of Causality', *Diogenes* 28 (1959) 63.

[16] The only way of escaping this conclusion would be to hope, rather unrealistically, that some future discovery would establish the existence of the velocities larger than that of light and gravitation; but this would falsify one of the central ideas of the relativity theory.

[17] Quoted by H. Poincaré, 'Science and Method' in *The Foundations of Science*, The Science Press, Ephrata, Pa., 1946 p. 400.

[18] P. Langevin, 'Le temps, l'espace et la causalité dans la physique moderne', *Bull. de la Société française de la philosophie* (October, 1911); 'L'évolution de l'espace et du temps', *Rev. de métaphysique et de morale* 19 (1911) 455; *La Physique depuis vingt ans*, Bibliothèque d'Histoire et de Philosophie des Science, Paris 1923, p. 265.

[19] Even such a careful thinker as Philip Frank makes the following slip: "We find that the spatial distance S, as well as the temporal distance T, of two events, depends on the system of reference. *Either of them can even disappear if we choose a certain system of reference*"; *cf. Philosophy of Science. The Link between Science and Philosophy*, Prentice-Hall, Englewood Cliffs, N.J., 1957, p. 161. Italics are mine. The italicized statement is certainly *not* generally true; S cannot be eliminated in the case *a* nor can T be eliminated in the case *c*! Professor Frank discussed both these cases *a* and *c* on the previous page (p. 160) without, however, perceiving the consequences drawn by Langevin in 1911. In fairness to him we must stress that he is as much opposed to the static interpretation of space-time as we are; *cf.* the whole chapter of the same book ("Is the World 'Really Four-Dimensional'"), p. 158, especially p. 162; also his polemic against Sir James Jeans' static interpretation in *Interpretations and Misinterpretations of Modern Physics*, Hermann and Cie, Paris, 1938, p. 46.

[20] The error in Laplace's assumptions underlying his calculations was pointed out by W. Wien, 'Über die Möglichkeit einer elektromagnetischen Begründung der Mechanik', *Wiedemann's annalen* (1901) 501.

[21] A. Einstein, 'Autobiographical Notes', in *Albert Einstein: Philosopher-Scientist, op. cit.*, p. 61.

[22] *Newtoni Opera* (ed. by Horsley), Vol. 3, p. 172: "unumquodque durationis indivisibile momentum ubique," Similarly Pierre Gassendi: "Et quodlibet Temporis momentum idem est in omnibus locis". *Opera omnia*, Florentia 1728), p. 198.

[23] A. S. Eddington, *The Nature of the Physical World*, Cambridge University Press, Cambridge, 1933, p. 42.

[24] A. N. Whitehead, *Science and the Modern World*, Macmillan, New York, 1926, p. 172.

[25] A. A. Robb, *The Absolute Relations of Time and Space*, Cambridge University Press, Cambridge, 1921, p. 12.

[26] Robb, *Geometry of Time and Space*, Cambridge University Press, Cambridge, 1936, p. 15.

[27] H. Reichenbach, *The Direction of Time*, University of Calafornia Press, Berkeley, 1956, p. 265.

[28] Whitehead, *The Concept of Nature*, Cambridge University Press, 1920, p. 116; B. Russell, *The Analysis of Matter*, Dover, New York, 1954, p. 61.

[29] '*Jetzt*' ist der Zeitmodus des erlebenden Ichs. *Cf.* Hugo Bergmann, *Der Kampf um*

das Kausalgesetz in der jüngsten Physik, Vieweg, Braunschweig (1929), p. 28; quoted approvingly by A. Grünbaum, *Philosophical Problems of Space and Time, op. cit.*, p. 323.
[30] H. Weyl, *Philosophy of Mathematics and Natural Science*, Princeton University Press, Princeton, 1949, p. 116.
[31] Grünbaum, *op. cit.*, p. 329.
[32] This is rarely sufficiently emphasized. *Cf.* my article 'Note on Whitehead's Definitions of Co-Presence', *Philos. of Sci.* **24** (1957) 79.
[33] This was in substance Hugo Bergmann's argument. *Cf.* Grünbaum, *op. cit.*, p. 323.
[34] See note 29.
[35] J. E. MacTaggart, *Studies in the Hegelian Dialectic*, 2nd ed., Russell and Russell, New York, 1964, p. 160.
[36] At least as long as we assume their interaction.
[37] Quoted by G. J. Whitrow, *The Natural Philosophy of Time*, Nelson, London, 1961, p. 311.

G. J. WHITROW

'BECOMING' AND THE NATURE OF TIME*

The failure of Boltzmann's ingenious attempt to invert the Second Law of Thermodynamics so as to provide a statistical definition of time and also of the later refinement of his theory by Reichenbach is further evidence for our thesis that the idea of time cannot be derived from prior concepts which involve no implicit appeal to it. At first the statistical theory of time had almost the force of a concealed tautology. But its subsequent history shows a striking similarity to that of the ingenious attempts that have been made this century to reduce pure mathematics to logic. Just as we are now obliged to conclude that mathematics is a subject *sui generis*, so we are similarly compelled to accept the view that the notion of earlier and later must be regarded as a primitive concept.[1]

The idea that time-relations are ultimate and irreducible is one which many philosophers and philosophically-minded scientists have been unwilling to accept. Even though it has seldom been denied that time is 'real' in the sense that it is a phenomenon of our experience – and, indeed, in Leibniz's phrase, a 'phenomenon *bene fundatum* – thinkers as diverse in their general outlook as, for example, Plato and Kant, and Bradley and Weyl have repeatedly argued that the temporal mode of our perception has no *ultimate* significance. Although this contention is primarily associated with the long line idealist philosophers going back to Parmenides, it has also been accepted by so empirically minded a thinker as Bertrand Russell. In his essay on 'Myticism and Logic', after dismissing the idealist arguments for the contention that time is unreal, he admitted that

Nevertheless, there is some sense – easier to feel than to state – in which time is an unimportant and superficial characteristic of reality. Past and future must be acknowledged to be as real as the present, and a certain emancipation from slavery to time is essential to philosophic thought.[2]

As a recent historian of philosophy has remarked when commenting on this passage, any philosopher who approaches philosophy through logic is likely to argue in this way,[3] since implication is not a temporal relation.[4]

Even Whitehead, a profound student of problems concerning time, who was greatly influenced by Bergson, felt obliged to regard the temporal extension of matter as a less significant characteristic than its spatial extension. For, he argued, if material has existed for a period of time, it has existed in each part of that period, so that dividing the time does not divide the material. On the other hand, dividing the space which it occupies does divide the material. Hence, "the fact that material is indifferent to the division of time leads to the conclusion that the lapse of time is an accident, rather than of the essence, of the material."[5] Against this line of argument,[6] however, we may set the following: any object can be at the same place at two or more different times, but normally it cannot be at the same time in two or more different places, that is, for a given object (for example, a clock) position is a single-valued function of time, but time need not be – and often is not – a single-valued function of position. From this point of view, time is the basic variable rather than any spatial coordinate.

Philosophers who deny the ultimate reality of time often claim that the idea is self-contradictory. Their arguments are either based, like those of Zeno, on objections to the extensive aspects of time, for example, its supposed infinity or continuity, or else on objections to its transitory aspect, that is, to the concept of 'becoming' and its relations to past, present, and future. These relations concern the very essence of time. Perhaps the most thorough scrutiny to which they have yet been subjected was that made early in the present century by McTaggart who maintained that the statements that an event E is now present, has been future, and will be past are mutually incompatible. McTaggart distinguished the changing A series, as he called it, of past, present, and future from the static B series in which events are related in the order 'earlier than' or 'later than'.[7] He argued, correctly I believe, that the A characteristics of events are an essential feature of the ideas of time and change. But he argued further, and incorrectly, that they involved a contradiction which could only be circumvented by an infinite regress. Therefore, he believed that in the final analysis the contradiction could *not* be circumvented.[8]

The foundation of McTaggart's detailed and intricate argument was his contention that an event can never cease to be an event.

Take any event – the death of Queen Anne, for example, and consider what changes can take place in its characteristics. That it is a death, that it is the death of Anne Stuart, that it has such causes, that it has such effects – every characteristic of this sort never changes. 'Before the stars saw one another plain', the event in question was the death of a Queen. At the last moment of time – if time has a last moment – it will still be the death of a Queen. And in every respect but one, it is equally devoid of change. But in one respect it does change. It was once an event in the far future. It became every moment an event in the nearer future. At last it was present. Then it became past, and will always remain past, though every moment it becomes further and further past.[9]

McTaggart argued that although past, present, and future are incompatible determinations, every event must have them all. If one makes the obvious retort that events do not have these characteristics simultaneously but successively, McTaggart counters with the argument that our statement that an event E is present, will be past, and has been future means that E is present at a moment of present time, past at some moment of future time, and future at some moment of past time. But each of these moments is itself an event in time and so is past, present, and future; in other words, the difficulty breaks out all over again and we are launched on a vicious infinite regress.

The answer to this ingenious conundrum has been clearly formulated by Broad who points out that we do *not* say that the Battle of Hastings preced*es* the Battle of Waterloo, but that it preced*ed* the latter, and that generally the copula in propositions which assert temporal relations between events is not the timeless copula of logic but the temporal copula 'is now', 'was', or 'will be'.

When I utter the sentence 'It has rained', I do *not* mean that, in some mysterious non-temporal sense of 'is', there *is* a rainy event, which momentarily possessed the quality of presentness and has now lost it and acquired instead some determinate form of the quality of pastness. What I mean is that raininess has been, and no longer is being, manifested in my neighbourhood. When I utter the sentence 'It will rain', I do *not* mean that, in some mysterious non-temporal sense of 'is', there *is* a rainy event, which now possesses some determinate form of the quality of futurity and will in course of time lose futurity and acquire instead the quality of presentness. What I mean is that raininess will be, but is not now being, manifested in my neighbourhood. [10]

The essence of McTaggart's argument is, in short, a philosophical fallacy of the same kind as St. Anselm's ontological argument for the existence of God. St. Anselm treated existence as if it were a predicate like goodness, and McTaggart treated absolute becoming as if it were a form of qualitative change.[11] *Time is not itself a process in time.*[12]

As McTaggart himself realized, if time cannot be explained without assuming time and we reject his contention that this proves that it is unreal, the only alternative is that time must be regarded as ultimate.[13]

And this is the view to which we must now adhere.[14] Events happen, and do not exist in any other sense. Moreover, the happening of an event is not itself a further event, and so there is no infinite regress of the type contemplated by McTaggart.

McTaggart's great merit, however, as compared with other idealist philosophers such as Bradley, was that, not content with merely denying the reality of time, he attempted to explain how we come by the illusion that makes us attribute temporal characteristics to existents. His explanation was based on the ingenious hypothesis that there is a third series, the C-series, which the percipient misperceives as a time-series, although it is in fact a real non-temporal series. The two basic relations of this series, like those of the B-series, are transitive and asymmetrical, and one is the converse of the other (just as 'earlier' in the B-series is the converse of 'later'). McTaggart decided that the relations 'included in' and 'inclusive of' fulfilled the intricate set of twelve conditions which he believed the C-series should satisfy. Be that as it may (and, without some implicit appeal to the idea of time, there does not seem to be a completely conclusive case for his correlation of 'inclusive of', rather than its converse, with 'later than'), the fact remains that the C-series is insufficient for a complete account of time, since it does not dispose of the A-series, which – as McTaggart himself originally insisted – is essential to time. For, although the terms which the B-series relates are events, that series is not itself a *temporal* series. Nevertheless, in the later parts of McTaggart's analysis the B-series does duty almost exclusively for time and the A-series is strangely neglected. As Miss Cleugh has commented:

the passage from the B-series to the C-series is successful in so far as the B-series is *not* temporal... As long as the B-series is taken as a series all is well; but as soon as reference is made to the specifically *temporal* connotation of the series, trouble begins. The ghost of time cannot permanently be laid.[15]

McTaggart's theory of time and the criticism to which it has been subjected are not matters of concern for professional philosophers alone. They have a direct bearing on the hypothesis of the 'block universe'. As we have already seen, this hypothesis has been powerfully reinforced by the space-time interpretation of the theory of relativity. From the point of view taken by Einstein and also by Weyl:

The objective world simply *is*, it does not *happen*. Only to the gaze of my consciousness, crawling upward along the life-line of my body, does a section of the world come to life as a fleeting image in space which continuously changes in time.[16]

In other words, the relativistic picture of the world recognizes only a difference between earlier and later and not between past, present, and future[17] Indeed, we can draw an analogy between the terms of McTaggart's C-series and the successive backwards-directed light-cones with vertices along the world line of an observer in the Minkowski diagram. As has been stressed by Eddington[18] and also by Reichenbach[19] the theory of relativity does not provide a complete account of the role of time, even in physics. Like McTaggart's theory it concerns *existence* but not *happening*.

Those who adhere to the hypothesis of the 'block universe' regard the present, to use an analogy due to Broad, like the spot of light from a policeman's bull's-eye traversing the fronts of the houses in a street. This tendency to reify time[20] as a serial order of events along which the quality of presentness moves from past to future was criticized by F. H. Bradley[21] He wrote:

We seem to think that we sit in a boat and are carried down the stream of time, and that on the bank there is a row of houses with numbers on the door. And we get out of the boat and knock at the door of No. 19, and re-entering the boat suddenly find ourselves opposite No 20, and having done the same we go on to No 21. And all this while the firm fixed row of the past and future stretches in a block behind us and before us.

Instead, he suggested the following analogy which is much closer to our actual experience of time.

If it is really necessary to have some image, perhaps the following will save us from worse. Let us fancy ourselves in total darkness, hung over a stream and looking down on it. The stream has no banks and its current is covered and filled continuously with moving things. Right under our faces is a bright illuminated spot on the water, which ceaselessly widens and narrows its area and shows us what passes away on the current, and this spot is our now, our present.

Although the theory of relativity has nothing significant to say about the question of 'becoming' and the role of the present nor about the associated question of the difference between past and future, light has been thrown on these problems by quantum theory. For, in quantum mechanics we find that the past history of an individual system does not determine its future in any absolute way but merely the probability distribution of possible futures. In general, there is no conceivable set of observations that can provide enough information about the past of a system to give us complete information as to its future. The future is a mathematical construction which can be changed by an observation.[22]

This indeterminism *in principle* of the future finally disposes of Laplace's claim[23] that

> An intelligent being who at a given instant knew all the forces animating Nature and the relative positions[24] of the beings within it would, if his intelligence were sufficiently capacious to analyse the data, include in a single formula the movements of the largest bodies in the universe and those of its lightest atom. Nothing would be uncertain for him: the future as well as the past would be present to his eyes.[25]

We now realize that such claims are completely baseless. *The past is the determined, the present is the moment of 'becoming' when events become determined, and the future is the as-yet undetermined.*

There is indeed a profound connection between the reality of time and the existence of an incalculable element in the universe. Strict causality would mean that the consequences pre-exist in the premises. But, if the future history of the universe pre-exists logically in the present, why is it not already present? If, for the strict determinist, the future is merely 'the hidden present', whence comes the illusion of temporal succession? The fact of transition and 'becoming' compels us to recognize the existence of an element of indeterminism and irreducible contingency in the universe.[26] The future is hidden from us – not in the present, but in the future. Time is the mediator between the possible and the actual.[27]

NOTES

* From *Natural Philosophy of Time*, Thomas Nelson and Sons, London, 1961, pp. 288–296.

[1] As an alternative to entropy-increase, Eddington (*New Pathways in Science*, Cambridge 1935, pp. 67–68) suggested that cosmical expansion could enable us to decide which of two epochs is the later by the criterion that the later epoch corresponds to the larger volume of the universe. This criterion presupposes, however, that the universe *always* expands, which is certainly not self-evident. Moreover, as Eddington recognized, it is inappropriate as a signpost for local time.

[2] B. Russell, *Mysticism and Logic*, London 1917, p. 21.

[3] An amusing story concerning the Russian philosopher Nicolas Berdyaev is told by Eugene Lampert (*The Listener* 60 (1958) 193): "I have heard him plead passionately for the insignificance and unreality of time, and then suddenly stop and look at his watch with genuine anxiety at the thought that he was two minutes late for taking his medicine!"

[4] J. Passmore, *A Hundred Years of Philosophy*, London 1957, p. 273.

⁵ A. N. Whitehead, *Science and the Modern World*, Cambridge 1926, p. 63.

⁶ Whitehead referred only to matter, but it should not be overlooked that the situation is effectively reversed at the mental level. For, as has been shown by the surgical removal of parts of the cortex, within limits spatial 'division' has comparatively little influence on thought, whereas temporal division reduces it to fragments. Corresponding to the significant concept of *density* of material objects, we have the equally significant concept of *rate of thought* (*and decision*) in mental processes.

⁷ J. M. E. McTaggart, *The Nature of Existence*, Vol. II, Cambridge 1927, p. 10.

⁸ McTaggart's view that the infinite regress is 'vicious' may be contrasted with the view held by J. W. Dunne (*An Experiment with Time*, 3rd edition, London 1934 (reprinted 1958), p. 197) that "an infinite regress is, after all, the proper and valid description of mind's relation to its objective universe."

⁹ J. M. E. McTaggart, op. cit., p. 13.

¹⁰ C. D. Broad, *Examination of McTaggart's Philosophy*, Vol. II, Part I, Cambridge 1938, p. 316.

¹¹ A similar mistake was made by J. W. Dunne in his theory of serial time.

¹² We are reminded of Zeno's paradox concerning place: for, if everything that exists has a place, it is clear that place too will have a place and so on *ad infinitum* (Aristotle, *Physics*, Book 4, 209a, 23).

¹³ J. M. E. McTaggart, *Philosophical Studies*, London 1934, p. 126.

¹⁴ Although the analysis of time (like that of infinity) is beset with logical perils, for example Schopenhauer's definition of time "as the possibility of opposite states in one and the same thing" (*On the Fourfold Root of the Principle of Sufficient Reason* (transl. by Mme Karl Hillebrand), London 1897, p. 32) – which echoes Leibniz's "le temps est l'ordre des possibilités inconsistentes" (*Die Philosophischen Schriften*, Berlin 1880, Vol. 4, p. 568) – and Miss Cleugh's definition, "The alogical element in the universe" (*Time*, London 1937, p. 280), we reject the idealist conclusion that time is illusory. Instead, we agree with Broad when he says that, *if* logic excludes time, "so much the worse for logic" (*Scientific Thought*, London 1923, p. 83).

In a recent penetrating analysis of McTaggart's argument, L. O. Mink (*Philosophical Quarterly* **10** (1960) 252–263) points out that McTaggart was arguing not about time itself but about arguments about time. Mink concludes that, although the attempt to "embalm the fact of transience" in language gives rise to "logical regresses, circles, paradoxes and reduplications", it does not follow that time itself is unreal unless it is automatically assumed that time must share *all* the characteristics of discourse.

¹⁵ M. A. Cleugh, *Time*, London 1938, pp. 164–165.

¹⁶ H. Weyl, *Philosophy of Mathematics and Natural Science*, Princeton 1949, p. 116.

¹⁷ At the *microphysical* level this may be justified; see R. P. Feynman (*loc. cit.*) who suggests that in studying the 'close collision' of elementary particles we should abandon the Hamiltonian method which considers the future as developing continuously out of the past. Instead, we should "imagine the whole space-time history laid out and that we just become aware of increasing portions of it successively." On the other hand, J. L. Martin (*Proc. Roy. Soc.*, A **251** (1959) 536) maintains that, in general, the Hamiltonian method is more fundamental than the Lagrangian approach for two reasons: "It seems more natural in the first instance to treat the behaviour of a system in time in terms of a continuously unfolding transformation, rather than in terms of a variational principle applying simultaneously to the whole of a time range. More practically, it will be found that the Hamiltonian approach is the wider of the two." Some Hamiltonian systems do not have Lagrangian forms.

[18] A. S. Eddington, *The Nature of the Physical World* (Everyman edition), London 1935, p. 76.
[19] H. Reichenbach, *The Direction of Time* (ed. by M. Reichenbach), Berkeley and Los Angeles 1956, passim.
[20] That it is almost instinctive is revealed by numerous instances, for example the action of those who rioted when the Gregorian calendar was introduced into England in September 1752 and demanded "Give us back our eleven days!"
[21] F. H. Bradley, *The Principles of Logic*, Oxford, Vol. 1, pp. 54–55.
[22] M. S. Watanabe, 'Réversibilité contre Irréversibilité en Physique Quantique', in *Louis de Broglie: Physicien et Penseur*, Paris 1953, pp. 385–400.
[23] P. S. Laplace, *Œuvres Complètes*, Vol. 7, Paris 1886, p. vi; cit. E. W. Barnes, *Scientific Theory and Religion*, Cambridge 1933, p. 578.
[24] Strictly speaking, from the point of view of Newtonian mechanics (which Laplace assumed) the velocities, as well as the relative positions, at the given instant must be known.
[25] In his famous lecture *On the Limits of Natural Knowledge*, delivered at Leipzig in 1872, E. du Bois-Reymond even maintained that the Laplacean calculator would be able to tell from his formula who was the Man in the Iron Mask and when England would have burnt her last piece of coal! Of only one problem would he be ignorant – the explanation of consciousness. On the other hand, in an important paper, published in 1950, K. R. Popper (*Brit. J. Phil. Sci.* **1** (1950) 117 et seq., 173 et seq.) argued that, even assuming that the future is completely subject to strict Newtonian determinism, the Laplacean calculator (regarded as a physical predicting machine which is itself part of the physical world) could not *predict* it. Instead, the calculator would only be able to 'predict' the state of its environment (including itself) at any specified future time *after* the arrival of the time in question! For, there is an inherent delay, which cannot be overcome, in deriving information from the environment about the environment; in particular, the calculator has to take into account the results of its own preceding calculation. (J. R. Platt (*American Scientist* **44** (1956) 183) has made the further point that we could never know the positions and velocities of all the particles in the universe, at a given instant, since we should require an impossibly large number of amplifiers – and these would have to be outside the universe! Indeed, the individual motions of the billions of molecules in a small quantity of gas are unknowable, *even in principle*. "The number of independently knowable particles must always be orders of magnitude less than the number of particles in the amplifiers."
[26] M. F. Cleugh, *Time*, London 1937, Chapter XII.
[27] A. Schopenhauer, *World as Will and Idea* transl. by A. B. Haldane and J. Kemp), Vol. 2, London 1883, p. 73.

TIME: CONTINUOUS OR DISCRETE*

Although psychological time is certainly discrete, in classical physics, as we have noted above, the time parameter is assumed to vary continuously over the range of all real numbers. This justifies the use of the time differential element dt in the application of the principle of elementary abstraction, and in so far as differential equations constitute a satisfactory way of expressing the fundamental principles of physical theories, the continuity of time would seem to be a convenient postulate. However, the recent stress on the element of *discreteness* involved in the quantum theory and its application to atomic structure has focused attention on the possible value of associating discreteness with conceptual time or space or both. The quantum concept of stationary state can be reconciled with a continuous time parameter only through the medium of quantum mechanics. If one tries to interpret this concept in terms of the usual space-time picture of classical physics, one is faced with the uncomfortable inconsistency that, although the behavior of a system in a stationary state can be described in terms of classical mechanics with its continuous t, the transitions from one state to another correspond to nothing in classical mechanics. This breakdown of the classical method has been stressed by Bohr ever since his founding of the quantum theory of atomic structure in 1913. As we have said, only the introduction of quantum mechanics with its new concept of *state* seems to be able to solve the difficulty, and this only by departing considerably from our primitive notions. The quantum theory lays great stress on the concept of frequency, related to the radiated quantum energy by the celebrated relation $E = h\nu$, where h is the Planck constant of action. It has been suggested[1] that there exists a maximum energy quantum corresponding to the largest possible radiation frequency which can exist. This would in turn correspond to a minimum period of time $T_{min} = 1/\nu_{max}$, which has been taken as a time-atom and called by some the *chronon*. The highest frequency radiation so far observed is that of cosmic rays, viz., approximately $1/4.5 \times 10^{24}$ sec^{-1}, which yields, for T_{min}, 4.5×10^{-24} sec. Now the theory of relativity suggests that the energy associated with any mass m is $E=mc^2$, where c is the velocity of light (3×10^{10} cm/sec). If one assumes that all this energy can be changed into radiation energy the frequency will clearly be mc^2/h. The substitution of the mass of a *proton* or *neutron* for m yields a value very close to v_{max} above. If the proton is the most massive atomic particle which can be, so to speak, annihilated with the change of its mass into energy in accordance with the Einstein relation $E=mc^2$, some meaning might be attached to the chronon as just defined. Since it is conceivable, however, that still more massive particles (i.e., atomic nuclei) may be subject to the same process, one is hardly justified at the present time in taking the chronon hypothesis too seriously.

It is of interest to observe, in any case, that if there exists a time quantum, all spectra must be line spectra and 'continuous' spectra are only those in which the lines are too close together to be resolved by ordinary optical instruments. The future alone can decide the fate of the assumption of discrete conceptual time. It may be in order, however, to remark briefly the profound alteration the introduction of discrete time would necessitate in all physical theories. Continuous conceptual time provides, so to speak, a continuous background against which to describe both discrete and continuous phenomena. If time itself is assumed discrete, this background is lost and the whole question of the use of time in physical description must be examined anew.

NOTES

* From *Foundations of Physics*, John Wiley and Sons, 1936, pp. 76–78.
[1] See, for example, G. I. Pokrowski, *Nature* **127** (1931) 667.

A. N. WHITEHEAD

THE INAPPLICABILITY OF THE CONCEPT OF INSTANT ON THE QUANTUM LEVEL*

In order to explain exactly how mathematics is gaining in general importance at the present time, let us start from a particular scientific perplexity and consider the notions to which we are naturally led by some attempt to unravel its difficulties. At present physics is troubled by the quantum theory. I need not now explain[1] what this theory is, to those who are not already familiar with it. But the point is that one of the most hopeful lines of explanation is to assume that an electron does not continuously traverse its path in space. The alternative notion as to its mode of existence is that it appears at a series of discrete positions in space which it occupies for successive durations of time. It is as though an automobile, moving at the average rate of thirty miles an hour along a road, did not traverse the road continuously; but appeared successively at the successive milestones, remaining for two minutes at each milestone.

In the first place there is required the purely technical use of mathematics to determine whether this conception does in fact explain the many perplexing characteristics of the quantum theory. If the notion survives this test, undoubtedly physics will adopt it. So far the question is purely one for mathematics and physical science to settle between them, on the basis of mathematical calculations and physical observations.

But now a problem is handed over to the philosophers. This discontinuous existence in space, thus assigned to electrons, is very unlike the continuous existence of material entities which we habitually assume as obvious. The electron seems to be borrowing the character which some people have assigned to the Mahatmas of Tibet. These electrons, with the correlative protons, are now conceived as being the fundamental entities out of which the material bodies of ordinary experience are composed. Accordingly if this explanation is allowed, we have to revise all our notions of the ultimate character of material existence. For when we penetrate to these final entities, this startling discontinuity of spatial existence discloses itself.

There is no difficulty in explaining the paradox, if we consent to apply to the apparently steady undifferentiated endurance of matter the same principles as those now accepted for sound and light. A steadily sounding note is explained as the outcome of vibrations in the air: a steady colour is explained as the outcome of vibrations in ether. If we explain the steady endurance of matter on the same principle, we shall conceive each primordial element as a vibratory ebb and flow of an underlying energy, or activity. Suppose we keep to the physical idea of energy: then each primordial element will be an organised system of vibratory streaming of energy. Accordingly there will be a definite period associated with each element; and within that period the stream-system will sway from one stationary maximum to another stationary maximum – or, taking a metaphor from the ocean tides, the system will sway from one high tide to another high tide. This system, forming the primordial element, is nothing at any instant. It requires its whole period in which to manifest itself. In an analogous way, a note of music is nothing at an instant, but it also requires its whole period in which to manifest itself.

Accordingly, in asking where the primordial element is, we must settle on its average position at the centre of each period. If we divide time into smaller elements, the vibratory system as one electronic entity has no existence. The path in space of such a vibratory entity – where the entity is *constituted by* the vibrations – must be represented by a series of detached positions in space, analogously to the automobile which is found at successive milestones and at nowhere between.

We first must ask whether there is any evidence to associate the quantum theory with vibration. This question is immediately answered in the affirmative. The whole theory centres around the radiant energy from an atom, and is intimately associated with the periods of the radiant wave-systems. It seems, therefore, that the hypothesis of essentially vibratory existence is the most hopeful way of explaining the paradox of the discontinuous orbit.

In the second place, a new problem is now placed before philosophers and physicists, if we entertain the hypothesis that the ultimate elements of matter are in their essence vibratory. By this I mean that apart from being a periodic system, such an element would have no existence. With this hypothesis we have to ask, what are the ingredients which form the vibratory organism. We have already got rid of the matter with its appearance of undifferentiated endurance. Apart from some metaphysical compulsion, there is no reason to provide another more subtle stuff to take the place of the matter which has just been explained away. The field is now open for the introduction of some new doctrine of organism which may take the place of the materialism with which, since the seventeenth century, science has saddled philosophy. It must be remembered that the physicists' energy is obviously an abstraction. The concrete fact, which is the organism, must be a complete expression of the character of a real occurrence. Such a displacement of scientific materialism, if it ever takes place, cannot fail to have important consequences in every field of thought.

Finally, our last reflection must be, that we have in the end come back to a version of the doctrine of old Pythagoras, from whom mathematics, and mathematical physics, took their rise. He discovered the importance of dealing with abstractions; and in particular directed attention to number as characterising the periodicities of notes of music. The importance of the abstract idea of periodicity was thus present at the very beginning both of mathematics and of European philosophy.

In the seventeenth century, the birth of modern science required a new mathematics, more fully equipped for the purpose of analysing the characteristics of vibratory existence. And now in the twentieth century we find physicists largely engaged in analysing the periodicities of atoms. Truly, Pythagoras in founding European philosophy and European mathematics, endowed them with the luckiest of lucky guesses or, was it a flash of divine genius, penetrating the inmost nature of things?

NOTES

* From *Science and the Modern World*, Macmillan, New York, 1926, pp. 52–56.
[1] *Ibid.*, ch. 8.

SPATIO-TEMPORAL CONTINUITY, QUANTUM THEORY AND MUSIC*

Quantum mechanics was a subject in mathematical physics which had originated in 1900 in the work of Max Planck on the equilibrium of radiation in a cavity. In plain language, the subject matter of quantum theory is the study of such light as we find inside of a hot furnace after light and hot matter have come to equilibrium so that if we look into a cavity with heated walls, such as a blast furnace, the light coming from inside the furnace changes in character as the temperature changes. This is a readily observable effect which we all know from the difference between a red-hot piece of metal and a white-hot piece of metal. The spectrum of the light coming from the red-hot furnace ceases somewhere in the red or yellow, but the light coming from the white-hot furnace may go far into the ultraviolet.

The nub of the difficulty in explaining this relation between light and heat, which Planck solved by a brutal new hypothesis, was that the traditional representation of light as a continuous phenomenon was not satisfactory. In light as in matter, he argued, there is a granular rather than continuous texture.

The earlier physics had not been able to conceive any mechanism by which the color distribution of light in a furnace could be determined by the furnace's temperature. Planck's eventual explanation of this easily observable phenomenon was, however, not simple. It was associated with ideas concerning mathematics and thought in general which go back to the end of the seventeenth century, during a period when an important intellectual battle was being fought between the atomists, who believed in the discreteness of matter, and those who believed that matter is continuous. There were various philosophical considerations which made this debate especially critical.

It was not, however, so much the general philosophical climate of the time as a technical innovation which brought the dispute to a head. This innovation was the discovery of the microscope by the Dutchman Leeuwenhoek, who had perfected his device to the point at which he could see something of the teeming life in a drop of stagnant water.

The discovery of a new instrument often leads immediately to a new insight. Before Leeuwenhoek, the study of living organisms had been limited to what could be seen by the naked eye or, at best, with a primitive hand lens. Thus, scientists, while they might have had the Democritean idea that the world exists of extremely small particles or atoms, had made no considerable progress in seeing phenomena smaller than, say, a grain of sand.

Leeuwenhoek's microscope showed by direct observation that a drop of pond water was a teeming mass of life suggestive of a crowded city. The new power lent to the eye engendered a new range of imagination, and everyone's thoughts turned to the fine structure of the world and to the philosophical implications suggested by the process of magnification. One result of this experience, perhaps, was Swift's famous jingle:

> So, naturalists observe, a flea
> Hath smaller fleas that on him prey;
> And these have smaller still to bite 'em;
> And so proceed *ad infinitum*.

The background of this little jingle is more interesting than the jingle itself may seem to us at this late date, for among the objects that Leeuwenhoek studied with his new microscope were the spermatozoa of man and the animals, which Leeuwenhoek quite reasonably interpreted as playing a part in conception. Through the imperfect microscopes of Leeuwenhoek and his followers, however, it was easy to imagine that the spermatozoon contained a small, rolled-up fetus. This theory gave a plausible interpretation to the act of conception, for it was believed conception consisted merely in the implantation of the spermatozoon in the womb, in which environment it could grow in size till it became an embryo of the sort which was already familiar to the doctors. The idea that the spermatozoon was the sole antecedent of the embryo led to some very interesting biological speculations.

If the spermatozoon was itself an early stage of the fetus, it was natural to think that it was a human being in miniature, with all the organs of the human being on a smaller scale, distorted but still essentially there. By this token, it should contain smaller spermatozoa, much as Swift's flea carried lesser fleas on a scale far smaller than the microscope of the day could show. These in turn could be thought to contain still smaller spermatozoa, and so on *ad infinitum*, so that the whole future of the human race actually lies preformed within the bodies of those now existing. This preformationism argued for an infinite subdivisibility of matter, and the philosophical consequences of this were eagerly studied, particularly by the great philosopher, Leibniz.[1]

Leibniz conceived the world after the analogy of the drop of water and the similarly teeming drop of blood as a plenum. That is, he conceived that all the apparent spaces between living beings and within living beings are themselves filled with living beings on a smaller scale. This theory led Leibniz to postulate the infinite subdivisibility of life and, consequently, the continuity of matter.

This opinion, which was generated as we have seen by the microscopic observations of his day as well as by the inner workings of his own philosophy, led Leibniz eventually to a new interpretation of mathematics. He was, we must remember, one of the co-inventors of the calculus, and he originated the notation which we use even now. For him not only are time and space infinitely subdivisible, but quantities distributed in time and space may have rates of change in all their dimensions. For example, one quantity distributed in time and space is temperature. When I say that the thermometer is dropping at the rate of ten degrees an hour, I am speaking of its time rate of change. When I say that it is dropping at the rate of three degrees per mile as I go west, I am giving one of its space rates of change. In discussing quantities which have a distribution both in time and in space, a natural mathematical law is the partial differential equation in which time rates of change and space rates of change are related to one another in a system where time and space are both infinitely subdivisible. Thus, Leibniz, in arguing for the continuity of the physical world, became the spokesman for a view in direct contradiction to atomism.

The development of physics since his time has brought both atomism and the continuistic theory to a perfection and to a sharpness of opposition which they did not possess in his day. The molecule has been all but seen, and the chemical evidence for the existence of the discrete atom is clear. Beyond the atom, new vistas of atomicity have been discovered in the electron, the proton, and the many new fundamental particles discovered in the atomic nucleus; while in the meantime the continuum theory has become a useful and almost indispensable tool for the study of the dynamics of gases, liquids, and solids and for the theory of light and electricity. That these two great directions of thought have come into head-on collision with one another has led to some of the chief problems of modern physics.

This collision began to take shape about a hundred years ago, when Clerk Maxwell developed what is now known as the kinetic theory of gases. A gas consists of particles called molecules which can move in several independent ways. A molecule can move up and down, to the right and to the left, and to and fro as a whole, besides which it can rotate about a vertical axis and two horizontal axes. All these motions belong to it as a rigid body, but it is often more than a rigid body and may have internal vibrations which appertain to it as an elastic system. We can count the number of modes of motion, or, as a physicist calls them, degrees of freedom, of a single particle. By adding up the number of the modes of motion of the different particles forming a gas, we can determine the number of modes of motion or degrees of freedom of the gas as a whole. Maxwell remarked that when a gas has settled down to an internal statistical equilibrium, each mode of motion will have on the average a certain energy and this average energy will be the same for all modes of motion. This is a most important theorem in connecting temperature with the other properties of a gas.

It results at once that the ability of a given volume of a gas to absorb energy depends on the number of degrees of freedom per unit volume. The measure of this ability is called the specific heat. It enables us to ascertain how much energy a body in heat equilibrium will contain at a given temperature. If the number of degrees of freedom per unit volume is infinite, the body will be able to take up an infinite amount of energy with a finite increase in temperature; or what is the same, a finite accretion to its energy will not make it any hotter. If we apply a similar argument to a continuous medium, which will naturally have an infinite number of degrees of freedom per unit volume, then this too will have infinite specific heat, and the notion of temperature will not be applicable to it.

Now, Clerk Maxwell was not only the originator of this theory we have just indicated, which is known as the kinetic theory of gases, but also of the theory that light and electricity are transmitted as oscillations of a continuous medium known as the luminiferous ether. This means that the ether can be heated indefinitely without getting any hotter. Since the motions of the luminiferous ether are known as radiation, taking the form of light, X rays, radiant heat, etc., the Maxwell theory of the ether is inconsistent with the existence of any temperature to radiation. Maxwell's theory of light, satisfactory as it is for free radiation in the absence of matter, makes it impossible for light to come to equilibrium with matter in temperature as it is actually known to do in furnace. Something more and different from the Maxwell theory was needed for the study of the radiation of light, and this something more was suggested by Planck.

Planck observed not only that there is a temperature to radiation, but that the relation between this temperature and the character of the radiation follows a definite law, which is known by Planck's name. In order to justify this law, he supposed that radiation was emitted according to certain small atomic quantities which he called quanta, and this work of his is the first form of the quantum theory of modern physics.

In general, 1900 represents a critical period in scientific thinking. It had not been many years since the most advanced scientists considered that future centuries would be devoted to determining already existing physical theory to further and further decimal places of accuracy. About 1900, however, the quantum theory was beginning to destroy some of the ideas of continuity in the field of radiation. The Gibbsian statistical mechanics was already well on the way to replacing determinism by a qualified indeterminism, and the optical experiment of Michelson and Morley, which showed the impossibility of measuring the velocity of the earth through the ether, had recently become an essential part of the chain of ideas which was to lead to Einstein's relativity.

Einstein's theory of relativity was formulated in 1905, and in the same year he made a critical contribution to quantum theory. He showed that certain of the constants involved in the photoelectric effect, which connects light absorption or emission and electricity, were numerically and dimensionally the same as a famous constant used by Planck in quantum theory. Seven years later, in 1912, Niels Bohr, of Copenhagen, discovered the same constant in the theory of the radiation of the hydrogen atom.

The theory of radiation which was put forward by Bohr was brilliantly although not perfectly successful. It was a curious hybrid in which features of a discontinuous theory were somewhat unnaturally grafted on to a continuous theory like that of planetary orbits. This quantized mechanics had important numerical successes and rather incomplete theoretical unity. By 1925, the year of my talk in Göttingen, the world was clamoring for a theory of quantum effects which would be a unified whole and not a patchwork.

Without being aware of the way in which interest in Göttingen was already concentrating about the difficulties of the quantum theory, my talk in Göttingen, like quantum theory, dealt with a field in which the laws of ordinary magnitudes do not continue down into the range of the very small. As I have said, my talk concerned harmonic analysis – in other words, the breaking up of complicated motions into sums of simple oscillations. Harmonic analysis, for all its many modern ramifications, has a history going back to Pythagoras and his interest in music and the vibrations of the strings of the lyre. There are many ways in which a string can vibrate, but the most elementary and simplest of all is known as the simple harmonic oscillation. The motion of the string of a musical instrument, if indeed it is not simply harmonic, is well known to be the most elementary sort of combination of simple harmonic motions. In fact, for a first very crude approximation, we can treat such a motion as simply harmonic.

Now, let us see what musical notation really is. The position of a note vertically on the staff gives its pitch or frequency, while the horizontal notation of music divides this pitch in accordance with the time. The time notation contains the indication of the rate of the metronome, the subdivision of sound into whole notes, half notes, quarter notes, etc., the various rests, and much else besides. Thus musical notation at first sight seems to deal with a system in which vibrations can be characterized in two independent ways, namely, according to frequency, and according to duration in time.

A finer assumption of the nature of musical notation was that things are not as simple as all this. The number of oscillations per second involved in a note, while it is a statement concerning frequency, is also a statement concerning something distributed in time. In fact, the frequency of a note and its timing interact in a very complicated manner.

Ideally, a simple harmonic motion is something that extends unaltered in time from the remote past to the remote future. In a certain sense it exists *sub specie aeternitatis*. To start and to stop a note involves an alteration of its frequency composition which may be small, but which is very real. A note lasting only over a finite time is to be analyzed as a band of simple harmonic motions, no one of which can be taken as the only simple harmonic motion present. Precision in time means a certain vagueness in pitch, just as precision in pitch involves an indifference to time.

The considerations are not only theoretically important but correspond to a real limitation of what the musician can do. You can't play a jig on the lowest register of the organ. If you take a note oscillating at a rate of sixteen times a second, and continue it only for one twentieth of a second, what you will get is essentially a single push of air without any marked or even noticeable periodic character. It will not sound to the ear like a note but rather like a blow on the eardrum. Actually, the complicated mechanism of the reflection of impulses which is necessary to make an organ pipe speak in a musical manner will not have a fair chance to get started. A fast jig on the lowest register of the organ is, in fact, not so much bad music as no music at all.

It was this paradox of harmonic analysis which formed an important element of my talk at Göttingen in 1925. At that time, I had clearly in mind the possibility that the laws of physics are like musical notation, things that are real and important provided that we do not take them too seriously and push the time scale down beyond a certain level. In other words, I meant to emphasize that, just as in quantum theory, there is in music a difference of behavior between those things belonging to very small intervals of time (or space) and what we accept on the normal scale of every day, and that the infinite divisibility of the universe is a concept which modern physics cannot any longer accept without serious qualification.

To see the relevance of my ideas to the actual development of quantum theory, we must step ahead a few years, to the time when Werner Heisenberg formulated his principle of duality or indeterminism. The classical physics of Newton is one in which a particle may have at the same time a position and a momentum – or, what is not very different, a position and a velocity. Heisenberg eventually observed that under the conditions under which a position can be measured with high precision, a momentum or velocity can be measured only with low precision, and vice versa. This duality is of exactly the same nature as the duality between pitch and time in music, and in fact Heisenberg came to explain it through the same harmonic analysis which I had already presented to the Göttingers at least five years before.

NOTES

* From *I am a Mathematician*, MIT Press, Cambridge, 1964, pp. 97–107.

[1] The step from Leibniz to Swift involves certain aspects of the history of the early eighteenth century which deserve comment. Leibniz was a great philosopher and physicist by avocation, but his official position had been that of archivist to the court of Hanover. In this position he showed himself to be not only a librarian but a diplomat of the first rank, eager for the welfare and the aggrandizement of his ruler. There is much to be said for the conjecture that he was active in the negotiations which put the house of Hanover on the throne of England. Since it was the Whig party in England which desired the coming of the Hanoverians, in order to terminate the reign of the unpopular Stuarts, Leibniz became identified with the Whig intrigues. His contact with England was greatly facilitated by his membership and his active share in the Royal Society.

Swift, on the other hand, was a Tory supporter of the Stuarts, and he took an active share in the attempted *coup d'état* which tried to put the Old Pretender on the throne after the death of Queen Anne. Thus, Leibniz and Swift were key figures, respectively, of the two conflicting parties in the English politics of the day. It is no wonder that a great antagonism grew up between them.

This antagonism is shown in the third of the four books of *Gulliver's Travels*, the voyage to Laputa. Many people have wondered at the virulence with which Swift lashes scientists, these impractical projectors who measure a man with a sextant to fit him with a suit of clothes, who extract sunbeams from cucumbers, and who attempt to attain all the learning of the ages by a process equivalent to Eddington's monkeys and typewriters. In fact, they represent nothing but the Royal Society and in particular the Leibnizian influence in the Royal Society. It is thus not astonishing that one of the targets of Swift's wit should be the essentially Leibnizian situation of the fleas upon the fleas and so on ad infinitum.

This was not the only place at which Swift showed himself fascinated with the problem of the variable scale of nature and of what would happen to the world and the individuals in it under a contraction or an expansion. It is likewise the theme of the first two books of *Gulliver's Travels*, the voyage to Lilliput, where the inhabitants are one twelfth the height of a normal man, and the voyage to Brobdingnag, where the inhabitants are giants seventy feet tall.

In both cases, Swift's imagination concerning the effect of change of size is keen but limited. It applies to the physical dimensions but not to their powers of motion. He is not aware that the Lilliputians, if they were made of human flesh and blood, should be able to jump a height several times greater than their own, nor of the similar fact that his Brobdingnagians would be so slothful and earth-bound that they would be scarcely able to stand up.

D. BOHM

INADEQUACY OF LAPLACEAN DETERMINISM AND IRREVERSIBILITY OF TIME*

1. On the Abstract Character of the Notion of Definite and Unvarying Modes of Being

Empirical evidence available thus far shows that nothing has yet been discovered which has a mode of being that remains eternally defined in any given way. Rather, every element, however fundamental it may seem to be, has always been found under suitable conditions to change even in its basic qualities, and to become something else. Moreover, the notion of the qualitative infinity of nature implies that every kind of thing not only can change in this very fundamental way but that, given enough time, conditions in its infinite background and substructure will alter so much that it *must* do so. Hence, the notion of something with an exhaustively specifiable and unvarying mode of being can be only an approximation and an abstraction from the infinite complexity of the changes taking place in the real process of becoming. Such an approximation and abstraction will be applicable only for periods of time short enough so that no significant changes can take place in the basic properties and qualities defining the modes of being of the things under consideration.

When we come to times that are long enough for the basic kinds of things entering into any specific theory to undergo fundamental qualitative changes, then what breaks down is the assumption that we can specify the modes of being of these things *precisely* and *exhaustively* in terms of the concepts that were applicable before this change took place. Indeed, the very fact that a thing is able to undergo a qualitative change is itself a property that is an essential part of the mode of being of the thing and yet a property that is not contained in the original concept of it. For example, the fact that the liquid, water, turns into steam when heated and ice when cooled, is a basic property of the liquid in question, without which it could not be water as we know it. Nevertheless, the original concept of water as nothing more than a liquid evidently does not contain these possibilities, either explicitly or implicitly, as necessary properties of this liquid. Hence, this concept does not give a precise and exhaustive representation of all the properties of the liquid in question.

Now the way one usually deals with this problem is to regard the transformations between solid, liquid, and vapour that take place at certain temperatures as part of the qualities defining the mode of being of a single broader category of substance; viz. water. But now the same kind of problem arises again at a new level. For the laws governing the transformations of these qualities are, in turn, being regarded as part of an eternal and exhaustive specification of the properties of the substance, water. On the other hand, in reality this law is applicable and has meaning only under limited conditions. For example, it will no longer have relevance at temperatures and denotes of matter so high that there can be no such things as atoms, and therefore no such a substance as water. Thus, we are led to include water as a special state of a still broader category of things (e.g. systems of electrons, protons, neutrons, etc.) and the laws governing the transformation of water into other kinds of substances as a part of the mode of being of this still broader category. But if *all* things eventually undergo qualitative transformations, then the process described above will never end. Thus we conclude that the notion that all things can become other kinds of things implies that a complete and eternally applicable definition of any given thing is not possible in terms of any finite number of qualities and properties.

If, however, we now start from the opposite side, viz. from the notion of the qualitative infinity of nature, we are then immediately able to arrive at a type of definition of the mode of being of any given kind of thing that does not contradict the possibility of its becoming something else. For the reciprocal relationships between all things then imply that no given things can be *exactly* and *in all respects* the kind of thing that is defined by any specified conceptual abstraction. Instead, it is always *something more* than this and, at least in some respects, *something different*. Hence, if the thing becomes something else, no unresolvable contradiction is now necessarily implied. For it is in any case never exactly represented by our original concept of it. Logically speaking, what this point of view towards the meaning of our conceptual abstractions does is, therefore, to create room for the possibility of qualitative change, by leading us to recognize that those aspects of things that have been ignored may, under suitable conditions, cease to have negligible effects, and indeed may become so important that they can bring about fundamental changes in the basic properties of the things under consideration.

We may illustrate the above conclusions by returning to a more detailed discussion of the transformations between steam, liquid water, and ice. Thus, the macroscopic concept of a certain state of matter (e.g. gaseous, liquid, or solid) leaves out of account an enormous number of kinds of factors that are not and cannot be defined in the macroscopic domain alone. Among these are the motions of the molecules constituting the fluid quantum fluctuations, field fluctuations, nuclear motions, mesonic motions, motions in a possible subquantum mechanical level, and so on. In short, we may say that the real fluid is enormously richer in qualities and properties than is our macroscopic concept of it. It is richer, however, in just such a way that these additional characteristics may, in a wide variety of applications, be ignored in the macroscopic domain. Nevertheless, when we come to the problem of understanding why transformations between gas, liquid, and solid are possible, we can no longer completely ignore the additional properties of the real fluid. Thus, the molecular motions are able to explain at least the essential features of the transformation in the system from a state in which one set of qualities (i.e. those corresponding to a gas) are the determinant, dominant, and controlling factors to a state in which these are replaced by another set of qualities (e.g. those corresponding to a liquid). Moreover, according to the notion of the qualitative infinity of nature, the same general kind of result is obtained for all things, including, for example, even the most fundamental entities that may have been discovered at any particular stage in the development of physics.

Not only is the notion of unvarying and exhaustively specifiable modes of being of things an abstraction that fails for periods of time that are too long (because of the possibility of fundamental qualitative changes), but it also fails for times that are too short. This is because the characteristic properties and qualities of a thing depend in an essential way on processes that are taking place in the background and substructure of the thing in question. Thus, for example, the properties of an atom (e.g. spectral frequencies, chemical reactivity, etc.) arise and are determined mainly in the process of motion of the electrons in the orbit, which take a period of time of the order of 10^{-15} seconds. Over shorter periods of time, however, the properties of an atom as a whole are so poorly defined that it is not even appropriate to consider them as such. A better conception of what the atom is can then be obtained by regarding it as a collection of electrons in motion around the nucleus. But as we shorten the period of time still further, the same problem arises with regard to electrons, protons, neutrons, mesons, etc. And if we go to a larger scale, the reader will readily see that a similar behaviour is obtained (e.g. the existence of a living being is maintained by inner metabolic and nervous processes that are fast in comparison with the period in which it makes sense to define the basic characteristics of such a being). Indeed, the notion of the qualitative infinity of nature implies that such behaviour is inevitable. For each kind of thing is maintained in existence by a balance of opposing processes in its infinite background and substructure, which are tending to change it in different ways. Thus, the properties of such a thing can be defined only over periods of time long enough so that the average of the effects of all these processes does not fluctuate significantly.

It is clear, then, that all our concepts are, in a great many ways, abstract representations of matter in the process of becoming. The choice of such abstractions is, however, limited by the requirement that they shall represent what is essential in a certain context to a suitable degree of approximation and under appropriate conditions.

The particular kind of abstraction that is used may evidently then vary, depending on what the context is. Thus, in theories of simple types of phenomena where things can be approximated as being in equilibrium, the modes of being of the basic entities and properties may be conceived of as completely static (e.g. as in statics and in thermodynamics). In the study of phenomena where motion is important, however, a higher level of abstraction is needed. For example, in mechanics one considers a system of particles which can change their positions without ceasing to be particles. In other words, the being of the particles is *indifferent* to their positions, and we can therefore consider them to be in motion through space. But the unvarying *laws* applying to these motions are now regarded as constituting an essential part of the modes of being of the particles in question. Thus, we have not escaped the necessity for considering unvarying and exhaustively specifiable modes of being. Of course, we could in principle go further and suppose, for example, that even the laws of motion of the particles were evolving with time. But then we would still be assuming that the higher laws applying to this process of evolution were themselves unvarying and in principle exhaustively specifiable in their form. On the other hand, according to the notion that everything takes part in the process of becoming, even these latter features of the laws could not ever really be completely unvarying and exhaustively specifiable in terms of a finite number of kinds of things.

We conclude, then, that we must finally reach a stage in every theory where we introduce the notion of something with unvarying and exhaustively specifiable modes of being, if only because we cannot possibly take into account all the inexhaustibly rich properties, qualities, and relationships that exist in the process of becoming. At this point, then, we are making an abstraction from the real process of becoming. Whether the abstraction is adequate or not depends on whether or not the specific phenomena that we are studying depend significantly on what we have left out. With the further progress of science, we are then led through a series of such abstractions, which furnish ever better representations of more and more aspects of matter in the concrete and real process of becoming.

Now, when we refer to the process of becoming by the word 'concrete', we mean by this to call attention to the quality of being special, peculiar, and unique that one always finds to be characteristic of real things when one studies them in sufficient detail. For example, if we consider any concept (e.g. apples), then this concept contains nothing in it that would permit us to distinguish one apple from another. We may then indicate other qualities which make such a distinction possible (e.g. red apples, hard apples, sweet apples, etc.). Evidently, no finite number of such qualities can ever give a complete representation of any specific example of a real apple. Of course, by going deeper (e.g. by giving the physical and chemical state of each part of the apple) we could come closer to our goal. But this process could never end. For even the modes of being of the individual atoms, electrons, protons, etc., inside the apple are in turn determined by an infinity of complex processes in their substructures and backgrounds. Thus, we see that because every kind of thing is defined only through an inexhaustible set of qualities each having a certain degree of relative autonomy, such a thing can and indeed must be *unique*; i.e. not completely identical with any other thing in the universe, however similar the two things may be.[1]

Carrying the analysis further, we now note that because all of the infinity of factors determining what any given thing is are always changing with time, *no such a thing can even remain identical with itself* as time passes. In certain respects, this brings us to a deeper notion of the process of becoming than we had before. For at each instant of time, each thing has, when viewed from one side, an enormous (in fact infinite) number of aspects which are in common with those that it had a short time ago. Indeed, if this were not so, it would not be a thing; i.e. it would not preserve any kind of identity at all. On the other hand, when viewed from another side, it has an equally enormous (in fact infinite) number of aspects that are not those that it had a short time ago. For typical sorts of things with which we commonly deal, however, these latter aspects are not essential in the normal contexts and conditions with which we work. In new contexts (e.g. a subatomic or a supergalactic time scale) or under new conditions (e.g. very high temperatures), these aspects may, however, take on a crucial importance.

We are in this way led to the conclusion that the process of becoming will necessarily have, at each moment, certain aspects that are concrete and unique. In other words, each thing in each moment of its existence must have certain qualities which, in some respects, belong uniquely to that thing and to that moment. The notion of unvarying and exhaustively specifiable modes of being is then an abstraction obtained, in general, by considering what is common to the same thing at different moments, or to many similar things at the same moment. In doing this, we evidently ignore the differences between these things, which are just as essential a side of them as are their similarities. By abstracting in more detail from these differences, we are then led to see newer but subtler aspects in which these differences contain common or similar relationships that apply to all of these things. Thus, the uniqueness of each thing at each instant of time is reflected in our abstract concepts by the limitless richness and complexity of the concepts that one needs to obtain a better and better abstract representation of matter in the process of becoming, or, in other words, by the inexhaustibility of the qualities that are to be found in nature.

2. Reasons for Inadequacy of Laplacean Determinism

We are now ready to see why the mechanistic determinism of Laplace does not apply if the notion of the qualitative infinity of nature is correct. For this kind of determinism implies that the laws of nature are such as to permit the super-being of Laplace to know them in their totality. On the other hand, according to the point of view that we have been presenting, this is impossible.

First of all, let us recall that no matter how far one goes in the expression of the laws of nature, the results will always depend in an unavoidable way on essentially independent contingencies which exist outside the context under investigation, and which are therefore undergoing chance fluctuations relative to the motions inside the context in question. For this reason, the causal laws applying inside any specified context will evidently not be adequate for the perfect prediction even of what goes on inside this context alone.

Secondly, however, the essential independence of different contexts implies that the processes taking place within a given context cannot provide a complete and perfect reflection of what goes on in the infinite totality of possible contexts. For example, because of the cancellation of chance fluctuations, the precise details of atomic motions are not usually reflected to any significant extent in the laws of the macroscopic level. The laws of each new context must then, in general, be discovered with the aid of new kinds of experiments, set up so as to create conditions in which the laws of the new context under investigation are significantly reflected in the behaviour of the apparatus. Hence, even to know what the totality of all the laws of nature is, the super-being would have to do an infinity of different kinds of experiments, each of which would give results that depended significantly on the laws of a different context, so that he could thereby obtain the necessary information. In doing this, he would have to be able to discover not only all the already operating kinds of laws, but also all the new laws that are expressible only in terms of the infinity of new qualities, new entities, and new levels that are going to come into being, all the way into the infinite future. It is evident, then, that if the Laplacean super-being resembles us to the extent of obtaining his knowledge through a series of investigations of partial segments of the universe, and not, for example, by Divine revelation or by *a priori* intuitions which he finds by plumbing the depths of his own mind, he will never be able to predict the entire future of the universe or even to approach such a prediction as a limit, no matter how good a calculator he may be. And if he did have such revelations or intuitions, a calculation would hardly be necessary, since the detailed prediction of the behaviour of the universe would then require a miracle only slightly greater than that by which he would learn the basic laws of the universe in the first place.

We see, then, that the behaviour of the world is not perfectly determined by any possible purely mechanical or purely quantitative line of causal connection. This does not mean, however, that it is arbitrary. For if we take any given effect, we can always in principle trace it to the causes from which its essential aspects came. Only as we go further and further back into the past, we discover three important points: viz. first, that the number of causes which contribute significantly to a given effect increases without limit; secondly that more and more qualitatively different kinds of causal factors are found to be significant; and finally, that these causes depend on new contingencies leading to new kinds of chance. For example, let us consider an eclipse of the moon. Over moderate periods of time this is a fairly precisely predictable event, which is determined mainly by the coordinates and momenta of the earth and the moon relative to the sun. But the longer the time that we consider, the more precise this determination must be, in order to make possible a prediction of the effect with a given accuracy. For the details of the motion become very sensitive to the precise initial conditions. As a result, perturbations arising from other planets, from tides in the earth, the moon, the sun, and still other essentially independent contingencies become significant. Over long enough periods of time, even the fluctuations arising from the molecular motions could in principle come to have significant effects; but before this could really become important, we should have gone so far into the past as to reach the qualitatively different phase of the gaseous nebulae from which the earth, moon, and sun came. Here we see that the random motions of the gas molecules in these nebulae contributed to making the eclipse eventually occur in the way that it did. If we go further back, we might reach the dense state of matter, that perhaps existed before the explosion that may have led to the present state of the part of the universe that is now visible to us. Then, the motions of the entities existing in this previous state, whatever they may have been, would have contributed to making the eclipse occur in the way that it did. But these motions would be contingent on something still earlier. And so on without limit. It is clear, moreover, that the eclipse of the moon is a phenomenon that is subject to an exceptionally simple type of determination, because of the approximate isolation of the earth and moon from other things. In other processes, where the degree of isolation is much less, the intertwining and fusion of the effects of more and more contingencies and more different qualities as we go further back is much greater. Thus, over an infinite period of time, the determination of even the essential features of an effect is evidently not purely mechanical, because it involves not only an infinite number of contingent factors but also an infinity of kinds of qualities, properties, laws of connection, all of which themselves undergo fundamental changes with the passage of time.

3. Reversibility versus Irreversibility of the Laws of Nature

In this section we shall make a few remarks concerning the implications of the qualitative infinity of nature with regard to the question of whether the laws of nature are reversible or irreversible.

It is well known that thus far the laws of microscopic physics have demonstrated themselves to be reversible. This follows from the fact that starting with any solution of the basic equations for the system (Newton's laws of motion, the laws of relativity, the laws of quantum theory), another possible solution can be found by replacing the time[2], t, by its negative, $-t$. Physically this means that given any motion, it is always possible, in principle at least, for a similar motion to take place, which is, however, executed in the reverse order. Of course, to obtain such a reversal of motion in reality, we would have to alter the boundary conditions appropriately (e.g. reverse all the velocities of the various particles, rates of change of the fields, etc.). Such a reversal does not, in general, occur spontaneously, at least within any practically significant periods of time. To show that this is so let us consider, for example, two boxes of gas, one containing hydrogen and the other containing oxygen, and let us imagine that we open a tube that connects them. As is well known, the gases will diffuse into each other. The reason is, of course, that the complicated and irregular motions of the hydrogen molecules will tend to carry them into the chamber originally containing oxygen, while similar motions of the oxygen molecules will tend to carry them into the chamber originally containing the hydrogen. Such processes can be treated in terms of the laws of chance, so that the theory of probability can be applied to them. Since over a long period of time it is equally probable that any particular molecule will occupy any given region of space, we conclude that on the average and in the long run we will obtain a practically uniform mixture of hydrogen and oxygen. It is characteristic of the laws of chance, however, that fluctuations away from the average can occur, although large fluctuations are very rare. A simple calculation, using the appropriate law of probability for these fluctuations, shows, for example, that a chance combination of motions that led all the hydrogen and oxygen back into their original containers would under typical conditions, not occur for $10^{10^{10}}$ years (i.e. 1 followed by then thousand million zeros). Clearly, then, although the motion may in principle reverse, the probability that this will happen is so small that we may for practical purposes ignore this possibility, especially considering the fact that, in any case, the containers of gas could not possibly last for such a long time.

It is possible by means of analysis described above to understand the observed irreversibility in various physical phenomena, such as the flow of heat, the establishment of thermal and mechanical equilibrium in fluids, etc. But this still leaves us with a disturbing problem. For the above reduction of the observed irreversibility of certain large-scale phenomena to the effects of chance does not alter the fact that the fundamental equations of motion are reversible, so that there is no inherent reason why processes in general must necessarily always take place in one direction only, since either direction would in principle be possible. Thus, if all the velocities and rates of change of fields did actually manage to be reversed for any reason whatever (e.g. by chance), then heat could go from a lower to a higher temperature, water could flow from the sea back to its sources in the mountains, etc. The fact that these events are so fantastically improbable does not detract from the problem of principle presented here, which is this: "Do the generally irrevocable effects of the passage of time in so wide a range of fields really come out of nothing more than the random mixing or shuffling according to the laws of chance of molecular and other types of motion, the reversal of which is in principle possible but in practice too improbable to be considered as having any real importance?"

If we take into account the character of the laws of physics implied by the qualitative infinity of nature, however, we can immediately answer this question in the negative. For, as we have seen, the notion of a law that gives a perfect one-to-one mathematical correspondence between well-defined variables in the past and in the future, is only an abstraction, good enough to describe limited domains of phenomena for limited periods of time, but, nevertheless, not valid for all possible domains over an infinite time. Thus, the very entities with which physics now works, satisfying the currently studied laws of physics, must have come into being at some time in the past, while changing conditions, brought about in part by the effects of just these laws and in part by chance contingencies, will eventually lead to a stage of the universe in which new kinds of entities satisfying new kinds of laws will come into being. On a smaller scale, we see also that new levels, such as that of living matter, have come into being, in which characteristic new qualities and new laws appear. Thus, we are not justified in making unlimited extrapolations of any specific set of laws to all possible domains and over infinite periods of time. This means that the description of the laws of nature as in principle completely reversible is merely a consequence of an excessively simple representation of reality. When we consider the mechanical laws in their proper contexts of ever-changing basic qualities, it becomes clear that irrevocable qualitative changes do take place, which could not even in principle be reversed. This is because, for systems of appreciable complexity, the fundamental character of the laws that apply cannot be completely separated from the historical processes in which these systems come to obtain their characteristic properties[3]. The possibility of such a behaviour is especially clear with regard to living matter, for here the very mode of being an organism and the basic qualities and laws which define this mode of being arise in the process by which the organism comes into existence, and passes through the various stages of its life. Thus, it is quite impossible that a human being could become a human being except by a process of growth, through embryo, childhood, adulthood, etc. But when one analyses processes taking place in inanimate matter over long enough periods of time, one finds a similar behaviour. Only here the process is so much slower that the abstraction in which we conceive of matter as having properties that are independent of its specific historical development is usually quite good as long as one considers periods of time which are measured in units smaller than billions of years.

The importance of considering the impact of qualitative changes on the basic modes of being of things is also seen when we consider the predictions of the 'heat death' of the universe, which were especially common towards the end of the nineteenth century. The 'heat death' refers to the prediction that eventually, because of random mixing and shuffling of molecules, the temperature of the universe would become uniform, and therefore, at least on the large scale, nothing could happen, so that the universe would be 'dead'. However, long before this comes about, it is evidently quite possible and indeed very likely that qualitatively new developments reflecting the inexhaustible and infinite character of the universal process of becoming will have invalidated predictions of the type described above. For example, just as there may have been a time before molecules, atoms, electrons, and protons existed, the further evolution of the universe could also lead to a new time in which they cease to exist, and are replaced by something else again. And new sources of energy coming from the infinite process of becoming may be made available even if atoms, molecules, etc., continue to exist. Thus, in the last century only mechanical, chemical, thermal electrical, luminous, and gravitational energy were known. Now we know of nuclear energy, which constitutes a much larger reservoir. But the infinite substructure of matter very probably contains energies that are as far beyond nuclear energies as nuclear energies are beyond chemical energies. Indeed, there is already some evidence in favour of this idea. Thus, if one computes the 'zero point' energy due to quantum-mechanical fluctuations in even one cubic centimetre of space, one comes out with something of the order of 10^{38} ergs, which is equal to that which would be liberated by the fission of about 10^{10} tons of uranium[4]. Of course, this energy provides a constant background that is not available at our level under present conditions. But as the conditions in the universe change, a part of it might be made available at our level.

Not only is the qualitatively and quantitatively infinite universal process of becoming too complex even to reverse itself or to come to some kind of final equilibrium, but it also cannot go in a cycle. For even if the laws applying in certain contexts and conditions should be consistent with a cyclical universe, such laws will always leave out an infinity of new kinds of factors, which will in the long run become important as conditions change sufficiently. Unless these new factors are exactly coordinated with those already existing in more limited contexts and set of conditions, they will eventually break the cycle and bring in fundamental qualitative changes. But because of their relative and approximate autonomy, these factors would not in general be coordinated in such a way. Hence a cyclical behaviour would also be inconsistent with the character of the universe that we have been considering here.

In conclusion, then, the notion of the qualitative infinity of nature implies that the development of the universe in time will lead to an inexhaustible diversity of new things.

NOTES

* From *Causality and Chance in Modern Physics*, Harper Torchbooks, 1957, pp. 153–164.
[1] According to the Pauli exclusion principle, any two electrons are said to be 'identical'. This conclusion follows from the fact that within the framework of the current quantum theory there can be no property by which they could be distinguished. On the other hand, the conclusion that they are *completely* identical in *all* respects follows only if we accept the assumption of the usual interpretation of the quantum theory that the present general form of the theory will persist in every domain that will ever be investigated. If we do not make this assumption, then it is evidently always possible to suppose that distinctions between electrons can arise at deeper levels.
[2] In the case of the quantum theory, we must also replace the wave function, ψ, by its complex conjugate, but this does not change any probabilities of physical processes, which depend only on $|\psi|$.
[3] Of course, this may not be the only reason or even the main reason for the observed irreversibility in nature, but in any case, for this reason alone, irreversibility would follow.
[4] Actually, according to present theories, this energy is infinite, but if one assumes that the theory is valid down to fluctuations having wavelengths of the order of 10^{-13} cm, then the above value of the energy is obtained. This wavelength was chosen, because it is generally believed that current theories of quantum electrodynamics break down for shorter wavelengths, and break down in such a way that the effects of quantum-fluctuations become finite. Thus, in a very rough estimate, we may ignore the effects of wavelengths shorter than 10^{-13} cm.

H. WEYL

THE OPEN WORLD*

The antinomy between freedom and determination takes its most acute form in the relation between knowing and being. Let us assume once more with Laplace that the state of the world at one moment, i.e., a three-dimensional section $t = $ const, defines by strict mathematical laws its course during all past and future time. Then we might suppose that I can calculate the future from what I know (or can know) here and now at the world point O. I should like to state with all emphasis that this antinomy, which formerly existed, disappears in the relativity theory. On another occasion I described the causal structure according to which a kind of conical surface issues from each point O of the four-dimensional world as vertex and separates the causal past and future. Causality is here not merely a methodological principle but becomes through this structure an objective constituent of the world. In the figure the section $t = $ const through O separates the past and the future sheets of the cone through O. But it is not this plane section, it is the surface of the backward light-cone which separates what is knowable at O from what is not. And it is a mathematical consequence of the classical physical laws that whereas the backward half of the world, cut off by $t = $ const, determines the whole, the interior of the backward light-cone does not. That is to say, only after a deed is done can I know all its causal premises.

If we regard, however, our problem as concerning reality alone and not concerning the relation of knowledge and reality, and if free action shall be possible in this real world, then we must demand that the content of the forward pointing cone through O shall not be completely determined by the rest of the world. This would contradict classical physics. But classical physics, after decades of invasion by statistical theories, is now finally superseded by the quantum theory, and a new situation has arisen.

In three grams of hydrogen there are about 10^{24} hydrogen molecules whirling about; it is of course impossible to calculate exactly their motion under the forces they experience from the walls of the container and from one another. Their average velocity determines the temperature, their bombardment of the walls, or rather the impulse per unit area it conveys, the pressure. Certain mean values are what our observations measure and these can be predicted by probability calculations, without detailed investigation of the motion. Consider, for example, a cubical container divided up into many small cubes, all of equal size, and suppose the chance of a given molecule to be in one of these is the same for each and that the space probabilities of the different molecules are independent in the statistical sense. Then we can show that the gas density in each of the small cubes differs with utterly overwhelming probability by less than, say, .01% from the mean density of the whole. Macroscopically speaking, the gas in equilibrium is uniformly dense. In the same way the kinetic theory of gases, first formulated by Bernoulli, leads to the other well-known gas laws.

The theory of probability not only tells us the mean value of a quantity, but also how great its deviation from this mean may be expected to be. The spontaneous variations in the density of the atmosphere which arise through the random motions of its molecules are the cause of the diffusion of the sun's rays in daylight, which makes a cloudless sky appear not black but blue. Small though they are individually, combined they produce a perceptible effect. Such variation-phenomena are the main supports of the statistical theory. The powerful researches of Maxwell and Boltzmann have made clear that the majority of physical concepts are not exact in the sense of classical physics, but statistical mean values, with a certain degree of indetermination, and that most of the familiar laws of physics, especially all those which concern the thermodynamics of atomic matter, are not to be regarded as strictly valid natural laws but as statistical regularities.

The first epistemological attitude toward statistical physics was to regard the probability theory simply as a short cut to certain consequences of the exact laws. For instance, strictly speaking, one would have to prove by means of the classical laws of motion that the time intervals during which the gas deviates noticeably from thermodynamic equilibrium were together vanishingly small as compared to the whole period of observation. Attempts at such proofs were indeed made, but it was always necessary to introduce an unproved hypothesis, the so-called ergodic hypothesis, at the critical point. If we adhere to the actual practice of physical research we are bound to admit that with the progress of the statistical theory and its continual increase in fruitfulness the attempts to base it on strict functional laws have gradually been abandoned. Historical evolution has spoken and demands that we recognize statistical concepts as equally fundamental with the concepts of law. I believe that such historical evolution can exert a more compelling pressure than any reasoning which pretends to be heaven knows how rigorous.

It should be remarked in this connection that in the world of exact laws time is reversible; changing t into $-t$ makes no difference. On the other hand, the definite direction of flow from past to future is perhaps the one outstanding mark of subjective time. This uniqueness of direction enters into physics not through its functional laws, but through our probability judgments; from a state at a given moment we deduce the probable state at a subsequent moment according to computed probabilities, and not the state at a previous one. Thus probability exposes a part of the causal idea which was quite suppressed in the exact laws.

Yet only the latest aspect of physics, quantum mechanics, has reduced the statistical nature of physical lawfulness to its ultimate foundations. This step became necessary in order to give an account of the double nature of physical entities, brought into evidence first in the case of light. Light is a spatially continuous undulatory process of electromagnetic nature. Only this conception enables us to understand diffraction and interference. But on the other hand a number of phenomena discovered in the last decades force us to conceive of light as consisting of single quanta, thrown out from the source of light in definite directions, and whose energy content is determined by the frequency, or the color of the light. I will describe here one of these phenomena. If a plate of metal is irradiated with ultra-violet light, electrons are emitted from the plate. Assuming the intensity of the light to be small, the energy of the wave which traverses an atom would not suffice to remove an electron from the atomic system. Even if we imagine some kind of a mechanism allowing the accumulation of wave energy within the atom, this effect could only begin after a long period of accumulation. Instead of this, it sets in immediately. The force with which the electrons are knocked out is totally independent of the intensity of the light; but it depends on its color. Only the number of electrons emitted in unit time increases with the intensity. This process can only be understood if light consists of single quanta. The energy content of such a light quantum, which hits an atom, is carried over to an electron, thus enabling this electron to break its bond with the nucleus of the atom, and furthermore imparting to the electron a certain kinetic energy. This energy depends on the energy content of the light quantum and hence on the color of the light. The dual nature of light – its being a wave capable of interference and also at the same time a light quantum striking suddenly here and there – we try to cover by assuming that the intensity of the wave field at a certain point represents the relative probability that a light quantum will be at that point. The more intense the light, the denser the accumulation of light quanta in unit time. The wave field obeys a strict functional law. But exactly the same condition prevails for the constituents of matter, the electrons. Everyday experiences suggest that their nature is corpuscular. But electrons have lately been shown to be susceptible of diffraction and interference. Hence there exist precise laws, but they deal with wave fields and therefore with quantities, which for real events have only the significance of probabilities. They determine the actual processes in the same way that *a priori* probabilities determine statistical mean values, frequencies – always containing a factor of uncertainty.

You know how it is possible with the aid of a prism or a grating to select monochromatic light from natural light. All light quanta in a ray of monochromatic light have the same definite energy and the same momentum. If we let the ray traverse a Nicol prism, we impress on it a certain direction of polarization. Let us describe this in terms of light quanta. A certain light quantum either will pass through the Nicol or it will not; hence there may be ascribed to the light quantum a certain quantity q_s corresponding to the position s of the Nicol, and taking on the values $+1$ or -1, according as the light quantum passes through or not. The monochromatic, polarized, plane light wave is the utmost in homogeneity that is obtainable. But we observe that such a homogeneous ray of light is again split up into a transmitted and reflected ray, when sent through a second Nicol in a position t different from s. The relative intensities are completely determined by the angle between the two positions s and t. They represent the probabilities that for a light quantum with $q_s = 1$, the quantity $q_t = +1$ or -1. The ray of light which passed through both Nicols is not more homogeneous than the ray which passed through only the first one: it is of exactly the same character as it would have been if we had omitted the first Nicol. Hence the selection due to the first Nicol is destroyed by the second one. It is legitimate to speak of the quantity q_s for a light quantum, because there exists a method of determining its value. We can also speak of the quantity q_t. But it is meaningless to ask for the values taken simultaneously by the quantities q_s, q_t for a light quantum, because measuring q_t by selecting the light quanta with $q_t = 1$ destroys the possibility of measuring q_s by selecting the light quanta with $q_s = 1$.

This impossibility is not due to human limitation, but must be regarded as an essential one. Another example will make this clearer. An atom of silver possesses a certain magnetic moment, it is a small magnet of definite strength and direction. It can be represented by an arrow, the vector of magnetic moment. This vector has, in any spatial direction z, a component m_z, capable of taking on only two values, ± 1, when measured in a certain unit, the magneton. By means of a magnetic field inhomogeneous in the direction z, it is possible to separate from a beam of atoms flying through the field the two component beams for which m_z equals $+1$ and -1 respectively. The same evidently applies in any other spatial direction. But a vector, whose components in every spatial direction are capable of taking on only the values ± 1, is geometrically absurd. The resolution of this paradox is this: if the component m_z is fixed by the separation, then no further component can be determined. Only probabilities can be calculated for their possible values ± 1.

Classical physics in attempting to establish conditions which would guarantee maximum homogeneity, assumed that for such a 'pure case' any physical quantity of the physical system considered took on a well-defined value, which under the same conditions would always be reproduced. Quantum mechanics also requires the experimenter to create a pure case whose homogeneity cannot be increased. But the ideal of classical physics is not realizable for quantum mechanics. We must not ask what value is taken on by a physical quantity in a certain pure case, but instead what the probability is that this physical quantity will take on a given value in this pure case. The idea that an electron describes a path cannot be upheld any longer. It is true that an electron's position at a certain instant can be measured; its velocity, too, is measurable, but not both at the same time. The measurement of position destroys the possibility of an exact measurement of speed. There is no human incapacity involved; the difficulty lies in the very nature of things. The meaning of a physical quantity is bound to the method by which it is measured. The attributes with which physics deals manifest themselves only through experiments and reactions which are based on postulated laws of nature. Formerly physicists took the point of view that these attributes were assigned to the physical bodies themselves, independently of whether or not the measurements necessary to establish them were actually carried out. It was proper to connect them by the logical 'and'; it was reasonable to postulate determinism and to satisfy this methodical postulate by introducing suitably chosen, concealed attributes. This epistemological position of constructive science is now submitted to an essential restriction in quantum mechanics.

We may try to escape this verdict by saying that the wave field, which obeys precise laws, is reality. Nevertheless it is a fact that this wave field cannot be observed directly, but only determines all observable quantities in the same way that *a priori* probabilities determine statistical frequencies. In this connection the uncertainty principle is unavoidable. We may say that there exists a world, causally closed and controlled by precise laws, but in order that I, the observing person, may come in contact with its actual existence, it must open itself to me. The connection between that abstract world beyond and the one which I directly perceive is necessarily of a statistical nature. This fact, together with the new insight which modern physics affords into the relation between subject and object, opens several ways of reconciling personal freedom with natural law. It would be premature, however, to propose a definite and complete solution of the problem. One of the great differences between the scientist and the impatient philosopher is that the scientist bides his time. We must await the further development of science, perhaps for centuries, perhaps for thousands of years, before we can design a true and detailed picture of the interwoven texture of Matter, Life and Soul. But the old classical determinism of Hobbes and Laplace need not oppress us any longer.

NOTE

* From *The Open World*, Yale University Press, New Haven, 1952, pp. 46–55.

INDEX OF NAMES

Adams 331
Aeschylus 10
Alembert d' 359
Anaxagoras 11
Anaximander xviii, 10, 15
Anaximenes 10
St. Anselm 43, 527
Apollodorus of Seleuca 164, 197
Archytas of Tarentum xix, xx, 14, 159, 160
Aristophanes 10
Aristotle xix, xx, xxi, xxiii, xxvii, xxviii, xxxi, xxxii, xxxiii, 6, 8, 9, 24–5, 27–9, 31, 41–2, 44, 59, 65, 69, 91, 93, 139, 141, 147, 159–61, 163, 167, 185, 187–9, 197–201, 207, 269, 428
Ascoli d', Gradazei xxxii, 185
St. Augustine xxix, xxxi, xxxii, lv, 179, 203

Bacon, Roger 44
Bailey, Cyril xxiv, 17, 33, 143
Barrow, J. xxviii, xxxi, xxxiii, xxxv, xxxvii, 203
Becquerel, Jean lii, 360, 433
Bergmann, Hugo 491
Bergson, Henri v, xxvi, xxix, xxxi, xxxvii, xxxviii, xxxix, xliii, lii, liii, 245, 359, 433, 490, 502, 505, 521, 526
Berkeley, G. xviii, xxxix, xl, xli, xlii, 267, 444
Bernoulli 562
Biran de, Maine 357
Boehme, Jakob 62–3
Bohm, David lii, liii, liv, lv, 451, 547
Bohr, Niels xxvi, 533, 543
Boltzmann xxxiv, 525, 562
Bolyai 411
St. Bonaventura 43–4
Bondi, H. 490–3
Bonnet, Nicolas xxxii, 185

Borel, E. 426–8
Boscovich, R. xviii, xxxvi, xl, 225, 289
Bradley, F. H. li, 251, 525, 528–9
Bradwardine, Thomas 47–9
Brahe, Tycho 62, 321
Bridgman, P. W. liii
Brillouin, M. 426
Broad, C. D. 521, 527, 529
Bruno, Giordano xix, xx, xxii, xxiii, xxv, xxviii, xxxii, xxxiii, 57–9, 69–71, 189
Brunschwicg, L. 359
Burleigh, Walter, 48
Burnet, John xxiv, 18

Caesar, Julius, 217
Calinon, A. xviii, xlii, xliii, 297, 320, 321, 415
Calvin 47
Campanella xxii, xxiii, 71
Campanus of Novara 44
Cantor 235–6, 240
Čapek, M. 484, 493–5, 501
Cardan 67
Carnap, R. xliv
Carnot, Sadi 357, 360
Carr, Wildon 441–5
Cassirer, Ernst 67, 69, 353–4
Chase 511
Chrysippus xxix, 39–41, 159–62, 168, 170, 199
Clarke, S. xix, xlii, 129, 273, 275, 285, 359
Clausius 360
Cleanthes 168
Cleomedes 39
Cleugh 528
Clifford, W. K. xviii, xlii, xliii, 291, 295
Cohen, Hermann 263
Columbus 322
Copernicus xxii, xxiii, 51–4, 57–61, 310, 394

INDEX OF NAMES

Cornford, F. M. 3, 137
Costa de Beauregard 504–7
Cournot 360
Crescas 65, 67, 69, 71
Cunningham 354–5
Cusanus, Nicolas xxii, 51, 65, 67

Dante, Alighieri 66
Darwin, Charles Robert 394
Democritus xxiv, xxvii, 5–6, 12, 21–4, 27, 36, 87, 269
Descartes, René xxiv, xxxiii, xxxiv, xlvii, li, 22, 24, 73, 85–6, 121–2, 359, 365–6, 428, 432, 487
Digges, Thomas xxii
Diogenes 17
Donne, John 51
Driesch, Hans 484
Duhem, P. v, xviii, 21, 27, 39, 41, 43, 185

Eddington, A. S. xliii, xlvi, li, liv, 4, 354–7, 365–6, 431–2, 463, 475, 490, 505, 514–5, 521, 529
Einstein, Albert xx, xliv, xlv, xlvi, li, 15, 329, 345, 354–8, 363, 390, 394, 397–9, 402, 404, 406, 409, 417, 420, 422, 426, 428–9, 431–2, 448, 452, 455, 458, 468, 485, 489, 501–6, 513–4, 517, 528, 534, 543
Empedocles 10, 15, 17, 139, 167, 200, 207
Empiricus, Sextus 48, 93, 144
Epicurus xxvii, xxix, xxxiv, 6, 33–5, 87, 91, 144–5, 179, 197
Euclid 3–7, 15, 293, 298, 410, 422
Eudemus of Rhodes 170
Eudoxus 57
Euler, L. xxvi, xl, 113, 305
Euripides 10
Evellin, M. 242

Fantappié 506
Faraday, Michael 336
Feynman 506
Fitzgerald, G. F. 374
Fizeau 330
Flammarion 509
Foucault 130
Frank, Philipp xlvii, 387
Fresnel, A. 334

Galen 160
Galilei, G. 60, 125–6, 233–4, 309, 312–3, 360, 428, 449, 483, 512
Galle 231
Gamov, G. liv
Gassendi, P. v, xxiii, xxiv, xxviii, xxxii, xxxiii, xxxiv, xxxv, xlii, 91, 195
Gauss, C. F. xliii, 415
Gilbert, William xxv, 65, 71
Giussani 34
Gödel, Kurt xlv, xlvii, liii, 455, 458–9, 503–7, 516
Gradazei, *see* Ascoli
Graves, John xlix
Grosseteste, Robert 41, 44
Grünbaum, Adolf xlvi, xlvii, xlix, l, li, 471, 505
Guericke, Otto von xxv, 277

Hegel, G. W. F. 235, 359
Heisenberg, Werner xxxv, 545
Helmholtz, H. von xliii, xlviii, 410
Heracleitus 164, 167–8
Hertz, Heinrich 331, 366
Heymans 129
Hipparchus 57
Hobbes, Thomas 477–8, 566
Höffding, H. 57
Huggins 231
Huygens, C. 52

Isidore of Sevilla 43

James, William v, xxix, xlvi, li, 384, 484, 521
Jammer, M. xviii, xx, 65
Jeans, James xlvii, 14, 388–9, 392–5, 456, 505
John of Salisbury 425

Kant, E. xxxv, xxxvi, xl, 129, 161, 305, 306, 414, 427, 455, 503, 525
Kelvin 468
Kepler, J. xxiii
Klein 411
Koyré, A. 47, 51

Lagrange, J. L. 257, 359
Langevin, P. 353, 356, 360, 427, 505, 510

INDEX OF NAMES 569

Laplace, P. S. li, 305, 393, 426, 507, 512, 514, 530, 547, 553, 561, 566
Larmor, J. 375
Lavoisier, A. L. 258–60
Le Sage 425–9
Leeuwenhoek, Anton van 539–40
Leibniz, G. W. xviii, xix, xxxi, xlii, 129–30, 236, 255, 257, 273, 277, 359, 425, 541
Lemaître liv
Lenzen, Victor lii, lv, 431
Leucippus 6, 7, 12, 17–9, 21, 24, 27
Leverrier, M. 231
Lindsay, R. B. lv, 533
Lipsius, F. 387
Lobachevsky, N. xxv, xlii, 411
Locke, John xix, xxxv, xxxvi, xxxvii, xliii, 14, 107, 211
Lombard, Peter 43
Lorentz, H. A. 330–5, 344–5, 374–5, 393, 437, 438, 451, 483, 508, 512
Lotze, l, 519
Lovejoy 54, 521
Lucretius xix, xx, xxiii, xxvii, xxxii, 6, 13–4, 33–7, 87, 143–5

Macaulay, W. H. 130, 132
Mach, E. xix, xxvi, xxxix, xl, xli, xlii, lii, 129, 133, 309, 334–5, 397–8, 400
Manzoli, Pietro xxii, 63
Marais, Henri 355–6
Margenau, H. lv, 533
Maurus, Rabanus 43
Maxwell, J. C. xxvi, xxxvii, xli, 121, 129, 231, 292, 330–2, 336, 366, 426, 542, 548, 562
McColly 52
McTaggart, J. M. E. xlviii, l, li, 521, 526–9
Mehlberg xliv
Melissus xix, 17, 19, 140
Metrodorus of Chios 13
Meyer, Louis 359
Meyerson, Emile xxvi, xxxix, xliii, xlv, xlvi, li, 255, 353, 363–7, 425, 501–5, 521
Michelson, A. A. xl, xliv, liii, 344, 373, 376, 513, 543
Mill xli

Minkowski, H. xlv, xlvii, xlix, 333, 339, 353–7, 387, 391, 477, 485–6, 501–3, 506–8, 529
Mirandola 71
Mits, T. C. 409
More, Henry xxiv, xxxiv, xxxv, 85, 282
Morley 373, 376, 513, 543
Müller, W. 447

Nemesius 91
Neumann, Carl xxvi, xxxvii, xl, 125, 131–3, 305–6
Newton, Isaac v, xii, xviii, xix, xxii, xxiv, xxv, xxvi, xxviii, xxix, xxxii, xxxv, xl, xli, xliv, 15, 97, 125–6, 129–30, 160, 209, 231, 234, 267, 287, 289, 305, 309–12, 325, 329–30, 334, 339, 357, 398, 400–2, 422, 425–7, 432, 489, 507, 509, 513–6, 545
Nicholas of Cusa, see Cusanus
Nietzsche, F. xxix, liv, 170
Noble 513
Noël, M. 241

Osiander xxiii

Palingenius, see Manzoli
Parmenides xix, xxvi, xxvii, li, 11–2, 15, 17, 19, 137–8, 140, 142, 264
Pascal, B. xxv, xxxvi, 89
Patritius, Franziscus xxii, xxiii, 67–71
Patrizzi, see Patritius
St. Paul 111, 286
Pearson, Karl 129
Peirce v
Philoponus, John xx, xxi, 39–41
Planck, M. 418, 474, 533, 539, 543
Plato xxvii, xxx, li, 4, 7, 13, 21–3, 27–9, 138, 140, 160, 163–4, 199, 491, 525
Plotinus xxix, xxx, xxxi, xxxii, 173, 199
Plutarch 165, 200, 207
Poincaré, H. xxxvii, xliii, xliv, 236, 257, 317, 417, 420
Poinsot 260
Porphyry 10
Possidonius 196
Priestley, Joseph 258
Ptolemy 53
Putnam, Hilary xlvi
Pythagoras 4, 7, 10, 170, 537, 544

Raleigh, Walter 54
Rankine xx
Reichenbach, Hans xlii, xliv, lii, liii, 367, 397, 447, 483, 490–1, 505, 517, 525, 529
Renouvier 264
Rey, Abel liv
Riccioli, Gianbattista 52
Riemann, B. xxv, xlii, 295, 411, 413
Robb, A. A. xliv, xlvi, 369, 514, 516
Robertson, H. P. xliii, li, 409
Rochot, Bernard xxxiii
Römer, Olaf xlvi, 512
Ross 13
Russell, Bertrand xxv, xxvi, xxxviii, xxxix, xliii, 129, 235, 251, 356, 410, 414, 415, 486, 518, 525

Sambursky, S. xxix, 31, 159, 167
Scaliger, C. 67
Scheele 258
Schlick, M. xliii, 367
Schopenhauer, A. xxxvii, 227
Schütz, J. R. 350
Schwarzschild, K. 415–7
Scot, Michael 43
Seneca 199
Sextus, *see* Empiricus
Silberstein 503
Simplicius 34, 160, 201
Smart, J. J. C. 483–6
Socrates xxix, xxx
Solomon 111
Sommerfeld, A. 353–4
Spinoza 122, 359
Stallo, J. B. xviii, xl, xli, xlii, 305
Stay, Benedict xxxvi
Stobaeus 197, 199
Strato 159–60, 163

Streintz 131
Swift, J. xxv, 540

Telesio, Bernadino xxii, xxiii, xxxii, 67–8, 71, 187
Tempier, Stephen 185
Thales 4
Theophrastus 86, 138–9
St. Thomas xxi, liii, 41–3
Timaeus 23
Tomaschek 513
Torricellius of Florence 277
Trouton 513

Vincent of Beauvais 43
Virgil 86

Ward 129
Weierstrass 236–7, 254
Wells, H. G. 255, 501, 503
Weyl, H. xlv, xlvi, lii, lv, 336, 353–4, 357, 365–6, 426, 432, 485, 502–6, 525, 528, 561
Whitehead, A. N. v, xxvi, xxix, xli, xliv, lii, liii, lv, 163–4, 410, 441, 505, 514, 515, 518, 521, 525, 535
Whitrow, G. J. l, li, 485–90, 495, 521, 525
Wien, W. 353, 357
Wiener, Norbert lv, 539
William of Auvergne 43
William of Ockham 185
Williams, D. 505
Wolfson 70

Xenocrates xxix, 162
Xenophanes xxiii

Zeno xxvi, xxvii, xxxviii, 40, 159–60, 168, 235–7, 245–7, 526

SYNTHESE LIBRARY

Monographs on Epistemology, Logic, Methodology,
Philosophy of Science, Sociology of Science and of Knowledge, and on the
Mathematical Methods of Social and Behavioral Sciences

Managing Editor:

JAAKKO HINTIKKA (Academy of Finland and Stanford University)

Editors:

ROBERT S. COHEN (Boston University)
DONALD DAVIDSON (The Rockefeller University and Princeton University)
GABRIËL NUCHELMANS (University of Leyden)
WESLEY C. SALMON (University of Arizona)

1. J. M. BOCHEŃSKI, *A Precis of Mathematical Logic*. 1959, X + 100 pp.
2. P. L. GUIRAUD, *Problèmes et méthodes de la statistique linguistique*. 1960, VI + 146 pp.
3. HANS FREUDENTHAL (ed.), *The Concept and the Role of the Model in Mathematics and Natural and Social Sciences, Proceedings of a Colloquium held at Utrecht, The Netherlands, January 1960*. 1961, VI + 194 pp.
4. EVERT W. BETH, *Formal Methods. An Introduction to Symbolic Logic and the Study of Effective Operations in Arithmetic and Logic*. 1962, XIV + 170 pp.
5. B. H. KAZEMIER and D. VUYSJE (eds.), *Logic and Language. Studies dedicated to Professor Rudolf Carnap on the Occasion of his Seventieth Birthday*. 1962, VI + 256 pp.
6. MARX W. WARTOFSKY (ed.), *Proceedings of the Boston Colloquium for the Philosophy of Science, 1961–1962*, Boston Studies in the Philosophy of Science (ed. by Robert S. Cohen and Marx W. Wartofsky), Volume I. 1973, VIII + 212 pp.
7. A. A. ZINOV'EV, *Philosophical Problems of Many-Valued Logic*. 1963, XIV + 155 pp.
8. GEORGES GURVITCH, *The Spectrum of Social Time*. 1964, XXVI + 152 pp.
9. PAUL LORENZEN, *Formal Logic*. 1965, VIII + 123 pp.
10. ROBERT S. COHEN and MARX W. WARTOFSKY (eds.), *In Honor of Philipp Frank*, Boston Studies in the Philosophy of Science (ed. by Robert S. Cohen and Marx W. Wartofsky), Volume II. 1965, XXXIV + 475 pp.
11. EVERT W. BETH, *Mathematical Thought. An Introduction to the Philosophy of Mathematics*. 1965, XII + 208 pp.
12. EVERT W. BETH and JEAN PIAGET, *Mathematical Epistemology and Psychology*. 1966, XII + 326 pp.
13. GUIDO KÜNG, *Ontology and the Logistic Analysis of Language. An Enquiry into the Contemporary Views on Universals*. 1967, XI + 210 pp.

14. ROBERT S. COHEN and MARX W. WARTOFSKY (eds.), *Proceedings of the Boston Colloquium for the Philosophy of Science 1964–1966, in Memory of Norwood Russell Hanson*, Boston Studies in the Philosophy of Science (ed. by Robert S. Cohen and Marx W. Wartofsky), Volume III. 1967, XLIX + 489 pp.
15. C. D. BROAD, *Induction, Probability, and Causation. Selected Papers*. 1968, XI + 296 pp.
16. GÜNTHER PATZIG, *Aristotle's Theory of the Syllogism. A Logical-Philosophical Study of Book A of the Prior Analytics*. 1968, XVII + 215 pp.
17. NICHOLAS RESCHER, *Topics in Philosophical Logic*. 1968, XIV + 347 pp.
18. ROBERT S. COHEN and MARX W. WARTOFSKY (eds.), *Proceedings of the Boston Colloquium for the Philosophy of Science 1966–1968*, Boston Studies in the Philosophy of Science (ed. by Robert S. Cohen and Marx W. Wartofsky), Volume IV. 1969, VIII + 537 pp.
19. ROBERT S. COHEN and MARX W. WARTOFSKY (eds.), *Proceedings of the Boston Colloquium for the Philosophy of Science 1966–1968*, Boston Studies in the Philosophy of Science (ed. by Robert S. Cohen and Marx W. Wartofsky), Volume V. 1969, VIII + 482 pp.
20. J. W. DAVIS, D. J. HOCKNEY, and W. K. WILSON (eds.), *Philosophical Logic*. 1969, VIII + 277 pp.
21. D. DAVIDSON and J. HINTIKKA (eds.), *Words and Objections: Essays on the Work of W. V. Quine*. 1969, VIII + 366 pp.
22. PATRICK SUPPES, *Studies in the Methodology and Foundations of Science. Selected Papers from 1911 to 1969*. 1969, XII + 473 pp.
23. JAAKKO HINTIKKA, *Models for Modalities. Selected Essays*. 1969, IX + 220 pp.
24. NICHOLAS RESCHER et al. (eds.), *Essays in Honor of Carl G. Hempel. A Tribute on the Occasion of his Sixty-Fifth Birthday*. 1969, VII + 272 pp.
25. P. V. TAVANEC (ed.), *Problems of the Logic of Scientific Knowledge*. 1969, XII + 429 pp.
26. MARSHALL SWAIN (ed.), *Induction, Acceptance, and Rational Belief*. 1970, VII + 232 pp.
27. ROBERT S. COHEN and RAYMOND J. SEEGER (eds.), *Ernst Mach; Physicist and Philosopher*, Boston Studies in the Philosophy of Science (ed. by Robert S. Cohen and Marx W. Wartofsky), Volume VI. 1970, VIII + 295 pp.
28. JAAKKO HINTIKKA and PATRICK SUPPES, *Information and Inference*. 1970, X + 336 pp.
29. KAREL LAMBERT, *Philosophical Problems in Logic. Some Recent Developments*. 1970, VII + 176 pp.
30. ROLF A. EBERLE, *Nominalistic Systems*. 1970, IX + 217 pp.
31. PAUL WEINGARTNER and GERHARD ZECHA (eds.), *Induction, Physics, and Ethics, Proceedings and Discussions of the 1968 Salzburg Colloquium in the Philosophy of Science*. 1970, X + 382 pp.
32. EVERT W. BETH, *Aspects of Modern Logic*. 1970, XI + 176 pp.
33. RISTO HILPINEN (ed.), *Deontic Logic: Introductory and Systematic Readings*. 1971, VII + 182 pp.
34. JEAN-LOUIS KRIVINE, *Introduction to Axiomatic Set Theory*. 1971, VII + 98 pp.
35. JOSEPH D. SNEED, *The Logical Structure of Mathematical Physics*. 1971, XV + 311 pp.
36. CARL R. KORDIG, *The Justification of Scientific Change*. 1971, XIV + 119 pp.

37. MILIČ ČAPEK, *Bergson and Modern Physics*, Boston Studies in the Philosophy of Science (ed. by Robert S. Cohen and Marx W. Wartofsky), Volume VII, 1971, XV + 414 pp.
38. NORWOOD RUSSELL HANSON, *What I do not Believe, and other Essays* (ed. by Stephen Toulmin and Harry Woolf), 1971, XII + 390 pp.
39. ROGER C. BUCK and ROBERT S. COHEN (eds.), *PSA 1970. In Memory of Rudolf Carnap*, Boston Studies in the Philosophy of Science (ed. by Robert S. Cohen and Marx W. Wartofsky, Volume VIII. 1971, LXVI + 615 pp. Also available as a paperback.
40. DONALD DAVIDSON and GILBERT HARMAN (eds.), *Semantics of Natural Language*. 1972, X + 769 pp. Also available as a paperback.
41. YEHOSHUA BAR-HILLEL (ed.), *Pragmatics of Natural Languages*. 1971, VII + 231 pp.
42. SÖREN STENLUND, *Combinators, λ-Terms and Proof Theory*. 1972, 184 pp.
43. MARTIN STRAUSS, *Modern Physics and Its Philosophy. Selected Papers in the Logic, History, and Philosophy of Science*. 1972, X + 297 pp.
44. MARIO BUNGE, *Method, Model and Matter*. 1973, VII + 196 pp.
45. MARIO BUNGE, *Philosophy of Physics*. 1973, IX + 248 pp.
46. A. A. ZINOV'EV, *Foundations of the Logical Theory of Scientific Knowledge (Complex Logic)*, Boston Studies in the Philosophy of Science (ed. by Robert S. Cohen and Marx W. Wartofsky), Volume IX. Revised and enlarged English edition with an appendix, by G. A. Smirnov, E. A. Sidorenka, A. M. Fedina, and L. A. Bobrova. 1973, XXII + 301 pp. Also available as a paperback.
47. LADISLAV TONDL, *Scientific Procedures*, Boston Studies in the Philosophy of Science (ed. by Robert S. Cohen and Marx W. Wartofsky), Volume X. 1973, XII + 268 pp. Also available as a paperback.
48. NORWOOD RUSSELL HANSON, *Constellations and Conjectures* (ed. by Willard C. Humphreys, Jr.), 1973, X + 282 pp.
49. K. J. J. HINTIKKA, J. M. E. MORAVCSIK, and P. SUPPES (eds.), *Approaches to Natural Language. Proceedings of the 1970 Stanford Workshop on Grammar and Semantics*. 1973, VIII + 526 pp. Also available as a paperback.
50. MARIO BUNGE (ed.), *Exact Philosophy – Problems, Tools, and Goals*. 1973, X + 214 pp.
51. RADU J. BOGDAN and ILKKA NIINILUOTO (eds.), *Logic, Language, and Probability. A selection of papers contributed to Sections IV, VI, and XI of the Fourth International Congress for Logic, Methodology, and Philosophy of Science, Bucharest, September 1971*. 1973, X + 323 pp.
52. GLENN PEARCE and PATRICK MAYNARD (eds.), *Conceptual Chance*. 1973, XII + 282 pp.
53. ILKKA NIINILUOTO and RAIMO TUOMELA, *Theoretical Concepts and Hypothetico-Inductive Inference*. 1973, VII + 264 pp.
54. ROLAND FRAÏSSÉ, *Course of Mathematical Logic – Volume I: Relation and Logical Formula*. 1973, XVI + 186 pp. Also available as a paperback.
55. ADOLF GRÜNBAUM, *Philosophical Problems of Space and Time*. Second, enlarged edition, Boston Studies in the Philosophy of Science (ed. by Robert S. Cohen and Marx W. Wartofsky), Volume XII. 1973, XXIII + 884 pp. Also available as a paperback.
56. PATRICK SUPPES (ed.), *Space, Time, and Geometry*. 1973, XI + 424 pp.
57. HANS KELSEN, *Essays in Legal and Moral Philosophy*, selected and introduced by Ota Weinberger. 1973, XXVIII + 300 pp.

58. R. J. SEEGER and ROBERT S. COHEN (eds.), *Philosophical Foundations of Science. Proceedings of an AAAS Program, 1969*. Boston Studies in the Philosophy of Science (ed. by Robert S. Cohen and Marx W. Wartofsky), Volume XI. 1974, X + 545 pp. Also available as a paperback.
59. ROBERT S. COHEN and MARX W. WARTOFSKY (eds.), *Logical and Epistemological Studies in Contemporary Physics*, Boston Studies in the Philosophy of Science (ed. by Robert S. Cohen and Marx W. Wartofsky), Volume XIII. 1973, VIII + 462 pp. Also available as a paperback.
60. ROBERT S. COHEN and MARX W. WARTOFSKY (eds.), *Methodological and Historical Essays in the Natural and Social Sciences. Proceedings of the Boston Colloquium for the Philosophy of Science, 1969–1972*, Boston Studies in the Philosophy of Science (ed. by Robert S. Cohen and Marx W. Wartofsky), Volume XIV. 1974, VIII + 405 pp. Also available as paperback.
61. ROBERT S. COHEN, J. J. STACHEL, and MARX W. WARTOFSKY (eds.), *For Dirk Struik. Scientific, Historical and Political Essays in Honor of Dirk J. Struik*, Boston Studies in the Philosophy of Science (ed. by Robert S. Cohen and Marx W. Wartofsky), Volume XV. 1974, XXVII + 652 pp. Also available as paperback.
62. KAZIMIERZ AJDUKIEWICZ, *Pragmatic Logic*, transl. from the Polish by Olgierd Wojtasiewicz. 1974, XV + 460 pp.
63. SÖREN STENLUND (ed.), *Logical Theory and Semantic Analysis. Essays Dedicated to Stig Kanger on His Fiftieth Birthday*. 1974, V + 217 pp.
64. KENNETH F. SCHAFFNER and ROBERT S. COHEN (eds.), *Proceedings of the 1972 Biennial Meeting, Philosophy of Science Association*, Boston Studies in the Philosophy of Science (ed. by Robert S. Cohen and Marx W. Wartofsky), Volume XX. 1974, IX + 444 pp. Also available as paperback.
65. HENRY E. KYBURG, JR., *The Logical Foundations of Statistical Inference*. 1974, IX + 421 pp.
66. MARJORIE GRENE, *The Understanding of Nature: Essays in the Philosophy of Biology*, Boston Studies in the Philosophy of Science (ed. by Robert S. Cohen and Marx W. Wartofsky), Volume XXIII. 1974, XII + 360 pp. Also available as paperback.
67. JAN M. BROEKMAN, *Structuralism: Moscow, Prague, Paris*. 1974, IX + 117 pp.
68. NORMAN GESCHWIND, *Selected Papers on Language and the Brain*, Boston Studies in the Philosophy of Science (ed. by Robert S. Cohen and Marx W. Wartofsky), Volume XVI. 1974, XII + 549 pp. Also available as paperback.
69. ROLAND FRAÏSSÉ, *Course of Mathematical Logic – Volume II: Model Theory*. 1974, XIX + 192 pp.
70. ANDRZEJ GRZEGORCZYK, *An Outline of Mathematical Logic. Fundamental Results and Notions Explained with all Details*. 1974, X + 596 pp.
71. FRANZ VON KUTSCHERA, *Philosophy of Language*. 1975, VII+305 pp.
72. JUHA MANNINEN and RAIMO TUOMELA, *Essays on Explanation and Understanding*. 1975, approx. 450 pp.
75. JAAKKO HINTIKKA and UNTO REMES, *The Method of Analysis. Its Geometrical Origin and Its General Significance*. Boston Studies in the Philosophy of Science (ed. by Robert S. Cohen and Marx W. Wartofsky), Volume XXV. 1974, XVIII+ 144 pp. Also available as a paperback.
76. JOHN EMERY MURDOCH and EDITH DUDLEY SYLLA, *The Cultural Context of Medieval Learning. Proceedings of the First International Colloquium on Philosophy, Science, and Theology in the Middle Ages – September 1973*. Boston Studies in the

Philosophy of Science (ed. by Robert S. Cohen and Marx. W. Wartofsky), Volume XXVI. 1975, X+566 pp. Also available as paperback.
77. STEFAN AMSTERDAMSKI, *Between Experience and Metaphysics. Philosophical Problems of the Evolution of Science*. Boston Studies in the Philosophy of Science (ed. by Robert S. Cohen and Marx W. Wartofsky), Volume XXXV. 1975, XVIII+193 pp. Also available as paperback.
80. JOSEPH AGASSI, *Science in Flux*. Boston Studies in the Philosophy of Science (ed. by Robert S. Cohen and Marx W. Wartofsky), Volume XXVIII. 1975, XXVI+553 pp. Also available as paperback.

SYNTHESE HISTORICAL LIBRARY

Texts and Studies
in the History of Logic and Philosophy

Editors:

N. KRETZMANN (Cornell University)
G. NUCHELMANS (University of Leyden)
L. M. DE RIJK (University of Leyden)

1. M. T. BEONIO-BROCCHIERI FUMAGALLI, *The Logic of Abelard.* Translated from the Italian. 1969, IX + 101 pp.
2. GOTTFRIED WILHELM LEIBNITZ, *Philosophical Papers and Letters.* A selection translated and edited with an introduction, by Leroy E. Loemker. 1969, XII + 736 pp.
3. ERNST MALLY, *Logische Schriften*, ed. by Karl Wolf and Paul Weingartner. 1971, X + 340 pp.
4. LEWIS WHITE BECK (ed.), *Proceedings of the Third International Kant Congress.* 1972, XI + 718 pp.
5. BERNARD BOLZANO, *Theory of Science*, ed. by Jan Berg. 1973, XV + 398 pp.
6. J. M. E. MORAVCSIK (ed.), *Patterns in Plato's Thought. Papers arising out of the 1971 West Coast Greek Philosophy Conference.* 1973, VIII + 212 pp.
7. NABIL SHEHABY, *The Propositional Logic of Avicenna: A Translation from al-Shifā': al-Qiyās*, with Introduction, Commentary and Glossary. 1973, XIII + 296 pp.
8. DESMOND PAUL HENRY, *Commentary on 'De Grammatico'. The Historical-Logical Dimensions of a Dialogue of St. Anselm's.* 1974, IX + 345 pp.
9. JOHN CORCORAN, *Ancient Logic and Its Modern Interpretations.* 1974. X + 208 pp.
10. E. M. BARTH, *The Logic of the Articles in Traditional Philosophy.* 1974, XXVII + 533 pp.
11. JAAKKO HINTIKKA, *Knowledge and the Known. Historical Perspectives in Epistemology.* 1974, XII + 243 pp.
12. E. J. ASHWORTH, *Language and Logic in the Post-Medieval Period.* 1974, XIII + 304 pp.
13. ARISTOTLE, *The Nicomachean Ethics.* Translated with Commentaries and Glossary by Hippocrates G. Apostle. 1975, XXI+372 pp.
14. R. M. DANCY, *Sense and Contradiction: A Study in Aristotle.* 1975, XII + 184 pp.
15. WILBUR RICHARD KNORR, *The Evolution of the Euclidean Elements. A Study of the Theory of Incommensurable Magnitudes and Its Significance for Early Greek Geometry.* 1975, XI+374 pp.
16. AUGUSTINE, *De Dialectica*, Translated with Introduction and Notes by B. Darrell Jackson. 1975, XII+151 pp.